Tolley's Basic Science and Practice of Gas Service

Gas Service Technology Volume 1

Tolley's Basic Science and Practice of Gas Service

Gas Service Technology Volume 1

Fifth edition

John Hazlehurst

LONDON AND NEW YORK

Second edition 1990
Third edition 1994
Fourth edition 2006
Fifth edition 2009

First published 1978 by Newnes

2 Park Square, Milton Park, Abingdon, Oxfordshire OX14 4RN
52 Vanderbilt Avenue, New York, NY 10017

Routledge is an imprint of the Taylor & Francis Group, an informa business

First issued in paperback 2020

British Library Cataloguing in Publication Data
Hazlehurst, John.
Tolley's basic science and practice of gas service. – 5th ed.
1. Gas appliances–Installation. 2. Gas appliances–Maintenance and repair.
I. Title II. Basic science and practice of gas service 683.8′8-dc22

Library of Congress Control Number: 2009925599

ISBN: 978-1-85617-671-2 (hbk)
ISBN: 978-0-367-65934-9 (pbk)

Table of Contents

Preface vi

1 Properties of Gases 1
2 Combustion 14
3 Liquefied Petroleum Gas 41
4 Burners 83
5 Energy 111
6 Pressure and Gas Flow 147
7 Control of Pressure 181
8 Measurement of Gas 207
9 Basic Electricity 240
10 Transfer of Heat 288
11 Gas Controls 311
12 Materials and Processes 361
13 Tools 387
14 Measuring Devices 439

Appendix 1 SI Units 461
Appendix 2 Conversion Factors 463
Index 466

Preface

Following comprehensive updates and revision of the two other volumes in this series 'Domestic Gas Installation Practice' and 'Industrial and Commercial Gas Installation Practice' (formerly Gas Service Technology 2 and 3), it was clearly essential that this, the first volume in the series, be brought up to date. 'Basic Science and Practice of Gas Service' leads the reader through the knowledge and understanding required to put into practice the safe installation and servicing procedures described in Volumes 2 and 3.

Changes to standards and legislation have been included, in particular the European gas directive relating to the prevention of products of combustion being released into a room in which an open-flued appliance is installed. Chapter 8 covers the devices used to ensure that these types of appliances conform to this directive. New types of combustion analysers and appliance testers which take advantage of the new technology available have also been included. Since the release of British Standards 7967 Parts 1, 2 & 3, on the 8th December 2005 and 7967 Part 4 on 29th June 2007 the industry has once again focused on Combustion Analysers. Compliance with the standards is of no value without a full working understanding of using your chosen Electronic Combustion Analyser. Chapter 14 covers requirements of British Standards 7967.

There have also been changes to the manner in which gas operatives are required to prove their competence. It is now a legal requirement that all gas operatives in the domestic field and most operatives working in industrial and commercial sectors currently be registered with the Confederation of Registered Gas Installers (CORGI). However, on 8th September 2008, HSE awarded a 10-year contract to the Capita Group Plc to provide a new registration scheme for gas installers from 1st April 2009. The current scheme has been in place for more than 17 years. During this time the number of domestic gas related fatalities has fallen significantly. However, a review in 2006 involving gas industry stakeholders (including gas installers and their representatives) and consumer groups identified no room for complacency and a strong case for change. To achieve membership all operatives must be successful in a series of initial gas safety assessments (Nationally Accredited Certification Scheme for Individual Gas Operatives ACS) in the areas of work in which they operate. This certification must be renewed every five years.

Scottish/National Vocational Qualifications are being amended to include these ACS assessments as part of the qualification process. This volume and the others in the series will prove invaluable to students studying for these qualifications and certificates, and for operatives wishing to improve their knowledge and understanding of natural gas and Liquefied Petroleum Gas (LPG) systems.

I would like to thank manufacturers for the use of photographs and diagrams, in particular S.I.T. Gas Controls (ODS devices) and BW Technologies (flue gas analysers), and also Blackburn College for the use of their facilities and resources.

Chapter | one

Properties of Gases

Chapter 1 is based on an original draft prepared by Mr E.W. Berry

INTRODUCTION

This first volume of the manual deals with the elementary science or 'technology' which forms the foundation of all gas service work. It outlines the principles involved and explains how they work in actual practice.

To do this it has to use scientific terms to describe the principles of things like 'force', 'pressure', 'energy', 'heat' and 'combustion'. Do not be put off by these words – they are simply part of the language of the technology which you have to learn. Every activity from sport to music has its own special words and gas service is no exception. While the football fan talks of 'strikers', 'sweepers' and 'back fours' the gas service man deals with 'calorific values', 'standing pressures' and 'secondary aeration'.

It is necessary for him to know about these things so that he can be sure that he has adjusted appliances correctly. He must also know what actions to take to avoid danger to himself or customers or damage to customers' property.

GAS: WHAT IT IS?

Every substance is made up of tiny particles called 'molecules' (see Chapter 2). In solid substances like wood or metal, there is very little space between the molecules and they cannot move about (Fig. 1.1).

In liquids, there is a little more space between the particles, so that a liquid always moves to fit the shape of its container. The molecules cannot get very far without bumping into each other, however, so they do not move very quickly and only a few get up enough speed to break out of the surface and form a vapour above the liquid (Fig. 1.2).

A gas has a lot more space between its molecules. So they are able to move about much more freely and quickly. They are continually colliding with each other and bouncing on to the sides of their container. It is this bombardment that creates the 'pressure' inside a pipe (Fig. 1.3).

Because the molecules are as likely to move in one direction as in any other, the pressure on all of the walls of their container will be the same. Gases must,

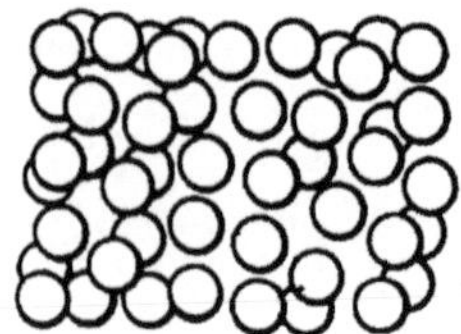

FIGURE 1.1 Molecules in a solid.

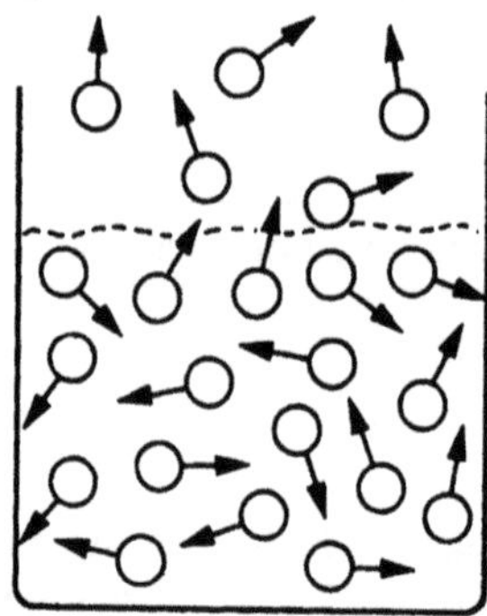

FIGURE 1.2 Molecules in a liquid.

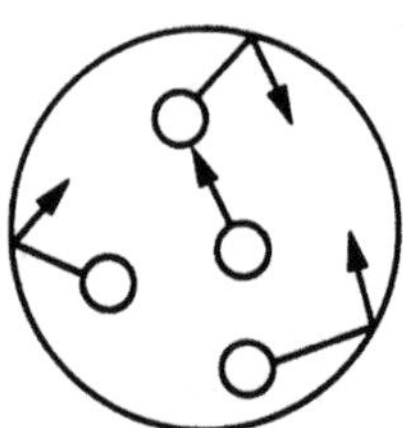

FIGURE 1.3 Molecules in a gas.

therefore, be kept in completely sealed containers otherwise the particles would fly out and mix, or 'diffuse', into the atmosphere (see section on Diffusion).

The word 'gas' is derived from a Greek word meaning 'chaos'. This is a good name for it, since the particles are indeed in a state of chaos, whizzing about, colliding and rebounding with a great amount of energy.

KINETIC THEORY

'Kinetic' means movement or motion, so kinetic energy is energy possessed by anything that is moving. A car has kinetic energy when it is moving along a road. The effect of this can clearly be seen if it collides with another object! Similarly gas molecules are in motion and possess kinetic energy at all normal temperatures. The amount of energy increases as the temperature increases. The Kinetic Theory states that:

1. The distance between the molecules of a gas is very great compared with their size (about 400 times as great).

2. The molecules are in continuous motion at all temperatures above absolute zero, −273 °C (see Chapter 5).
3. Although the molecules have an attraction for each other and tend to hold together, in gases at low pressures the attraction is negligible compared with their kinetic energy.
4. The amount of energy possessed by the molecules depends on their temperature and is proportional to the absolute temperature (see Chapters 5 and 8).
5. The pressure exerted by a gas on the walls of the vessel containing it is due to the perpetual bombardment by the molecules and is equal at all points.

DIFFUSION

If a small amount of gas is allowed to leak into the corner of an average-sized room, the smell can be detected in all parts of the room after a few seconds. This shows that the molecules of gas are in rapid motion and because of this, gases mix or 'diffuse' into each other.

Graham's Law of Diffusion

If two different gases at the same pressure were put into a container separated by a wall down the centre and a small hole made in the wall as shown in Fig. 1.4 then, because the molecules are in continuous motion, some molecules of each gas would pass through the hole into the gas on the other side. The faster and lighter molecules would pass more quickly through the hole into the other gas.

After studying the rates at which gases diffuse into each other, Graham discovered that the rates of diffusion varied inversely as the square root of the densities of gases. Or,

$$\text{Diffusion rate} \propto \frac{1}{\sqrt{\text{density}}}$$

Thus a light gas will diffuse twice as quickly as a gas of four times its density (see section on Specific Gravity).

The effect can be demonstrated experimentally by filling a porous pot, made from unglazed porcelain, fitted with a pressure gauge, with a dense gas such as

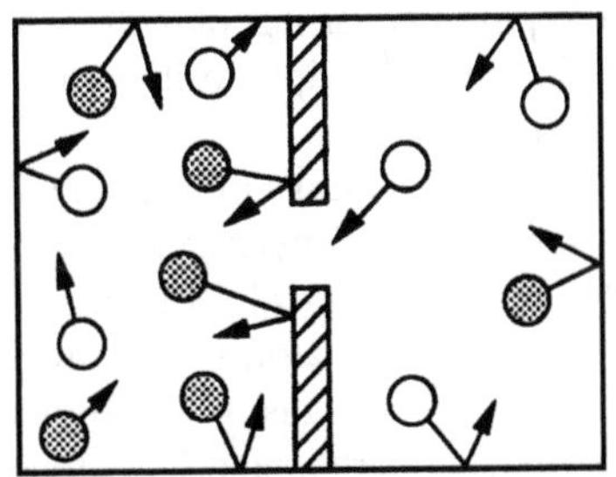

FIGURE 1.4 Diffusion of two gases.

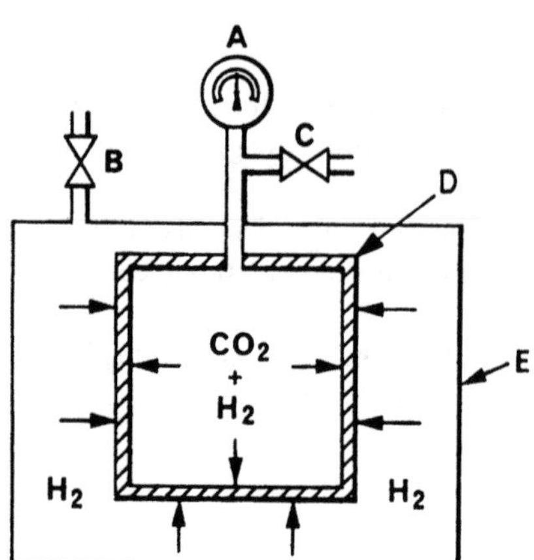

A: Pressure gauge
B: Inlet valve for hydrogen
C: Inlet valve for carbon dioxide
D: Inner porous pot
E: Outer container

FIGURE 1.5 Experiment to demonstrate the effect of diffusion.

carbon dioxide. Then place it in another vessel and fill that with a lighter gas such as hydrogen (see Fig. 1.5). The pressure inside the inner pot will be seen to rise, proving that the lighter gas is getting into the pot faster than the heavier gas that is getting out.

CHEMICAL SYMBOLS

Symbols are often the initial letters of the name of the substance, like H for hydrogen, O for oxygen. These symbols are not only a form of shorthand and save a lot of writing, they also show the amount of the substance being considered.

Each single symbol indicates one 'atom', which is the smallest chemical particle of the substance (see Chapter 2). So H indicates one atom of hydrogen, O is one atom of oxygen and so on.

It has previously been said that substances are made up of tiny particles called 'molecules'. This is true, but the molecules themselves consist of atoms. Sometimes a molecule of a substance contains only one single atom, like carbon, which is denoted by C. Often the molecules have more than one atom, like those of hydrogen and oxygen which both have two atoms. So while an atom is indicated by H, the smallest physical particle of hydrogen gas which can exist is shown by H_2. The '2' in the subscript position indicates that the molecule of hydrogen is made up of two atoms.

Some substances are made up of combinations of different kinds of atoms. Water is an example. Water is composed of hydrogen and oxygen and its formation is described in Chapter 2. The chemical formula for water is H_2O. This shows that a molecule of water has two atoms of hydrogen and one of oxygen combined together. Similarly, methane gas, which forms the main part of natural gas, is made up of one atoms of carbon and four atoms of hydrogen. So its formula is CH_4.

Table 1.1 Chemical Symbols and Formulae for Substances Commonly Used During Gas Service Work

Gases		Metals	
Substance	Chemical Formulae	Substance	Chemical Formulae
Hydrogen	H_2	Iron	Fe
Oxygen	O_2	Copper	Cu
Nitrogen	N_2	Lead	Pb
Carbon monoxide	CO	Tin	Sn
Carbon dioxide	CO_2	Zinc	Zn
Methane	CH_4	Antimony	Sb
Ethane	C_2H_6	Platinum	Pt
Propane	C_3H_8	Nickel	Ni
Butane	C_4H_{10}	Chromium	Cr
Propylene	C_3H_6	Tungsten	W
		Mercury	Hg

Table 1.1 shows the chemical symbols and formulae for some of the substances met with in gas service work.

ODOUR

Gas can, of course, be dangerous. It can burn and it can explode. Some gases are 'toxic' or poisonous. But all fuels are potential killers if not treated properly. Coke and oil both burn and can produce poisonous fumes. Electricity causes more domestic fires than any other fuel and the first indication you get of its presence could be your last!

Gas has the advantage of having a characteristic smell or 'odour' so it is easily recognisable. Several of the combustible gases, including hydrogen, carbon monoxide and methane, are colourless and odourless and could not easily be detected without elaborate equipment. To make it possible for customers to find out when they have a gas escape or have accidentally turned on a tap and not lit the burner, a smell or odour, is added to the gas.

Gas manufactured from coal has its own smell, natural gas does not. But suppliers of natural gas are required to add a smell to it before sending the gas out to the customers. So an 'odorant' is used, originally a chemical called tetrahydrothiophene. Only a very small amount is added, something like 1/2 kg to a million cubic feet of gas. Odorants now in use contain diethyl sulphide and ethyl and butyl mercaptan.

TOXICITY

A number of gases are 'toxic' or poisonous and inhaling them can result in death. Newspaper reports of people being 'gassed' are usually referring to carbon monoxide poisoning.

Carbon monoxide, CO, is the ingredient which causes the problem. By replacing oxygen in the bloodstream it prevents the blood from maintaining life and so the organs of the body become poisoned.

CO is one of the constituents of gas made from coal or oil, and inhaling the unburnt gas can prove fatal. Natural gas does not contain CO and so it is 'non-toxic'.

This means, of course, that people can no longer commit suicide by gassing themselves. There is another hazard, however. All fuels which contain carbon can produce carbon monoxide in their flue gases if the carbon is not completely burned. So people can still be gassed if the appliances are not flued or ventilated correctly (see Chapter 2). There is always a risk of suffocation if the presence of the gas reduces the amount of oxygen in the air.

CALORIFIC VALUE

All gases which burn give off heat (energy) and the 'calorific value', or CV, indicates the heating power. It is the number of heat (energy) units which can be obtained from a measured volume of the gas.

To measure CV in SI units, megajoules per cubic metre are used, written as MJ/m^3. The CV of natural gas in the UK is about 39.3 MJ/m^3 but does vary slightly from district to district. Because customers pay for the gas they use measured in heat units, The Gas supplier has to declare the calorific value of its supply and this is printed on every gas bill. The actual CV is monitored at official testing stations by gas examiners, appointed by the Department of Trade and Industry.

Meters presently used by The Gas supplier measure gas in cubic feet ($100\ ft^3 = 2.83\ m^3$) and until April 1992 customers were charged, based on the number of 'therms' used (1 therm = 105.506 MJ).

EC directive 80/181 required Britain to change their method of billing from imperial to metric units and The Gas supplier implemented this change from April 1992 using the metric kilowatt hour (kWh) as a basis for charge. It is the common unit used in Europe and is of course the basis of charge for electricity.

The total amount of heat obtained from gas is, in fact, the Gross CV. If however, the water vapour in the products of combustion is not allowed to condense into water, the amount of heat obtained is the Net CV.

SPECIFIC GRAVITY (RELATIVE DENSITY)

Every substance has weight or 'mass', including gas. Some complicated scientific equipment would be needed to do the weighing, but it can be weighed. It is necessary, for various reasons, to compare weights of gases and to do this a comparison is made of their 'densities'. The density of a substance is the weight of a given volume. In Imperial units it is the number of pounds per cubic foot (lb/ft^3), and in SI units it is the number of kilograms per cubic metre (kg/m^3).

Densities of substances vary very considerably. Lead is heavier than wood and wood is lighter than water. In order to compare densities they are related to a standard substance. For solids and liquids the standard is water. For gases the standard is air.

This relationship between the density of a substance and the density of the standard is known as the 'relative density' or 'specific gravity'. Let us take the example of mercury. The specific gravity (or SG) of liquid mercury is 13.57. So it is about 13 V2 times as heavy as the same bulk of water, or 1 l would weigh 13.57 kg.

The specific gravity of natural gas is in the region of 0.5. So it is about half the weight of the same volume of air. The specific gravity of liquefied petroleum gas (LPG) is greater than that of air.

WOBBE NUMBER

The Wobbe number of Wobbe 'index' gives an indication of the heat output from a burner when using a particular gas (the terms Wobbe number and Wobbe index are used interchangeably in this book).

The amount of heat which a burner will give depends on the following factors:

1. The amount of heat in the gas as given by its calorific value.
2. The rate at which the gas is being burned. This rate depends on the following.
   - 2.1 The size of the 'jet' or 'injector'.
   - 2.2 The pressure in the gas, pushing it out of the injector.
   - 2.3 The relative weight of the gas. This affects how easily the pressure can push it out and is indicated by the specific gravity.

Looking at these factors it can be seen that they divide up into two groups.

1. Factors depending on the gas:
   - calorific value (1),
   - specific gravity (2.3).
2. Factors depending on the appliance:
   - size of injector (2.1),
   - pressure of the gas (2.2).

Since the factors in group 2 are fixed by the design or the adjustment of the appliance, the only alterations in heat output would be brought about by changes in the group 1 factors. That is, changes in the characteristics of the gas, the CV and the SG. The Wobbe number links these two characteristics and is obtained by dividing the CV by the square root of the SG, thus:

$$\text{Wobbe number} = \frac{\text{CV}}{\sqrt{\text{SG}}}$$

It is essential to ensure that the heat outputs of appliances are kept reasonably constant. To do this the Wobbe number of the gas must be maintained within fairly close limits.

Natural gas with a CV of 39.33 MJ/m^3 and a SG of 0.58 would have a Wobbe number of 51.64.

FAMILIES OF GASES

To ensure that appliances operate correctly the gas quality must be maintained within close limits. In practice, it is kept to a quality range indicated by Wobbe numbers.

There are three ranges or 'families' which have been agreed internationally (Table 1.2). Family 1 covers manufactured gases, family 2 covers natural gases and family 3 covers liquefied petroleum gas (LPG).

Table 1.2 Families of Gases

Family	Approximate Wobbe Number Range	Type of Gas
1	22.5–30	Manufactured (inc. LPG/air)
2 L	39.1–45	Natural
2 H	45.5–55	
3	73.5–87.5	LPG

Appliances are designed to operate on gas of a particular family.

Manufactured gases are generally made from coal, oil feed-stocks or naphthas and also include LPG/air mixtures.

The demand for natural gas has historically been supplied from the North Sea reserves, with excess gas being supplied to Continental Europe. However, more recently the UK have become a net importer of gas. It is estimated that additional supplies will be required by 2010; the UK is projected to import up to half of its gas demand and that by 2020 imports may be as high as 90%.

LIQUEFIED NATURAL GAS (LNG)

The reasoning behind storing and transporting LNG as opposed to gaseous natural gas is the physical property of a reduction in volume of 600 times. Although there may be some modifications of the make up of the revapourised gas which change its combustion characteristics.

Liquefied petroleum gases include propane, butane and mixtures. They will be dealt with in more detail in Chapter 3.

AIR REQUIREMENTS

Everything that burns must have oxygen in order to do so. Fuel gases contain carbon and hydrogen compounds which burn when they are lit and allowed to combine with oxygen (see Chapter 2).

Fortunately the atmosphere consists of about 21% oxygen and 79% nitrogen (with a tiny amount of other gases). This means that if a gas flame is allowed to burn freely in the open, it can get the oxygen it needs from the surrounding air.

For each cubic metre (m^3) of gas burned, the amount of air required is:

- 4.89 m^3 for butane/air,
- 9.75 m^3 for natural gas,
- 23.8 m^3 for commercial propane.

AERATION

Gas can be burned straight out of a jet, getting the air it requires from the atmosphere around the flame. This happens with a 'neat' or 'non-aerated' burner.

For a number of uses a different kind of flame is necessary and to get this a proportion of the air is allowed to mix with the gas before it is burned. This happens in the 'aerated burner' and the air added first is called 'primary air', as distinct from 'secondary air' which is the remaining air required and obtained from around the flame.

The proportion of air mixed before 'combustion' or burning is called the 'primary aeration'. It is usually expressed as a percentage. For example, a burner with 50% primary aeration would have half the total air requirement mixed with the gas before it was burned.

GAS MODULUS

The 'gas modulus' is a numerical expression which relates the heat output from a burner with the pressure required to provide a satisfactory amount of aeration. It gives a figure which indicates how aeration and heat loading conditions may be maintained when changing from one gas to another.

The modulus is obtained by dividing the square root of the pressure by the Wobbe index. Thus,

$$\text{Gas modulus} = \frac{\sqrt{\text{pressure}}}{\text{Wobbe index}}$$

Using the modulus shows that to change from a manufactured gas with a Wobbe index of 27.2 supplied at a pressure of 6.23 mbar to a natural gas with a Wobbe index of 49.6 required the pressure to be increased to 20.62 mbar to maintain the same operating conditions (see Chapter 4, section on Modifying Appliances to Burn Other Gases).

FLAMMABILITY LIMITS

Mixtures of gas and air will burn, but only within limits. If there is either too much gas or too much air, the mixture will not burn. The 'flammability limits' are those air and gas mixtures at each end of the range which will just burn. For natural gas the range is from 5% up to 15% of gas in the mixture. So the limits are 5% gas for the lower explosive limit (LEL) and 15% gas for the higher explosive limit (HEL). For commercial propane the limits are 2.0–10.0% and for butane/air the limits are narrower being from 1.6 to 7.75% gas.

FLAME SPEED

All flames burn at particular rates. You can watch a flame burning its way along a match or taper. Similarly a gas flame is burning along the mixture (Table 1.3) as it comes out of the jet or burner at a particular speed. The speed at which the mixture is coming out has to be adjusted so that the flame will stay on the tip of the burner. If it came out too fast it would blow the flame off and if it was too slow the flame could burn its way back inside the tube!

The flame speed of gas is measured in metres per second. Typical flame speeds are:

- natural gas: 0.36 m/s,
- butane/air: 0.38 m/s,
- commercial propane: 0.46 m/s.

Table 1.3 Constituents of Gases

Constituent	Formulae	Typical Percentage by Volume		
		Natural Gas	Commercial Propane	Butane/Air
Oxygen	O_2			17.01
Nitrogen	N_2	2.7		63.99
Carbon dioxide	CO_2	0.6		
Methane	CH_4	90.0		
Ethane	C_2H_6	5.3	1.5	
Propylene	C_3H_6		12.0	
Propane	C_3H_8	1.0	85.9	2.5
Butane	C_4H_{10}	0.4	0.6	16.5
		100	100	100

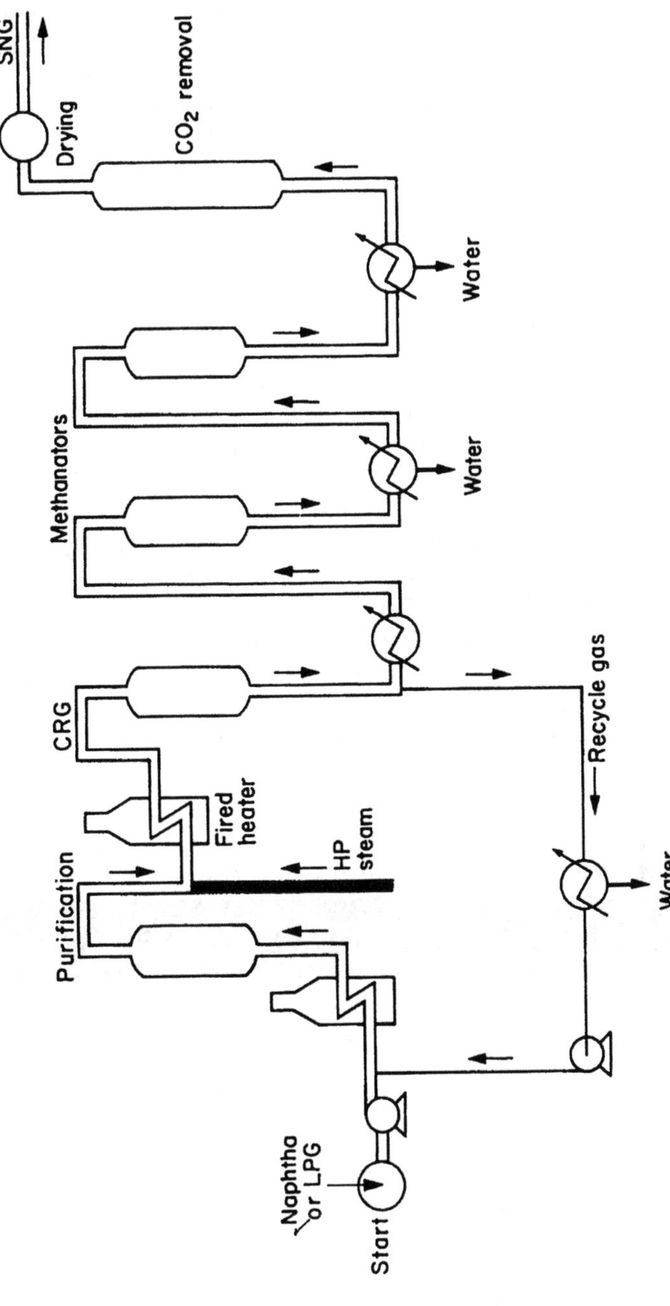

FIGURE 1.6 Substitute natural gas (SNG) plant. A catalytic rich gas producer with a double methanation process.

IGNITION TEMPERATURES

Gases need to be lit or 'ignited' before they will burn. This is done by heating the gas until it reaches a sufficiently high temperature to burst into flame and keep burning.

Heating the gas to the required temperature may be done with a match, a small gas flame or 'pilot', an electric spark, or a coil of wire or 'filament' made red hot by an electric current. Ignition temperatures of the common gases are:

- natural gas: 704 °C,
- butane/air: 500 °C,
- commercial propane: 530 °C.

SUBSTITUTE NATURAL GAS

Substitute natural gas (SNG) is manufactured either as a direct substitute for natural gas or as a means of providing additional gas to meet peak loads. It can be made from a range of feed-stocks in a number of different types of plant. The feed-stocks commonly used are LPG or naphthas, which are light petroleum distillates. The feed-stock is mixed with high pressure steam and passed over a catalyst to produce a Catalytic Rich Gas (CRG). After this it may pass through additional processes to increase its percentage of methane and to remove carbon dioxide. An example of an SNG plant is shown in Fig. 1.6.

For combustion to be useable, it must be controlled; uncontrolled it will be dangerous and inefficient. To control combustion and achieve fuel efficiency, and complete combustion process, it is necessary to understand the characteristics of the fuel and the way it burns. Table 1.4 below provides the comparison of properties of typical gases.

Table 1.4 Comparison of Properties of Typical Gases

Property	Units	Natural Gas	Commercial Propane	Butane/Air
Caloric value	MJ/m^3	39.3	97.3	23.75
Specic gravity	air = 1	0.58	1.5	1.19
Wobbe number	MJ/m^3	51.64	79.4	21.79
Air required	vol/vol	9.75	23.8	4.89
Flammability limits	% gas in air	5–15	2–10	1.6–7.75
Flame speed	m/s	0.36	0.46	0.38
Ignition temperature	°C	704	530	500

Table 1.5 Characteristics of Typical SNG

Constituent	Formulae	Percentage by Volume
Methane	CH_4	98.5
Hydrogen	H_2	0.9
Carbon monoxide	CO	0.1
Caloric value		38 MJ/m^3
Specic gravity		0.555 (air = 1.0)
Wobbe number		51 MJ/m^3

The characteristics of SNG from the 'double methanation process' are shown in Table 1.5. The gas produced is a non-toxic, high-methane, low-inert gas interchangeable with natural gas, and it can be supplied directly into pipelines.

Chapter | two

Combustion

Chapter 2 is based on an original draft prepared by Mr E.W. Berry

COMBUSTION

Fuel gases burn when they are ignited and allowed to combine with oxygen, usually taken from the air. This burning or 'combustion' is in fact a chemical reaction taking place, a reaction which produces heat and changes the gas and air into other gases. In order to understand what exactly is taking place and what new substances are being produced by the combustion, it is necessary to study a little very simple chemistry. Start by looking at atoms and molecules in a little more detail.

ATOMS

The atom was introduced in Chapter 1 as the smallest chemical particle into which a substance may be divided by chemical means. Atoms are, however, made up of three components. There are over 100 different kinds of atoms, each containing different numbers of the three basic components.

All atoms consist of a relatively heavy central core or 'nucleus' with very light 'electrons' revolving or orbiting round it at a little distance. The nucleus has a positive (or +) electrical charge and the electrons have negative (or –) electrical charges. The positively charged particles in the nucleus are called 'protons'. Usually the number of electrons and protons in an atom is the same so that the negative and positive charges balance out and the atom is electrically neutral. The simplest of all atoms is the common hydrogen atom (Fig. 2.1). It has a single proton round which revolves a single electron.

Some atoms have additional particles in the nucleus which are similar to protons but are electrically neutral (i.e. without charge). These particles are called 'neutrons'. A few hydrogen atoms have neutrons as well as protons, and the nucleus of 'heavy hydrogen' is thought to have one proton and one neutron (Fig. 2.2).

The weight or mass of an atom depends on the number of protons and neutrons in the nucleus. The electrons are so tiny, by comparison, that they do

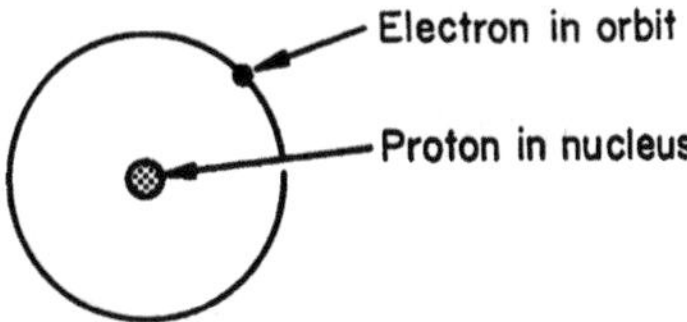

FIGURE 2.1 Structure of the hydrogen atom.

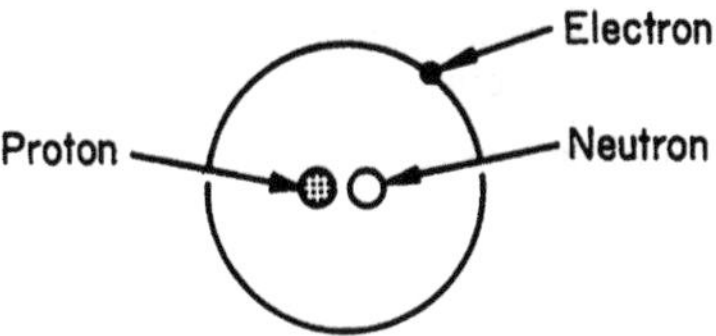

FIGURE 2.2 Structure of the deuterium (or 'heavy hydrogen') atom.

not affect the overall mass of the atom. So the three components of an atom are:

1. electrons,
2. protons,
3. neutrons.

Electrons move around the nucleus of an atom in fixed orbits. There can be up to seven orbits of electrons in the more complex atoms. The inner orbit or 'shell' can contain up to two electrons. The next holds up to eight electrons. Each shell holds a fixed number of electrons.

The chemical behaviour of an atom depends largely on the number of electrons in its outer shell. This particularly affects the ability of the atom to combine with others to form molecules. Stable substances are those whose outer shell contains its full complement of electrons. Other substances, with outer shells which are not completely full, are less stable and so combine more readily together to form more stable substances. Carbon, for example, has two electron shells. With two electrons in its inner shell, but only four in its outer shell, it combines readily with other substances.

MOLECULES

A molecule is the smallest particle of a substance which can exist independently and still retain the properties of that substance.

A molecule of methane, CH_4, consists of one atom of carbon and four atoms of hydrogen (Fig. 2.3). You can see that, by sharing electrons with the hydrogen atoms, the carbon atom can fill up its outer shell to the full eight electrons and so form a stable substance.

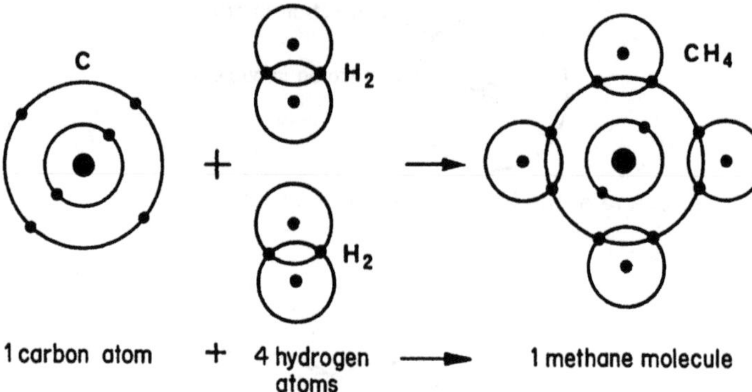

FIGURE 2.3 Structure of a molecule of methane.

ELEMENTS

An element is a substance whose molecules contain only one kind of atom. Examples of elements are hydrogen, H_2, oxygen, O_2, nitrogen, N_2, iron, Fe, copper, Cu.

COMPOUNDS

Compounds are substances whose molecules contain more than one kind of atom. For example, water, H_2O, contains the atoms of hydrogen and oxygen. Similarly methane, CH_4, is a compound since it contains both carbon and hydrogen atoms.

MIXTURES AND COMPOUNDS

It is possible for substances to be brought together either as compounds or as mixtures and there are important differences between the two. Table 2.1 shows the main differences.

Table 2.1 Differences Between Mixtures and Compounds

Compounds	Mixtures
There are fixed proportions of constituents	May have variable proportions
Produced by a chemical reaction usually associated with heat	Made by adding constituents together in some container
Cannot be separated by physical means	Can be separated by physical means
Has properties often very different from those of its constituents	Has properties related directly to its constituents, each contributing its own particular property to the whole

As an example, suppose that iron filings and sulphur were stirred together in any proportions. The result would be a grey powder which would be slightly magnetic. This would be a mixture.

It would be possible to separate the components of this particular mixture by removing the iron filings with a magnet or by dissolving the sulphur in carbon disulphide.

Now suppose that a mixture of iron filings and sulphur was heated. It would become red hot as if burning and, when it cooled, a black brittle solid would be found which was not magnetic or soluble in carbon disulphide. This would be ferrous sulphide, FeS, in which each atom of iron had combined with an atom of sulphur. This would be a compound.

It would not be possible to remove either the iron or the sulphur from the ferrous sulphide.

COMBUSTION EQUATIONS

Chemical reactions may be written in the form of equations. Take, for example, the combustion of methane.

Methane, CH_4, burns in combination with oxygen, O_2, in the air. In doing so it produces carbon dioxide, CO_2, and water vapour, H_2O (see Fig. 2.4). Written as an equation this becomes:

$$CH_4 + 2O_2 = CO_2 + 2H_2O$$

Notice that, as with all equations, both sides balance.

- There is one carbon atom on each side.
- There are four hydrogen atoms on each side.
- There are four oxygen atoms on each side.
- Altogether there are nine atoms on each side.

Since there is a direct relationship between the number of molecules and the volume of gases, the equation shows the volumes of gases involved. So,

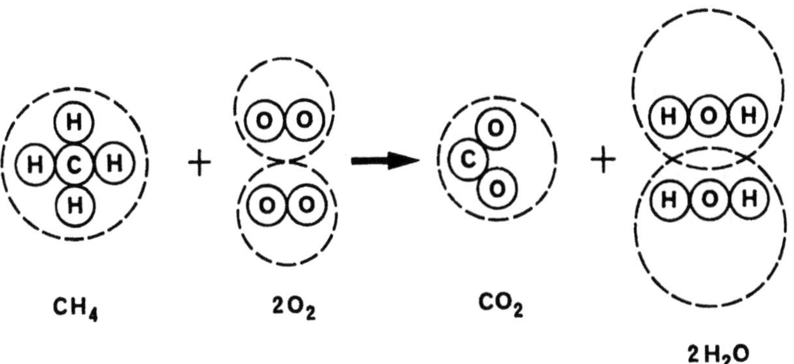

FIGURE 2.4 Combustion of methane.

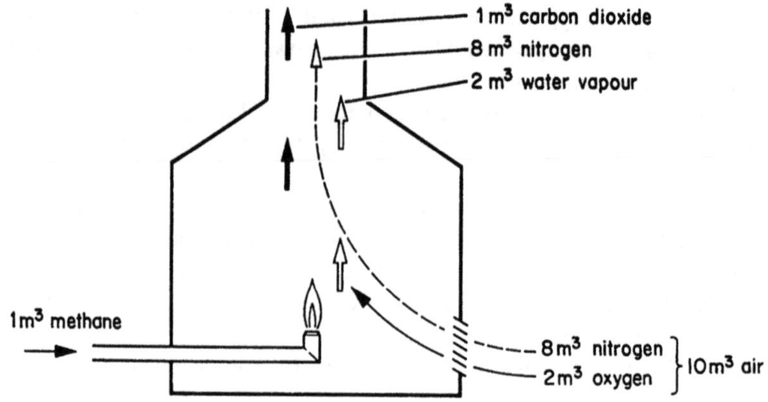

FIGURE 2.5 The volumes of gases involved in the combustion of methane.

CH_4	+	$2O_2$	=	CO_2	+	$2H_2O$
1 volume methane	requires	2 volumes oxygen	and gives off	1 volume carbon dioxide	and	2 volumes water vapour

This is true for any volume of methane. Figure 2.5 shows the theoretical volumes involved when one cubic metre of methane is burned in an appliance.

Both natural gas and LPG are mixtures of gases and their constituents are given in Table 2.3. The chemical equations for their combustion are as follows.

- Natural gas
  Nitrogen: This is not flammable and takes no part in combustion, Hence it is called 'inert'.
  Carbon dioxide: This is not flammable. It is actually a product of combustion.

  Methane : $CH_4 + 2O_2 = CO_2 + 2H_2O$

  Ethane : $2C_2H_6 + 7O_2 = 4CO_2 + 6H_2O$

  Propane : $C_3H_8 + 5O_2 = 3CO_2 + 4H_2O$

  Butane : $2C_4H_{10} + 13O_2 = 8CO_2 + 10H_2O$

- LPG gases

  Ethane, Propane, Butane } as for natural gas

  Propylene : $2C_3H_6 + 9O_2 = 6CO_2 + 6H_2O$

AIR REQUIREMENTS

It is possible to find out the combustion air requirements for a gas by first calculating the amount of oxygen required, as indicated by the combustion equation.

Take methane again as an example. Table 2.3 shows that in natural gas there is 90 m^3 of methane in each 100 m^3 of the gas.

The equation is:

$$\begin{array}{lccccccccc} & CH_4 & + & 2O_2 & = & CO_2 & + & 2H_2O \\ \text{or} & 1\text{ volume} & + & 2\text{ volumes} & = & 1\text{ volume} & + & 2\text{ volumes} \end{array}$$

So each 1 m^3 of methane requires 2 m^3 of oxygen and produces 1 m^3 of carbon dioxide and 2 m^3 of water vapour. Therefore, 90 m^3 of methane will require 180 m^3 of oxygen and produce 90 m^3 of carbon dioxide and 180 m^3 of water vapour. The oxygen requirements are therefore as shown in Tables 2.2–2.4.

Since the atmosphere consists of 21% oxygen the air requirements for the complete combustion of 1 m^3 of gas are as follows:

- *Natural gas*

$$2.06 \times \frac{100}{21} = 9.8 \text{ m}^3 \text{ of air or roughly } 10 \text{ m}^3$$

- *Commercial propane*

$$4.926 \times \frac{100}{21} = 23.45 \text{ m}^3 \text{ of air or roughly } 24 \text{ m}^3$$

Table 2.2 Oxygen Requirement – Natural Gas

Constituent	Percentage by Volume	Chemical Equation for Combustion	Volumes of Oxygen Required to Burn
N_2	2.7	–	–
CO_2	0.6	–	–
CH_4	90.0	$CH_4 + 2O_2 = CO_2 + 2H_2O$	$\frac{90}{1} \times 2 = 180.0$
C_2H_6	5.3	$2C_2H_6 + 7O_2 = 4CO_2 + 6H_2O$	$\frac{5.3}{2} \times 7 = 18.6$
C_3H_8	1.0	$C_3H_8 + 5O_2 = 3CO_2 + 4H_2O$	$\frac{5.3}{2} \times 7 = 18.6$
C_4H_{10}	0.4	$2C_4H_{10} + 13O_2 = 8CO_2 + 10H_2O$	$\frac{5.3}{2} \times 7 = 18.6$
Total	100		206.2

So 1 m^3 of natural gas requires 2.062 m^3 of oxygen – near enough 2 m^3.

Table 2.3 Oxygen Requirement – Commercial Propane

Constituent	Percentage by Volume	Chemical Equation for Combustion	Volumes of Oxygen Required to Burn
C_3H_8	85.9	$C_3H_8 + 5O_2 = 3CO_2 + 4H_2O$	$\frac{85.9}{1} \times 5 = 429.5$
C_3H_6	12.0	$2C_3H_6 + 9O_2 = 6CO_2 + 6H_2O$	$\frac{12}{2} \times 9 = 54.0$
C_2H_6	1.5	$2C_2H_6 + 7O_2 = 4CO_2 + 6H_2O$	$\frac{1.5}{2} \times 7 = 5.25$
C_4H_{10}	0.6	$2C_4H_{10} + 13O_2 = 8CO_2 + 10H_2O$	$\frac{0.6}{2} \times 13 = 3.9$
Total	100		492.65

So 1 m^3 of propane requires 4.926 m^3 of oxygen – near enough 5 m^3.

Table 2.4 Oxygen Requirement – Butane/Air

Constituent	Percentage by Volume	Chemical Equation for Combustion	Volumes of Oxygen Required to Burn
O_2	17.01	–	−17.01
N_2	63.99	–	
C_3H_8	2.5	$C_3H_8 + 5O_2 = 3CO_2 + 4H_2O$	$\frac{2.5}{1} \times 5 = 12.5$
C_4H_{10}	16.5	$2C_4H_{10} + 13O_2 = 8CO_2 + 10H_2O$	$\frac{16.5}{2} \times 13 = 107.25$
Total	100		102.74

So 1 m^3 of butane requires 1.0274 m^3 of oxygen – near enough 1 m^3.

- *Butane/air*

$$1.027 \times \frac{100}{21} = 4.89 \text{ m}^3 \text{ of air or roughly } 5 \text{ m}^3$$

PRODUCTS OF COMBUSTION

It does not follow that because 1 m^3 of natural gas requires 9.8 m^3 of air, the volume of products will be $1 + 9.8 = 10.8$ m^3 of products of combustion (Tables 2.5–2.7). For example,

Table 2.5 Products of Combustion – Natural Gas

Constituents	Percentage by Volume	Chemical Equation for Combustion	Products of Combustion	
			Carbon Dioxide	Water Vapour
N_2	2.7	–	–	
CO_2	0.6	–	0.6	
CH_4	90.0	$CH_4 + 2O_2 = CO_2 + 2H_2O$	$\frac{90}{1} \times 1 = 90.0$	$\frac{90}{1} \times 2 = 180.0$
C_2H_6	5.3	$2C_2H_4 + 7O_2 = 4CO_2 + 6H_2O$	$\frac{5.3}{2} \times 4 = 10.6$	$\frac{90}{1} \times 2 = 180.0$
C_3H_g	1.0	$C_3H_8 + 5O_2 = 3CO_2 + 4H_2O$	$\frac{1}{1} \times 3 = 3.0$	$\frac{1}{1} \times 4 = 4.0$
C_4H_{10}	0.4	$2C_4H_{10} + 13O_2 = 8CO_2 + 10H_2O$	$\frac{0.4}{2} \times 8 = 1.6$	$\frac{1}{1} \times 4 = 4.0$
Total	100.0		105.8	201.9

Note that the 0.6 m^3 of carbon dioxide present in the 100 m^3 of gas before being burned is included in the total volume. The 2.7 m^3 of nitrogen will need to be added into the final total.

Table 2.6 Products of Combustion – Commercial Propane

Constituents	Percentage by Volume	Chemical Equation Forcombustion	Products of Combustion	
			Carbon Dioxide	Water Vapour
C_3H_8	85.9	$C_3H_8 + 5O_2 = 3CO_2 + 4H_2O$	$\frac{85.9}{1} \times 3 = 257.7$	$\frac{85.9}{1} \times 4 = 343.6$
C_3H_6	12.0	$2C_3H_6 + 9O_2 = 6CO_2 + 6H_2O$	$\frac{12.0}{1} \times 6 = 36.0$	$\frac{12.0}{1} \times 6 = 36.0$
C_2H_6	1.5	$2C_2H_6 + 7O_2 = 4CO_2 + 6H_2O$	$\frac{1.5}{2} \times 4 = 3.0$	$\frac{1.5}{2} \times 6 = 4.5$
C_4H_{10}	0.6	$2C_4H_{10} + 13O_2 = 8CO_2 + 10H_2O$	$\frac{0.6}{2} \times 8 = 2.4$	$\frac{0.6}{2} \times 10 = 3.0$
Total	100.0		299.1	387.1

Table 2.7 Products of Combustion – Butane/Air

Constituents	Percentage by Volume	Chemical Equation for Combustion	Products of Combustion	
			Carbon Dioxide	Water Vapour
O_2	17.01			
N_2	63.99			
C_3H_8	2.5	$C_3H_8 + 5O_2 = 3CO_2 + 4H_2O$	$\frac{2.5}{1} \times 3 = 7.5$	$\frac{2.5}{1} \times 4 = 10.0$
C_4H_{10}	16.5	$2C_4H_{10} + 13O_2 = 8CO_2 + 10H_2O$	$\frac{16.5}{2} \times 8 = 66.0$	$\frac{16.5}{2} \times 10 = 82.5$
Total	100.0		73.5	92.5

C_3H_8	+	$5O_2$	=	$3CO_2$	+	$4H_2O$
1 volume	+	5 volumes	=	3 volumes	+	4 volumes
	6 volumes		=		7 volumes	

Obviously the volumes are not equal whether we use O_2 or air volumes. But what the equation does show is that 1 m^3 of propane combined with 5 m^3 of oxygen will produce 3 m^3 of carbon dioxide and 4 m^3 of water vapour. The amount of the products of combustion produced in each case can be calculated from the combustion equations.

Total volumes

To provide the 2.06 m^3 of oxygen required to burn 1 m^3 of natural gas, we needed 9.8 m^3 of air. The difference, $9.8 - 2.06 = 7.74$ m^3, is nitrogen. So, when 1 m^3 of natural gas is burned, the total volume of gases which are leaving the combustion chamber of the appliance is:

Carbon dioxide	1.058
Water vapour	2.019
Nitrogen from gas	0.027
Nitrogen from air	7.74
Total	10.844 m^3 or roughly 11 m^3

For commercial propane the total volume of products

Carbon dioxide	2.991
Water vapour	3.871
Nitrogen from air	18.524
Total	25.386 m^3 or roughly 26m^3

For butane/air the total volume of products is:

Carbon dioxide	0.735
Water vapour	0.925
Nitrogen from gas	0.639
Nitrogen from air	3.863
Total	6.162 m^3 or roughly 6 m^3

EXCESS AIR

In practice, slightly more air than is theoretically required for combustion is allowed to pass over the burner to take care of any slight variations in the gas rate which may occur and to provide a factor of safety. This additional air is known as 'excess air' and allowance is made for this when designing the flue for a gas appliance.

The method of measuring and calculating excess air will be dealt with later.

FLAMMABILITY

In Chapter 1 the limits of flammability of a gas were given as the percentages of gas in air at which combustion would just take place.

If a very small amount of gas was mixed with air in a closed container, the mixture would not ignite. It would be too weak or too 'lean' to burn until the amount of gas reached the lower limit for flammability, when it would just ignite.

As the amount of gas increased, it would still be possible to light it until a point was reached when there was not enough oxygen in the mixture for the gas to burn. This would be the upper limit of flammability when the mixture would be too 'strong' to ignite.

Between these limits gas would burn at a speed depending on the percentage of gas in the mixture. The fastest speed is usually obtained when the mixture contains slightly less air than the amount theoretically required to burn the gas (see Fig. 2.6).

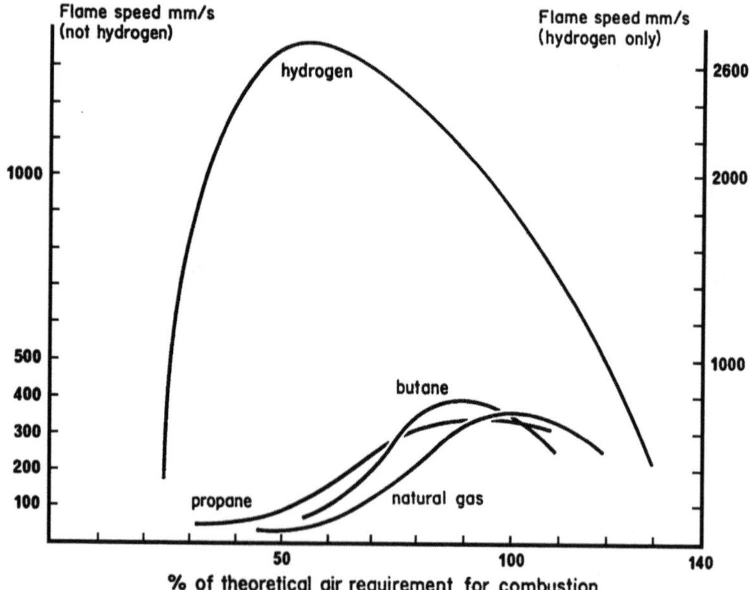

FIGURE 2.6 Effect of aeration on flame speed.

EXPLOSION

Other factors affect flame speed. If a flammable gas/air mixture is contained in a tube open at one end and is lit at that end, the flame will move fairly slowly and quietly down the tube until all the gas is burned.

If a similar mixture is contained in a closed box and lit, it will burn with a much faster speed than in the open tube. The heat generated causes an increase in the volume of the gas and its products, and results in a sudden increase in pressure and an explosion.

The larger the container, the faster the flame speed. The more turbulent the air/gas mixture, the greater the pressure developed and the more violent the explosion.

The force of an explosion does not seem to be affected by the way in which the mixture is lit, whether by a flame or a spark.

Vapour from liquids such as methylated spirit and petrol can diffuse into the atmosphere and can cause explosions. If lit in an enclosed space, 3.5% by volume of methylated spirit vapour or 1.5% of benzene will produce explosive conditions.

Clouds of dust in the air have been known to cause explosions in factories and mines.

Paraffin and other oils generally are less likely to cause explosions since they do not vaporise until they are heated. They can cause fires which may, of course, supply the necessary heat!

Extreme care is necessary when working with any flammable liquid or gas, particularly in an enclosed space.

Remember that it is not the strongest smelling mixture that is the most explosive.

FLAMES

When a mixture of gas and oxygen is burned it produces a flame. The flame itself is a zone, or space, in which chemical reactions are taking place. These reactions produce heat and, when combustion is complete, carbon dioxide and water vapour. In any flame, there are intermediate stages of combustion during which other chemicals are produced. Providing that the flame can burn freely without interference and with an adequate supply of oxygen, combustion will continue until it is complete.

As the gas and air mix, or diffuse, in the flame, they are heated. This heating causes the original constituents of the gas to break down or 'dissociate' and other different compounds of carbon, hydrogen and oxygen are formed. These may be alcohols, CH_3OH, or aldehydes, HCHO, and there may be some free carbon and carbon monoxide present. By the time the flame has taken in its full requirement of oxygen all the intermediate substances will have been oxidised to form the final products, CO_2 and H_2O.

There are many different sizes and shapes of flames, each suitable for a particular purpose. But they fall into two main types, 'post-aerated' and 'pre-aerated' flames. 'Post' means 'after' and 'pre' means 'before'. So a post-aerated flame gets all its air after it has left the burner and the pre-aerated gets some (or all) of its air before it leaves the burner.

Post-aerated flames

The simplest way to burn gas is to let it come straight out of a pipe or a jet and get its oxygen from the surrounding air. This produces a post-aerated flame, also variously known as a 'neat flame', 'luminous flame' or 'non-aerated flame'. Since all flames must have air, 'non-aerated' is hardly a true description. It could apply to the burner but not to the flame.

At low gas pressures the flame is ragged and shapeless with a large luminous zone (Fig. 2.7). It is not suitable for use in most gas appliances. Increasing the pressure can make the air mix with the gas more quickly and give a neatly shaped, stable flame. Unfortunately this can only be done for gases with a high flame speed, like manufactured gas. With a gas of low flame speed, like natural gas, the flame is blown off the jet and disappears when the pressure is increased.

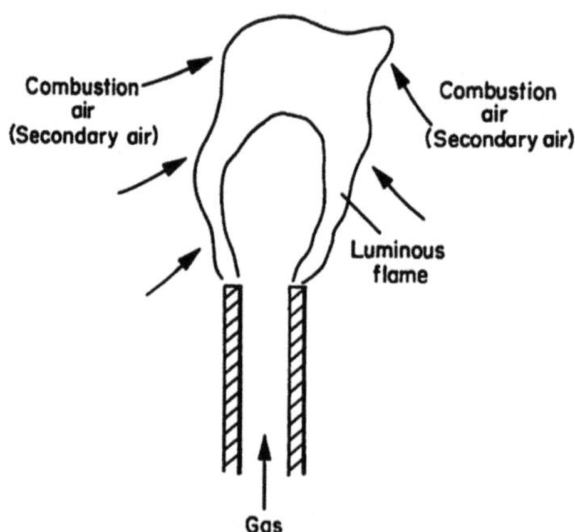

FIGURE 2.7 A post-aerated flame.

Post-aerated burners have been designed specially for natural gas and are discussed in Chapter 4. Essentially they incorporate some means of keeping the flame alight on the jet, usually by means of small 'retention flames' supplied with gas at a lower pressure than the main flame (see section on Retention Flames).

Pre-aerated flames

The flame most commonly used in domestic gas appliances is the one produced by a burner in which some of the air required for combustion is mixed with the gas before it is burned. This is the 'aerated' or 'pre-aerated' flame.

The air which is added before combustion is called 'primary air' or 'primary aeration'. The air needed to complete the combustion is obtained from around the flame itself and is called 'secondary air'. In most domestic burners about 40–50% of the total air requirement is added as primary air. So about half the air required to burn the gas is mixed before burning. There are some burners in use in industrial equipment where all the air is provided as primary air. These are dealt with in Chapter 4.

The pre-aerated flame is smaller and more concentrated than a post-aerated flame burning the same amount of the same gas. Both flames will give off the same total amount of heat, which depends only on the calorific value of the gas and not on the way in which it is burned.

Although the total heat output does not change, the flame itself gets smaller and hotter as more and more primary air is added. When all the air required is added as primary air it produces a very small intensely hot flame.

The characteristic feature of the pre-aerated flame is its 'inner cone'. A simple, stable flame appears to have two parts, an inner and an outer cone (Fig. 2.8). The inner cone is usually a bright green–blue colour and the outer flame a darker bluish-purple. To understand why the flame takes on this shape it is necessary to look at a simple burner as shown in Fig. 2.8.

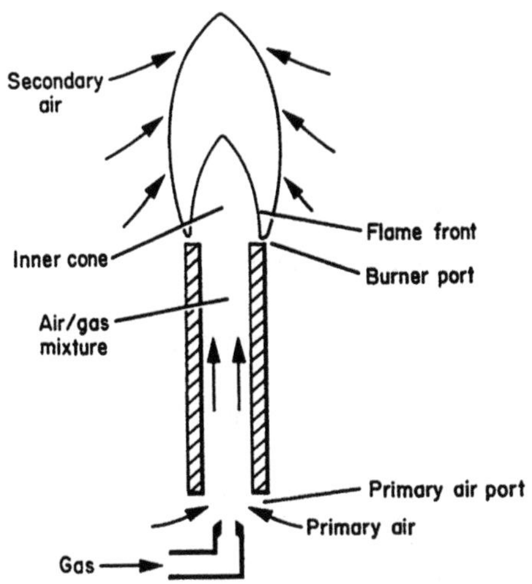

FIGURE 2.8 A pre-aerated flame.

Gas is forced, by its pressure, out of a jet placed centrally at the end of a tube. The stream of gas injected into the tube draws in the primary air and pushes it up the tube, mixing it with the gas on the way. The mixture is lit at the top end of the tube. All the holes in burners are called 'ports'. So air is drawn in at the 'primary air ports' and the mixture burns at the 'burner port'. The tube of the burner is called the 'mixing tube' and the jet supplying the gas is an 'injector'.

The boundary between the air/gas mixture emerging from the tube and the actual flame itself is called the 'flame front'. It is this flame front which takes on the cone shape and is the boundary of the inner cone. The cone occurs because the mixture flowing up the tube is slowed down at the sides where it is in contact with the walls of the tube. So it is faster towards the centre, where it reaches its

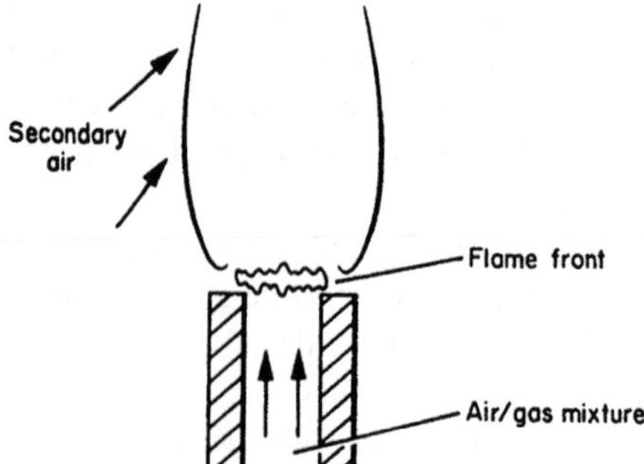

FIGURE 2.9 Flame structure with high primary aeration.

top speed. This means that it tends to push the flame front away from the burner port much more at the centre than at the sides. It follows, therefore, that the inner cone contains unburnt gas.

If the amount of primary air is small, the mixture reaches a high speed at its centre and a long inner cone is formed. As more primary air is added, the cone becomes shorter and brighter until a flat, ragged and noisy flame front is formed (Fig. 2.9).

Zones of pre-aerated flame

A pre-aerated flame has four zones.

1. Inside the inner cone is an unburnt air/gas mixture.
2. Between the mixture and the flame is the flame front where the speed of the gases passing through the surface of the cone is equal to the flame speed of the mixture.
3. The reaction zone is where gases are dissociated by the heat and partially burned.
4. The outer mantle is where combustion of the gas is completed by air diffusing into the flame.

The process is shown in Figs 2.10 and 2.11.

1. A to B is the top of the burner and the burner port. Just inside the opening the temperature of the mixture begins to rise, partly because the burner gets hot at the top and partly from radiant heat from the flame.
2. B to C is the zone of the flame up to the top of the inner cone which forms the flame front. Here the temperature of the mixture rises more quickly. Because air is drawn into the sides the average mixture strength falls although there is still unburnt gas inside the cone.
3. C to D is the reaction zone where the temperature continues to rise and the mixture strength continues to fall as more air is drawn in.
4. D represents the outer mantle of the flame where combustion is completed. It is the hottest part of the flame. Beyond this the temperature begins to fall as heat is lost to the surrounding air.

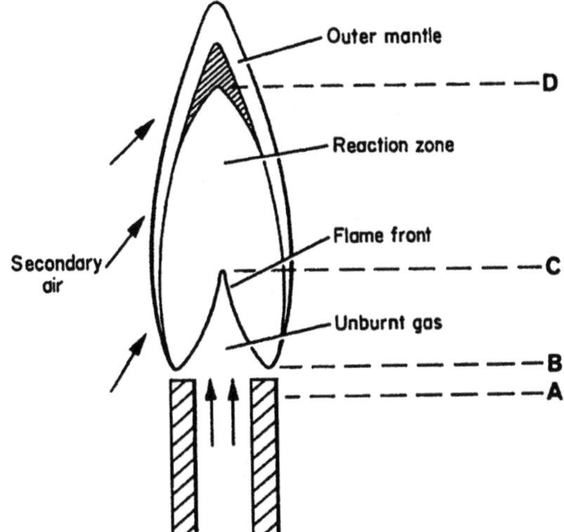

FIGURE 2.10 Zones of a pre-aerated flame. If the flame has a yellow tip, this is formed in the shaded area at the top of the reaction zone.

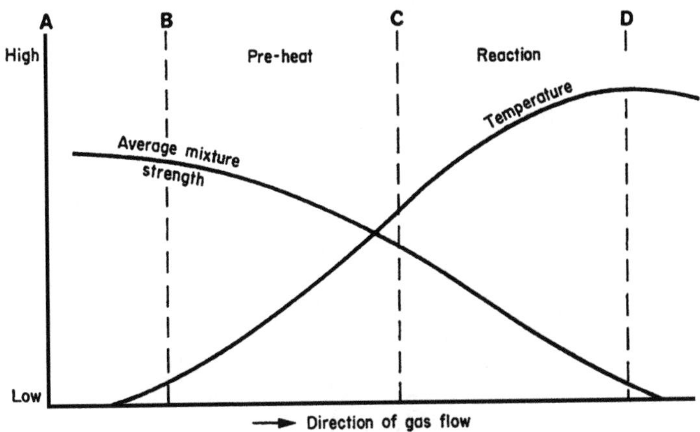

FIGURE 2.11 Graph of the air/gas mixture and temperature through the four flame zones of a pre-aerated flame.

LIGHTING-BACK

You have seen that, if the air/gas mixture flows up the tube at the same speed at which the flame can burn it up, then the flame will stay at the end of the tube. If, however, the speed of the mixture is reduced, the flame will burn its way down the tube to the injector. This is called 'lighting-back' or 'striking back'.

Since only about half the air required to burn the gas can enter through the air ports, the flame cannot burn completely inside the tube. So some gas continues to burn at the burner port and often combustion is not satisfactorily completed. Experiments using glass tubes can actually show the inner cone leaving the flame mantle and moving down the tube to burn on the injector.

The flame speed of an air/gas mixture is low at the flammability limit and increases as more primary air is added. You can see the effect of this on the flame. With only a little primary air the inner cone is long and the flame has a yellow tip. As more primary air is added the inner cone becomes shorter, brighter and more clearly defined. With even more primary aeration the inner cone becomes ragged, noisy and flat (Fig. 2.12). Finally the flame lights-back (Fig. 2.13). Slow-burning gases, e.g. natural gas, propane and butane, are unlikely to have lighting-back problems.

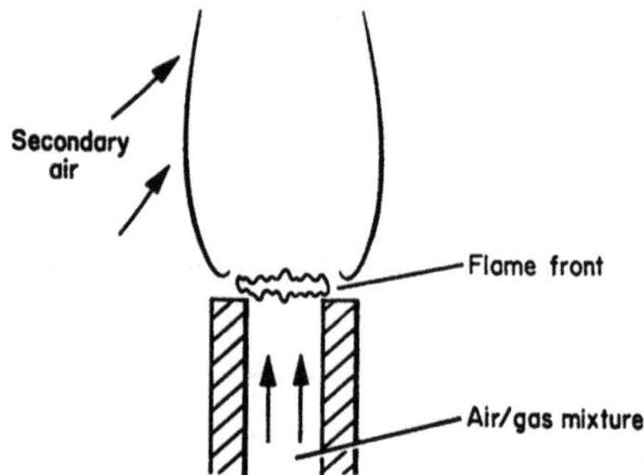

FIGURE 2.12 Flame structure with high primary aeration.

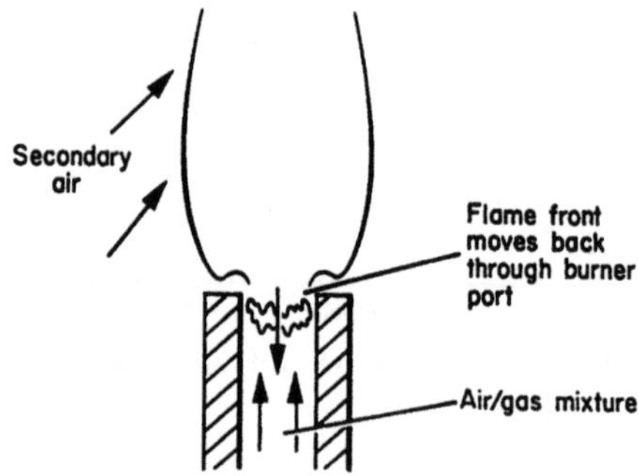

FIGURE 2.13 Lighting-back.

FLAME LIFT

This is the opposite of lighting-back. If the speed of the air/gas mixture up the tube is greater than the speed at which the flame can burn, then the flame will be pushed away from the burner port. If the mixture's speed is only slightly higher than the flame speed of the gas, the flame will continue to burn with its flame front just a little distance from the end of the tube. This is called 'flame lift' (Fig. 2.14).

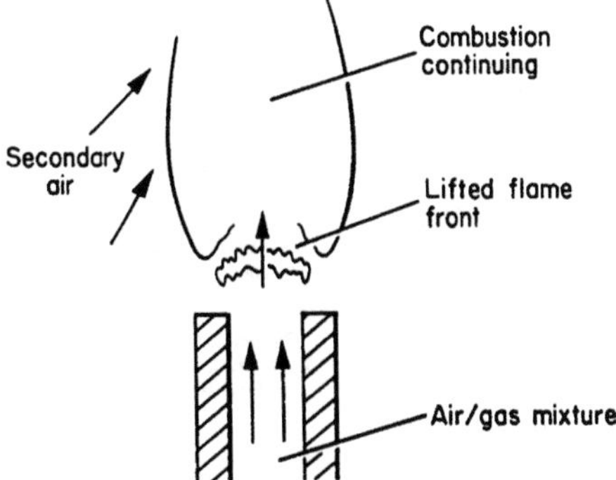

FIGURE 2.14 Flame lift.

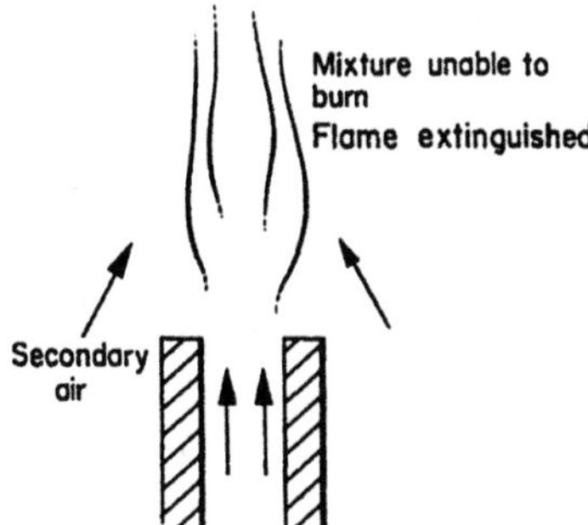

FIGURE 2.15 Blow-off.

If the speed of mixture is much greater than the flame speed, the flame front will be pushed away from the burner port completely and the flame will disappear. This is because the gas diffuses into the surrounding air and the mixture becomes too weak to burn. This is known as 'blow-off' (Fig. 2.15).

Flame lift or blow-off is particularly likely to occur with natural gas because of its low flame speed. Substitute natural gases contain hydrogen which increases their flame speed and widens the flammability limits. So the risk of lift off is reduced.

Retention flames

Special burners are used for natural gas (see Chapter 3) and they have small flames to keep the main flame alight at the burner port. These 'retention flames' are supplied with the air/gas mixture at a lower speed than the main flame (Figs 2.16 and 2.17).

The retention flames continuously relight the main flame as it tries to lift off. In addition they raise the temperature of the gas in the mixture and increase its flame speed so that a stable flame can form.

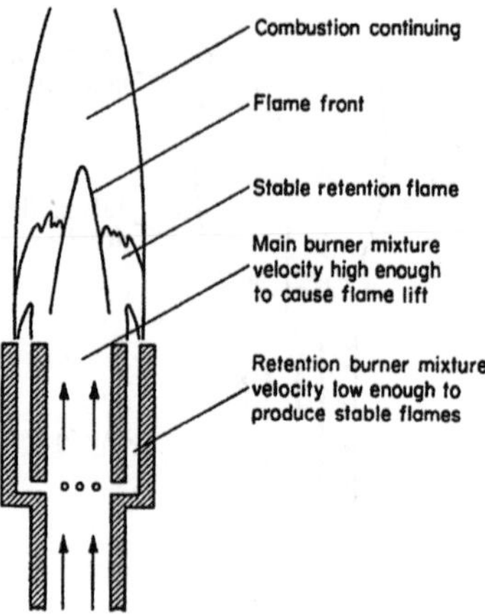

FIGURE 2.16 Pre-aerated burner with retention flame.

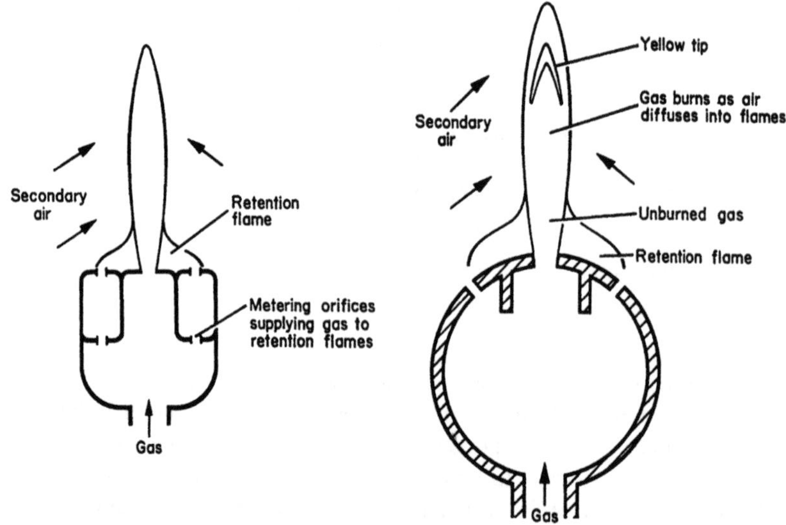

FIGURE 2.17 Post-aerated burners with retention flames.

LIFT AND REVERSION PRESSURES

For a particular burner and a particular gas the speed at which an air/gas mixture will flow up the mixing tube depends on the pressure of the gas. Increasing the pressure will increase the speed of the mixture and so can cause the flame to lift off the burner ports. If the pressure is decreased then the speed of the mixture will slow down again and the flame will revert back to its original position on the burner port. This 'reversion' takes place when the mixture speed is slightly less than that speed which causes flame lift.

It is usual to relate flame lift and reversion to the pressures at which they occur.

INCOMPLETE COMBUSTION

Earlier in the chapter it was explained that, in the reaction zone of the flame, the constituent gases were broken down and recombined to form other compounds before being burned to the final products of combustion, CO_2 and H_2O, or carbon dioxide and water vapour. These intermediate substances included carbon, carbon monoxide, alcohols and aldehydes.

If combustion is interrupted at this incomplete stage then we have 'incomplete combustion' and these products can be released from the flame.

Of the products of incomplete combustion, carbon monoxide is the most dangerous. It is a toxic gas. A concentration of only 0.4% in air can be fatal within a few minutes. Unfortunately carbon monoxide combines much more readily with the haemoglobin in the blood stream than does oxygen, which is what the blood needs. The blood normally picks up oxygen from the lungs and carries it to the tissues of the body, collecting the waste products in return. If this haemoglobin (red cells) becomes saturated with carbon monoxide, the tissues cannot take it in or get rid of their waste matter. The body is quickly poisoned and the person dies. Even if death does not occur, there may be very serious brain damage due to the lack of oxygen (see Table 2.8).

Carbon itself is not dangerous. It can be seen in the flame as a yellow tip or a yellow zone. If this part of the flame touches a cold surface the carbon is deposited either as hard lumps on the surface or in the form of fine soot throughout the combustion chamber.

Soot is a particular nuisance because it makes the servicing of appliances difficult. It is often experienced in the form of a fine dust, which blows about the house as soon as an attempt is made to remove it. It is difficult to get it off your hands or clothes. The formation of soot in narrow passages within a combustion chamber can restrict the flow of the products of combustion which, in turn, will reduce the flow of fresh air to or through the flames. The shortage of air may result in a more serious deposit of soot and carbon monoxide can be formed.

Alcohols are quickly oxidised to become aldehydes which have a characteristic smell and often give the indication that incomplete combustion is taking place. Although they are poisonous they are a nuisance rather than a danger because they cause irritation of the eyes which becomes unbearable before a dangerous concentration is reached. The amounts produced by incomplete combustion would not, in any case, be dangerous.

Causes of incomplete combustion

Almost anything that interferes with the combustion process may result in the release of products of incomplete combustion from the flame. The two main causes, however, are chilling the flame and starving it of oxygen.

CHILLING

Chilling occurs when a flame touches a cold surface. This does not mean a cold surface in the ordinary sense. It means a surface which is cold compared to the

Table 2.8 Effects of Carbon Monoxide on Adults

% Saturation of Haemoglobin with Carbon Monoxide	Symptoms
0–10%	No symptoms
10–20%	Tightness across the forehead, yawning
20–30%	Flushed skin, headache, breathlessness and palpitation on exertion. Slight dizziness
30–40%	Severe headache, dizziness, nausea. Weakness of the knees, irritability. Impaired judgement. * Possible collapse
40–50%	Symptoms as above with increased respiration and pulse rates. Collapse on exertion*
50–60%	Loss of consciousness, coma
60–70%	Coma, weakened heart and respiration
70% and above	Respiratory failure and death
% Volume of carbon monoxide in air	*% Saturation of haemoglobin with carbon monoxide*
0.01%	4% in 1½ hours About 15% maximum with indefinite exposure
0.03%	10% in 1½ hours 20% in 4–5 hours
0.05%	20% in 1½ hours 40% in 4–5 hours
0.4%	60% in a few minutes

* These two factors explain why CO poisoning is frequently fatal.

1. It impairs mental ability so that a person may be brought to the verge of collapse without realising that anything is wrong.
2. Any sudden exertion would then cause immediate collapse, and an inability to escape from the situation.

flame. If the surface cools the gases in the flame to below their ignition temperature, some of them will not be completely burned and may be released as incomplete combustion products.

The effect can be particularly bad if the inner cone of a pre-aerated flame is 'broken' by the cold surface. This is because this inner cone contains unburned gases.

Chilling can also result if the flame is exposed to a stream of cold air or to a draught. This may also tend to lift the flame off the burner. If the flame is under-aerated and yellow-tipped and touches a cold surface, soot will be deposited and, if the conditions are not corrected, the deposit will continue to grow, causing more interference.

LACK OF OXYGEN

Lack of oxygen produces carbon monoxide and may be caused by restricting the amount of air to the burner or by a shortage of oxygen in the air.

The air supply to the burner may be restricted either by a blockage of the air inlet to the appliance or by a blockage of the flue outlet. Both have the same effect on the flow of air over the burner.

Even when the total amount of air required can enter the appliance the required amount may not be able to pass through to the flames if there are deposits of dirt, fluff or soot on or in the burner or in the mixing tube.

A shortage of oxygen reaching the burner may be due to a lack of oxygen in the air. Normally air has about 20% oxygen and a negligible amount of carbon dioxide. If a gas appliance burns in a room it uses up oxygen and gives out carbon dioxide and water vapour. So the proportion of oxygen in the room is reduced and a condition can be reached at which combustion is affected.

This lack of oxygen is called 'vitiation'. Air is 'vitiated' (made impure) by reducing its proportion of oxygen compared to the other gases. Adequate ventilation should be maintained to prevent the air becoming vitiated.

If the flame is burning freely in the air, combustion will continue satisfactorily until there is not enough oxygen for the gas to burn completely.

If the flame is inside an appliance, it will become longer in its attempt to reach out for more oxygen. This can cause it to touch the combustion chamber and bad combustion will result.

When flames touch each other or a cold surface, they are said to 'impinge'. Flame 'impingement' will generally cause poor combustion either by chilling or by one flame robbing another of oxygen.

COMBUSTION STANDARDS

Because of the dangers of poor combustion, minimum standards of combustion have been laid down by the British Standards Institution. All 'approved' appliances must conform to these specifications, British Standard Specifications (BS) 5258, BS EN 483 and BS EN 625.

Standards, in Britain, are based on the ratio of the amount of carbon monoxide produced to the amount of carbon dioxide produced. The ratio $CO:CO_2$ must not exceed 0.02 under conditions which have been laid down. Tests are carried out using a range of test gases which simulate the worst conditions for light-back, sooting, flame lift and poor combustion likely to occur on the district.

The most important test is carried out at a 20% heat input overload rate. The ratio must still not exceed 0.02 at this rate.

In some countries the acceptance standard depends on the rate of production of CO in air-free products.

IGNITION: PILOT FLAMES

When a flammable mixture of gas and air is lit by another flame, ignition is usually very simple. A small gas flame, situated beside the main burner and called a 'pilot' flame, supplies much more heat energy than is required for ignition (Fig. 2.18). Although other methods are more complex, in every case it is essential that the ignition source comes into contact with an air/gas mixture within the flammability limits.

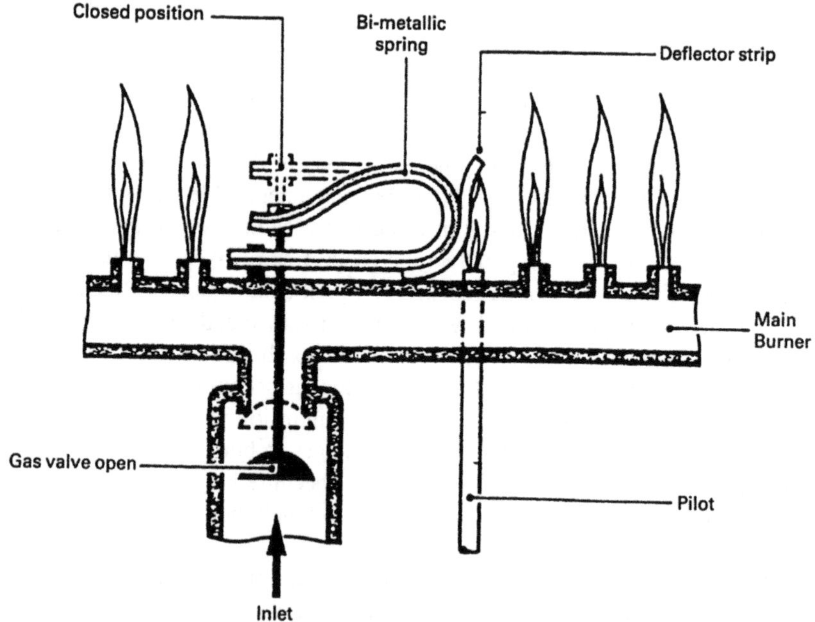

FIGURE 2.18 Pilot ignition.

In the example shown, the pilot flame also activates the bimetallic device (see Chapter 11, section on Bimetallic Devices).

IGNITION: HOT WIRE FILAMENTS

Lighting a gas flame by means of a small spiral coil of thin wire or 'filament' presents more problems. If, however, the gas to be lit contains free* hydrogen,

---

* The term 'free' applied to hydrogen (H_2) refers to hydrogen on its own or in a mixture of gases, but not when in a compound e.g. CH_4 or C_3H_6.

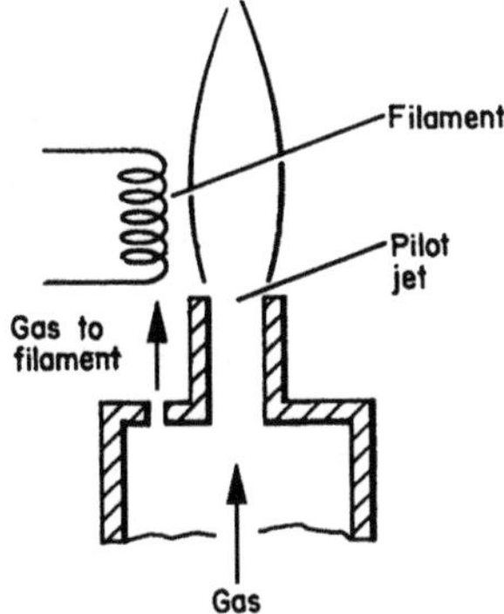

FIGURE 2.19 Filament ignition.

ignition by means of a platinum filament is made easier because of the 'catalytic effect' of platinum. A 'catalyst' is something which helps a reaction to take place without itself being changed. In this case, if the temperature of the filament is raised slightly by passing an electric current through it, hydrogen in the gas and oxygen in the air will combine on the surface of the filament to form water vapour. This reaction gives out heat which raises the temperature of the filament up to the ignition temperature of the gas (Fig. 2.19).

To ignite natural gas, which does not itself contain free hydrogen, it is necessary to use a higher voltage to raise the filament to the temperature required to light the gas. The minimum voltage needed is in the region of 3 V. A filament which does not operate by catalytic effect is known as 'glow coil'. Most natural gas appliances use spark ignition to light the pilot or main burner.

SPARK IGNITION

A spark ignition head is shown in Fig. 2.20. The 'points' which form the spark gap are positioned in the path of the air/gas mixture as it emerges from the

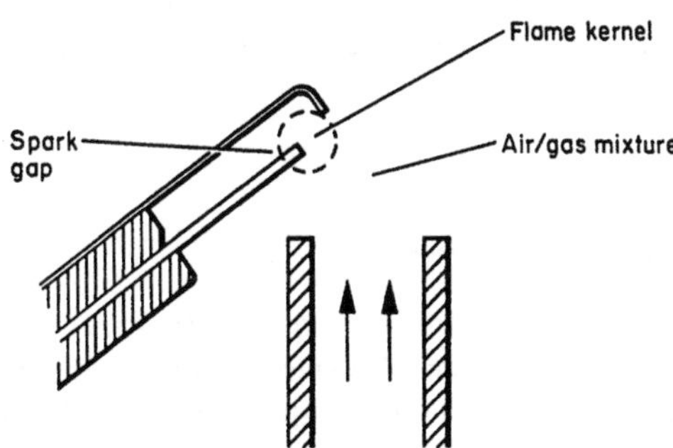

FIGURE 2.20 Spark ignition.

burner port. When the igniter is operated, sparks will be formed in the gap as the electricity flows across. A very high voltage is needed to break down the resistance of the air/gas mixture in the spark gap and the width of the gap requires careful adjustment. If the gap is too wide, no spark will occur.

An average spark gap on gas igniters is 4 mm and the applied voltage is approximately 12000 V.

When a spark is formed it raises the temperature of the gases in its path and, for a moment, lights a small sphere of gas known as a 'flame kernel'. Combustion will then take place on the surface of this sphere, which is at the same time losing heat to its surroundings. If the kernel is sufficiently large, it will ignite the surrounding mixture. If it is too small, it will lose heat quicker than heat is being generated by the combustion on its surface, and the flame will disappear.

There is, therefore, a minimum size of flame kernel required to produce the amount of energy for successful ignition. This size depends on the gases used and on their movement through the spark gap.

OXIDISING AND REDUCING FLAMES

When a substance becomes combined with oxygen it is said to be 'oxidised'. So 'oxidising' or 'oxidation' is simply adding oxygen to the substance concerned. A familiar example of natural oxidation is the rusting of iron. When unprotected iron or steel is left in contact with air or water the metal reacts with the oxygen in the air or water and forms rust. This rust is a kind of iron oxide, and so rusting is a type of oxidation.

Just for the record, oxidation can also be brought about by the removal of hydrogen, but that is another story.

'Reduction' is the opposite of oxidation. So 'reducing' is carried out by adding hydrogen or removing oxygen.

Gases may have oxidising or reducing effects and a study will show that the oxidising gases are found in the products of complete combustion, while the reducing gases are those within the reaction zone where there is a shortage of air.

It follows, therefore, that the outer part of the flame, where there is complete combustion and an excess of oxygen, produces oxidising conditions. Within the flame envelope, where there is a lack of oxygen, there are reducing conditions. The knowledge and control of such conditions is important for a number of industrial applications.

CONTROLLED ATMOSPHERES

During the heat treatment of metals and ceramics (pottery) in industry it is often necessary to control the condition of the air surrounding the material within the furnace. If this is not done, the finish on the product can be spoiled.

A great advantage of using gas to heat a furnace is that the combustion can be regulated to give the best conditions for the process. We can have 'controlled atmospheres' surrounding the material.

If gas is burned with an excess of air, the resulting atmosphere will be 'oxidising'. The oxygen not used for combustion is then available for use in combination with the surface of the material.

If there is a lack of air, the atmosphere will be 'reducing'. Because there is insufficient oxygen to burn the gas, it will try to get some from elsewhere and if metallic oxides are present in the material within the furnace, it will remove the oxygen from them. There are industrial processes in which this is a desirable effect.

It is possible to have a 'neutral' atmosphere which is neither oxidising nor reducing. That is, one where there is sufficient air for combustion without producing an excess of oxygen in the products.

This is an over-simplification of a complicated subject, and an atmosphere which is oxidising to one material may be neutral to another. But if the gases in and around flames are considered, their general effects on iron, as an example, are as follows:

Oxygen: A very active oxidising agent, it combines with the iron to form ferrous oxide, a black powdery scale, on the surface of the iron.

$$O + 2Fe = 2FeO$$

Carbon dioxide: At low temperatures this is a neutral gas but at high temperatures it can be oxidising.

$$CO_2 + Fe = FeO + CO$$

Water vapour: Water vapour acts as an oxidising gas.

$$H_2O + Fe = FeO + H_2$$

Carbon monoxide: This usually acts as a reducing gas, combining with oxygen to form carbon dioxide.

$$2CO + O_2 = 2CO_2$$

With low-carbon steels it can increase carbon at the surface of the metal, forming iron carbide. This process is called 'carburising'.

$$2CO + 3Fe = Fe_3C + CO_2$$

Hydrogen: Hydrogen is a powerful reducing agent.

$$H_2 + FeO = Fe + H_2O$$

Hydrocarbons: Gases such as methane, ethane etc. are fairly unreactive at low temperatures; at high temperatures they act as carburising gases.

To sum up, the causes of incomplete combustion and the actions to remedy are given below.

Cause of Incomplete Combustion	Action to Remedy
Incomplete combustion caused by blockage of air inlet grill of open flued appliance.	Remove blockage from air inlet grill, then check operation of appliance as further blockages may have occurred within the appliance.
Incomplete combustion caused by blockage of primary air port of appliance.	Remove blockage from primary air port, then check operation of appliance.
Incomplete combustion caused by blockage of heat exchanger with soot or debris.	Remove blockage from heat exchanger, then check operation of appliance as further blockages may have occurred within the appliance burner or flueways.
Incomplete combustion caused by over-gassing attributed to excessive burner pressure.	Reset burner pressure to manufacturer's instructions and if necessary check gas rate using the gas meter.
Incomplete combustion caused by over-gassing attributed to excessive gas rate.	If burner pressure is correct check that injectors are correct and if necessary change.
Incomplete combustion caused by flame impingement.	Check burner alignment. If incorrec fit correctly.
Incomplete combustion caused by flame impingement.	Check burner pressure and adjust as necessary.
Incomplete combustion caused by flame impingement.	If burner alignment & pressure are correct check that injectors are correct and if necessary change.
Incomplete combustion caused by appliance being designed for other gases e.g. LPG.	Check if conversion kit is available if not, replace appliance.

Chapter | three

Liquefied Petroleum Gas

Chapter 3 is based on an original draft by Mr Don Barber

WHAT IS LPG?

Before it is possible to understand the safe storage, handling and utilisation of any commodity, it is first essential to comprehend the nature of the material under consideration.

No person shall carry out any LPG 'work' in relation to gas fitting, gas storage vessels or gas installations unless they are competent to do so. They must be registered with an approved HSE registration body and hold a valid certificate of competence for each work activity that they wish to undertake. The certificate must have been issued under the Nationally Accredited Certification Scheme (ACS) for individual gas fitting operatives or an LPG Aligned Scottish/ National Vocational Qualification.

LPG is the normal abbreviation used to describe 'Liquefied Petroleum Gas' which is itself the term used to describe a group of hydrocarbons existing as vapours under ambient conditions of temperature and pressure, but which can be liquefied by the application of moderate pressure and/or refrigeration.

Such liquefaction is always accompanied by a considerable decrease in the volume occupied by the vapour so that the liquid formed requires much less storage space. The material is therefore normally stored and distributed in the liquid phase, in pressurised containers and systems, and is finally allowed to revert to the vapour phase before the point of eventual consumption. This property is the basis of one of the principal advantages of LPG as a convenient premium energy source.

SOURCES OF LPG

Liquefied petroleum gases originate from two main sources. LPGs are found at the oil and/or gas fields where they are removed as condensable products from natural gas (such LPGs are often referred to as Natural Gas Liquids – NGLs),

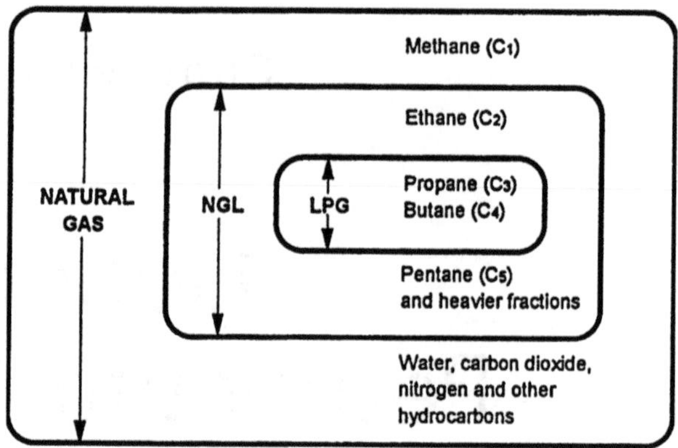

FIGURE 3.1 Constituents of natural gas.

and LPGs are also removed from crude oils during the 'stabilisation' process applied in order to reduce the vapour pressure prior to shipment (Fig. 3.1).

The second normal source of LPGs is found at the refineries where the crude oil is processed. LPGs are produced by many refinery activities, but the actual composition and quality vary considerably and is being determined by the process involved and the type of crude oil or feedstock being used.

Large quantities of LPG are used by oil refineries, particularly butane which is added to gasolene to improve the volatility. LPGs are also used as feedstock for petrochemical production.

CHEMICAL PROPERTIES OF LPG

Whatever the source of the LPG, one common factor is found, namely that the bulk of the material comprises hydrocarbons containing three or four carbon atoms per molecule. It is only these hydrocarbon gases, often referred to as C_3 or C_4 gases that are easily liquefied by moderate compression at ambient temperature. By comparison, natural gas (methane – CH_4) can be liquefied only by refrigeration. The normal constituents of LPG are:

Propane C_3H_8
Propylene C_3H_6
Butane C_4H_{10}
Butylene C_4H_8

The actual composition of an LPG will vary and may also include traces of ethane (C_2H_6) and ethylene (C_2H_4), together with impurities. It is therefore important to bear in mind that the LPGs in normal use are commercially available products rather than pure chemical hydrocarbons. The significance is that any properties quoted will represent typical or average values.

It is necessary, for many applications, to know the actual analysis and the nature of any impurities so as to determine the properties of the mixture, or to be assured by the producer that it has these properties within the limits of established acceptable specification standards.

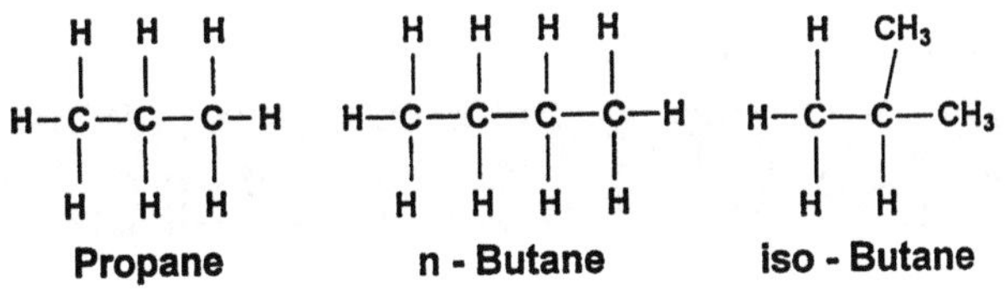

Propane n - Butane iso - Butane

FIGURE 3.2 Commercial LPG mixtures.

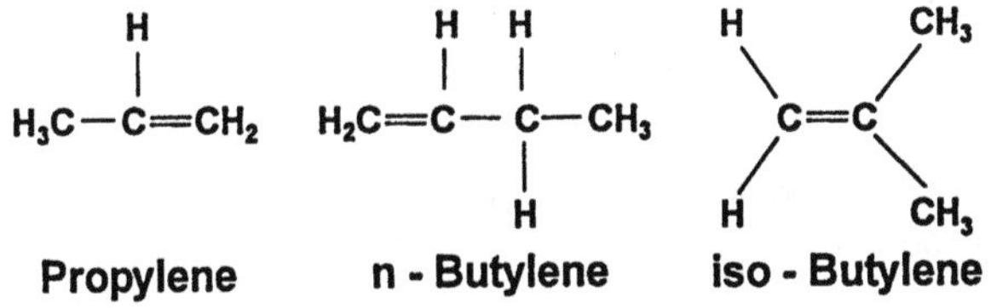

Propylene n - Butylene iso - Butylene

FIGURE 3.3 Unsaturated LPG mixtures.

Saturated hydrocarbons generally constitute the bulk of commercially available LPG, these being in the form of propane (C_3H_8) and butane (C_4H_{10}), the latter being found as isobutane and normal butane (see Fig. 3.2).

Whilst saturated hydrocarbons are found in LPGs produced from refinery pipe stills and distillation units, other processes, particularly cracking plants, will result in unsaturated C_3 and C_4 hydrocarbons being found in LPG mixtures. These unsaturated LPGs are usually propylene (C_3H_6), normal and iso-butylene and transbutylene (C_4H_8) (see Fig. 3.3).

In view of the variations which occur in the hydrocarbons found in LPG, national standards have been introduced which specify the requirements of the commercial product. These refer to butane, propane, or mixtures of the two hydrocarbons. Most of these national standards will in future be influenced by the ISO standard 9162. In the United Kingdom the quality specifications are defined by BS 4250 'Specifications for Commercial Propane and Commercial Butane'.

As the requirements of these standards are normally only of interest to LPG supply and marketing organisations, or to equipment manufacturers, full details have not been included in this chapter. The typical properties of LPG grades are given in Table 3.1.

SPECIAL GRADES OF LPG

LPG is also used for a range of specialised applications where standards more rigorous than those defined by the commercial standard are applied.

Examples of such applications are chemical feedstocks, food processing, cigarette lighter refills and aerosol propellants.

In the majority of these cases the material is also required in an odourless condition thus introducing the need for gas detection equipment to indicate the

Table 3.1 Typical Properties of Commercial LPG Grades

	Commercial Butane	Commercial Propane
Relative density of liquid at 15.6 °C (60 °F)	0.57–0.58	0.50–0.51
Litres per tonne at 15.6 °C	1723–1760	1966–2019
Relative density of gas compared to air at 15.6 °C and 100 kPa	1.90–2.10	1.40–1.55
Volume of gas (l) per kg of liquid at 15.6 °C and 100 kPa	406–431	537–543
Ratio of gas volume to liquid volume at 15.6 °C and 100 kPa	233	274
Boiling point (°C) at atmospheric pressure	−2	−45
Specified maximum vapour pressure kPa–Temperature (°C)		
−40.0	–	138
−17.8	–	311
0	193	552
37.8	586	1550
45.0	689	1860
Sulphur content – % weight	Negligible to 0.02	Negligible to 0.02
Limits of flammability – % volume of gas in air to form a flammable mixture – at atmospheric pressure.	Upper 9.0 Lower 1.8	Upper 10.0 Lower 2.2
Calorific values		
Gross		
(MJ/m^3) dry	121.8	93.1
(MJ/kg)	49.3	50.0
Nett		
(MJ/m^3) dry	112.9	86.1
(MJ/kg)	45.8	46.3
Air required for combustion (m^3 to burn 1 m^3 of gas)	30	24

presence of minor leaks which, with the commercial product, would be normally detected by smell as, like natural gas, an odorant is normally added.

Many of the special applications also require a closely controlled vapour pressure and an absence of residues. Special LPG used as hydrocarbon aerosol propellant requires very rigid quality control to achieve the production of an odour-free butane–propane blend and near pure products having little or no unsaturated hydrocarbons. This particular application has grown rapidly as a direct result of environmental pressure to limit, or ban, the use of chlorofluorocarbons in aerosols.

COMBUSTION OF LPG

The combustion of LPG may be considered using the following basic equations from which it will be recognised that correct combustion yields carbon dioxide and water vapour (see also Chapter 2).

$$C_3H_8 + 5O_2 = 3CO_2 + 4H_2O + \text{HEAT}$$

$$2C_4H_{10} + 13O_2 = 8CO_2 + 10H_2O + \text{HEAT}$$

Both combustion reactions are associated with an increase in volume of products plus further expansion due to the generation of heat. Each volume of propane and butane vapour requires 24 and 30 times its own volume of air, respectively, for complete combustion and, at the same time, yields three to four times its own volume of carbon dioxide.

It is essential, therefore, that adequate ventilation must be provided when LPG is burnt in enclosed spaces (e.g. caravans, boats, small bathrooms etc.), otherwise asphyxiation, directly due to the depletion of oxygen and the formation of carbon dioxide, will rapidly occur.

The formation of incomplete combustion products such as carbon monoxide will also occur when there is insufficient air for correct combustion. Whereas carbon dioxide is an asphyxiant gas, carbon monoxide is toxic.

EFFECT OF LPG ON MATERIALS

Commercial LPG does not affect metals but the use of aluminium is usually confined to vapour system equipment components and domestic size cylinders, unless stringent precautions are taken to ensure the removal of any traces of caustic soda, chlorides or moisture which may be present. These compounds have a vigorous chemical action when brought into contact with aluminium.

Many non-metallic substances are chemically attacked by LPG or its impurities – possibly the most notable being natural rubber which rapidly becomes spongy when exposed to LPG. For this particular reason, natural rubber or other non-LPG resistant materials must never be used for items such as LPG hoses, gaskets or valve seals.

One significant influence of the effect of LPG on materials is the solvent action on many of the traditional jointing compounds used for sealing screwed and flanged joints in gas pipework. It is important to use PTFE tape or one of the special compounds available which are impervious to LPG.

Other materials, notably some plastics, can become either soft or brittle, particularly if components of the plastic material are dissolved out by the LPG. If plasticisers are leached out from unsuitable material they may then cause problems by being deposited in the LPG system, particularly in regulators and control valves.

The solvent action of LPG on materials in service is frequently unpredictable and can therefore only be accurately ascertained, with any degree of certainty, from laboratory testing. Thorough laboratory evaluation should always be made before any untried or unproven materials are introduced into any equipment where it will come into contact with LPG.

PHYSICAL PROPERTIES OF LPG

Whilst the physical properties of the product owe their characteristics to the chemical composition, it is in the physical state that LPG is normally encountered (Table 3.2).

Vapour pressure

Having subjected the vapour to moderate pressure so as to achieve liquefaction, the resultant liquid must be contained within a pressurised system (a pressure vessel), until the gaseous state LPG is required by the consumer.

Table 3.2 Molecular Weight, Boiling Points and Critical Properties of LPGs

	Propane	iso-Butane	iso-Butane	Commercial Propane	Commercial Butane	Ethylene
Molecular weight	44.1	58.1	58.1	44	58	2.1
Boing point (°C) at 1 atm	−42.1	−0.5	−11.7	−45	−2	−103.7
Critical temperature (°C)	96.8	152.0	135.0	95	150	9.9
Critical pressure (atm)	41.9	37.5	36.0	40	35	50.7

Vapour pressure is a measure of the volatility of the gas. Where vapour exists, in an enclosed system, in conjunction with the liquid phase the pressure exerted is referred to as the 'saturated vapour pressure'. At the boiling point this pressure is equal to atmospheric pressure and it increases as the temperature rises to the critical. Propane, with its lower boiling point thus exerts, under identical conditions, a greater vapour pressure than butane.

Knowledge of the vapour pressure of LPG is essential so as to be able to specify the design conditions for the pressurised system, e.g. tank, tanker, pipework or cylinder (Fig. 3.4). It is also necessary to enable the gas offtake rates by natural vaporisation to be calculated and controlled. Control is effected using a 'regulator' which gives a steady outlet pressure of gas even when the inlet pressure changes.

In practical terms, systems are often specifically designed to be suitable for the pressure exerted by either butane or propane. This, of course, precludes a butane system from being used for propane but enables the propane system to be classified as 'dual purpose'.

Pressure vessels are normally protected by safety relief valves. These are designed to protect the system against overpressure brought about by fire exposure and are set to discharge in accordance with the designed working conditions. Very small cylinders are excluded from this requirement.

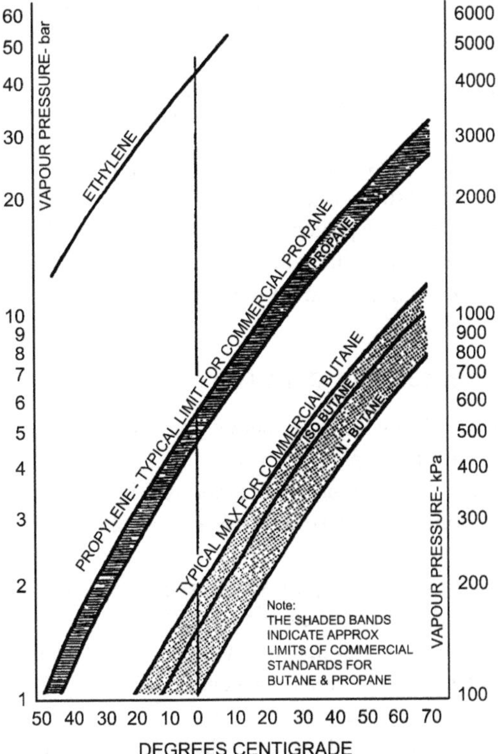

FIGURE 3.4 Liquefied petroleum gas typical vapour pressures.

Boiling point

The constituent gases found in the commercial LPG mixture all have very low boiling points and therefore normally exist in the vapour phase, under atmospheric conditions.

Where the gases are held at a temperature at, or below, their boiling point the vapour pressure will be equal to, or less than, atmospheric. This property has encouraged the development of large-scale storage at marine terminals, where the product is held in refrigerated form in what is, essentially, a non-pressurised system. Large shipments, to and from such terminals, by marine tanker are also transported in a refrigerated condition.

Above ambient temperature, the gases exert an increasing vapour pressure thus increasing the pressure required to achieve liquefaction. This pressure continues to increase until the 'critical temperature' is reached (96.67 °C for propane: 152.03 °C for *n*-butane), above which temperature the gases cease to exist in the liquid state even if further pressure is applied.

The vapour pressure of the product at the 'critical temperature' is known as the 'critical pressure'.

Latent heat

The latent of the liquid product is the quantity of heat required to enable vaporisation to occur. As such it is essential data required during the design of vaporiser systems.

When liquid phase LPG vaporises naturally the latent heat required is taken from the liquid itself, and its immediate surroundings, at the same time causing a drop in temperature. This is known as 'autorefrigeration'.

With propane under such conditions very low temperatures can be achieved (down to −40 °C), therefore protective clothing is essential to prevent operators receiving severe cold burns. All equipment used to handle LPG where these low temperatures can occur must also be made from suitable materials.

Specific volume, relative density

In the vapour phase, LPGs exist as heavy gases being approximately 1.5–2.0 times the density of air (propane 1.40–1.55; butane 1.90–2.10).

On liquefaction, LPG reduces considerably in volume: the ratio of gas volume to liquid volume at 15.6 °C/1016 mbar is 233 for butane and 273 for propane.

Both gases exist as a clear liquid which is approximately half the density of water – (propane 0.50–0.51; butane 0.57–0.58).

LPG vapours, being heavier than air, will cling to the ground seeking to enter trenches, drains or other low areas. Dispersion of these vapours will also take longer than would be the case with gases lighter than air. For this reason, LPG should not be stored or used in locations where any escape could concentrate in low lying areas or enter drains, trenches or basement areas.

Coefficient of cubical expansion of liquid

Liquid phase LPG expands considerably as its temperature increases. The coefficients of cubical expansion at 15 °C are approximately 0.0029/°C for propane and 0.0020/°C for butane, these values being, for example, four times the equivalent for fuel oil, ten times that for water and 100 times that for steel.

This high rate of expansion has to be fully considered when specifying the maximum quantity of LPG permitted to be filled into any pressure vessel. The maximum quantity is determined by 'the filling ratio' (as defined by Codes of Practice or Standards) for different LPG specifications under differing maximum ambient conditions.

All types of LPG container, from a butane cigarette lighter to a depot storage tank, have a design filling ratio which requires that they are never completely filled with liquid. The free space allowed, up to about 20%, allows for the liquid to expand if the temperature rises.

Note: Because the filling ratio precautions taken to prevent the hydraulic filling of storage systems cannot be extended to the connecting liquid phase pipework, these parts of the system are protected by the provision of small over-pressure relief valves situated in all areas where the liquid LPG can be trapped between closed valves. Failure to provide such valves can result in extremely high hydraulic pressures being generated which will eventually lead to the failure of joints, gaskets, seals or connecting hoses.

Vapour lines do not require hydrostatic relief valves to be fitted.

Odorisation of LPG

In their refined state, LPGs are normally almost odourless. To detect leaks easily, using the sense of smell, the published standards for commercial butane and propane call for the gas to have a characteristic odour, if necessary to be achieved by the addition of a stenching agent. When a stenching agent is used, this could be ethyl mercaptan (alternatively thiophane) in the approximate ratio of 1.0 kg of odorant per 100,000 l of liquid phase gas.

The odour required has to be sufficiently strong to ensure detection, by smell, of a leakage down to a concentration in air of one-fifth of the lower limit of flammability.

Limits of flammability

A mixture of LPG and air is highly combustible within certain concentrations, known as the flammable range. This range is bounded by the lower and upper limits of flammability which are approximately 1.8–8.4% for butane and 2.4–9.5% for propane. These are approximate values at atmospheric pressure. At higher pressures, or with oxygen, the flammability limits are different (Table 3.3).

The term 'limits of flammability' assumes that the gas and air are thoroughly mixed. In practice, LPG escaping into the atmosphere without ignition tends to settle and is often too rich to burn, i.e. the mixture is above the upper limits concentration.

Table 3.3 Approximate Limits of Flammability of Gas–air Mixtures (at 15 °C and 100 kPa)

Gas	Lower Limit	Upper Limit	% Gas in Stoichiometric Mixture
Commercial butane	1.8	8.4	3.2
Commercial propane	2.2	9.5	4.2
Methane	5.3	14.0	9.5
Hydrogen	4.0	74	29.6
Acetylene	2.5	80	7.75

On the fringe of such a 'gas cloud' mixing with air occurs and if the resultant combustible mixture reaches an ignition source, it will ignite. Heat generated by the burning gas causes turbulence and further mixing results such that, eventually, most of the gas will become mixed within the flammable range. A serious hazard may then result.

The table of properties (Table 3.1) shows that when butane liquid is vaporised under standard conditions of temperature and pressure, its free vapour volume is 233 times that of the liquid.

Because of this high liquid–vapour expansion ratio, 1 l of liquid butane will, for example, produce 233 l of neat vapour which, at 5% concentration in air, would yield some 4660 l of highly flammable and potentially explosive mixture.

The practical significance is that LPG must be securely held in its pressurised system, and that any liquid phase leak is significantly more hazardous than a similar sized leak in the vapour phase. (It follows that should a leak occur in an LPG system the gas should be allowed to disperse safely. This is achieved by ensuring that systems are always sited at a safe distance from possible sources of ignition.) The safety distance tables in the Codes of Safe Practice have been established to avoid this type of occurrence during normal operations Table 3.4 and Fig 3.5.

Calorific value

The calorific value of commercial LPG grades is considerably higher than natural gas. Details are given in the table of typical properties, Table 3.1.

Dew point

Whereas LPG is stored and distributed in the liquid phase for most applications, it is normally used in the vapour phase under controlled pressure. The vapour in

Table 3.4 Minimum Recommended Separation Distances for LPG Storage Vessels (Distances from Buildings, Boundaries and Sources of Ignition) (Also See Fig. 3.5)

Maximum water capacity					*Minimum separation distances*					
Of any single vessel in a group			Of all vessels in a group		Above ground vessels			Buried or mounded vessels*		
Litres	Gallons (approx.)	Nominal LPG Capacity (tonnes)	Litres	Gallons	From buildings boundary, property line or fixed source of ignition (a)	With fire wall (b)	Between vessels (c)	From buildings etc. to		Between vessels * (f)
								Valve Assembly (d)	Vessel (e)	
					m (ft)	m	m	m	m	m
150–500	28–100	0.05–0.25	1500	330	2.5 (8)	0.3"	1	2.5	0.3	0.3
>500–2500	100–500	0.25–1.1	7500	1650	3 (10)	1.5	1	3	1	1.5
>2500–9000	500–2000	1.1–4	27,000	6000	7.5 (25)	4	1	7.5	3	1.5
>9000–135,000	2000–30,000	4–60	450,000	100,000	15 (50)	7.5	1.5	7.5	3	1.5
>135,000–337,500	30,000–75,000	60–150	1,012,500	225,000	22.5 (75)	11	¼ of sum of the dia. of 2 adjacent vessels	11	3	***
>337, 50	>75000	150	2,250,000	500,000	30 (100)	**15**		**15**	3	***

*To qualify as buried or mounded vessels the cover shall not be less than 0.5 m. Mounded vessels with exposed ends shall be located to meet the above-ground distances around the exposed areas.

**For vessels up to 500 l, the fire wall needs to be no higher than the top of the vessel and may form part of the site boundary. The fire wall for a vessel up to 2500 l water capacity may form part of a building wall in accordance with Fig. 3.1. Where part of the building is used for residential accommodation the wall, including overhanging but excluding the eaves, against which the LPG is stored should be imperforate and of 60 min fire-resisting construction (to BS 476, Part 8).

***The spacing between adjacent vessels should be determined by the site conditions and the needs for safe installation, testing, maintenance and removal of vessels.

Source: LP Gas Association.

a system will be fed under pressure direct from the storage container (or in the case of many industrial applications via an indirectly heated vaporiser unit). Within the system the temperature must be maintained above the dew point so that the vapour does not re-condense. This is of particular importance for butane and butane–air systems (Fig 3.6).

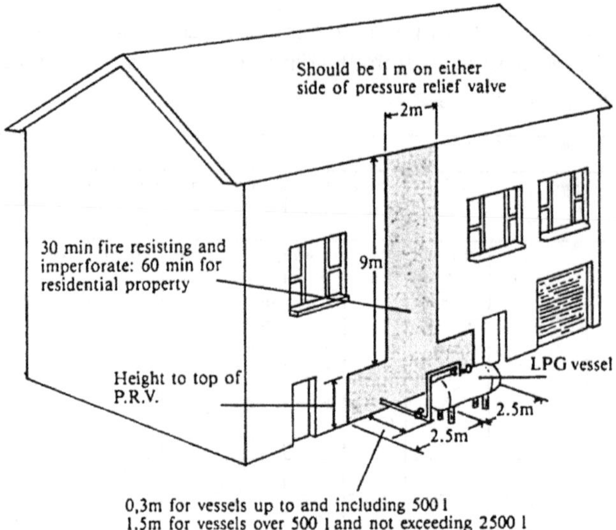

FIGURE 3.5 Vessel Siting and fire precautions. The LP Gas Association (LPGA) has produced Codes of Practice which deal with fired installations (Bulk Vessels) stored in vessels larger than 150 l (approximately 75 kg). The safety distances are laid down in the LPGA COP1 Part 4 (amended February 2008) (see Table 3.4).

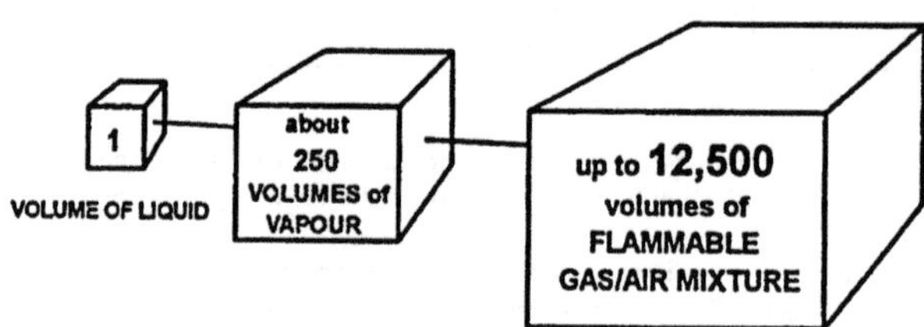

FIGURE 3.6 Expansion and flammability of LPG.

LPG–Air systems

Butane–air and propane–air systems are used to provide a gas which has comparable combustion characteristics to natural gas or to manufactured gas.

This is achieved by mixing LPG with air in a specialised venturi or carburettor mixer system and then distributing the gas through normal distribution pipework to the consumer.

This mixture is safe to handle as it is always outside the flammable range. Combustion is completed by adding the remainder of the required air at the burners.

The LPG–air mixtures have higher dew points than the neat LPG vapour and can therefore be distributed without any re-condensation occurring. The production of LPG–air mixtures requires specialised knowledge but the application of the mixture is the same as other gas service practice.

The LPG–air plants are normally used to supply a district in advance of a natural gas supply or for standby purposes, e.g. to replace natural gas supplied to industry on an interruptible basis.

Viscosity of LPG

The absolute viscosity of LPG, held as a liquid under vapour pressure, is considerably less than that of water thus requiring a very high integrity in the pressurised system if leakage is not to occur.

With the liquid phase LPG, an increase in temperature will be accompanied by a drop in viscosity whereas the same temperature rise will increase the vapour phase viscosity.

A knowledge of and familiarity with the viscosity characteristics of LPGs are therefore required before designing pumping and pipework systems.

THE SUPPLY OF LPG TO CONSUMERS

Unlike natural gas, which is piped to the consumer, LPG has to be transported from the storage terminal or secondary distribution depot. This transportation is always in the liquid phase at ambient temperature and therefore requires to be handled in specially designed pressure vessels.

A typical LPG distribution chain is shown in Fig. 3.7.

At the start of the distribution chain the LPG is handled in large road or rail tankers and in some countries of Europe by river barge. At the secondary distribution depot there is often a cylinder filling plant and loading facilities for small road tankers used for delivering to the consumer.

The typical consumer is the same as those using natural gas, in addition to which there are those outlets which are beyond the normal mains distribution network.

CYLINDER SUPPLIES OF LPG

The LPG cylinder

LPG cylinders are transportable pressure vessels, specially designed to handle the liquefied gas safely. They are constructed to one of a number of special design codes, which in the United Kingdom is **BS 5045 Part 2** for steel cylinders up to 130 l water capacity.

Note: The size of all LPG containers is defined by the water capacity, which is based on the total volume of the vessel. The actual quantity of LPG which can be filled into a container is determined from this figure, the density of the LPG being filled and the filling ratio.

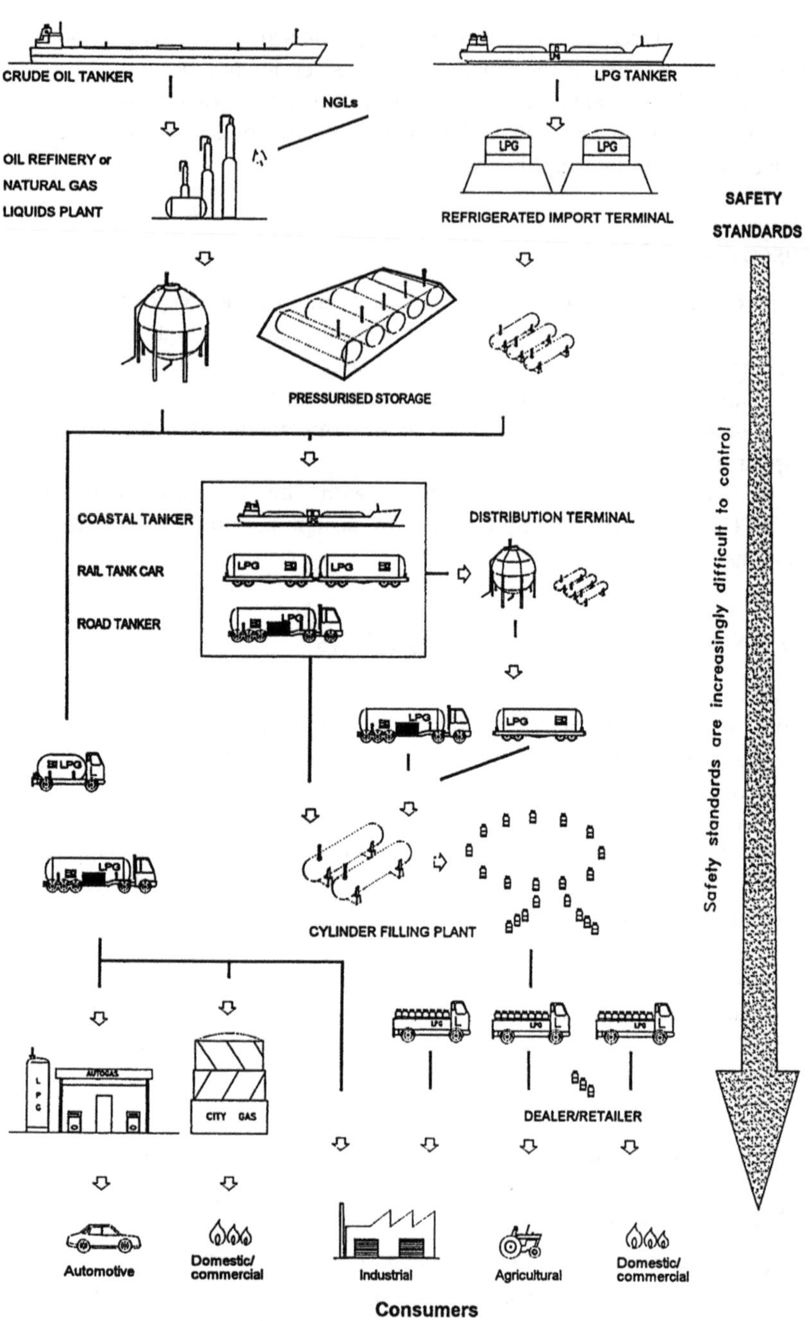

FIGURE 3.7 The LPG distribution chain.

Refillable cylinders are used in a range of semi-standard sizes, from the small camping and blowtorch cylinder of 0.35 kg capacity to the large industrial propane cylinder of around 50 kg.

In the United Kingdom both Commercial Butane and Commercial Propane are available in refillable exchange cylinders. For safety reasons the valves are different and the contents can often be determined by the cylinder colour. The convention is blue for butane and red for propane, but this is not universal.

Most of the cylinders in use are owned by the gas supply companies who are responsible for the safe condition of the cylinders and their regular inspection and testing, but certain cylinders are owned by the consumers.

Cylinder design and construction

The design code to which a cylinder is constructed specifies the type of materials which can be used and the test procedures which have to be carried out during and after manufacture. Details giving the name of the manufacturer, the date of manufacture, the design code and the cylinder test pressure, together with the maximum filling weight, are marked on all cylinders. There is also space on the data panel for the addition of subsequent test and repair dates.

Most cylinders are fabricated in what is described as two-piece construction. This is shown in Fig. 3.8. The body of the cylinder is in two parts, each deep pressed from sheet steel. After preparation of the central joint the two halves are welded together to form the pressure container.

In the centre top of the cylinder a threaded 'bung' is welded to the body. This enables the cylinder valve to be fitted.

The cylinder is completed by the attachment of a heavy base ring on which the cylinder stands, and a shroud, or collar, which protects the cylinder valve during transportation. The shroud also acts as a carrying handle. Some small cylinders do not have a shroud fitted.

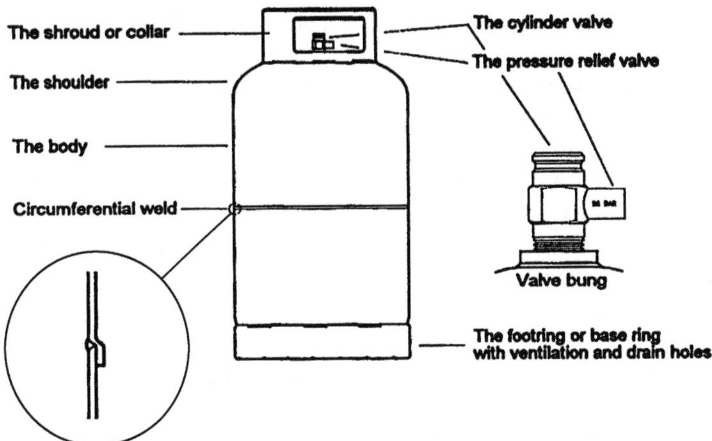

FIGURE 3.8 The LPG cylinder – construction details.

After fabrication most cylinders are heat treated and shot blasted before being given a protective and decorative coating. Some cylinders are zinc metal sprayed before painting as an extra anti-corrosion measure. After hydraulic testing the valve is fitted.

The cylinder valve

The cylinder valve serves two purposes which have very different requirements. These are to enable the cylinder to be filled with liquid phase LPG at a fast rate, and for vapour to be withdrawn for consumption at a slow rate in the vapour phase. Throughout these operations the valve should not leak across its main seat or at the connection between the valve and the filling head or consumer attachment.

The cylinder valve is a robust precision unit but it can easily be damaged by rough handling or abuse, and it may leak if subjected to the ingress of dirt.

There are many designs of LPG cylinder valve currently used worldwide. Those in common use, particularly in the United Kingdom, are described here.

The *Compact* valve was developed in Denmark but is today manufactured by many companies. It is a top entry clip on system which is used with matched low- and high-pressure regulators. The valve is made in a range of top diameters, the popular sizes being 20 and 21 mm. The matching regulators for each of these valves are not interchangeable and no attempts should be made to modify either valves or regulators to make them fit.

Inside the top of the valve is a soft seal which ensures a gas tight connection to the matching regulator. This seal must always be present and it is important that it is not shrunken, cracked, damaged or hard. Replacements should only be fitted by the gas supplier at the filling plant to enable leak testing to be carried out.

The latest developments of the 'Compact' system incorporate a self-sealing dust cap at the top of the valve.

Details of the 'Compact' valve system are shown in Fig. 3.9(a) and 3.9(b).

The 'Jumbo' valve originated in Germany and has been improved since the original version. When fitted with its matched regulator it becomes a two-stage regulation system. The latest versions incorporate a patented dust cap at the top of the valve.

Details of the 'Jumbo' valve system are shown in Fig. 3.10(a) and 3.10(b).

The 'POL' valve originated in the United States of America. Two versions of the valve exist today. These are basically the same but the American and European valves have slightly different dimensions. For this reason, care should be taken to ensure that compatible fittings are always used to connect to these valves.

The USA is the world's largest consumer of LPG therefore many items of LPG equipment are of American origin. Their suitability for use in the UK should always be checked carefully.

Details of the 'POL' valve system are shown in Fig. 3.11.

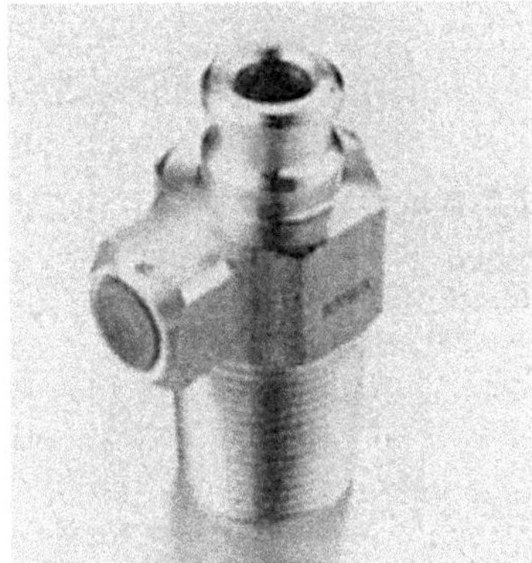

Compact valve

Low pressure regulator

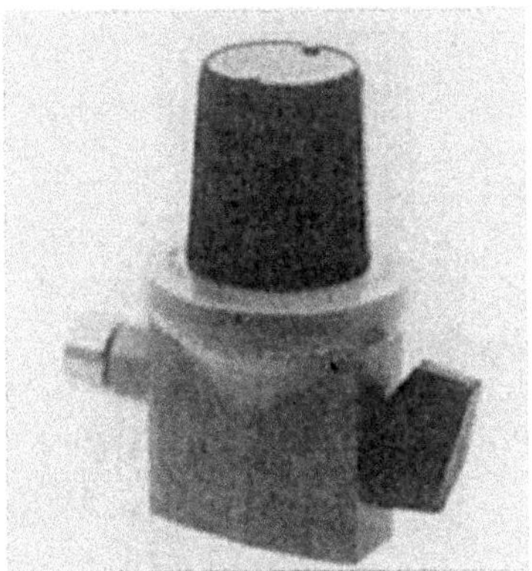

High pressure regulator

FIGURE 3.9(a) Compact valve system.

SPECIAL CYLINDER VALVES

There are a number of special cylinder valves in service, mostly to be found fitted to special propane cylinders. Cylinders designed to supply LPG as an engine fuel for fork lift trucks, or similar applications, provide LPG in the liquid

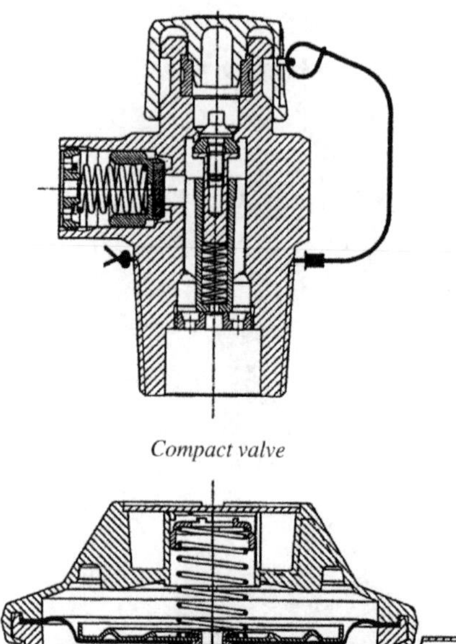

Compact valve

Kosangas

Low pressure regulator

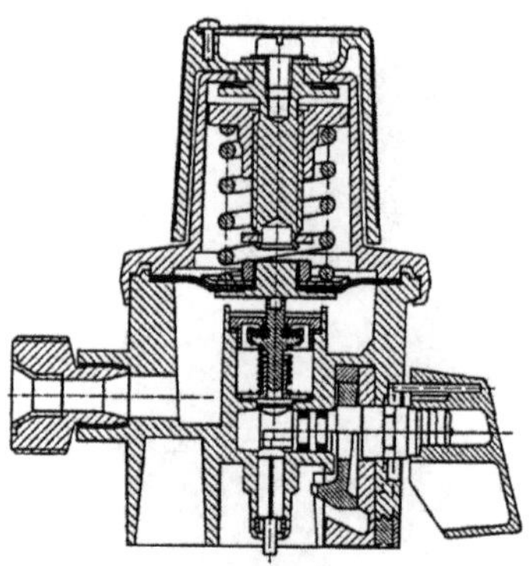

High pressure regulator

FIGURE 3.9(b) Compact valve system.

phase. The valves are of the self-sealing, quick-coupling type; large propane cylinders are also supplied with liquid offtake valves, for use in specialised applications, e.g. the liquid feed burners used for grain drying.

For safety reasons the special valves fitted to these cylinders have a different coupling system to the normal vapour offtake valves – it would

a

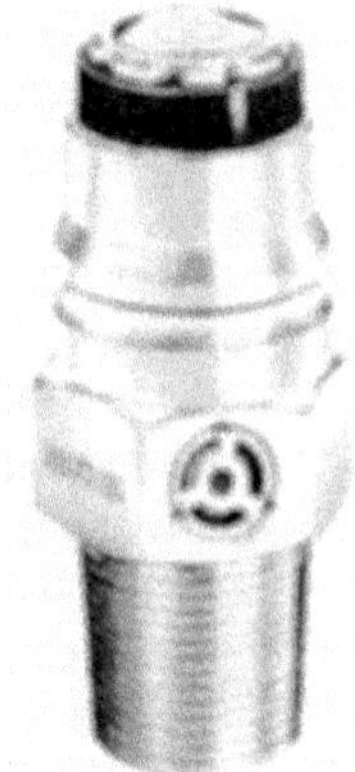

Jumbo value

Low pressure regulator for Jumbo valve

High pressure regulator for Jumbo valve

FIGURE 3.10(a) Jumbo valve system.

b

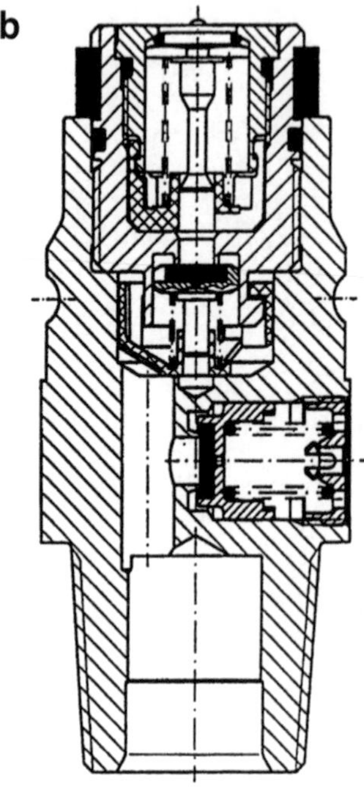

Jumbo valve

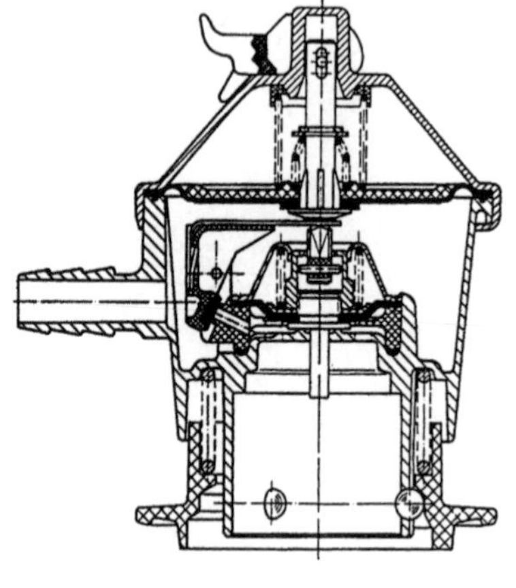

Low pressure regulator for Jumbo valve

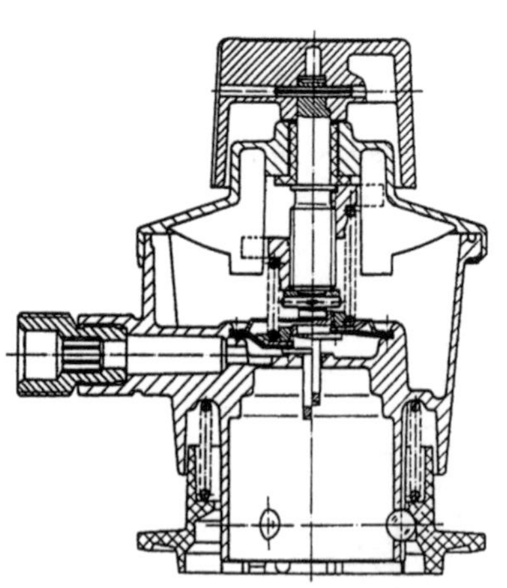

High pressure regulator for Jumbo valve

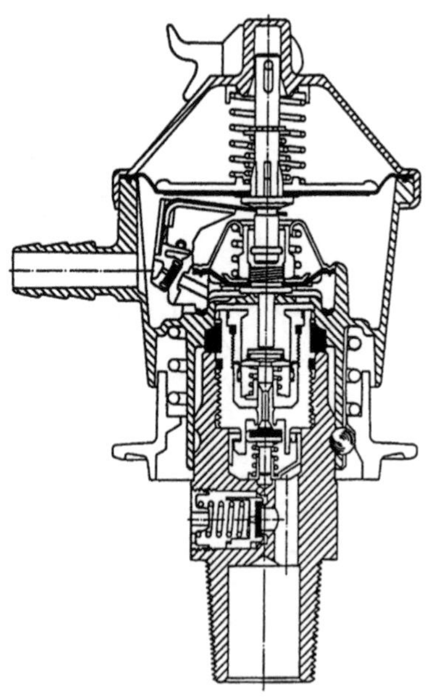

Jumbo operating system (low pressure regulator in use)

FIGURE 3.10(b) Jumbo valve system.

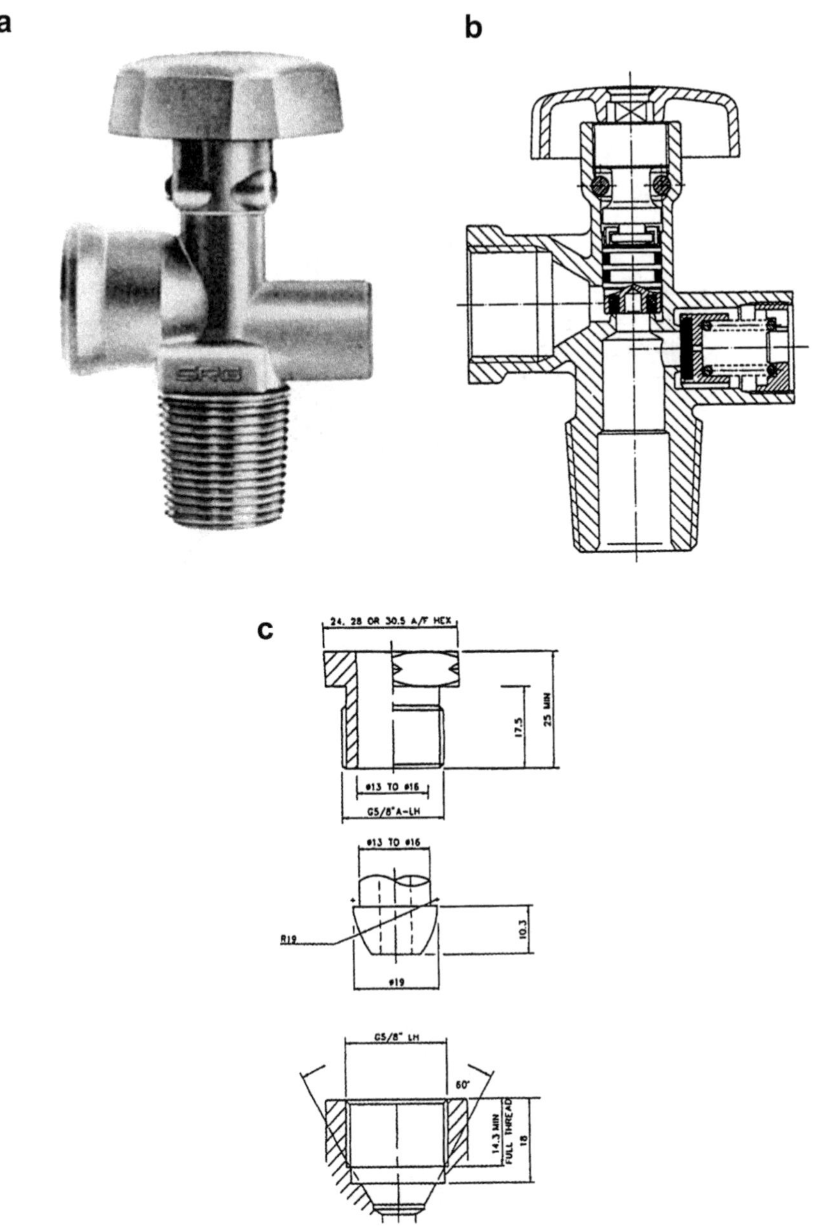

FIGURE 3.11 POL side entry valve system for propane: (a) POL valve with safety relief valve fitted; (b) section through POL valve showing pressure relief system; (c) POL bullnose connecting system for propane cylinders.

be very dangerous to be able to fit a liquid offtake cylinder to a vapour offtake system.

The ‘Camping Gaz’ range of refillable cylinders and disposable cartridges is shown in Fig. 3.12(a) and 3.12(b). This equipment, and a wide range of camping

FIGURE 3.12(a) Camping Gaz refillable cylinders and non-refillable cartridges.

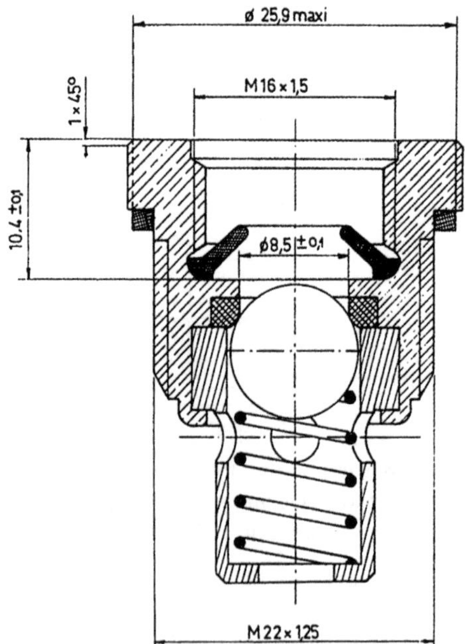

FIGURE 3.12(b) Section through Camping Gaz cylinder valve.

and light industrial appliances, are produced and sold by Camping Gaz which is part of the international ADG organisation who markets the same systems world-wide. This is the only LPG cylinder system which is truly international.

The refillable Camping Gaz cylinders have a valve which is fitted internally at the top of the cylinder. The connection is made by screwing a regulator, or one of a range of specially designed appliances, into the valve. No pressure relief valve is fitted. A plug with a carrying handle protects the thread of the valve when the cylinder is not in use.

Unlike most LPG cylinders, Camping Gaz cylinders are sold outright and become the property of the purchaser. They are refilled by Camping Gaz and users are guaranteed an exchange cylinder which has been correctly filled and is in a sound and safe condition.

The Camping Gaz system operates on butane.

Primus also manufactures a range of small cylinders with an internal screw valve, and also a very well known range of torch sets and burners. Many of these torches, some of which are illustrated in Fig. 3.13, are used by gas service technicians in their daily work (see also Chapter 13).

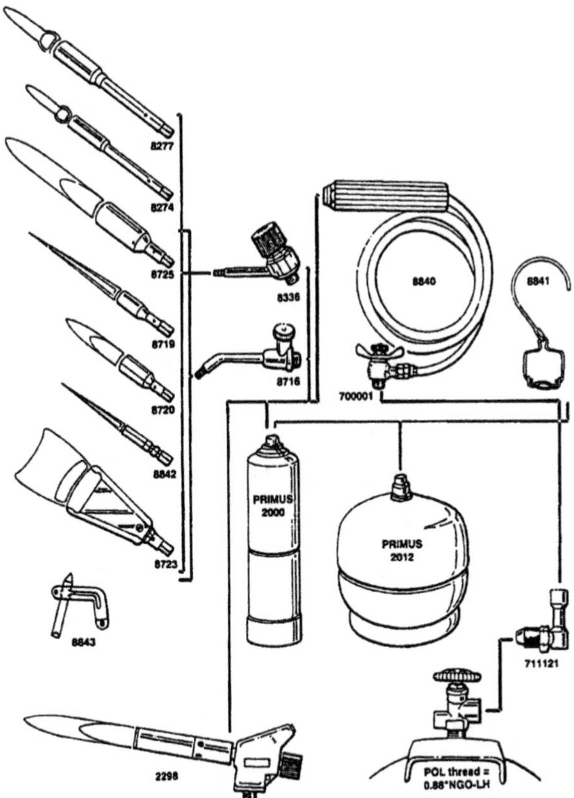

FIGURE 3.13 The Primus cylinder and torch system.

In addition to the equipment shown, the Primus range of equipment includes a large number of heavy duty burners and heaters, mostly designed for special applications.

The Primus equipment operates on propane.

Pressure relief valves on cylinders

In the 1970s the LPG industry in the UK undertook to fit pressure relief valves to all but the smallest refillable cylinders. These valves are an integral part of the cylinder valve and have a capacity in accordance with the formula found in the LP Gas Association Code of Practice 15, Part 1 – Safety Valves.

It should be noted that the requirement calls for a minimum flow rate according to the water capacity of the cylinder. For this reason valves are available with different capacities, the larger ones being used for the large propane cylinders. Typical LPG cylinder capacities and dimensions are shown in Fig. 3.15.

The Code also recommends the following set pressures:

- 21 bar for butane service,
- 26 bar for propane service.

Cylinders are filled at special centres which are equipped with the necessary equipment to ensure that the correct grade and quantity of gas is dispensed and that the cylinder and its valve are checked, weighed and leak tested to ensure that they are in a safe condition before being filled or despatched. After filling the cylinder valves may also be fitted with a commercial guarantee or safety seal to ensure that the cylinder is not tampered with before use.

Gas service technicians should note that LPG cylinders cannot be serviced in the field, and in the event of problems being experienced the defective cylinder should be immediately returned to the dealer, or gas company, with a label attached describing the fault.

Some of the common reasons for rejecting cylinders are shown in Fig. 3.14.

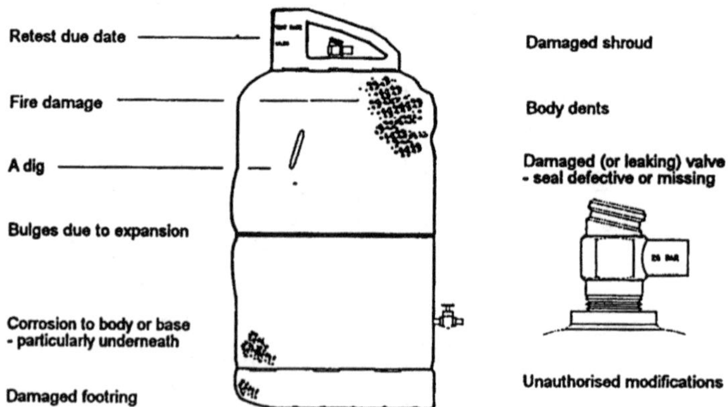

FIGURE 3.14 The LPG cylinder – rejection criteria.

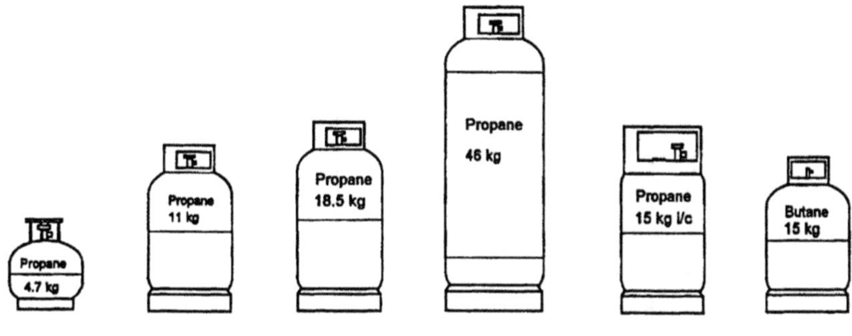

Cylinder type	Propane	Propane	Propane	Propane	Propane liquid -fork lift truck	Butane
Product capacity / kg	4.7	11	18.5	46	15	15
Total weight (approx). / kg	13.4	28.3	43	107	43.2	31
Diameter - typical / cm	28	33	33	38	33	33
Overall height - typical / cm	37	64	76	125	71	61

FIGURE 3.15 Typical LPG cylinders – capacities and dimensions.

Vapour offtake of LPG from cylinders

The diagrams in Fig. 3.16 show what happens when LPG vapour is withdrawn from a cylinder and fed to an appliance.

Diagram A shows a nearly full cylinder in static equilibrium as no gas is being used. There is vapour in the upper part of the cylinder above static liquid. The Vapour Pressure in the cylinder (VP) will be determined by the composition of the LPG and its temperature (VT and LT) – refer to the vapour pressure graphs in Fig. 3.4 to establish this pressure. If the cylinder has been standing for some time without gas being consumed, the temperatures of the vapour and the liquid will be similar and they will also be close to the Atmospheric Temperature (AT).

The liquid in the cylinder will be still.

Diagram B shows what happens when gas is withdrawn from the cylinder for use by an appliance.

The vapour pressure in the cylinder is considerably greater than the pressure required at the appliance so that, when the appliance control tap is opened, gas flows to the burners from the cylinder.

The pressure required by the burners is controlled by the regulator. In a single cylinder system there is normally only a single-stage regulator connected directly to the valve.

As gas is consumed the pressure in the vapour space falls (VP-1), and some of the liquid vaporises in order to restore the equilibrium.

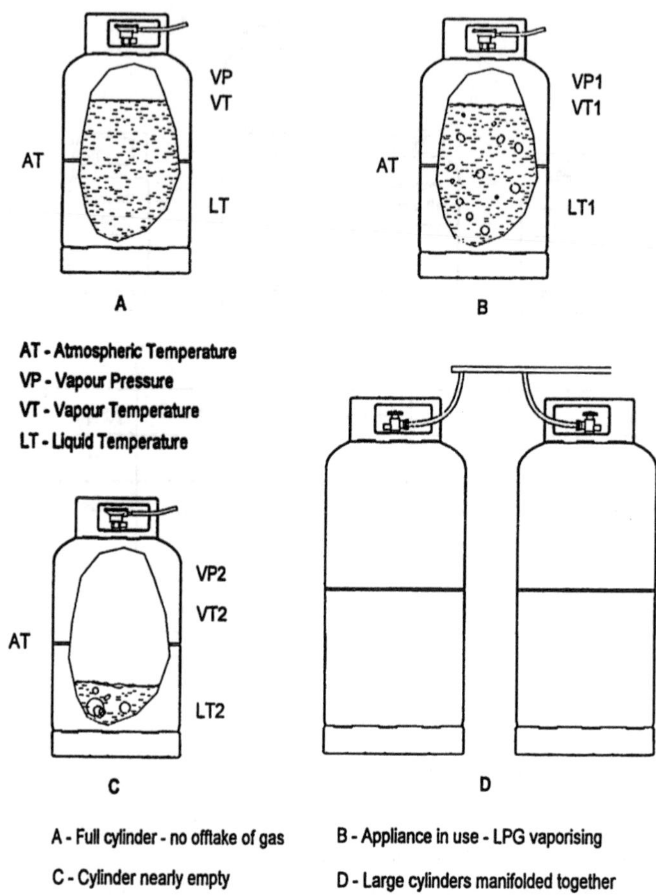

FIGURE 3.16 Vapour offtake of LPG from cylinders.

The latent heat required to convert the liquid to vapour is taken from the liquid itself causing it to fall in temperature (LT-1).

As more gas is consumed the liquid boils to generate the additional vapour and the temperature falls further. The cylinder walls take up the temperature of the cooled liquid and are now at a lower temperature than the atmospheric temperature. This temperature difference causes heat to be transferred from the atmosphere to the cylinder to supply the required latent heat.

When the gas to the appliance burner is turned off the system slowly returns to the equilibrium state described in A. Domestic LPG systems rely on intermittent use to ensure that gas is readily available at the required volume and pressure when needed.

Diagram C represents the condition when the cylinder is nearly empty. The pressure and temperature in the cylinder have fallen further (VP-2, VT-2, LT-2).

In this condition the volume of liquid which can supply latent heat is much reduced and the surface area of the cylinder wetted by liquid is smaller. The liquid in the cylinder will boil vigorously. The net result is that the amount of vapour which the cylinder can supply to the appliance is considerably less than when the cylinder was full.

The low temperature in the cylinder can often be seen, indicated by heavy condensation (on butane systems) or frost (on propane systems).

In the United Kingdom the LPG supplied in domestic sized cylinders is commercial butane. The most popular cylinder size is around 13 kg capacity, with 7 and 15 kg also available. Propane is supplied in cylinder sizes from 4.7 to 46 kg capacity.

When a cylinder is unable to supply the volume of gas needed by the appliance, the pressure may fall below that required at the burners. Under these conditions the regulator cannot function correctly as it is a pressure reducing/ control device.

To ensure that the required volume of gas is available, at the required pressure, it may be necessary to use a larger cylinder, or to combine two or more cylinders by connecting them to a manifold (Diagram D).

For such requirements, and the even larger gas consumption of commercial systems (e.g. catering), propane is used in preference to butane because of its higher vapour producing capability.

Where a constant supply of gas is essential the manifolded cylinders are used in two banks which are connected together via an automatic changeover device. This system permits one bank of cylinders to be the 'service' side and the other bank the 'reserve'. When the service cylinders are empty, or in the event of an extra heavy demand, the automatic changeover device switches the supply to the reserve cylinders. At the same time the unit displays a signal indicating that the service cylinders require replacing.

The automatic changeover device comprises two regulators which are set at slightly differing pressures. After replacing the empty cylinders a lever on the changeover device is operated to switch the former reserve cylinders to the service mode.

Details of a typical automatic changeover system are shown in Fig. 3.17(b).

From the above information it can be seen that the volume and pressure of LPG available from cylinders are not constant, and are affected by the specification of the gas, the type and size of the cylinder, the method of utilisation and the ambient conditions.

In countries where the LPG sold in cylinders is a mixture of butane and propane, the propane will vaporise preferentially when the cylinder is brought into service. As the contents of the cylinder are consumed the percentage of propane in the vapour will decrease and the percentage of butane will increase. This variation in gas composition can affect the combustion performance of some appliances.

The major LPG companies will supply information on the performance which can be expected from the cylinders they supply.

Regulation 14(2) of the Gas Safety (Installation and Use) Regulations 1998 states that 'No person shall cause gas to be supplied from a gas storage vessel

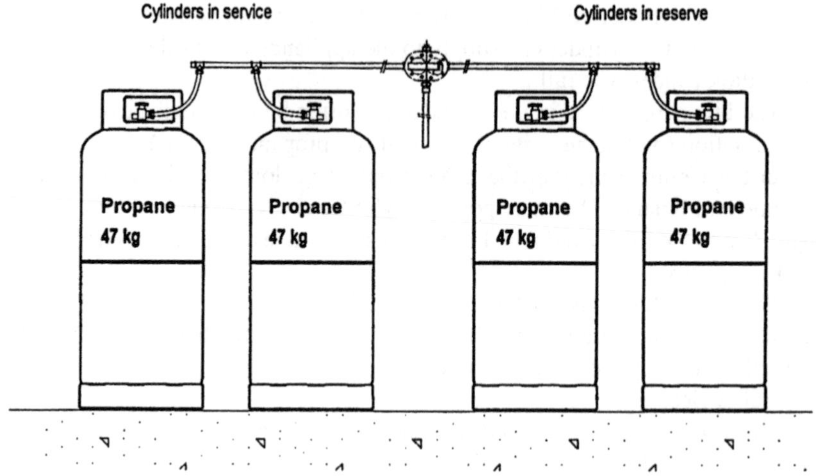

FIGURE 3.17(a) Multiple cylinder installation with automatic changeover system.

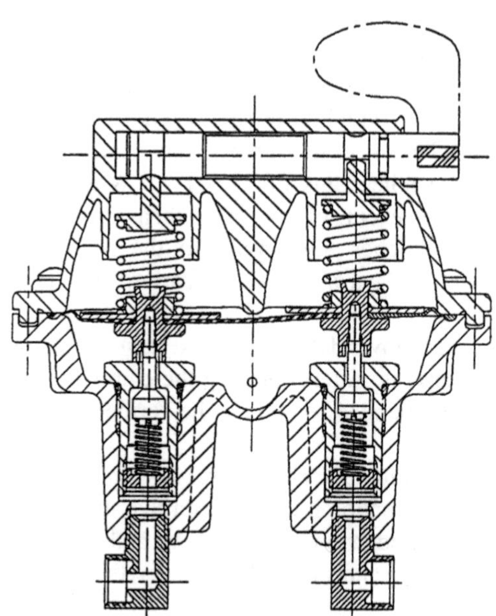

FIGURE 3.17(b) 'Rego' automatic changeover device type.

(other than a refillable cylinder or a cylinder or cartridge to be disposed of when empty) to any service pipework or fitting unless:

- there is a regulator installed which controls the nominal operating pressure of the gas
- there is adequate automatic means for preventing the installation pipework and fittings downstream of the regulator from being subjected to a pressure different from that for which they were designed for

- there is an adequate alternative automatic means for preventing the service pipework from being subjected to a greater pressure than that for which it was designed should the regulator fail'.

CUSTOMER BULK STORAGE FOR LPG

When the consumption of LPG is above that which can be effectively supplied by the use of manifolded large propane cylinders the gas is supplied in a different manner.

One or more consumer tanks are installed at the customers' premises and propane is delivered to these tanks by means of a road tanker.

Customer tanks vary in size from the small domestic and commercial tanks typically holding between 200 and 2000 kg, to the larger industrial tanks which can have capacities up to 100 tonnes or greater.

The following section relates to the smaller tanks where liquid phase LPG is pumped, via a meter, into the tank until the safe maximum level is reached. The meter records the amount of gas delivered which is then charged to the consumer. This type of delivery is often described as split bulk, to differentiate it from the full tanker loads delivered to the large industrial customers.

Small customer tanks are normally installed above the ground but there is increasing action from the safety and environmental authorities for more of these tanks to be buried or semi-buried below the ground. These systems are known as underground and mounded, respectively. The three systems are shown in Fig. 3.18.

The manner in which tanks should be installed is described in standards and codes of practice produced by the Health and Safety Executive and the LP Gas Association. Technicians involved in this type of activity should receive specialised training for the work and should always have access to the relevant codes. These are:

- *Health and Safety Executive*
  HS/G34 – The storage of LPG at fixed installations
- *LP Gas Association*
  Code of Practice 1
  Part 1 – Design, installation and maintenance of bulk LPG storage at fixed installations
  Part 2 – Small bulk LPG installations for domestic purposes
  (a more comprehensive list of the Codes and Guidance Notes published by the LP Gas Association and the Health and Safety Executive is included later in this chapter).

From mid-1994 this type of LPG installation must be installed and operated in accordance with the Gas Safety (Installation and Use) Regulations, and the Pressurised Systems and Transportable Gas Containers Regulations – 1989.

Because an LPG tank remains in situ with the customer, it is equipped with a range of special fittings to enable it to be safely filled and operated.

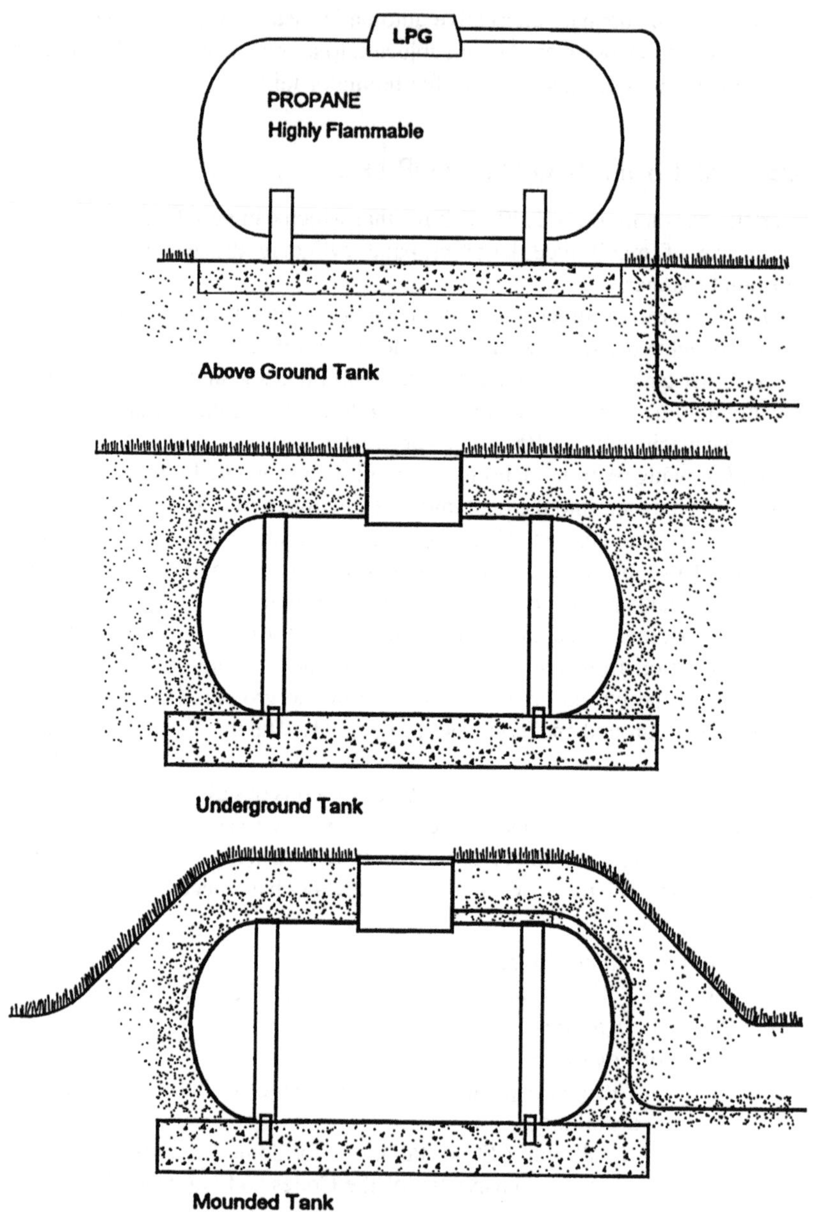

FIGURE 3.18 Methods of installing small LPG tanks.

These fittings, which are required by the codes, are described in outline here. (For full details please refer to the LP Gas Association Code 1.) Each tank shall have at least one of each of the following:

- Safety relief valve, connected to the vapour space.
  These valves may be either 'external' or 'internal' in design, of the spring loaded type, as shown in Fig. 3.19.

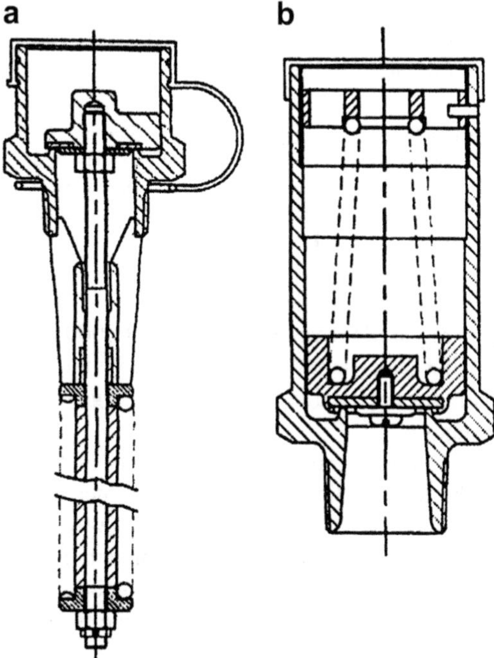

FIGURE 3.19(a) Internal pressure relief valve and (b) external pressure relief valve.

- Drains, or an alternative means of removing the contents in the liquid phase. The drain connection should be provided with a shut-off valve of the quick acting type and preferably not >50 mm in diameter. The outlet should be blanked off when the drain is not in use.
- A fixed maximum liquid level device.
  This is normally a small capillary tube the end of which is fixed at the maximum liquid level permitted in the tank. The tube is connected to a special bleed valve which has an orifice not >1.4 mm diameter. Gas is allowed to escape from this valve during the filling operation and the transition from vapour to liquid is clearly indicated by a white cloud, caused by the expanding LPG freezing the water vapour in the atmosphere. The maximum fill levels permitted are:
  - Commercial propane to BS 4250 86.6%,
  - Commercial butane to BS 4250 89.9%.
- A filling connection.
  The filling connection is required to be fitted with:
  - an excess flow or back check valve,
  - a shut-off valve or,
  - a remotely controlled valve of the fail safe type.

  Special couplings are used for connecting the tanker hose to the tank filler valve. These are described in the LP Gas Association Code 1.

- An outlet, or service, connection.
  For the smaller tanks this is normally a POL vapour outlet of the same type used on the larger cylinders.

  Note: The American POL coupling (specified by CGA Standard No. 510) is smaller than the UK equivalent coupling (BS 341) and is not compatible. Some of the larger tanks are fitted with a liquid outlet, which should be independent of the drain, connected to a draw off point slightly above the bottom of the tank. This type of outlet must have an excess flow valve, fitted internally between the tank and the valve.

- A pressure gauge fitted to the vapour space.
  This gauge is required for vessels in excess of 5000 l water capacity (~2.5 tonnes). Smaller tanks do not have to be fitted with a gauge but provision must be made for a valved tapping to which a gauge can be fitted.
- Contents gauges.
  The type of contents gauge normally fitted to the small LPG tanks determines the level of liquid in the tank by means of a float. The float is linked to a magnetically operated dial gauge by means of a lever arm and a crown wheel and pinion gear. All contents gauges are required to clearly indicate whether they read the contents in % of the water capacity, or % or fraction of the LPG capacity or actual contents in gallons, litres or tonnes.
- Temperature gauges are fitted only to the larger tanks.

  Note: On some designs of small tanks the above fittings may be combined in a special valve assembly known as a 'Multivalve'. This permits only one opening in the pressure vessel. On larger tanks the fittings may be found as individual components.

  The amount of vapour which can be taken from an LPG tank is determined by the level of LPG in the tank and the ambient conditions. Figure 3.20 shows the approximate quantity which can be withdrawn and will also be dependent on the nature of the consumption, i.e. intermittent or continuous.

When it is impractical to obtain all the LPG vapour required by natural vaporisation a vaporiser is installed. Liquid phase LPG is passed through an externally heated heat exchanger supplying the latent heat of vaporisation required. Vaporisers are made in a range of sizes and can be heated by electricity, hot water, steam or hot oil. The design of vaporiser fed systems should only be undertaken by experienced personnel.

A typical small vaporiser is shown in Fig. 3.21.

PRESSURE REGULATION

As described in the early part of this chapter, the pressure available from an LPG system is not constant and can vary considerably according to the type of gas and the operating conditions.

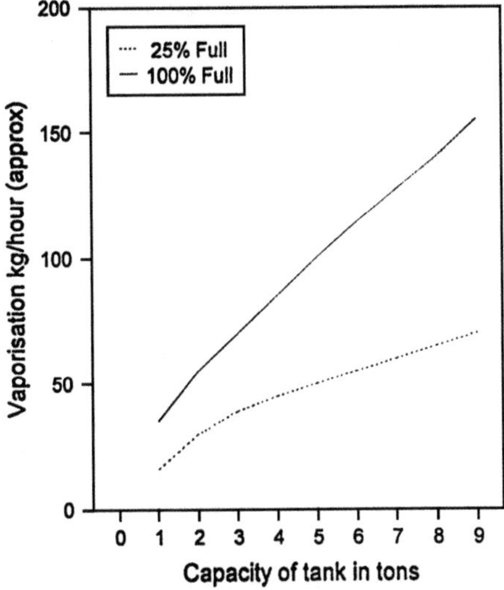

FIGURE 3.20 Vaporisation rates from propane vessels.

FIGURE 3.21 Kogangas vaporiser – electric.

In order that the eventual appliance can operate correctly, it is very important that it receives gas at a constant pressure. This is achieved by the use of a regulator between the cylinder or tank and the point of eventual consumption.

Most LPG regulators are purpose designed. It is most important that the correct regulator system is chosen, taking into account the total offtake of gas, and the outlet pressure, required. This is achieved by selecting an appropriate model from a manufacturer's catalogue and data sheets.

- Cylinder regulators for butane

  The most common regulators in use are of the clip-on type, used with top outlet cylinder valve systems (refer to section on Compact and Jumbo valve systems). These are normally designed to give a low pressure of 28–30 mbar at the outlet nozzle. The pressure is pre-set at the time of manufacture and is not adjustable in the field. The regulators have a maximum capacity of 1 or 2 kg per hour.

  The regulator for the 'Compact' system is a single-stage regulator designed for easy one-hand operation and incorporates special safety features. The connecting latch automatically engages when the regulator is pressed down on to the valve in the coupling position. At this stage there is no flow of gas through the regulator. Only when the regulator control tap is turned to the 'on' position, the gas can flow. This operation also locks the regulator to the valve.

  Different versions of the regulator are available with the relative positions of the operating tap and the outlet located according to the application, e.g. a model designed for cabinet heaters has the tap and nozzle adjacent so that control is easy through the gap in the cylinder shroud.

  The 'Jumbo' regulator is larger than the 'Compact' regulator and is designed for two stage pressure control when combined with the valve. Regulators for the 'Compact' and 'Jumbo' systems, and for the different valve diameters of the 'Compact' system, are not interchangeable.

  These regulators are designed for connection to a flexible hose and have special outlet nozzle profiles of either 8 or 10 mm bore. The appropriately sized LPG hose must be used. Some regulators from different manufacturers have features in excess of those required by the basic standard (BS 3016). These include locking taps, extra seals on the regulator inlet to prevent gas passing in the closed position and anti-corrosion coatings to the regulator's internal components.

  Special regulators designed for high pressure are also available. The outlet pressure is normally variable up to 2 bar. The hose outlet is normally a special $\frac{3}{8}''$(9.5 mm) LH thread.
- Cylinder regulators for propane

  Propane cylinders are fitted with the 'POL' valve which has a female bull nose outlet with a left hand thread. The matching regulators, designed for direct connection to the valve, have a male bullnose thread with a metal to metal seal. This type of regulator requires a spanner to ensure that the connection is tightened correctly. The special cylinder connection spanner should be kept near to the cylinders so that it is always available to make the connection, and so that in the case of an emergency the cylinders can be quickly disconnected.

Propane regulators normally have an outlet pressure of 37 mbar. This is factory set, but can be adjusted in the field on some designs. Different capacities of regulator are available.

A range of adjustable high pressure regulators are available for direct connection to the POL valved propane cylinders. These are designed to supply gas to those special appliances which are designed to be supplied with high pressure gas. These include heating torches, propane for oxy-propane cutting, and a variety of portable industrial heaters.

Notes:

- It is most important that the correct regulator is chosen for any application. This applies to both the capacity and the pressure. Particular care should be taken not to use a high pressure regulator to feed gas to a low pressure application.
- LPG regulators are not designed to be serviced in the field. Many units are completely factory sealed and require replacement when they are faulty or when they have reached the end of a normal service life. Those regulators that can be dismantled for servicing will require testing and recalibrating if refurbishment is carried out. This can only be done at specially approved reconditioning centres and is often not economic when compared to the cost of a new unit.

- Manifolded propane cylinders
  When cylinders are manifolded together an automatic changeover device is normally used. This unit acts as a first stage regulator and may also incorporate a second stage regulator which gives an outlet pressure of 37 mbar. Those units which do not incorporate a second stage regulator have a typical outlet pressure of between 0.4 and 0.75 bar. A separate, in line, second stage regulator is then used to give the line pressure of 37 mbar.
- Consumer tank regulators
  The pressure from a consumer tank is controlled in two stages. The first stage regulator, which is mounted on the tank at the vapour offtake valve, controls the variable tank pressure to a constant medium pressure of 0.75 bar. This pressure is reduced by the second stage regulator to 37 mbar. Safety features are required by the codes to protect the consumer. These incorporate under pressure shut off (UPSO) and over pressure shut off (OPSO) controls. The manner in which these controls operate is described in the following section.

SAFETY REGULATION

A safety regulator comprises four devices integrated into a single module, the purpose being to protect the installation against abnormally high or low outlet

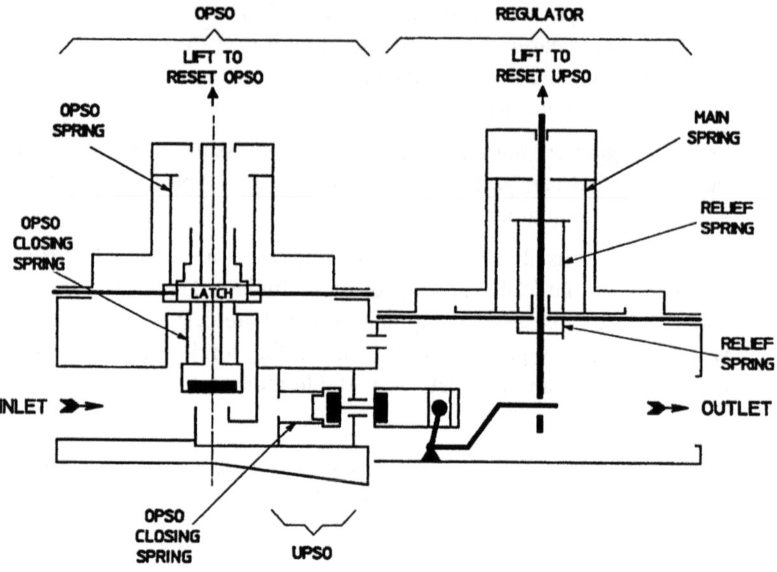

FIGURE 3.22 Safety regulator.

pressures. The schematic of the typical safety regulator is shown in Fig. 3.22 and the four devices are:

1. Regulator,
2. Over Pressure Shut Off (OPSO),
3. Under Pressure Shut Off (UPSO),
4. Limited Capacity Relief Valve.

An OPSO device must be totally independent of the regulator, although the two can be connected.

An UPSO device must be totally independent of the regulator mechanism, although it may be activated by the mechanism.

A relief valve may be fully integrated into the regulator mechanism.

In a typical safety regulator, all four devices operate on classic principles, i.e. they contain a reference element (spring), a sensing element (diaphragm) and a control element (valve). During operation, the diaphragms sense the outlet pressure, creating forces that are compared to those generated by forces that are in equilibrium. All devices operate normally, but when they are out of balance motion of the valves is triggered to initiate the appropriate action.

Typical operation of a safety regulator is as follows:

Regulator

In response to changes in outlet pressure, the diaphragm motion is transmitted to the control valve causing it to open or close as appropriate to restore and maintain the desired outlet pressure determined by the setting spring.

UPSO Device
Abnormally low outlet pressure causes the regulator diaphragm to assume a low position. This in turn allows the UPSO device to seat against its orifice, closing the supply. Reset is manual.

OPSO Device
Outlet pressure from the regulator is fed to the under-side of the OPSO diaphragm via an internal impulse tube. Abnormally high outlet pressure causes the OPSO diaphragm to lift against the force of the setting spring. This releases the latch and causes the OPSO to seat against supply. Reset is manual.

Relief Valve
If the regulator fails to lock-up, downstream pressure creeps, causing the regulator diaphragm to rise. Eventually the force generated by the relief spring will be overcome and the relief valve will open to restore normal conditions.

Typical settings of a second stage safety regulator are:

Regulator: 37 mbar, operating tolerance ±5 mbar
OPSO: Set to close at 75 mbar, tolerance ±5 mbar
UPSO: Set to close between 25 and 32 mbar
Relief Valve: Set to vent at 55 mbar, tolerance ±7 mbar/5 mbar.

SAFETY CONSIDERATIONS AND EMERGENCY ACTION

(an example taken from a typical suppliers safety leaflet)

Safety precautions for the storage and use of cylinders

GENERAL INFORMATION

Liquefied Petroleum Gases are stored in cylinders, as propane and butane, in liquefied form under pressure. When the cylinder valve is opened and gas is withdrawn from the cylinder the pressure falls and the liquid boils.

Both propane and butane, as gases, are heavier than air and any leaking gas will tend to collect at a low level. The gases have a strong and unpleasant smell enabling leakage to be easily detected.

The gases are highly flammable and a small quantity of gas mixed with air can form an explosive mixture.

STORAGE

Never store cylinders below ground or near to drains.

Always store cylinders upright in a cool well ventilated outdoor area away from other flammable materials and naked flames, and secure from vandalism.

Always handle cylinders with care.

CONNECTION

Always stand cylinders upright on a firm level base.

Never use a cylinder tilted or on its side.

Never change or connect cylinders near a source of ignition – i.e. cigarette, open fire, electric fire etc.

Always check that the male 'bullnose' connection on propane regulators or pipework is not damaged or dirty before making connection. For clip on butane connections check that the sealing washer is in position and in good condition.

Always use the correct spanner to tighten screwed connections (left-hand thread).

Never rely on finger-tight joints.

Never check for leaks with a naked flame. The general area of a leak can often be detected by smell and its exact position determined by brushing leak detection fluid over the suspected area.

Always use special LPG quality flexible hose, conforming to British Standards Specification. Natural rubber and most plastics are not suitable for use with LPG.

Always check hoses for wear before making a connection.

Always use a hose clip for securing hoses to regulators and appliances. On ridged nozzles position the hose clip on the plain section.

USE

Never use gas appliances without adequate ventilation. All gas appliances require a plentiful supply of air for correct operation – carbon monoxide (a very poisonous gas) will be produced if adequate combustion air is not available.

Always read and observe equipment manufacturers' instructions.

Always use the correct type of pressure regulator. Never tamper with the pressure regulator.

Always have a means of ignition ready before turning on the gas – if the flame goes out accidentally, do not attempt to re-light it until the escaped gas has been dispersed.

Never apply external heat to a cylinder even if it is showing signs of frosting.

Always keep the number of cylinders in a work area to the absolute minimum and remove empty cylinders to a secure storage area.

EMERGENCY PROCEDURES

LPG suppliers should provide all users with instructions on the action to be taken in the event of an emergency, such as a gas leak or a fire in the vicinity of LPG cylinders, as follows:

In the event of a gas leakage or suspected leakage

- extinguish all naked flames and sources of ignition
- turn off all gas appliances
- do not operate any electrical switches
- turn off gas at cylinders
- if leak is in a building, ventilate by opening doors and windows
- call in a competent person, notify gas supplier.

Do not restore gas supply until it has been verified as safe by a competent person.

In the event of fire

- call the fire brigade immediately, warn them that LPG cylinders are involved
- do not approach any cylinders in the vicinity of the fire
- turn off gas supply if practical and safe to do so.

Do not restore gas supply until it has been verified as safe by a competent person.

The following illustration, Fig. 3.23 is an example of an Emergency Card provided by an LPG supplier to customers using a small bulk tank installation. These installations are normally owned by the LPG company who also ensure that they are inspected and maintained and serviced correctly.

THE LP GAS ASSOCIATION

The LP Gas Association is a non-profit making association, founded on 30 June 1970 as the Liquefied Petroleum Gas Industry Technical Association. The current name was adopted in 1991. Its objects are:

(a) To promote for the benefit of the public of the United Kingdom the advancement of the scientific and economic development and efficiency of those branches of the commerce and industry of the United Kingdom concerned with or involved in the manufacture and safe utilisation of liquefied petroleum gas.
(b) To promote study and research into the principles, methods and techniques concerned with the safe development, manufacture, use, application, storage and transportation of liquefied petroleum gas and to publish the results of all such study and research.

The Association publishes a wide range of Codes of Practice covering many aspects of LPG activity. The list of Codes available (December 2008) is as follows.

COP1 Bulk LPG Storage at Fixed Installations
Part 1: Design, Installation and Operations of Vessels Located Above Ground 2004
Part 2: Small Bulk Installations for Domestic Purposes 2000 *(incorp Amendment 1: 01/03)*

Consumer Emergency Procedures

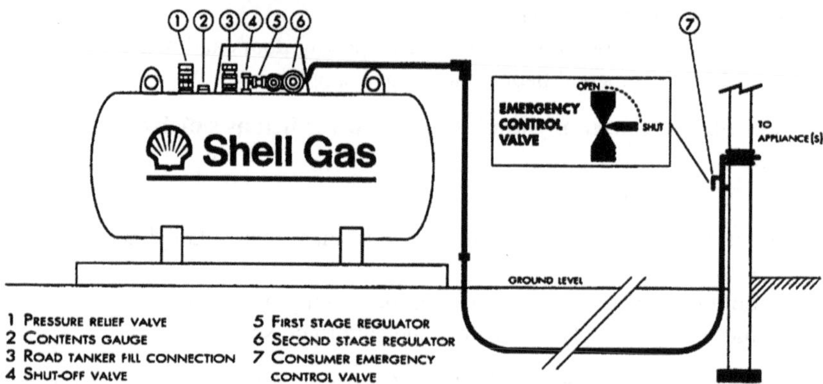

In the event of a suspected gas leak identified either by smell, frost on the pipe or misting of air around the leaking point take the following action:

- Extinguish all naked flames and ignition sources.
- Turn off all gas appliances.
- Do not switch on or off electrical equipment.
- Turn off gas supply at emergency control valve and service take off valve.
- If the leak occurs indoors - open doors & windows to increase ventilation.
- Immediately notify Shell Gas on the emergency telephone number provided.

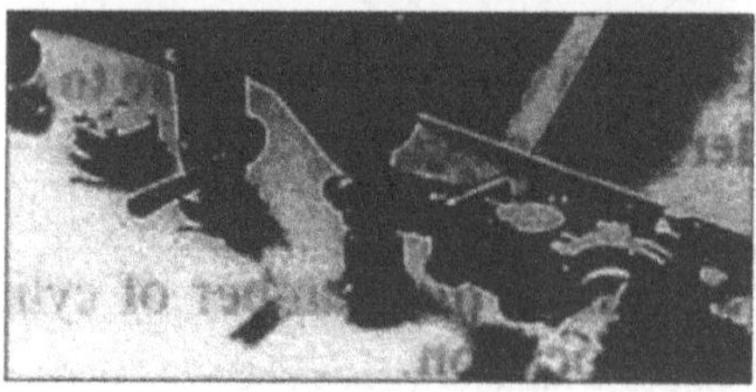

Gas Emergency Control

In the event of an escape

Turn off supply at emergency control valve. Open windows. Do not search with naked light and if gas persists immediately contact Shell Gas.

Do not turn on gas until escape has been repaired.

FIGURE 3.23 Shell safety card.

Part 3: Examination and Inspection 2006
Part 4: Buried/Mounded LPG Storage Vessels 2008

COP2 Safe Handling and Transport of LPG in Road Tankers and Tank Containers by Road 2007

COP3 Prevention or Control of Fire Involving LPG 2006

COP4 Safe and Satisfactory Operation of Propane-Fired Thermoplastic and Bitumen Boilers, Mastic Asphalt Cauldrons/Mixer, Hand Tools and Similar Equipment 2004

COP7 Storage of Full and Empty LPG Cylinders and Cartridges 2004

COP9 LPG–Air Plants 2005

COP10 Containers attached to Mobile Gas Fired Equipment 2005

COP11 Autogas Installations 2001 *(incorp Amendment 3: 11/03)*

COP12 Recommendations for Safe Practice in the Design and Operation of LPG Cylinder Filling Plants 2005

COP14 Hoses for the transfer of LPG in Bulk: Installation, Inspection, Testing and Maintenance 2002 *(incorp Amendment 1: 08/06)*

COP15 Valves and Fittings for LPG Service
Part 1: Safety Valves 1998 *(incorp Amendment 1: 11/03)*
Part 2: Valves for Transportable LPG Cylinders 1998) *(incorp Amendment 1: 11/03)*

COP17 Purging LPG Vessels and Systems 2001 *(incorp Amendment 1: 11/05)*

COP18 Safe Use of LPG as a Propulsion Fuel for Boats, Yachts and Other Craft 2003

COP19 Liquid Measuring Systems for LPG
Part 1: Flow Rates up to 80 litres per minute in Installations Dispensing Road Vehicle Fuel 2001
Part 2: Transfers between Mobile Equipment and Fixed LPG Storage at Flow Rates above 80 litres/minutes 2003

COP20 Automotive LPG Refuelling Facilities 2001 *(incorp Amendment 1: 2/04)*

COP21 Guidance for Safety Checks on LPG Appliances in Caravans *(1997) (incorp Amendment 1: 1/2000)*

COP22 LPG Piping System Design and Installation 2002

COP24 The Use of LPG Cylinders
Part 1: The Use of Propane in Cylinders at Residential Premises 2006
Part 2: The Use of LPG in Mobile Catering Vehicles and Similar Commercial Vehicles 2000
Part 3: The Use of LPG for Catering and Outdoor Functions 1999
Part 4: The Storage and Use of LPG on Construction Sites 2000
Part 5: The Use of Propane in Cylinders at Commercial and Industrial Premises 2000

COP25 LPG Central Storage and Distribution Systems for Multiple Consumers 1999 *(incorp Amendment 1: 02/08)*

COP26 Uplifting of Static LPG Vessels from Site, and their Carriage to and from Site by Road 2004 *(incorp Amendment 1: 07/08)*

COP27 Carriage of LPG Cylinders by Road 2004 *(incorp Amendment 1: 12/07)*
COP29 Hazard Information and Packaging Labelling for Commercial LPG Cylinders *2003 (incorp Amendment 1: 11/04)*
COP30 Gas Installations for Motive Power on Mechanical Handling and Maintenance Equipment 2004
GN2 A Guide to Servicing Cabinet Heaters 2002
GN3 A Guide to the Preparation of Major Accident Prevention Policies (MAPP's) 1999

THE HEALTH AND SAFETY EXECUTIVE

Regulations

There are a number of regulations which affect the way in which LPG installations are installed and the way the gas is used. Those which technicians should be aware of are listed below:

The Highly Flammable Liquids and Liquefied Petroleum Gases Regulations 1972 (SI 1972/917)
The Gas Safety (Installation and Use) Regulations
Notification of Installations Handling Hazardous Substances Regulations 1982 (SI 1982/1357)
The Control of Industrial Major Accident Hazard Regulations (CIMAH) 1984 (SI 1984/1902)
The Dangerous Substances (Notification and Marking of Sites) Regulations 1990 (SI 1990/304)
The Pressure Systems and Transportable Gas Containers Regulations 1989 (SI 1989/2169)

The Health and Safety Executive are responsible for the drafting and interpretation of these regulations and their inspectors control the safety requirements at industrial premises. For non-industrial premises administration is the responsibility of local authority Environmental Health Departments.

To assist in the interpretation and implementation of the various regulations the Health and Safety Executive has produced a range of publications which should be used for reference. These include:

- HS(G)34 Storage of LPG at Fixed Installations
- CS4 The Keeping of LPG in Cylinders and Similar Containers
- CS6 The Storage and Use of LPG on Construction Sites
- CS8 Small Scale Storage and Display of LPG at Retail Premises
- CS11 The Storage of LPG at Metered Estates
- Liquefied Petroleum Gas – Storage and Use. An Open Learning Course.

Chapter | four

Burners

Chapter 4 is based on an original draft prepared by Mr E.W. Berry

INTRODUCTION TO BURNERS

Gas can be used for many purposes. In the home it can provide facilities for cooking, space heating, water heating and refrigeration.*

In industry and commerce, gas is used in ovens, furnaces, baths and as working flames for the cutting, melting and heat treatment of metals, for the production of glass and pottery, for manufacturing foods and for drying paints and enamels. Many manufacturing processes require heat and gas can easily be applied to provide it.

Many different shapes and sizes of flames are required for these different applications, each of which is produced by its own gas burner. Burners may, however, be divided into two main classes, aerated and non-aerated, perhaps better described as pre-aerated and post-aerated burners. As its name implies, the pre-aerated burner has some, or all, of the air for combustion mixed with the gas before it is burned.

Pre-aerated burners may be further subdivided into those which obtain their primary air supply directly from the atmosphere and are called 'atmospheric' or 'natural draught' burners and those whose air is supplied by an electric fan and which are called 'forced draught' burners or sometimes 'air blast' or 'pre-mix' burners. The following nine sections deal with the principles of natural draught pre-aerated burners.

PARTS OF A BURNER

Figure 4.1 shows a cross-section through a simple, atmospheric ring burner. The parts of the burner are lettered as follows:

(a) Injector,
(b) Primary air port,
(c) Aeration control shutter (primary air adjustment),

---

*Although refrigerators for use with natural gas are not now generally on sale, many are still in use.

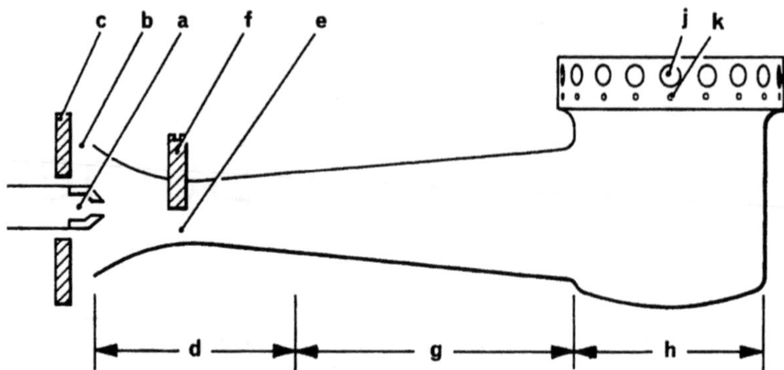

FIGURE 4.1 Pre-aerated ring burner.

(d) Venturi,
(e) Venturi throat,
(f) Throat restrictor (alternative primary air adjustment),
(g) Mixing tube (burner tube),
(h) Burner head (burner body),
(i) Burner ports (flame ports),
(j) Retention ports (retention flame ports).

In some burners, there is no venturi and the mixing tube is parallel or slightly tapered. To ensure good mixing of the air and gas and even flame heights throughout its length the burner may contain:

(k) Deflectors (baffles),
(l) Internal gauze.

Figure 4.2 shows a cross-section of a bar burner with gauze and deflectors. A burner will generally have an air shutter or a throat restrictor, but not both.

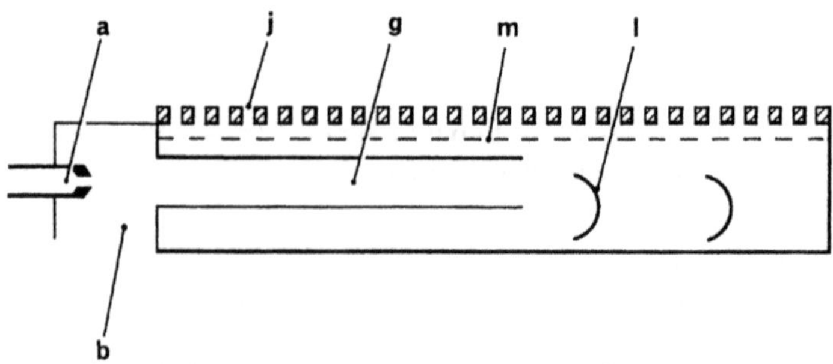

FIGURE 4.2 Pre-aerated bar burner.

INJECTORS

The injector is so-called because it injects a stream of gas into the burner. It is usually made of brass and is normally situated in line with the central axis of the burner tube, as in Fig. 4.1. Because the gas behind the injector is at a pressure several millibars higher than the air, it possesses energy. When it leaves the injector the gas moves at about 45 m/s (162 km/h) and draws air into the burner just as a fast-moving lorry will draw dust in its wake. This drawing of air into the burner is called 'entraining' the air.

For the most efficient primary air entrainment, the injector should be aligned so that the gas stream is directed down the centre of the venturi. If the gas impinges on the side of the throat, momentum will be lost and the aeration will be reduced.

Some burners were made so that the direction of the gas stream could be adjusted to flow in the centre or to one side of the throat axis. This 'eccentric' injector provided a means of varying the aeration without the use of an air shutter or a throat restrictor (Fig. 4.3).

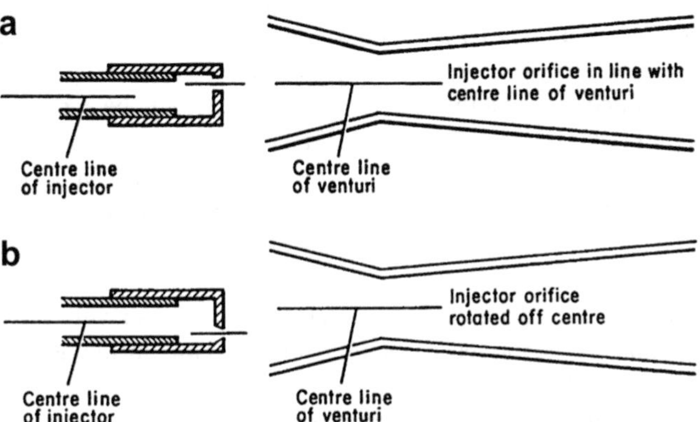

FIGURE 4.3 Aeration adjustment by eccentric injector: (a) orifice central with venturi and (b) orifice off centre.

Injectors may have a single hole or several holes. They may produce one or more cylindrical jets or a flat stream of gas. As a general rule, for a given heat input and pressure, single-hole injectors give a higher air/gas ratio than multi-hole injectors. But they are more noisy, producing a hiss which can cause annoyance at times.

The size of the injector controls the gas rate to the burner at a particular pressure. So the heat input rate to the burner for gas of a particular Wobbe Number and supplied at a specific pressure will depend on the area of the hole or 'orifice' in the injector.

The shape of the inside of the injector has also some effect. The injector shown in Fig. 4.4a will cause a greater loss of pressure than Fig. 4.4b where the approach to the orifice is at an angle.

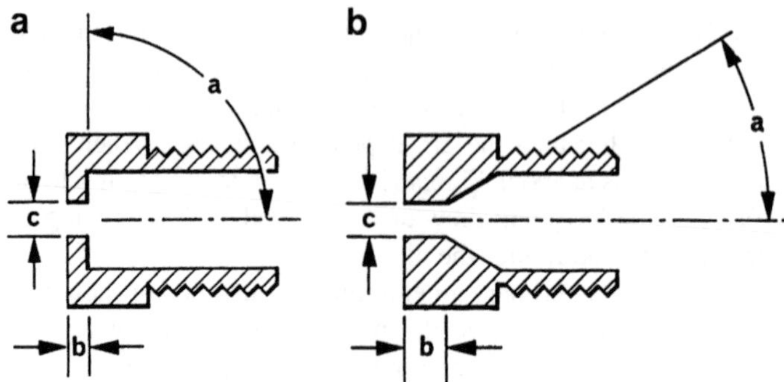

FIGURE 4.4 Comparison of injector characteristics: (a) low coefficient of discharge, high pressure loss and (b) high coefficient of discharge, low pressure loss.

Injector orifices are usually calibrated according to a measured flow rate rather than their physical dimensions to take into account manufacturing tolerances. They may be marked with a rate at a standard pressure or with the drill size of a perfect hole which would give the actual rate obtained.

It is most important that the injector orifice size should be correct for the gas rate of its particular burner. The supply of an excessive gas rate to an appliance could cause incomplete combustion and might result in a fatal accident. Great care must be taken when cleaning injectors to ensure that the holes are not enlarged, because a small increase in orifice diameter can make a big difference in gas rate.

For example, consider a gas fire which has an injector with a diameter of 0.5 mm. If this was enlarged to 0.6 mm the increase in diameter would be only 0.1 mm, a 20% increase. But a 20% increase in diameter results in a 44% increase in area, so the gas rate would also increase by 44% or nearly half as much again! (This is because the area increases as the square of the diameter.)

If required, it is possible to calculate the volume of gas which will pass through any size or orifice. In metric units this is:

$$Q = 0.0467 \times A \times C \frac{\sqrt{p}}{s}$$

usually written as

$$Q = 0.0467AC \frac{\sqrt{p}}{s}$$

$$Q = 0.036d^2C \frac{\sqrt{p}}{s}$$

where Q = volume of gas in cubic metres per hour (m^3/h),
A = area of orifice in square millimetres (mm^3),
d = diameter of orifice in millimetres (mm),
P = pressure behind the injector in millibars (mbar),
s = specific gravity of gas,
C = the 'coefficient of discharge'.

The coefficient of discharge, C, varies with the shape of the injector orifice and depends mainly on the angle of approach and the orifice length. As the angle of approach decreases from 90° (Fig. 4.4a), the value of C increases until the angle is about 30° (Fig. 4.4b). After that it begins to decrease again.

As the length of the injector channel increases the value of C increases until the length is about 1.5–2 times the orifice diameter. After that it begins to fall.

The actual value of C is usually between 0.85 and 0.95.

PRIMARY AIR PORTS

Primary air ports can be situated at the beginning of the venturi or mixing tube or at the side or bottom. They are rarely at the top on a horizontal burner because of the risk of gas escaping from the port at low injector pressure as, for example, when the burner is turned down to a low gas rate.

The size of the opening at the primary air port may be fixed in the case of some burners. This is called 'fixed aeration' and the correct area of air port is calculated when the burner is designed.

Most burners have some method of aeration control so that the burners may be adjusted to the correct aeration for different applications or for different gases.

AERATION CONTROL

The control of primary aeration to a natural draught burner is carried out by one of two methods.

Air shutters

Opening and closing some form of sliding or rotating air shutter has the effect of increasing or decreasing the area of the primary air port so increasing or decreasing the amount of air entrained.

Various types of shutter are shown in Fig. 4.5 and the effect of altering the shutter opening on the gas/air ratio of the burner is shown in Fig. 4.6.

Correct aeration for a particular burner is usually measured by the height of the inner cone. As you saw in Chapter 2, increasing the primary aeration causes the flame speed of the air/gas mixture to increase and this makes the inner cone of the flame to become shorter. The effect of altering the air shutter on the cone height is shown, for a cooker burner, in Fig. 4.7. With natural gas flames the inner cones are sometimes indistinct and not easy to measure.

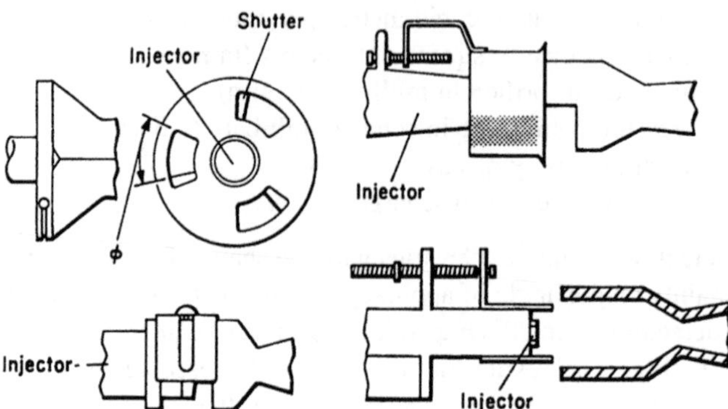

FIGURE 4.5 Types of aeration control.

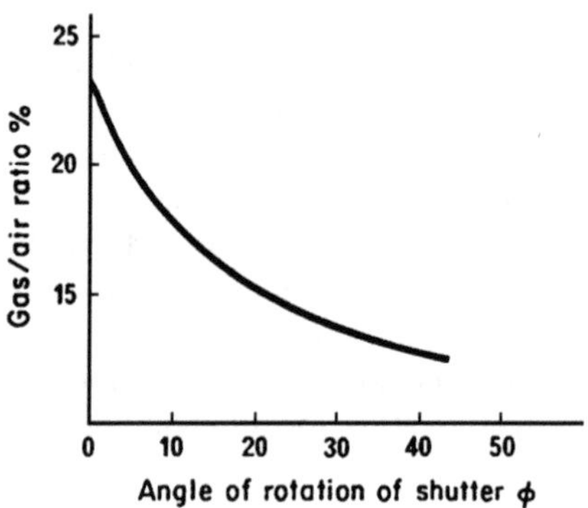

FIGURE 4.6 Graph showing the effect of air shutter rotation on the gas/air ratio of a burner.

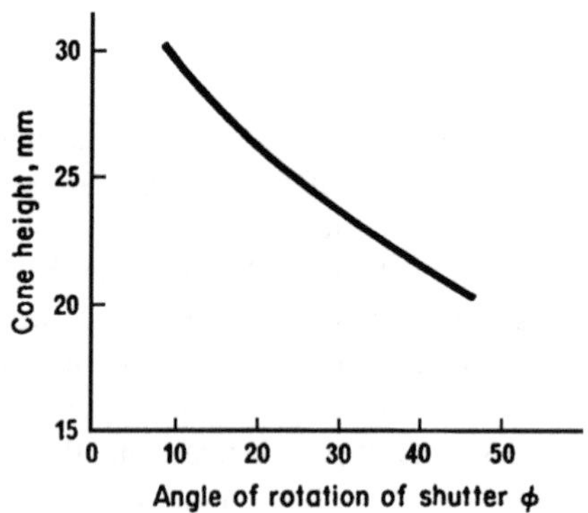

FIGURE 4.7 Graph showing the effect of air shutter rotation on the cone height of aflame.

Throat restrictors

Throat restrictors control the amount of primary aeration by changing the resistance within the burner. Anything which increases the resistance to the flow of the air/gas mixture through the burner will reduce the speed of the flow and so reduce the amount of primary air entrained.

Usually restrictors take the form of a screw or a vane located in the throat of the venturi as shown in Fig. 4.8. They are designed so that, even when fully closed, the restriction produced is not sufficient to cause gas to escape from the primary air ports.

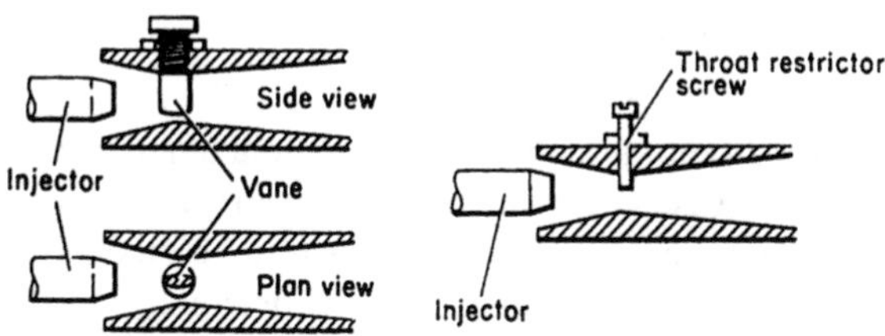

FIGURE 4.8 Throat restrictors.

VENTURI

This is a tube which tapers down in the first third of its length to a narrow 'throat'; the tapering is shaped like a curved 'bell' on a trumpet or trombone. From the throat the venturi gradually widens out again to its original diameter over the final two-thirds of its length (see Fig. 4.1).

The venturi has a low resistance to flow, gives good mixing and a reasonably constant air/gas ratio over a wide range of inlet pressures. Figure 4.9 shows that although the inlet pressure to a burner varies from 2 to 16 mbar the air/gas ratio remains fairly constant at around 2.1. A venturi also gives better pressure recovery at the burner head.

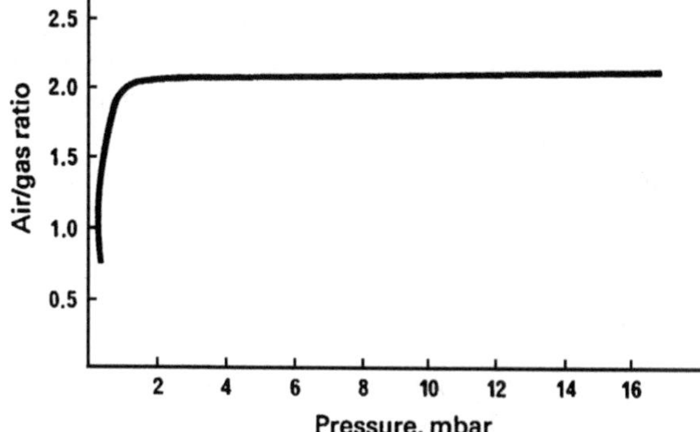

FIGURE 4.9 Graph of variation of air/gas ratio with inlet pressure to burner with venturi.

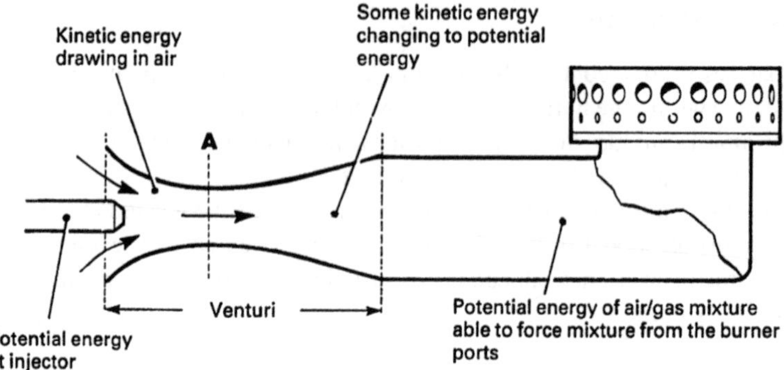

FIGURE 4.10 Energy changes in burner venturi.

For these reasons venturis are usually incorporated into burners where there is likely to be some resistance to the flow of the air/gas mixture due to bends, roughness inside the burner, small flame ports or long mixing tubes.

If the burner resistance is low, then a parallel or slightly tapered mixing tube is adequate.

Venturis possess their advantages because they act as energy convertors. They change potential energy into kinetic energy and then partially reconvert the kinetic energy into potential energy again (Fig. 4.10). What actually happens is this.

Gas behind the injector is under pressure and it is this pressure which gives the gas its potential energy and which forces it out and into the mouth of the venturi.

As the stream of gas goes forward, it begins to draw air with it. The mixture gains speed as the area narrows down until, at the throat, A, it has reached its maximum velocity. This gain in speed results in a change of the potential energy into kinetic energy.

As the potential energy is reduced, so the gas pressure falls until, at the throat, it is very close to that of the air outside the burner. By this time the gas has pulled in all the air it can.

The air/gas stream moves out of the throat and through the gently widening end of the venturi and into the mixing tube, gradually slowing down as it does so. This reduction in the speed of the mixture causes a change back of some of the kinetic energy into potential energy. Consequently there is some increase in pressure and that serves to overcome the resistance of any bends or roughness in the burner and to push the mixture out of the burner ports.

BURNER HEAD

The purpose of a burner is to provide flames with the right heat input and shape for a particular application. This is achieved by using the right size and shape of burner head and the correct design of the flame ports. Saucepans are round, so the burners which heat them have circular heads. Gas fires must have a long flat area of radiants, so their burners are also long and flat to give even heating over the whole radiant surface.

In some of the long 'box' or 'bar' burners with flame ports along the top or to one side, there is no separate burner head (see Fig. 4.2). Very often the air/gas mixture leaves the mixing tube at a point about half-way along the inside of the tube. Some of the mixture travels to the far end, the rest turns back towards the injector end.

With other bar burners the mixing tube is outside the burner and there is a tendency then for the pressure in the burner to vary from one end to the other. Variations in pressure can affect the flame length and result in an uneven distribution of heat along the burner. There are some cases where this is desirable, for example, on a grill burner where a reduced heat output from the middle of the burner prevents overheating in the centre of the grill. Generally, however, even heating is required and some action has to be taken to prevent the flame length varying along the burner.

There are three common methods of ensuring even flame distribution with a box-type burner.

1. insert baffles,
2. fit an internal gauze,
3. shape the burner body.

Baffles

Baffles fitted inside a burner by its manufacturer should prevent more gas getting to one part of the burner than to another. They must not cause too high an internal resistance because this would reduce the primary aeration.

Gauzes

When fitted below the burner ports, gauzes will even out the internal pressure. A gauze can also have the effect of making a burner 'flameproof', that is incapable of lighting-back.

If a sheet of metal gauze is held above a lighted burner, the flame will not pass through it until it becomes red-hot. Then the gas will light and burn above the gauze (Fig. 4.11).

If the gauze is held above an unlit burner and the gas above the gauze is lit, the gas will continue to burn above the gauze. This is because the gauze conducts heat away from the centre to the edge and the cold gas and air passing through the gauze together keep the temperature of the gauze low enough for the gas below not to ignite (Fig. 4.12).

So a gauze fitted inside a burner to even out the flames can also be used to prevent lighting-back. But the burner must be very carefully designed. If the burner ports themselves are large, the flames may pass through them and burn on the gauze, causing overheating and possibly distorting the burner. If the gauze is fitted close to the burner ports, its effective area will be reduced and it

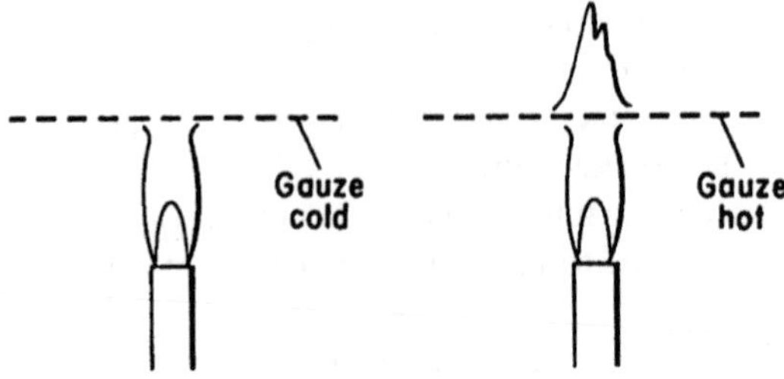

FIGURE 4.11 Effect of gauze above aflame.

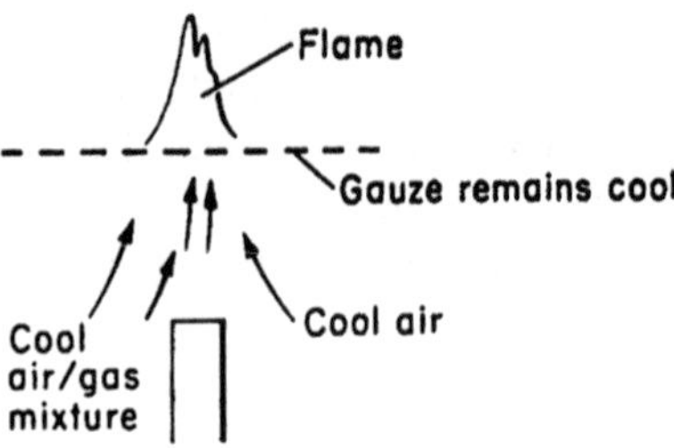

FIGURE 4.12 Effect of gauze below aflame.

will quickly be blocked with fluff and 'lint' carried in by the primary air. This will increase the resistance of the burner and reduce the aeration.

Gauzes in burners are particularly useful when the flame ports are not completely flame-proof by themselves, but are designed so that at the normal gas rate the flames burn correctly. The only risk of lighting-back is at the time of ignition. The gauze will prevent this happening and will also prevent the mixture lighting noisily in the burner when the gas is turned off.

When gauzes are used in burners to even out the mixture flow there is generally no need to fit baffles.

Shaped burner body

It is possible for the burner body to be tapered in such a way that the velocity and the pressure of the air/gas mixture are constant throughout the entire length of the burner. An example is shown in Fig. 4.13.

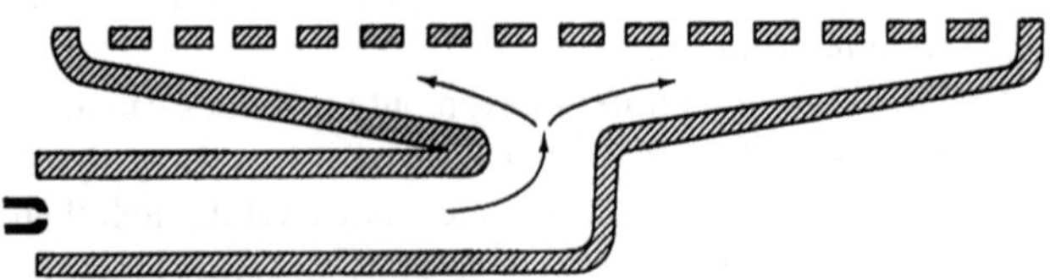

FIGURE 4.13 Tapered burner body.

BURNER PORTS

Flames must be of such size and shape that they will transfer the maximum amount of heat to whatever has to be heated, whether it is food in a pan, water in a boiler or air in a room. The shape and sizes of flames are, to a great extent, controlled by the ways in which they are able to obtain their secondary air. This depends on the size and position of the burner ports.

If a burner has several rows of ports close together, the inner cones of the flames at the outer rows may be short and well defined. But in the middle rows they may be long with cones that disappear completely. Secondary air is not able to reach the centre of the flame, which lengthens until sufficient oxygen to complete combustion can be obtained from the surrounding atmosphere (Fig. 4.14).

Many burners for Family 2 and 3 gases (see Chapter 1) are made with the ports in rows across the head so that secondary air has free access to the centre of the flames (Fig. 4.15).

Although allowing space between the flames for access of secondary air, the flame ports must be close enough for one flame to light the next when an igniter is operated at one end of the burner. Burners must be capable of 'cross-lighting' at their normal operating gas rates.

The diameter of the burner port can affect the ability of a burner to resist lighting-back. In simple terms, a burner with a large number of small burner ports is unlikely to light-back. Another burner with fewer, larger ports, burning the same amount of gas with the same aeration, will be much more prone to lighting-back.

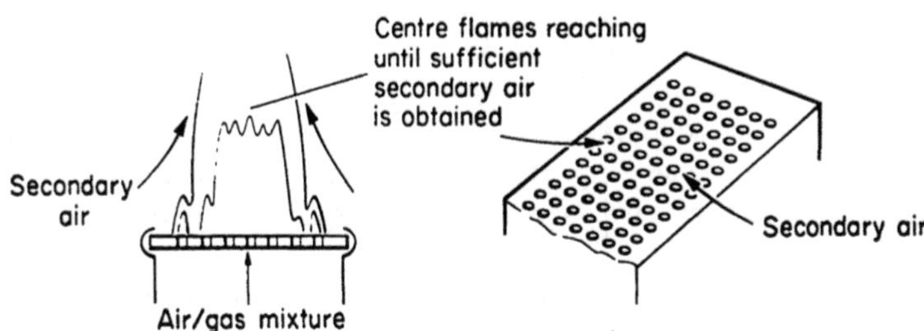

FIGURE 4.14 Burner with insufficient secondary air.

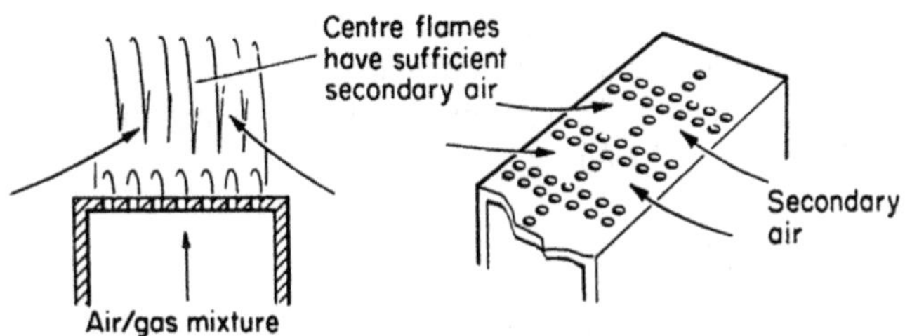

FIGURE 4.15 Burner with adequate secondary air.

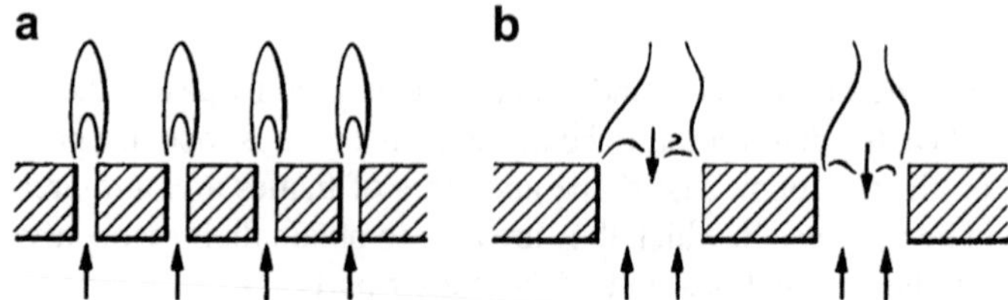

FIGURE 4.16 Effect of burner port diameter on resistance to lighting-back: (a) small ports, high resistance and (b) large ports, low resistance.

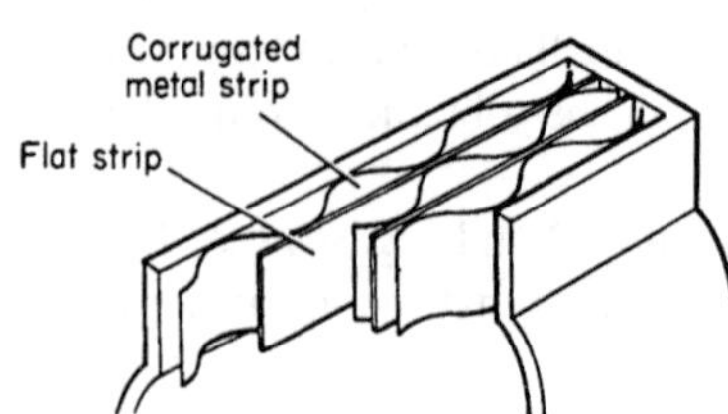

FIGURE 4.17 Ribbon burner.

The reason for this is that, in the case of the narrow port (Fig. 4.16a), gas at the centre of the flame is close to the walls of the port. If it attempts to pass back through the port, it will lose sufficient heat for the temperature to drop below the ignition temperature of the gas. So the flame will not be able to travel through the hole and will continue to burn at the top of the port. This is called 'flame quenching'.

With the wide port (Fig. 4.16b), the gas at the edge of the hole will be cooled but that at the centre of the flame will be much less affected and the flame may be able to travel down through the port and cause light-back.

For ports to be flame-proof, those in thin metal must not be more than 1 mm in diameter: in thicker metal they may be bigger.

Deep ports can be formed by sandwiching together alternate strips of corrugated and flat metal strip. Burners made in this way are called 'ribbon burners' and possess the advantage of having a large burner port area for the size of the burner (Fig. 4.17).

The heat output from the flame ports can be calculated, if required. The 'flame port loading' is the heat output for each square centimetre of flame port area and is measured in watts per square centimetre (W/cm^2) or megajoules per square centimetre hour (MJ/cm^{2}h).

For an aerated burner without retention flames, to prevent flame lift the flame port loading must be about 900 W/cm^2 (3.24 MJ/cm^{2}h).

If the burner has retention flames, the flame port loading may be increased to about 4.5 kW/cm^2 (16.2 MJ/cm^{2}h).

RETENTION PORTS

Burners which are designed to operate on more than one family of gases by changing the injector, the operating pressure usually have retention flames adjacent to the main flames. Gas for the retention flames is often controlled by small internal orifices between the main burner head and the passage carrying gas to the retention ports. These are known as 'metering orifices' (Fig. 4.18).

There are methods for stabilising main flames other than by the use of separate retention flames. The general principle is to slow down at least part of the air/gas stream as it leaves the burner port or immediately after it leaves the port. Alternatively, a zone of high temperature may be created close to the burner port.

The two methods described may be combined. Some of the methods used are shown in Figs 4.19 and 4.20.

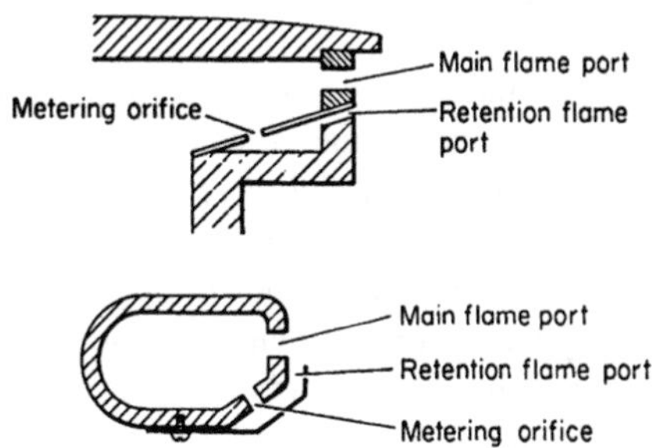

FIGURE 4.18 Metering orifices.

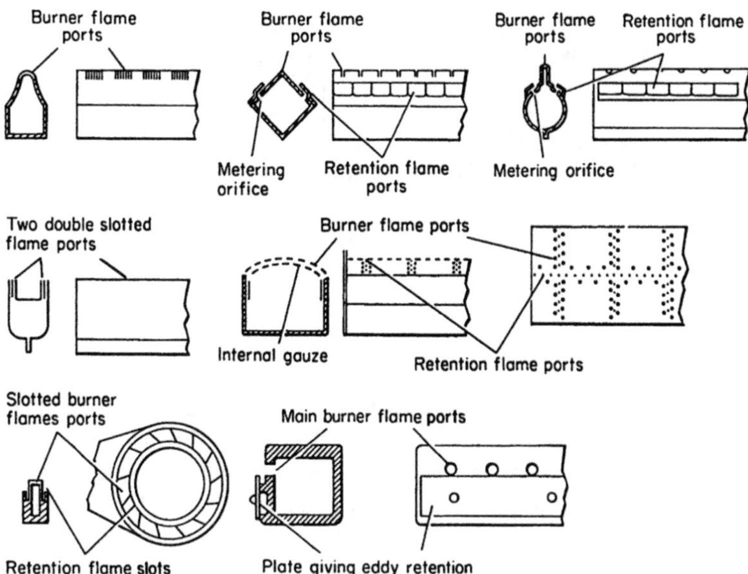

FIGURE 4.19 Methods of flame retention.

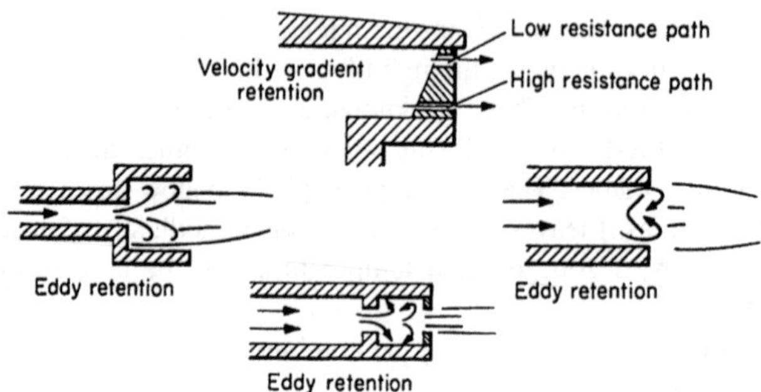

FIGURE 4.20 Methods of flame retention.

PRE-AERATED JETS AND PILOTS

The jets are small burners made to be screwed into a gas 'manifold' or burner bar. Jets may be obtained to give a range of heat input rates with various flame shapes. The most popular are those which give flat, fan-shaped flames.

Pre-aerated pilots are designed to have low gas rates so that their running cost is low. They may give a flame or flames in one or several directions.

Jets and pilots usually take in primary air at the side. It may pass straight up to the flame port or down to the injector and then up to the port. This gives better mixing and allows the overall height of the jet to be kept short. Alternatively, gauzes or conical or spherical baffles may be fitted in the gas stream to help mixing of the air and gas (Fig. 4.21).

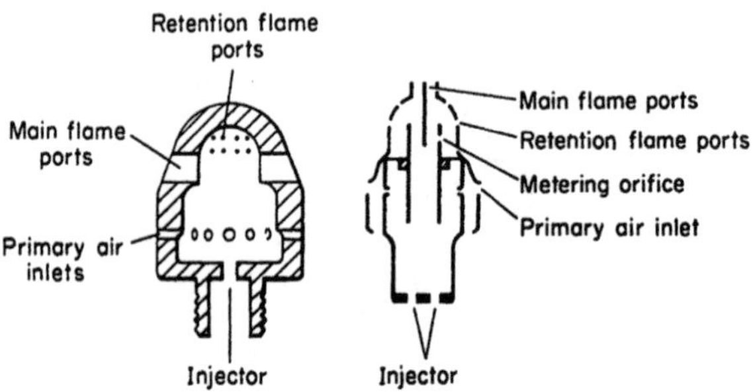

FIGURE 4.21 Pilot jets.

POST-AERATED BURNERS

With Family 1 gases, which had a high flame speed and a wide flammability range, the burner in common use was the 'Bray' jet. This had a ceramic top which produced two opposing streams of gas, one from each side of the head. The two streams impinged on each other and gave a flat, fan-shaped flame which was non-luminous and stable over a wide range of gas rates.

If these burners are used with gases which have a lower flame speed and narrow flammability range, the flame blows off unless the pressure is very low,

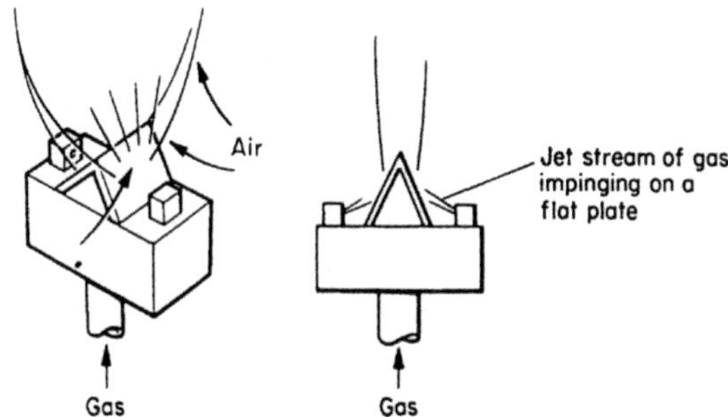

FIGURE 4.22 Target burner.

when a small, sooty flame forms. Attempts to provide a simple form of flame retention have not been successful.

The type of jet used with natural gas usually has thin, cylindrical flames known as 'rat-tail' flames which can be stabilised. When these are used with faster-burning substitute natural gases the flames become shorter and this can cause the burner to overheat. As the burner gets very hot the gases inside can break down and soot may be deposited inside the burner.

Burner 'modules' are made in various lengths and have one or two rows of holes in line along the length of the burner, forming the main burner ports. There are rows of smaller holes at each side at which gas burns at a low pressure. The small, outer flames retain the main flames.

Another form of post-aerated burner is the 'target' burner. In this, a stream of gas strikes a flat surface at an angle of about 40°. This produces a flat, fan-shaped flame in which the air has been rapidly mixed by the turbulence so that the flame, although lifted, is well aerated and stable. This burner is too noisy to be used in domestic appliances except for low-rated pilots. It is used for commercial and industrial equipment and has the advantage that it is not subject to 'linting'. An example using opposing streams of gas is shown in Fig. 4.22.

The 'matrix' burner distributes gas over a flat, horizontal, perforated surface through a very large number of very small ports. Air is passed up through the perforations and the air/gas mixture burns as a thin, flat sheet of flame. Matrix burners are particularly suitable for use in compact heat exchangers when air can be blown through the appliance by a fan (Fig. 4.23).

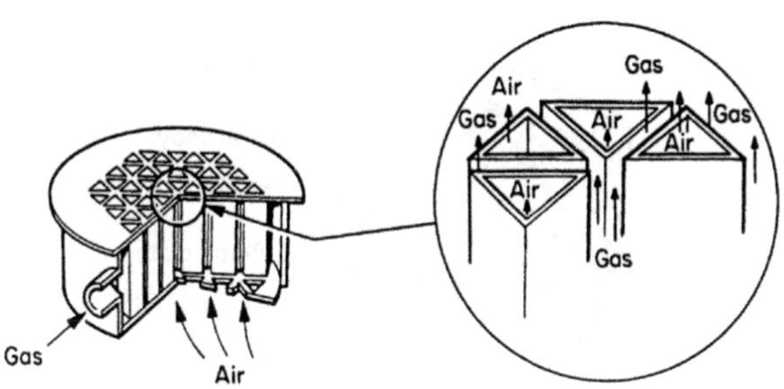

FIGURE 4.23 Matrix burner.

As with the other post-aerated burners, if fast-burning substitute gases are used, the shortened flame can cause overheating of the burner.

FULLY PRE-MIX BURNERS

This type of burner is used when a considerable amount of heat is needed to be released in a small space. In the past they were usually found on industrial appliances. Significant developments in design, materials used in manufacture and the associated control systems have led to more domestic appliances adopting this type of system.

The amount of gas required to provide the heat input is mixed with slightly more air than is required to burn it completely (usually about 10–15% more). When burned on a suitable burner head a short, very hot flame is produced with almost no visible outer mantle.

Gas may be mixed with air either before or after the centrifugal fan which forces the mixture into the burner head. The head of the burner shown in Fig. 4.24 is made from perforated metal and is shaped to suit the appliance heat exchanger. The ribbon burner head shown in Fig. 4.25 is made from stainless steel; burners made from ceramic materials are also in use. No secondary air is required, all the air for combustion being supplied as primary air. The burner

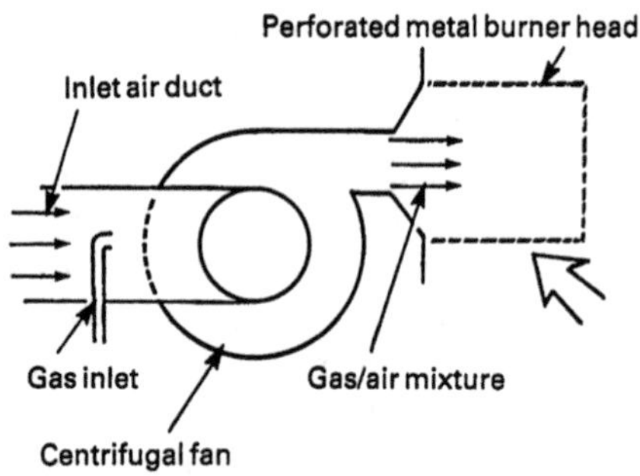

FIGURE 4.24 Fully pre-mix burner.

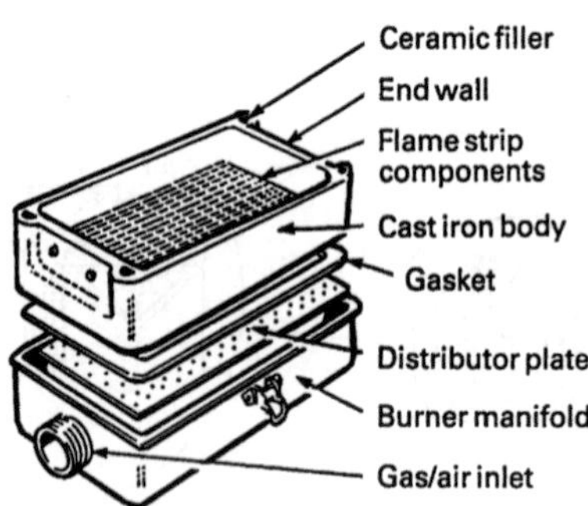

FIGURE 4.25 Fully pre-mix ribbon burner.

port loading must be low enough for a stable flame to form without 'lift' and the temperature of the burner head must remain cool enough to prevent 'light-back'.

By using a fully pre-mix burner, appliances can be made more efficient, flexible and compact. Using a fan to supply the combustion air and to remove the products of combustion makes it possible for smaller flue pipes and terminals to be used. This helps to simplify the installation of the appliance and reduce costs.

FORCED-DRAUGHT BURNERS

With some large domestic and commercial boilers an electric centrifugal fan is used to blow air past nozzles supplying gas to a single, large burner port. Flame retention is provided, usually by eddies of hot gases around the main flame. This type of burner is usually controlled automatically so that unburned gas can never reach the combustion chamber.

Figure 4.26 shows examples of forced-draught burners. In one type the gas flows up the central tube to a number of ports at the end, the ports are arranged so that the gas emerges in the form of a cone. Air is blown up the outer tube and mixes with the gas in the conical metal burner head at the end of the tube. Holes in the metal cone can be moved relative to the gas ports, so varying the rate of mixing and altering the length of the flame.

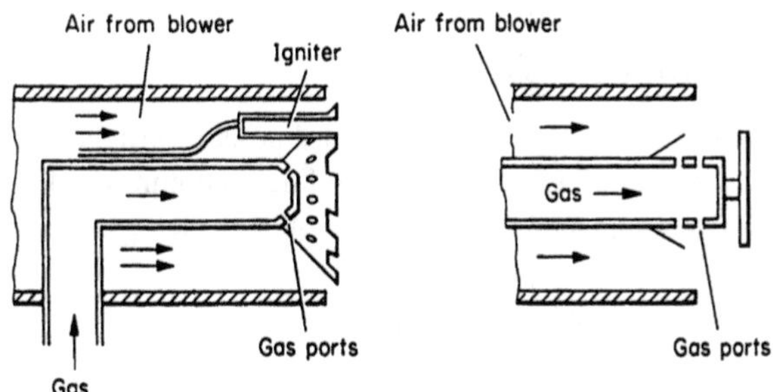

FIGURE 4.26 Forced-draught burners.

In the second type the air/gas mixture flows past a circular plate. Turbulent eddies are formed at the edge of the plate, retaining the main flame.

RADIANT BURNERS

There are two main forms of radiant burners. In the first gas and air are mixed by a centrifugal fan and in the second a conventional atmospheric burner is used (Fig. 4.27).

With the pre-mix type the burner head is formed of granules of refractory (fire-proof) brick bonded together so that there are tiny spaces through which the air/gas mixture can pass. When lit, the flame burns on the surface causing the temperature of the brick to rise. As the temperature rises the flame can burn

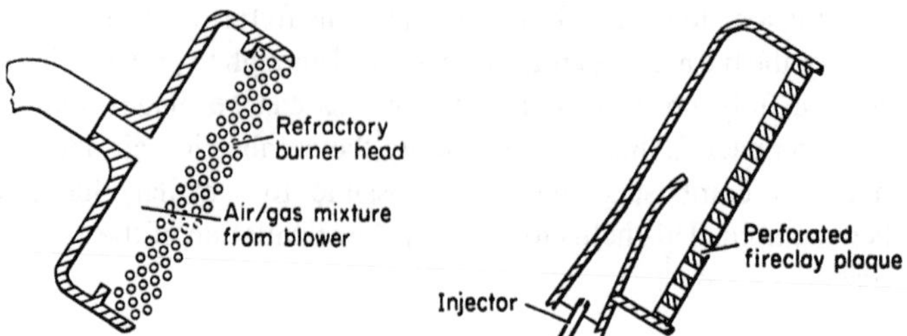

FIGURE 4.27 Radiant burners.

just inside the surface, which becomes red-hot, or 'incandescent'. This is called 'surface combustion'.

Because the refractory head is a poor conductor of heat and because the incoming air/gas mixture is cool, the inner side of the head remains at a temperature below the ignition temperature of the gas and so lighting-back is prevented.

With the atmospheric burner, the head consists of a thick slab of refractory material with many small burner ports through which the air/gas mixture can pass easily.

Radiant burners are used for overhead heating in buildings, for cooker grills and for some industrial processes.

FLAMELESS COMBUSTION

A form of radiant burner (Fig. 4.28) has for its burner head a flat pad of refractory fibre with one layer impregnated with very fine platinum powder. If the pad is heated to the operating temperature and a suitable air/gas mixture is passed

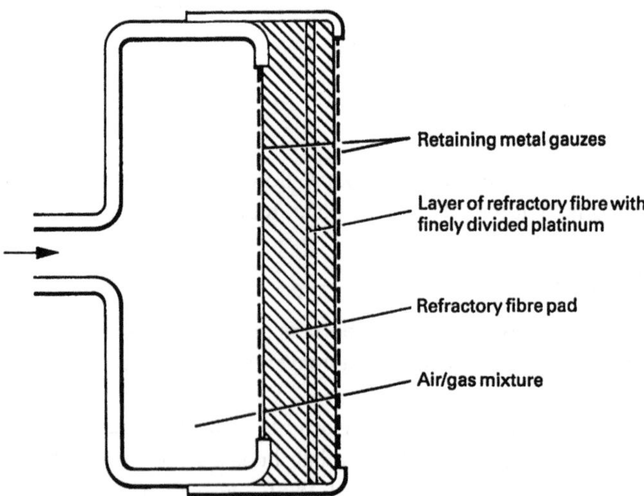

FIGURE 4.28 Catalytic combustion burner.

through the pad, a flameless reaction will take place in the platinum layer, just as if the gases were burning. This reaction is referred to as 'catalytic combustion'.

This means of raising the temperature of the pad initially may be by an electric element embedded in the surface. Alternatively gas can be lit on the surface and will burn there until the pad has reached the temperature at which combustion can continue flamelessly within the pad.

This system has been used for outdoor heating and for buildings where flammable liquids or vapours may be present, because there is no actual flame and the surface temperature is too low to ignite vapours. It is more likely to be found on LPG than natural gas appliances.

FAULT DIAGNOSIS

An experienced engineer can often tell what is wrong with an appliance simply by looking at the flames on the burner. The size, shape, colour and general appearance of the flame is called the 'flame picture'. The flame picture tells the story of any faulty adjustment or the presence of any obstructions in the appliance. The things to look for are as follows.

Flames

In pre-aerated burners, flames and inner cones must be of the correct length. Installation and servicing instructions usually give these lengths but the thing to do is to remember what a correctly adjusted flame should look like. If you carry in your mind mental images of the correct flames for different types of appliances, you will be able to compare them with the actual flame pictures and identify any faults.

In addition to its length, the general appearance of the flame shows what is happening. On domestic appliances flames are usually stable, quiet and with a firm, distinct shape. Cooker hotplate burners and some water heater burners may be noisy but oven and grill burners have 'softer' flames and gas fires must burn as quietly as possible.

Gas rate

For an appliance to work properly the burner gas rate must be right. It must have an adequate gas supply at the correct pressure. 'Pressure points' are provided on most burners and the method of checking is described later. The gas rate depends not only on the pressure but also on the size of the injector or, in the case of post-aerated burners, the jet or flame port. These are sometimes marked with the size or rate.

Where jet sizes are not indicated or if an accurate check is required, the gas rate may be measured by the meter (see Chapter 7). However, it is often sufficient to check the pressure and then look at the flames. The flame picture will show when the gas rate is incorrect. 'Over-gassing', or too much gas, has the same effect as too little air. The flames become longer and 'soft' and if they come into contact with cool surfaces, poor combustion will result. The Gas Safety (Installation and Use) Regulations 1998 impose legal requirements regarding the setting of appliance gas rates.

Air supply

There must be an adequate, but not excessive, supply of air in the form of primary or secondary air, as required. Air inlets must be clear of any external obstructions or blockage by lint or dirt. Too little air causes flames to lengthen and become 'soft', they may 'float' away from the burner ports and could be extinguished completely.

The same effect can be caused by the heat exchanger or the flue-ways being blocked or by vitiation of the air. If a gas cooker oven flue outlet is choked by grease and dirt, the oven flame will be unable to get enough oxygen. It will lengthen, wave about in its search for oxygen and finally float off the burner. This is called 'smothering'.

Other examples of floating flames may be seen on some eye-level grill burners when all the hot plate burners below are alight. The products of combustion from the hot plate burners vitiate the combustion air for the grill burner.

Some excess air is necessary to most burners, but too much will cool down the heat exchanger and reduce the efficiency of the appliance.

Too little primary aeration results in long or non-existent inner cones and soft, yellow-tipped flames. This may be due to incorrect adjustment of the primary air control or it may be caused by deposits in the burner (or 'linting') or by blockage of the burner ports. On cooker burners this blockage is caused by spillage from the cooking pots and on gas fires by debris from the radiants or the heat exchanger.

Over-aeration can produce the same effect as lowering the position of the burner. Combustion will generally improve but the efficiency of the appliance will fall.

The Gas Safety (Installation and Use) Regulations 1998 impose legal requirements regarding the need to ensure an adequate air supply to appliances.

Burner location

The burner or jets must be situated correctly in relation to the heat exchanger or other object to be heated. If a burner is too high, the flames may touch the heat exchanger and poor combustion will result. If the burner is too low, combustion will be improved but, at the same time, efficiency may be reduced as the flame is not sufficiently close to the heat exchanger.

Where one flame impinges on another they produce larger flames into which secondary air cannot easily flow. This causes long, pointed streaks of yellow flame where the flames touch. Usually this occurs on burners made up of jets screwed into a burner bar and can be corrected by altering the angle of the jets causing the trouble. Partial blockage of burner ports by debris may also cause flame impingement and can be cured by cleaning the burner.

Linting

'Lint' has already been mentioned as a substance which chokes up pre-aerated burners. It is worth studying in a little more detail.

Air in a room usually contains dust and fibres from the mats, carpets, furnishings and the clothing of the occupants. This is lint. Its composition is given in Table 4.1. When this air is drawn into the primary air ports of pre-aerated burners the lint goes in with it.

Lint collects at small holes such as burner ports, primary air ports or gauzes. It is also deposited where the gas flow changes direction, such as at baffles or deflectors.

So it is just those features of burner design which overcome lighting-back and poor flame distribution which themselves can cause linting. If the holes in gauzes are large enough to allow lint to pass through, they will neither be flame-proof nor will they be effective in evening out the pressures within the burner.

The problem can be overcome in several ways. Some burners are made with primary air inlets close to the flames so that the lint is incinerated before it enters the burner. Gauzes can be fitted outside the primary air ports but they need

Table 4.1 Composition of Lint

Analysis of Samples Extracted From Burners	Average %
Constituent	
Fats and oils	5.4
Rayon	3.1
Nylon	14.8
Terylene	2.3
Wool and hair	30.7
Cotton	21.7
Dust	22.0
Distribution of dust size (microns)*	
0–1	30.0
1–5	53.0
5–10	8.5
Above 10	8.5
Lengths of fibres (microns)	
7–15	100

*A micron, abbreviated to μ, is one millionth part of a metre.

regular cleaning. Otherwise the effect of them becoming blocked is just the same as when the gauzes are restricted in the burner.

All burners should, in any case, be cleaned regularly. It is a simple matter in those appliances where the external gauze is accessible without dismantling the appliance, or where internal gauzes are removable. All the lint must be removed. If any is left, it provides a base on which more lint will build up very much more quickly than it can on a clean surface.

The amount of lint in the room air is greater close to the floor. So if the combustion air openings to an appliance or to a cupboard containing an appliance are at least 200 mm above floor level, it will take a lot longer for the burner to lint up.

Linting can, of course, be avoided by using post-aerated burners where this is possible. Another alternative is to have appliances which draw their combustion air from outside the building through a pipe in the wall. These are called 'room-sealed' appliances and are far less likely to become linted. Some cases have been reported of their burners getting choked with fluffy seeds, like dandelion, or even spiders' webs, but these are pretty rare.

MODIFYING APPLIANCES TO BURN OTHER GASES

Most appliances which were designed to burn Family 1 gases can be 'converted' to work on Family 2 gases, and vice versa. As an example, two typical gases are:

1. Family 1 calorific value (CV) $= 18.9\ \text{MJ/m}^3$
   specific gravity (SG) $= 0.475$

   $$\text{Wobbe number} = \frac{18.9}{\sqrt{0.475}} = 27.4$$

   Air required to burn 1 m^3 $= 4.3\ \text{m}^3$
   Volume of products from 1 m^3 $= 5.0\ \text{m}^3$
2. Family 2 calorific value $=38.5\ \text{MJ/m}^3$
   specific gravity $=0.601$

   $$\text{Wobbe number} = \frac{38.5}{\sqrt{0.601}} = 49.7\ \text{m}^{31}$$

   Air required to burn 1 m^3 $= 9.8\ \text{m}^3$
   Volume of products from 1 m^3 $= 10.8\ \text{m}^3$

Comparison between 1 and 2

CV: The calorific value of Family 2 is about twice that of Family 1 so only half the volume of Family 2 gas is required to give the same heat input.

SG: The specific gravity of Family 2 is slightly higher than that of Family 1, so the Family 2 Wobbe number is not quite twice that of Family 1. This affects the gas modulus figure and means that a much higher pressure will be required at the injector to maintain satisfactory aeration.

Air Requirement: Family 2 needs about twice as much air as Family 1. But since only half the volume of Family 2 is used then the air inlets should be adequate.

Products of Combustion: Family 2 produces about 20% greater volume of burnt gases than twice the amount produced by Family 1. So if half the volume of Family 2 is used, it will result in about 10% greater volume of products than the original Family 1.

If the flueways of the appliance were designed to allow at least this amount of excess air, combustion would still be satisfactory. If the allowance for excess air were less, then the appliance must have its gas rate reduced to ensure satisfactory combustion. This is called 'downrating' the appliance.

The actual pressure and the new injector diameter required can be calculated, if necessary, using the formulae given in Chapter 1.

If the appliance has an injector diameter of 3.5 mm working on a pressure of 5 mbar on Family 1 gas, the new pressure and diameter are as follows:

The gas modulus is obtained from

$$\frac{\sqrt{\text{pressure}}}{\text{Wobbe number}} \text{ or } \frac{\sqrt{p}}{W}$$

So

$$\frac{\sqrt{p_1}}{W_1} = \frac{\sqrt{p_2}}{W_2}$$

to maintain the same conditions
or

$$\frac{\sqrt{5}}{27.4} = \frac{\sqrt{p_2}}{49.7}$$

$$\sqrt{p_2} = \frac{\sqrt{5} \times 49.7}{27.4} = 4.06$$

Therefore $p_2 = 16.5$ mbar.

To find the injector size use the formula found earlier in this chapter and assume that the coefficient of discharge is the same for both injectors. Then, since the quantity of gas, Q, must remain the same:

$$Q = 0.036Cd_1^2\frac{\sqrt{p_1}}{s_1} = 0.036Cd_2^2\frac{\sqrt{p_2}}{s_2}$$

Divide both sides by $0.036C$, to obtain

$$3.5^2 \frac{\sqrt{5}}{0.475} = d_2^2 \frac{\sqrt{16.5}}{0.601}$$

which gives

$$d_2^2 = 7.57$$

$$d_2 = 2.75 \text{ mm}$$

So with Family 2 gas an injector with a diameter of 2.75 mm and a working pressure of 16.5 mbar is required.

This assumes that the burner is unaltered except for an injector change. In practice it might be possible to adjust a primary air control or completely change the burner. Either might allow conversion to be carried out at a lower pressure than that given by the calculation.

TYPICAL BURNERS

Figures 4.29–4.36 illustrate typical gas appliance burners.

Figure 4.29 shows a hob unit with four burners. The burners are sealed to the spillage tray to prevent spillage entering the underside of the hotplate.

Figure 4.30 shows the removable enamelled cap and cast burner head. The spark ignition electrode is secured to the sealed base of the burner and is protected from spillage by the enamel cap.

Figure 4.31 shows the mixing tubes which are secured to the underside of the spillage tray. The spark generator is connected by individual wires to each electrode.

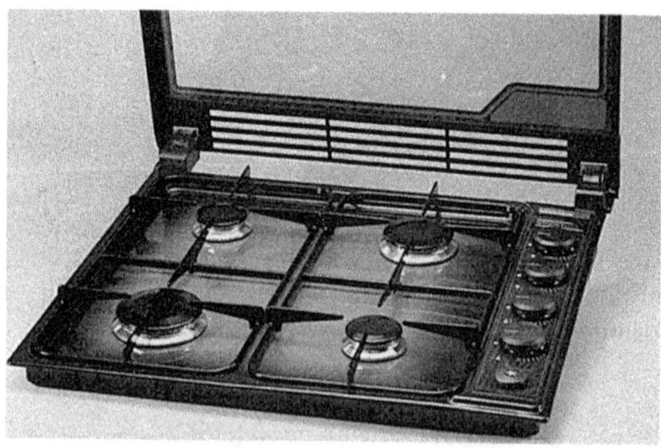

FIGURE 4.29 Hob unit (New World).

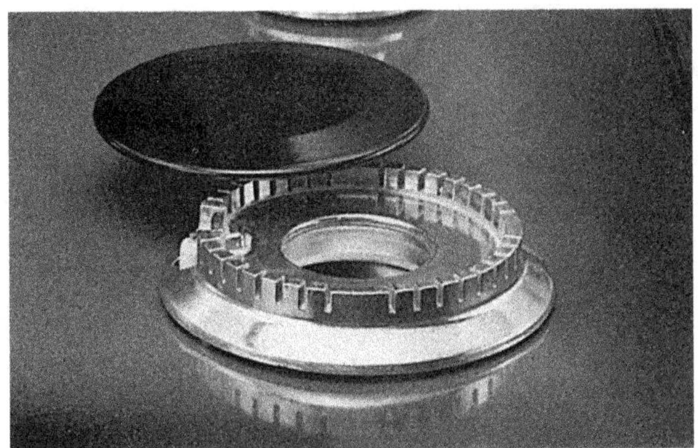

FIGURE 4.30 Hotplate burner.

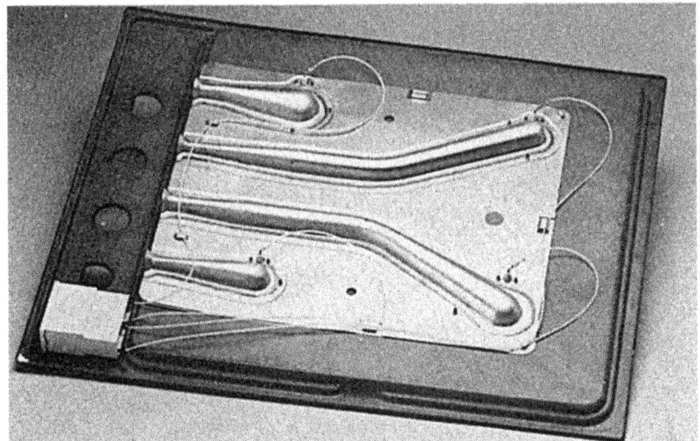

FIGURE 4.31 Burner mixing tubes.

FIGURE 4.32 Burner taps and injectors.

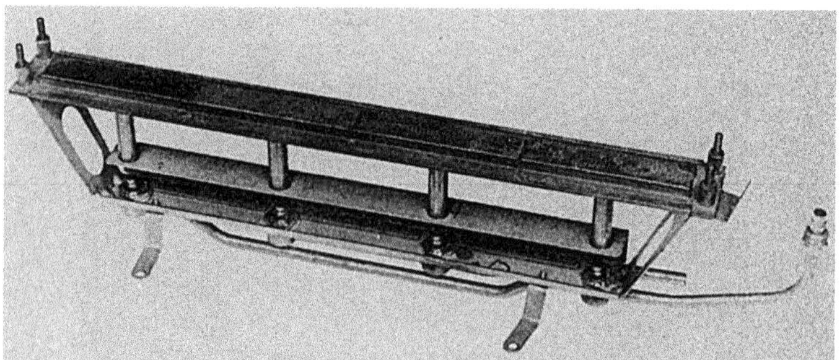

FIGURE 4.33 Gas fire burner, (four sections and four injectors).

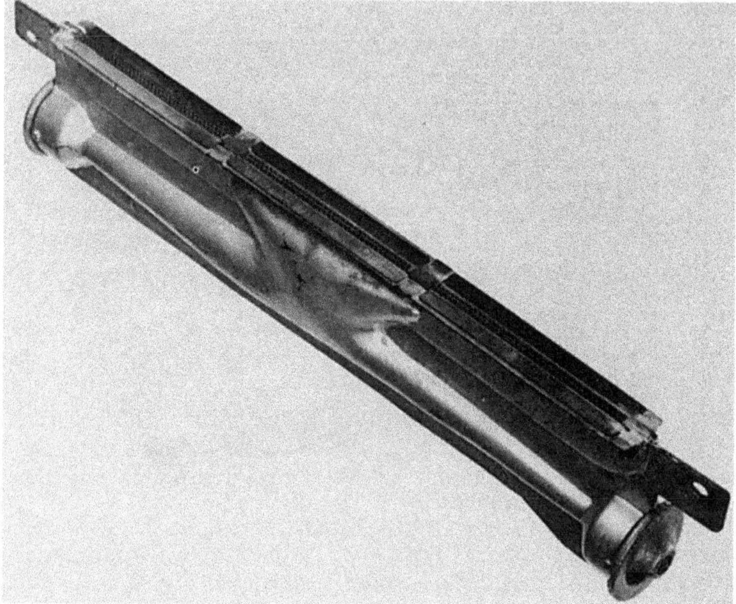

FIGURE 4.34 Gas fire burner, (three sections and two injectors).

Figure 4.32 shows the taps and injectors in the hotplate well.

Figure 4.33 is a gas fire burner made from pressed steel with stainless steel gauze flame ports. The burner is in four separate sections, each served by its own injector. This allows the side burners to be turned off to reduce the heat output.

Figure 4.34 burner is also made from pressed steel and is used in a gas fire. It has an injector at each end, one supplying the centre section and the other the two outer sections. The burner ports are made from stainless steel ribbons and the mixing tubes are tapered through their length.

Figure 4.35 shows a multi-point water heater burner. The burner manifold is made up of six stainless steel burner tubes, each supplied by an injector

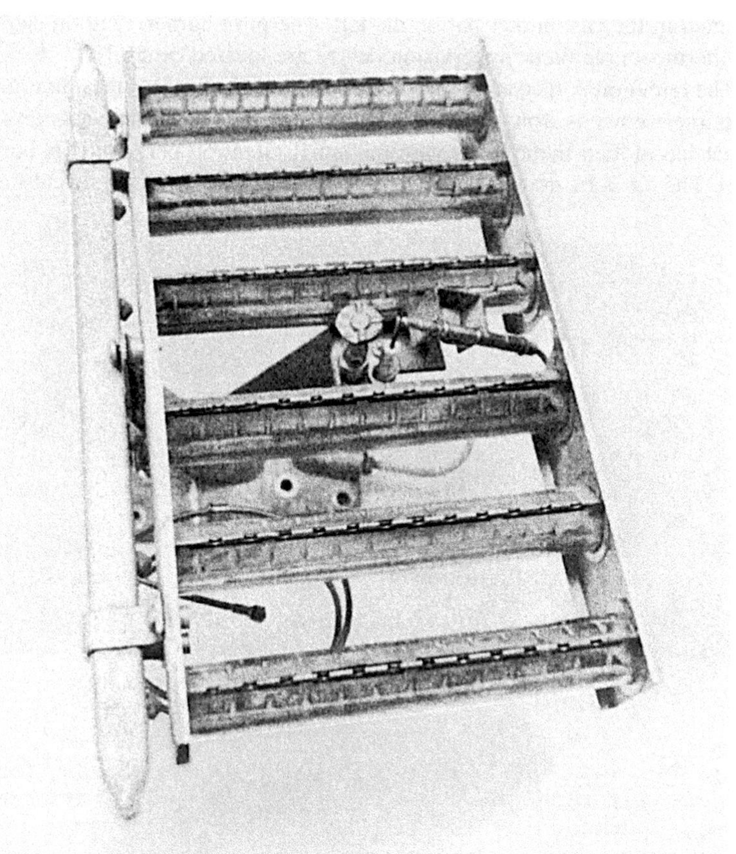

FIGURE 4.35 Multipoint water heater burner.

FIGURE 4.36 Central heating unit burner.

mounted in the cast burner bar on the left. The pilot burner, ignition electrode and thermocouple flame supervision device are located centrally.

The retention ports can be seen on the sides of the burner tubes. In Fig. 4.36 the burner shown is from a central heating unit. The cylindrical, stainless steel burner has slotted flame ports with the small retention port drillings between them. The air ports are covered by rectangular gauze boxes to prevent linting.

Chapter | five

Energy

Chapter 5 is based on an original draft prepared by Mr L. Howson

INTRODUCTION TO ENERGY

The term 'energy' is used in everyday life to mean vigour, forcefulness and activeness. To the sportsman or athlete it means the ability to go on for that longer period which wins the match or race.

In science the word 'energy' has a similar but very precise meaning. It means the ability or capacity for doing work.

Some mention has already been made of energy in the earlier chapters, and both potential and kinetic energy have been referred to. The purpose of this chapter is to show not only how it may be recognised, but how it may be measured.

FORMS OF ENERGY

There are a number of different forms of energy. If energy is due to position or the static storage of energy as in a compressed spring, it is called 'potential' energy.

The energy of a cistern full of water up in a roof or a volume of gas compressed in a pipe is potential energy. When the spring is released or the water or gas is allowed to flow, they can be made to do work.

If the energy is due to motion, then the energy is 'kinetic'. The motor car going along a road, gas or water flowing in a pipe or a rotating flywheel all have kinetic energy.

Apart from these mechanical forms, energy may also be in the form of chemical, electrical, heat, light and sound energy.

The Gas Industry is in the energy business. It markets a fuel which is a source of energy. Up to 1920 gas had been principally used for lighting, from then on it was sold on its heating value and the amount of heat used by customers was measured in 'therms'. One therm is equal to 100,000 British thermal units (Btu).

Under the Imperial system different units are used for different kinds of energy. Heat energy is measured in Btu or therms. Electrical energy is measured in kilowatt-hours (kWh) and mechanical energy is measured in horse-power hours (hph).

In the SI system, energy is measured in the same units as those used for measuring work, that is joules (named after a British physicist, Dr James Prescott Joule). A joule is a small unit and the megajoule is a more useful size. For comparison, 1 British thermal unit equals 1055.06 joules, or 1 Btu = 1055 J (approximately).

The unit used for gas charges in the United Kingdom is the kWh (see Calorific Value, Chapter 1). Although not an SI unit, it is a useful unit of charging and is an ideal method of comparing gas and electricity.

$$\begin{aligned} 1 \text{ therm} &= 105.506 \text{ MJ} \\ 1 \text{ kWh} &= 3.6 \text{ MJ} \\ 1 \text{ therm} &= 29.3071 \text{ kWh} \end{aligned}$$

CONSERVATION OF ENERGY

Energy can be changed from one form into another. A lake of water high in the mountains can change from potential energy when the water flows down through the turbine of a hydroelectric plant. It first becomes kinetic energy and finally electrical energy. To demonstrate an energy change for yourself, rub your hands together briskly for a few moments. The palms of your hands will begin to feel warm, and the energy used to obtain this effect was mechanical energy.

We talk of 'losing energy' and 'using up energy', but in fact energy is never really lost or destroyed. What happens is that it changes from one form to another. Unfortunately not all the energy we start with ends up in the form that is finally wanted. Quite a bit is often changed into unwanted heat or sound and dissipated into the atmosphere.

The Law of Conservation of Energy states that energy can neither be created nor destroyed, it may simply be changed from one form to another. For greatest efficiency, however, we need to end up with most of the energy we start with changed into the required end product and only a small percentage changed into unwanted forms of energy. So it is necessary to choose carefully the original form of energy to be used and the process for its change. Otherwise we may squander the readily available forms of energy at our disposal. Here gas has an obvious advantage since it needs only one change to produce heat, whereas electricity requires three or four energy changes for the same result.

The energy in natural gas, oil and coal comes from the light and heat energy given off by the sun many millions of years ago. But supplies of these 'fossil fuels' are limited and the best use needs to be made of them. It is now necessary to conserve energy generally as an economic measure and, in particular, to ensure that gas, the most refined natural source of heat energy, is used for those applications where its advantages can be most usefully employed.

The sun's energy can be used to produce electricity by means of solar cells. These form the power supply for space satellites, and on earth they can be used in those countries where the sun shines for most days in the year. Solar heating panels, fixed on roofs, can provide a proportion of a household's domestic heat requirements even in the United Kingdom. They pick up radiation from the sun and use it to heat circulating water.

MASS AND WEIGHT

Since energy in scientific terms is the capacity to do work, the next subject for study is 'work'. In order to start at the beginning and build up to this, it is necessary to look first at 'mass'.

In everyday speech 'mass' and 'weight' are often confused. We commonly use the term 'weight' to describe what is actually 'mass'.

- The 'mass' of an object is the amount of matter or substance that it contains.
- The 'weight' of an object is the pull of the earth's gravity on the object.

So the mass of an object is always the same whereas the pull of gravity will depend on the distance between the object and the centre of the earth.

Actually,

$$\text{mass} = \frac{\text{weight}}{\text{gravity}} \text{ or } \frac{W}{g}$$

A kilo of sugar always has the same mass, but its weights at, say, the top of a high mountain or down a deep mine shaft are fractionally different, because the pull of gravity reduces with increasing distance from the earth's centre.

From a practical point of view the difference in gravitational force between one place and another is so small that it can be ignored. Only when considering something like space travel does the difference become apparent. Outside the earth's gravitational field objects become 'weightless', but of course, they still have mass.

The reason why mass and weight get confused is simple. To find out the mass of an object the gravitational force on it is compared with the gravitational force on a standard mass. In other words we weigh it!

We use the same unit for both weight and mass, that is, the kilogram (kg). It is useful to remember that 1 l of water has a mass of 1 kg approximately. While it may be necessary in an examination to write '1 l of water has a mass of 1 kg', in real life we will probably continue to say '1 l of water weighs 1 kg'.

SPEED AND VELOCITY

Speed

Speed is rate of movement and it is measured by the distance which has been covered in a particular time. The difference between speed and velocity is that when measuring speed, changes of direction are ignored, whereas velocity refers to movement in a straight line (or in the case of angular velocity, movement in a circle).

Speed is measured in metres per second (m/s), flame speed of gases in millimetres per second (mm/s), and for practical purposes the units used are kilometres per hour (km/h).

Velocity

Velocity in a straight line is called 'linear' velocity. Linear means simply 'consisting of a line' and you will find it used to describe movements in a straight line when dealing with a number of other subjects. Velocity is the distance travelled in a straight line in a unit of time.

In other words:

$$\text{Velocity} = \frac{\text{distance}}{\text{time}} = \frac{\text{metres}}{\text{seconds}} = \text{m/s}$$

(The word 'per' in the terms 'metres per second' simply means 'divided by'. So

$$\text{metres per second} = \frac{\text{metres}}{\text{seconds}} = \frac{\text{distance}}{\text{time}})$$

As with speed, the practical units for linear velocity are km/h.

Angular velocity relates to objects which are rotating, and in a definition similar to that for linear velocity, angular velocity is the angle through which an object in circular motion moves in a given time. It is more relevant to mechanical engineering than to gas technology, and is often measured in revolutions per minute (rpm). The unit known as the radian is also used in expressions for angular velocity, in terms of radians per second (rad/s). The radian measure is the circular distance which has been travelled divided by the radius of the circle. In the case of one complete revolution of a circle of radius r the distance covered is the circumference of the circle, $2\pi r$. So:

$$\text{angle travelled (in radians)} = \frac{\text{distance}}{\text{radius}} = \frac{2\pi r}{r} = 2\pi$$

Thus one complete revolution, in angular measure, is either 360° or 2π radians, and:

$$1 \text{ rpm} = 2\ \pi \text{ radians per minute}$$

$$= \frac{2\pi}{60} \text{ or } \frac{\pi}{30} \text{ rad/s}$$

ACCELERATION

The velocity of a moving object may change over a period of time. When velocity is increasing it is called 'acceleration'. When velocity is decreasing it is 'deceleration' (deceleration is simply a negative form of acceleration).

Acceleration is the change in velocity over a time period, or the rate of change of velocity. So:

$$\text{acceleration} = \frac{\text{velocity}}{\text{time}} = \frac{\text{metres per second}}{\text{second}} = \frac{\text{m/s}}{\text{s}}$$

$$\frac{\text{m/s}}{\text{s}} = \text{metres per second per second, or}$$
$$\text{metres per second squared} = \text{m/s}^2$$

An example of acceleration can be seen in the force of gravity. If an object is allowed to fall freely towards the earth, from a height, it will accelerate at a steady rate of approximately 9.81 m/s^2 (assuming that the air offers no resistance to its fall) (Fig. 5.1). So, at the beginning, the velocity would be nil.

$$\text{at the end of 1 s, velocity} = 9.81 \text{ m/s}$$
$$\text{at the end of 2 s, velocity} = 19.62 \text{ m/s}$$
$$\text{at the end of 3 s, velocity} = 29.43 \text{ m/s}$$

and so on.

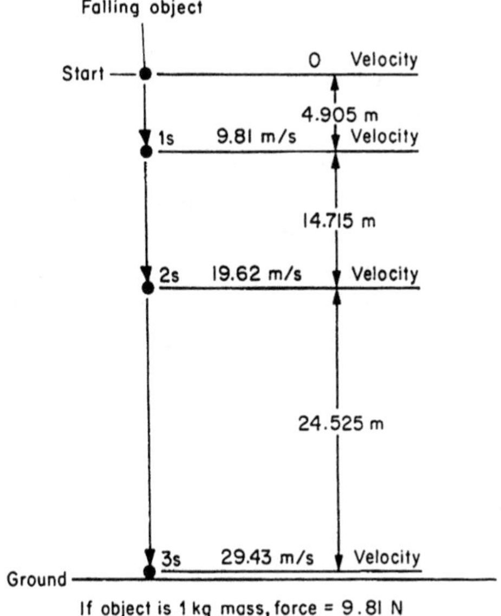

FIGURE 5.1 Acceleration due to gravity.

The distance that the object falls can be calculated from the formula: distance = velocity × time. During the first second the average velocity is mid-way between 0 and 9.81 or

$$\frac{0 + 9.81}{2} = 4.905 \text{ m/s}$$

In the one second the distance fallen will be:

$$4.905 \text{ m/s} \times 1 \text{ s} = 4.905 \text{ m}$$

Similarly, in each of the other two seconds, the distances will be:

$$\begin{aligned} \text{2nd second average velocity} &= \frac{9.81 + 19.62}{2} = 14.715 \text{ m/s} \\ \text{Distance} &= 14.715 \text{ m} \end{aligned}$$

$$\begin{aligned} \text{3rd second average velocity} &= \frac{19.62 + 29.43}{2} = 24.525 \text{ m/s} \\ \text{Distance} &= 24.525 \text{ m} \end{aligned}$$

FORCE

Having now established the definitions of mass and acceleration it is possible to move on to 'force', which is the product of both of them.

Newspaper reports of court proceedings sometimes mention that the police had to use force in order to make an arrest. This means that a blow was struck or that the offender was dragged or pushed into a police vehicle. In some way or another, force was applied to make something move.

Consider what happens when a blow is delivered to the jaw. The effect of the blow depends on two things:

1. The mass of the fist and its velocity when it strikes the jaw.
2. The rate at which the fist's velocity falls to zero on hitting the jaw, that is its deceleration. If the jaw is firm, deceleration is very rapid and the force is considerable. If, however, the punch is ridden, that is the head is moved away in the direction of the punch as the blow is landed, then deceleration is much slower and the force is very much reduced.

So: force = mass × acceleration (or deceleration).
In terms of SI units,

$$\begin{aligned} \text{force} &= \text{kilograms} \times \text{metres per second squared} \\ &= \text{kg} \times \frac{\text{m}}{\text{s}^2} \\ &= \frac{\text{kg m}}{\text{s}^2} \text{ or kg m/s}^2 \end{aligned}$$

This unit of force is called the 'newton' (N) in honour of Sir Isaac Newton, who first defined force and recognised gravitational force.

A force of 1 N exerted on a mass of 1 kg would cause it to accelerate at 1 m/s^2.

In Fig. 5.1 the falling object has a mass of 1 kg, so gravity exerts a force of 9.81 N.

WORK

'Work', in the mechanical sense, is done when a force displaces an object through a distance in the direction of the force. When a force applied to an object moves it over a distance then the amount of work done is measured by the force multiplied by the distance moved in the direction of the force. So:

$$\text{work} = \text{force} \times \text{distance} = \text{newtons} \times \text{metres} = \text{newton metres} = \text{N m}$$

This unit is called the 'joule'.

As you saw earlier in the chapter the joule is used not only to measure work but also as the unit for measuring all forms of energy.

POWER

When work is done energy may be used or produced at different rates. In a race the various competitors cover the same distances in different times. A piece of solder can be melted more quickly by using a larger blowlamp nozzle than by using a smaller one which will only give a small flame. The large flame gives out more heat energy than the small one and is said to be more 'powerful'. It can do the same job in much less time. 'Power' is the rate at which work is done, or the rate at which energy is produced or consumed.

$$\text{power} = \frac{\text{work}}{\text{time}} \text{ or } \frac{\text{energy}}{\text{time}} = \frac{\text{joules}}{\text{seconds}} = \mathrm{J/s}$$

This unit is called the 'watt' (W) after another British pioneer.

Electrical power is always measured in watts.

It is very important to understand clearly the difference between kilowatt and kilowatt-hour. What is confusing is that 'kilowatt' does not look as though it refers to time – but it does! And 'kilowatt-hour' looks like a rate – but it is not!

A kilowatt is 1000 J/s, so it is a rate of doing work, a unit of power. A kilowatt means that energy is used at the rate of 1000 J/s.

A kilowatt-hour is obtained from kilowatts × hours. It is the total amount of energy used when energy at the rate of 1000 joules per second is used for the period of 1 h.

$$1 \text{ kWh} = \frac{1000 \text{ joules}}{1 \text{ s}} \times 3600 \text{ s} = 3{,}600{,}000 \text{ joules or } 3.6 \text{ MJ}$$

If in doubt remember that kilowatts equate to MJ/h and kilowatt-hours equate to MJ.

TEMPERATURE

People buy gas generally to raise the temperature of something. This might be water for washing or bathing, air in the rooms for comfort or food to provide appetising meals. In each case the temperature of the substance will rise as more heat energy is applied. The effect of the substance absorbing heat is measurable by 'taking its temperature'. That is, by measuring the degree of hotness of the substance.

Temperature is measured by various instruments. For the lower temperatures a number of different kinds of 'thermometers' are used, for the very high temperatures 'pyrometers' are required. Both thermometers and pyrometers are described in detail in Chapter 14. The main differences between them are that thermometers are usually in direct contact with the substance to be measured and are operated by the expansion of a liquid or the change of a liquid into a gas. Pyrometers may not necessarily be in actual contact with the heated object and often employ electricity in their function. This may be by using an electric current to heat a wire until its incandescence matches that of the furnace or by employing a thermo-electric device which generates a current when it gets hot. The electrical meters which measure the amount of current passing are graduated directly in degrees of temperature.

The simplest form of thermometer is the one which consists of a fine bore glass tube containing mercury or alcohol in a bulb at one end and with the space above the liquid made into a vacuum. The tube is marked off in degrees of temperature and 'calibrated' (set to give a correct standard reading) to the freezing and boiling points of water (Fig. 5.2).

The scales in common use are the Fahrenheit and Centigrade or Celsius scales. The Fahrenheit scale is being superseded by the Centigrade scale which is now being given its continental name 'Celsius', after the Dutchman who devised it.

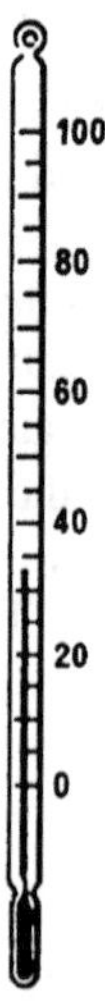

FIGURE 5.2 Glass thermometer.

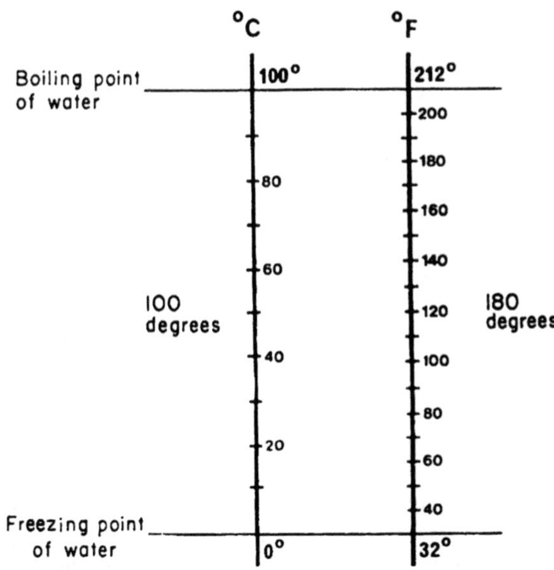

FIGURE 5.3 Comparison between Celsius and Fahrenheit scales.

Both Fahrenheit and Celsius scales are calibrated to the freezing and boiling points of pure water. If the pressure is the same, then the changes from ice to water (solid to liquid) and from water to steam (liquid to vapour) always take place at the same temperatures.

Figure 5.3 shows a comparison between the two scales. On the Fahrenheit scale the freezing point of water is at 32° (32°F) and the boiling point at 212° (212°F). So there are 180° between the two points.

On the Celsius scale freezing point is at 0°(0°C) and boiling point at 100°(100°C). So there are 100° between the two points.

If 100° on the Celsius scale equals 180° on the Fahrenheit scale, then:

$$\begin{aligned} \text{1 degree on the Celsius scale} &= \frac{180}{100}\,\text{degrees on the Fahrenheit scale} \\ &= \frac{9}{5}\,\text{degrees on the Fahrenheit scale} \end{aligned}$$

So to convert a Celsius temperature to the Fahrenheit equivalent (°C to °F):

$$°F = \left(°C \times \frac{9}{5}\right) + 32$$

To convert °F to °C:

$$°C = (°F - 32) \times \frac{5}{9}$$

Another, possibly quicker method of converting is:

°C to °F – Add 40, multiply by 9/5 take away 40,
°F to °C – Add 40, multiply by 5/9 take away 40.

Note: −40 °C and −40 °F are the same temperature.

Table 5.1 gives a selection of comparative temperatures.

The SI unit of temperature is the 'kelvin' (K), named after another British scientist. This unit is the same size as the degree Celsius from which it was developed. So:

$$1 \text{ degree on the Celsius scale} = 1 \text{ K}$$

The difference is that the Kelvin scale is linked to the lowest possible temperature obtainable.

You will discover, later on, that gases expand when heated and contract when cooled. If a volume of gas was gradually cooled, the volume would get smaller and smaller. If it was cooled until it disappeared, then the temperature at which this would happen would be 'absolute zero'. This point is −273 °C and 0 K.

There is absolutely nothing colder than absolute zero!

In fact gases generally turn into liquids or solids as they are cooled and a 'permanent gas' does not exist. It is, however, possible to calculate absolute zero temperature and the Kelvin scale is used in some calculations of the expansion or compression of gases. For all normal purposes, Celsius is the scale used and this can be converted to Kelvin by adding 273.

When using symbols it is usual to use '°C' for an actual temperature and 'deg C' for a temperature difference. For example:

The temperature of domestic hot water is 60 °C. This indicates a temperature rise, from 13 °C (cold), of 47 deg C, or

$$60\ ^\circ\text{C} - 13\ ^\circ\text{C} = 47 \text{ deg C}$$

The symbol for kelvin is K (*not* °K).

HEAT

Although a thermometer indicates the degree of hotness of a substance, it does not indicate the amount of heat which the substance contains. As an example, take a bucket containing 10 l of water and a bath containing 120 l. If both had been heated from the same cold water temperature up to the same hot water temperature, then it is fairly obvious that the bath water would have needed much more heating than the water in the bucket. In fact it needs twelve times as much heat and it would contain twelve times as much heat energy than the water in the bucket.

Table 5.1 Comparative Temperatures

Process		Temperature	
		°C	K
Absolute zero		−273.2	0
Home freezer cabinet		−21 to −24	252–255
Refrigerator temperatures			
Frozen food storage			
3 months		−18	255
1 month		−12	261
1 week		267	267
Cabinet		4–7	277–280
Freezing point of water		0	273
Room temperatures			
Hall, stairs and landing		16	289
Bedrooms, kitchen, toilet		18	291
Living and dining rooms		21	294
Bathroom		22	295
Body temperature		36.9	309.9
Water temperatures			
Warm bath		40.5	313.5
Washing-up		49–60	322–333
Water heater thermostats		60	333
Central heating circulation		60–82	333–355
Boiling point of water		100	373
Cooker oven			
Slow	mark 1 to 2	135–150	408–423
Moderate	mark 3 to 4	165–175	438–448
Moderately hot	mark 5 to 7	190–220	463–493
Hot	mark 8 to 9	232–246	505–519
Melting points			
Solder			
Grade A		185	458
Grade 99C		228	501

Continued

Table 5.1 Comparative Temperatures—cont'd

Process		Temperature	
		°C	K
Tin		232	505
Lead		325	598
Zinc		430	703
Copper		1083	1356
Steel		1600	1873
Glass		1450	1723

SPECIFIC HEAT AND SPECIFIC HEAT CAPACITY

The amount of heat in a substance depends upon its temperature rise, its weight (or mass) plus its ability to absorb or give off heat. This ability to absorb or give off heat is known as the 'specific heat' of the particular substance and compares the heat-absorbing properties of the substance to those of water.

The specific heat of a substance is the amount of heat required to produce a given temperature rise in the substance, divided by the amount of heat required to produce the same temperature rise in the same mass of water (so the specific heat of water = 1).

Specific heats vary considerably between different substances. Copper, which absorbs heat readily and quickly gives it out, has a specific heat of about 0.1. Brick or stone, which absorbs heat slowly, is 0.2 and wood about 0.5.

When SI units are used, calculations of thermal or heat energy, must be based on 'specific heat capacity'. This is because the unit of energy is the joule, which has no connection with water.

Specific heat capacity is the amount of heat energy required to raise the temperature of a given amount of the substance by 1 degree. Or, alternatively, the amount of heat given off when a specific amount of the substance is cooled through 1 degree.

$$\text{Specific heat capacity} = \frac{\text{energy}}{\text{mass} \times \text{temperature change}}$$

$$\text{Specific heat capacit} = \frac{\text{joules}}{\text{kilograms} \times \text{degrees change}}$$

$$= \text{joules per kilogram } ^{\circ}\text{Celsius} = \text{J/kg } ^{\circ}\text{C}$$

For example, at 15 °C the specific heat capacity of water is equal to 4.186 kJ/kg°C. This means that to raise temperature of 1 kg of water from 15 to 16 °C we require 4.186 kJ of heat energy. The reverse is also true. When 1 kg of water cools from 16 to 15 °C it gives out 4.186 kJ of energy. Using the previous section's example of a bucket of water containing 10 l and a bath containing 120 l we can prove that we need twelve times as much heat to raise the water temperature in the bath than the bucket:

$$\text{Heat energy} = \text{mass (kg)} \times \text{temperature rise } (^\circ\text{C}) \times \text{specific heat capacity}$$

Assume a temperature rise of 27 °C

$$\begin{aligned} \text{Heat energy in the bath} &= 120 \times 27 \times 4.186 = 13{,}562.64 \text{ kJ} \\ \text{Heat energy in the bucket} &= 10 \times 27 \times 4.186 = 1130.22 \text{ kJ} \\ 13{,}562.64 \div 1130.22 &= 12 \end{aligned}$$

therefore the water in the bath contains twelve times the amount of heat than that in the bucket.

At 12 °C the specific heat capacity of air = 1.012 kJ/kg °C. Quantities of air are usually measured in m^3, so when dealing with air it is usual to use the heat capacity on a volume basis rather than on mass. That is, to use cubic metres, rather than kilograms.

$$\text{Heat capacity of air} = 1.34 \text{ kJ/m}^3 \ ^\circ\text{C}$$

Table 5.2 gives a list of the specific heat capacities of common substances.

Table 5.2 Specific Heat Capacities.

Material	Specic Heat Capacity (kJ/kg °C)
Water	4.2 (4.186)
Air	1.0 (1.012)
Ice	2.1
Mercury	0.14
Methylated spirit	2.5
Brick	8.4
Concrete	9.2
Glass	6.7
Iron	4.6
Copper	0.4
Lead	0.13

SENSIBLE HEAT AND LATENT HEAT

If a piece of ice was placed in a pan and heated, its temperature would gradually rise until it reached 0 °C. Then the ice would begin to melt. If the heating was continued until the ice was melted, the pan would then contain water at 0 °C. There would be no increase in temperature while the ice was melting.

If heating was still continued, the temperature of the water would rise until it reached 100 °C. Then the water would begin to boil. The temperature would stay at 100 °C until all the water was turned into steam. If the heating process was continued with the steam, the temperature would begin to rise again.

Figure 5.4 shows, on a graph, what is happening.

From point A to point B the temperature is rising steadily. The heat being absorbed during that period is called 'sensible heat' because it can be 'sensed' by the thermometer.

From B to C it is apparent that heat is still being absorbed but it is not visible on the thermometer, so it is called 'latent heat' because it cannot be seen.

From C to D the thermometer shows the second increase in sensible heat and from D to E shows latent heat again.

The same sort of thing happens with most substances. When a solid melts into a liquid or a liquid cools to a solid, the heat absorbed or given off is called the 'latent heat of fusion'.

When a liquid evaporates into a vapour or a vapour condenses into a liquid the heat involved is called the 'latent heat of vaporisation'.

Different substances require different amounts of heat to bring about these changes in state. For example:

Latent heat of fusion of ice = 334 kJ/kg

Latent heat of vaporisation of water = 2250 kJ/kg.

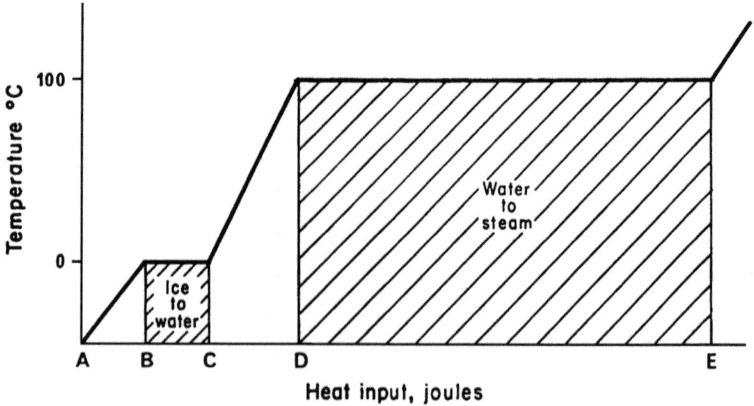

FIGURE 5.4 Graph of sensible and latent heat of water (not to scale).

CONDENSATION AND EVAPORATION

When a vapour or a gas changes into a liquid it is said to have 'condensed' and the change and its product are called 'condensation'.

When the reverse happens and the liquid becomes a vapour or a gas, it has 'evaporated' and the process is called 'evaporation'.

During condensation, heat is given up and during evaporation heat is absorbed.

The atmosphere contains water in vapour form and the amount which it can hold depends on its temperature. Warm air can hold more moisture than cold air. When the air has as much water vapour as it can hold at that temperature it is said to be 'saturated' at that particular temperature. If the temperature of the air is raised, then the air could hold more vapour. So, at the new, higher temperature it is only 'partially saturated'.

The relationship between the amount of water vapour and the temperature of the air is called the 'humidity'.

The relative humidity of the air is the relationship between the amount of water vapour present and the total amount that the air could hold at that temperature. It is expressed as a percentage and, for comfort, it should not be above 70%. When the atmosphere is more than 70% saturated, it feels 'muggy' and 'sticky'.

If air containing water vapour is cooled, then a temperature will be reached at which the air is saturated. If cooling continues below this then water vapour will begin to condense out. The temperature at which this condensation begins is called the 'dew point', literally the point at which dew begins to form. The dew point of a gas also depends on the pressure to which the gas is subjected.

A familiar example of condensation occurring is when warm, moist air in a room comes in contact with a cold window pane. The air in contact with the glass is quickly cooled to below the dew point and tiny droplets of water are condensed on to the window in the form of a fine mist. The window 'mists' or 'steams up'. Similarly condensation can cause damp patches on cold walls. Because the products of combustion of gas contain water vapour, precautions must be taken to avoid condensation taking place within the appliance or its flue. The products are often diluted with excess air to provide a larger volume of gases to hold the water. The heat exchanger is designed to heat up quickly and the flue is kept short, warm and as straight as possible.

COMFORT CONDITIONS

The human body is probably at its most comfortable when it is doing the least amount of work. Certainly this is true so far as thermal comfort is concerned. A healthy body must maintain its temperature at a steady 36.9 °C.

To do this the body has to work. Food is digested and turned into a source of heat. The average person gives out about 425 kJ when sitting down and up to 800 kJ when digging the garden.

If the body starts to get cold, then the pores close and shivering produces rapid movement in an endeavour to raise the body temperature to normal. If, on the other hand, the body becomes too hot, then the pores open and sweating occurs. This is so that the moisture, in being evaporated, will take heat from the body and lower the temperature again.

The body is comfortable when, for any activity, the heat it is producing exactly balances the amount of heat being lost. In other words, the body does not have to do any work to keep its temperature steady. The factors which affect comfort are:

1. air temperature,
2. air movement,
3. radiation from the environment,
4. humidity.

1. Air temperature
   This needs to be a 'comfortable' temperature for our activities. It is recommended that, in a house, the temperature should be:

Room	Room temperature (°C)	Air change per hour
Living room	21	1
Dining room	21	2
Bedsitting room	21	1
Bedroom	18	½
Hall	16	1½
Bathroom	22	2
Kitchen	18	2
Toilet	18	1½

Source: British Standard Code of Practice 5449: Part 1.

2. Air movement
   There needs to be some air movement in a room to give a feeling of freshness. But excessive movement or draughts make us feel cold even when the air temperature is still quite high. In summer, of course, a fan is used to help us keep cool although it does not actually reduce the air temperature. What does happen is that air movement helps to increase the rate at which perspiration is being evaporated and so increases the rate at which heat is taken from the body. Recommended air changes are shown in the table giving recommended room temperatures.
3. Radiation from the environment
   Heat always flows from a hot object to a colder one and any hot object gives out heat rays or 'radiation'. The sun is the main source of radiant

heat but anything which is hotter than something else within range will give off radiant heat (see Chapter 10).

This means that, for comfort, our surroundings should ideally be at about the same temperature as the air. If they are a good deal colder, we will lose heat to them. If they are much hotter, we will be heated up.

4. Humidity
   Since we control our temperature by the evaporation of perspiration from the skin, we can do this more easily when the atmosphere is only partially saturated with water vapour and so is able to accept some more.
   If the humidity is high, we will not easily be able to get rid of the perspiration and so will not be comfortable. The reverse is also true. If the atmosphere is too dry we will be equally uncomfortable.
   Air conditioning controls not only the air temperature by making it cool in summer and warm in winter, but it also maintains a comfortable humidity by adding or removing moisture.

HEAT ENERGY RATES

The rates at which gas is used and heat is produced in a gas appliance can be expressed in a number of ways.

Gas meters measure the volume of the gas passing so it is easy to check the gas rate of the appliance by the meter. This gives the gas rate in cubic feet per hour or cubic metres (not meters!) per hour.

The heat input rate can then be calculated by multiplying the gas rate in m^3 or ft^3 by the calorific value (CV) in the relevant units.

$$\text{Heat input rate} = \text{gas rate} \times \text{calorific value}$$

$$\text{MJ/H} = \text{m}^3/\text{h} \times \text{MJ/m}^3$$

or

$$\text{Btu/h} = \text{ft}^3/\text{h} \times \text{Btu/ft}^3$$

MJ/h may be converted to kW for convenience.

For example, assume:

$$\text{calorific value} = 39.3\ \text{MJ/m}^3 \text{ or } 1035\ \text{Btu/ft}^3$$

$$\text{gas rate} = 1.7\ \text{m}^3/\text{h or } 60\ \text{ft}^3/\text{h}$$

Then,

$$\begin{aligned}\text{heat input rate} &= 1.7 \times 39.3\ \text{MJ/h or } 60 \times 1035\ \text{Btu/h} \\ &= 66.81\ \text{MJ/h or } 62{,}100\ \text{Btu/h} = \frac{66{,}810{,}000}{3600}\ \text{J/s} \\ &= 18{,}558.3\ \text{J/s(or watts)} = 18.56\ \text{kW}\end{aligned}$$

Given a heat input rate it is also possible to convert this into a gas rate in terms of m^3/h or ft^3/h

$$\text{gas rate} = \frac{\text{heat input rate}}{\text{calorific value}}$$

$$(\text{in units})\ \text{m}^3/\text{h} = \frac{\text{MJ/h}}{\text{MJ/m}^3}$$

or

$$\text{ft}^3/\text{h} = \frac{\text{Btu/h}}{\text{Btu/ft}^3}$$

For example, assume:

$$\begin{aligned}\text{Heat input rate} &= 18\ \text{MJ/h or } 17{,}060\ \text{Btu/h}\\ \text{Calorific value} &= 39.3\ \text{MJ/m}^3 \text{ or } 1035\ \text{Btu/ft}^3\end{aligned}$$

Then,

$$\text{Gas rate} = \frac{18\ \text{m}^3/\text{h}}{39.3} \text{ or } \frac{17{,}060\ \text{ft/h}}{1035} = 0.46\ \text{m}^3/\text{h or } 16.5\ \text{ft}^3/\text{h}$$

If the heat input rate had been in kW, it would have needed to be converted to MJ/h before doing the calculation.

For example,

$$\begin{aligned}\text{Heat input} &= 5\ \text{kW} = 5000\ \text{J/s} = 5000 \times 3600\ \text{joules/hour}\\ &= 18{,}000{,}000\ \text{J/h} = 18\ \text{MJ/h}\end{aligned}$$

Alternatively, given the conversion factor of
1 kW = 3.6 MJ/h

$$5\ \text{kW} = 5 \times 3.6 = 18\ \text{MJ/h}$$

THERMAL EFFICIENCY

When energy changes from one form to another the change is never 100% efficient. Some of the original form of energy always ends up in a useless form and so can be said to be wasted. Efficiency is the comparison (or ratio) between what we get out and what we put in. It is usually expressed as a percentage, so the ratio has to be multiplied by 100.

$$\%\ \text{efficiency} = \frac{\text{output}}{\text{input}} \times 100$$

The efficiency of any process, mechanical, electrical or chemical can be calculated in this way. The units in which the output and input are measured must be the same.

The relationship between the potential heat energy in a fuel and the useful heat obtained from an appliance is the 'thermal efficiency'.

$$\text{Thermal efficiency} = \frac{\text{heat output}}{\text{heat input}} \times 100$$

The output and input could be measured in kW or MJ/h. For any calculation the output and the input must be in the same units.

Example: A gas water heater has a gas rate of 2.67 m^3/h and delivers 7 l of water per minute when raised to 45 °C. Find the percentage efficiency. Assume CV = 39.3 MJ/m^3.

$$\begin{aligned}\text{Heat output} &= 7\,\text{l} = 7\,\text{kg/min} = 7 \times 60 = 420\,\text{kg/h} \\ &= 420 \times 45 \times 4.186\,\text{kJ/kg/deg C} \\ &= 79,115.4\,\text{kJ/h or } 79.11\,\text{MJ/h}\end{aligned}$$

$$\text{Heat input} = 2.67\,\text{m}^3/\text{h} \times 39.3 = 104.93\,\text{MJ/h}$$

Therefore,

$$\text{efficiency} = \frac{79.11}{104.93} \times 100 = 75\%$$

When a number of quantities are related, as in this case, it is always possible to calculate one missing one, if all the others are known.

If

$$\%\ \text{efficiency} = \frac{\text{output} \times 100}{\text{input}}$$

then

$$\text{output} = \frac{\%\ \text{efficiency} \times \text{input}}{100}$$

and

$$\text{input} = \frac{\text{output} \times 100}{\%\ \text{efficiency}}$$

MECHANICS

'Mechanics' is the study of the effects of forces on objects. It is divided into two sections, statics and dynamics.

1. Statics
   In this category the motion of the objects is not affected by the forces. So the objects include beams, joists, structural steelwork and things that are stationary, or 'static' (it would also include an aircraft in straight level flight at constant speed, but that is another story).
2. Dynamics
   Here the forces are changing the motion of the objects concerned and the objects may be loads being lifted, valves, diaphragms or pistons moving (or the aircraft when taking off and landing). This chapter has already covered a number of aspects of mechanics including energy, force, work and power. It remains for us to consider the effects of loads on the structure of a house or building and to see how the principles of mechanics are applied to mechanisms, tools and lifting aids.

MOMENT OF A FORCE

The 'moment' of a force is simply its turning effect. Figure 5.5 shows a heavy door in an open position. To close the door you will have to apply a force.

Try at first to close it by pushing with one finger on the handle at A. The door should move relatively easily.

Now try to close it with the same finger at a point about 50 mm from the hinged edge at B. It takes a lot more force to move it now.

If you were to measure the forces applied at A and B and then multiply each force by the distance from the hinge at which it was applied, the answer in each case would be the same. The moment of a force in relation to its turning point is the size of the force multiplied by the perpendicular distance of its line of action from the turning point.

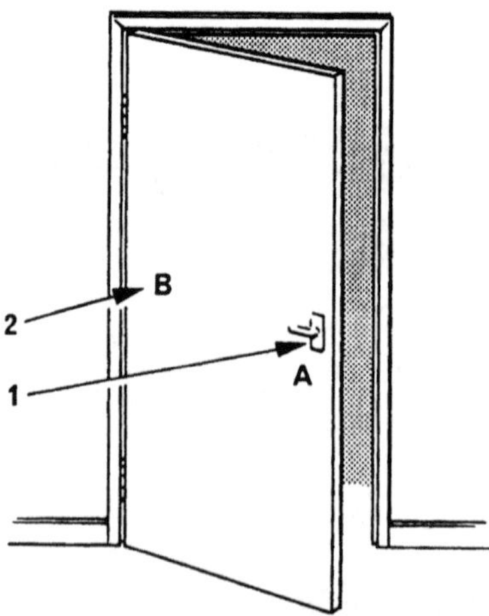

FIGURE 5.5 Forces on a door.

For example, if the force which would just cause the door to move, when applied at a distance from the hinge of 750 mm, was 5 newtons, then the force required at any other known distance could be calculated (Fig. 5.6).

Force 1 × Distance C = Force 2 × Distance D. In other terms:

$$F_1 \times \mathrm{C} = F_2 \times \mathrm{D}$$

or

$$F_2 = \frac{F_1 \times \mathrm{C}}{\mathrm{D}}$$

If distance D was 75 mm the force required would be:

$$F_2 = \frac{750 \times 5}{75}\,\mathrm{N} = 50\,\mathrm{N}$$

Similarly, if any two forces and one distance, or any two distances and one force were known, the other missing quantity could be calculated.

if

$$F_1D_1 = F_2D_2$$

then

$$F_2 = \frac{F_1D_1}{D_2} \qquad D_2 = \frac{F_1D_1}{F_2}$$

$$F_1 = \frac{F_2D_2}{D_1} \qquad D_1 = \frac{F_2D_2}{F_1}$$

From Fig. 5.6 it can be seen that the forces exerted on the door are attempting to turn it in an anticlockwise direction. That is, in the opposite direction to that in which the hands of a clock rotate. So the forces are causing an 'anticlockwise moment'.

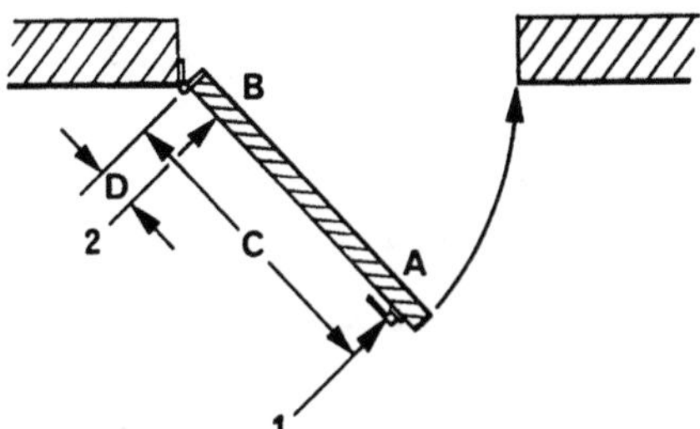

FIGURE 5.6 Calculation of forces.

If the door was to be held open against the forces trying to close it, then it would stay in one position when the anticlockwise moments attempting to close it were balanced by an equal and opposite clockwise moment. At this point the door is said to be in a state of rest or 'equilibrium'.

Since anticlockwise moments are measured as negative moments (or minus, –) and clockwise moments are measured as positive moments (or plus, +), then an object will be in equilibrium when the total of all the moments acting on it equals zero.

In other words, when the moments are equal and opposite, as on the door, this principle makes it possible to calculate the forces on levers and structures.

LEVERS

There are three forms or 'orders' of levers shown in Fig. 5.7a–c. In these diagrams the forces on the levers are shown as L for Load and E for Effort.

The distances are:

Effort to pivot $= e$

Effort to load $= a$

Load to pivot $= l$

By the principle of moments:

$$E \times e = L \times l \quad \text{or} \quad \frac{E}{L} = \frac{l}{e}$$

$$E = \frac{L \times l}{e} \qquad L = \frac{E \times e}{l}$$

$$e = \frac{L \times l}{E} \qquad l = \frac{E \times e}{L}$$

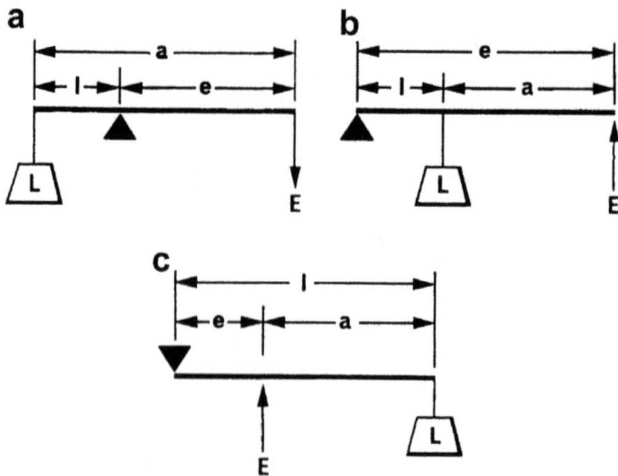

FIGURE 5.7 Forms of lever.

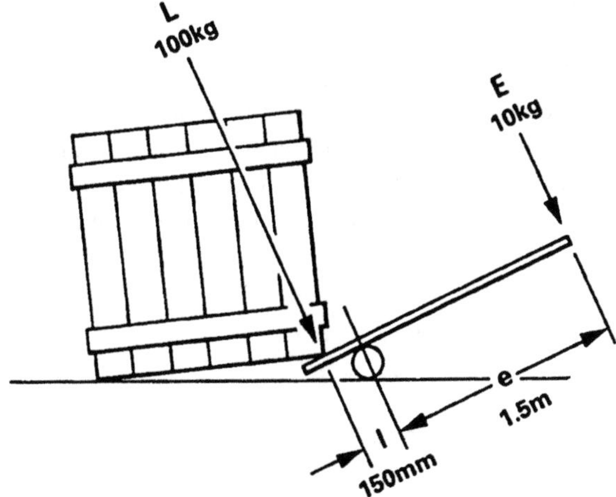

FIGURE 5.8 Lifting a crate.

The distance from effort to load varies:

In Fig. 5.7a, $a = l + e$

In Fig. 5.7b, $a = e - l$

In Fig. 5.7c, $a = l - e$

As a practical example, Fig. 5.8 shows a crate being lifted by a first order lever. If the edge of the crate is exerting a weight of 100 kg at right angles to the lever, what weight (or effort) would be needed to support it?

Now

$$E = \frac{L \times l}{e}$$

So

$$E = \frac{100 \text{ kg} \times 0.15 \text{ m}}{1.5 \text{ m}} = 10 \text{ kg}$$

STRESS AND STRAIN

All the materials you will meet and use in the course of your work are subjected to loads of one kind or another. Steelwork and joists in a building are designed to carry the floors, furniture and occupants. Tools are subjected to loads when in use.

Building and engineering materials must behave satisfactorily under loads. This means that they must not break under the load and also that they must not stretch or squash beyond an acceptable limit. A manufactured article must be strong enough for the purpose for which it will be used. So there is a need to study the 'stress' which will be imposed on the material.

'Stress' may be 'tensile' or pulling stress, or it may be 'compressive' or pushing stress. In both cases the same rules apply. Stress is the relationship between the load and the area on which the load is applied, more precisely stress = load ÷ area.

Figure 5.9 shows concrete blocks (which might be part of a building) carrying loads. You can see that, if one block at (a) can carry a load of 1 tonne, then the four blocks at (b) will each carry a load of 1 tonne, making a total of 4 tonnes. If the four blocks were replaced by one larger block having the same total area (c), then the large block would carry a load of 4 tonnes (4 tonnes = 39.25 kN).

In each case, the stress would be the same.

(a)

$$\text{area} = a \times a = a^2$$

$$\text{load} = L$$

$$\text{strees} = \frac{L}{a^2}$$

(c)

$$\text{area} = 2a \times 2a = 4a^2$$

$$\text{load} = 4L$$

$$\text{stress} = \frac{4L}{4a^2} = \frac{L}{a^2}$$

So,

$$\text{Stress} = \frac{\text{load}}{\text{area}} = \frac{\text{newtons}}{\text{square metres}} \text{ or N/m}^2$$

The 'ultimate stress' is the stress required to break the particular material. For structural steel, this is the same amount for both tensile and compressive

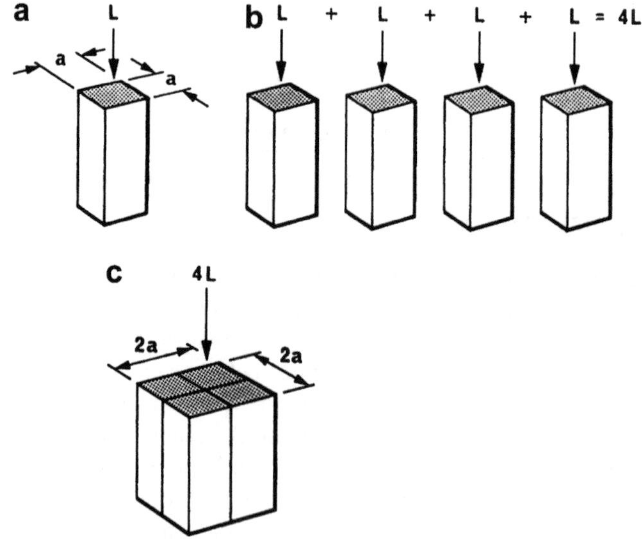

FIGURE 5.9 Stress.

stress, at 414 MN/m^2. Bricks and concrete will withstand greater loads in compression (about 14 MN/m^2) than in tension (about 2 MN/m^2).
Because materials may have to stand up to unexpected overloads which could not be accurately estimated, some allowance has to be made to ensure that the material will not reach its ultimate stress in use. This is called the 'factor of safety'. The factor of safety is the number of times by which the area of the material must be increased to provide an adequate safety margin. It will depend on whether the load is steady or varying and on the possibility of shocks. For structural steel the factor for a steady load is about 3. For Wood, it is about 8 and brick 15.

When materials are being stressed they become distorted or deformed. They are stretched or compressed, depending on the type of stress. This distortion is called 'strain'.

Strain is the relationship between the amount of stretch, or 'elongation', and the original length of the piece of material, thus:

$$\text{strain} = \frac{\text{stretch}}{\text{length}} \quad \text{or} \quad \frac{\text{millimetres}}{\text{millimetres}}$$

so it has no units, it is simply a number.

For example, if a steel rod 5 m in length extended a distance of 2.5 mm under a load of 20 kg, then a bar of 1 m length would stretch $1/5 \times 2.5 = 0.5$ mm.

But 0.5 mm = 0.0005 m, so each 1 m stretches 0.0005 m, or each 1 inch stretches 0.0005 inches and so on.

Although this example refers to a tensile load it is equally true for compression loads and strain might be better expressed more generally as:

$$\text{strain} = \frac{\text{increase or decrease in length}}{\text{original length}}$$

HOOKE'S LAW

Robert Hooke discovered that, for some materials, strain is directly proportional to stress, up to a limiting value, the 'limit of proportionality' or 'elastic limit'. If a graph is drawn of stress to strain for one of these materials the result is a straight line, as shown in Fig. 5.10. Beyond the limit of proportionality the graph becomes curved showing that the material is becoming permanently distorted and will finally break. Up to the elastic limit the material will regain its original length if the stress is removed. Materials which obey Hooke's Law include most metals and wood.

The ratio of stress to strain for a material obeying Hooke's Law is called the 'modulus of elasticity' or 'Young's modulus'. It is a measure of the stiffness of the material. Steel is the stiffest with a modulus of elasticity of about 30,000,000, brass is about 15,000,000 and aluminium 10,000,000 (1,000,000 is often written as 1×10^6).

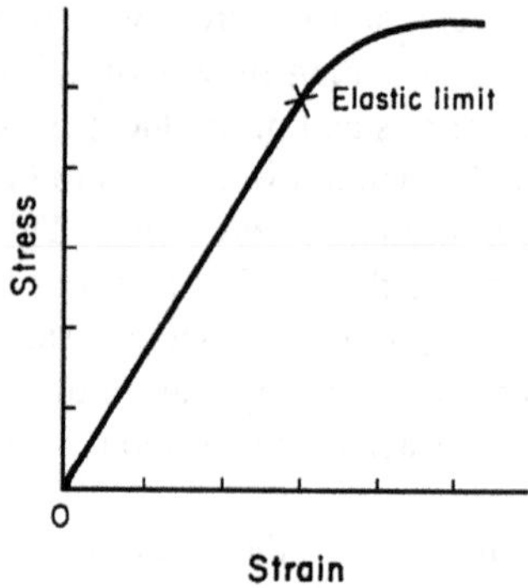

FIGURE 5.10 Graph of the ratio of stress to strain, Hooke's Law.

STRESSES IN BEAMS

The stresses in beams with which we are principally concerned are those due to bending. Figure 5.11 shows a wood floor joist supported on two walls. If the weight or load (*W*) is uniformly distributed over the whole length of the joist then each support will carry half the load (1/2*W*) (the same would be true if the weight was concentrated at the centre).

It is possible to calculate the 'bending moment' using the principle of moments previously described, and to draw a graph showing the value of the bending moment at any point along the beam. The graph in Fig. 5.11 shows that there is zero bending at the supports and maximum in the centre.

Figure 5.12 shows the same beam with each of the supports moved in a distance *l* from the ends. In this case the graph shows that the bending moment reaches its maximum over the supports and again in the centre. But the maximum is now considerably less than in the previous example in Fig. 5.11 and smaller beams can support the same load.

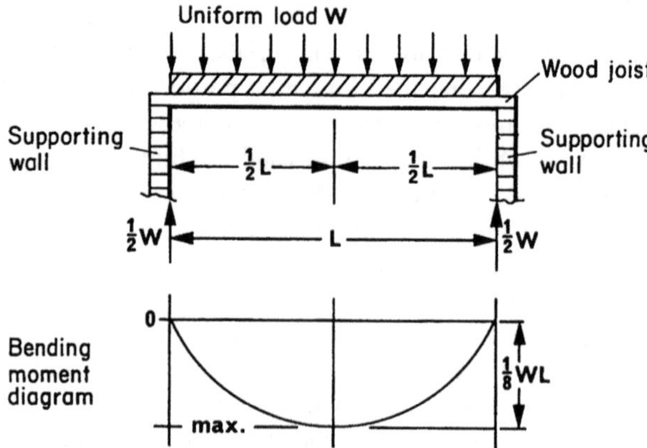

FIGURE 5.11 Bending moment in a floor joist.

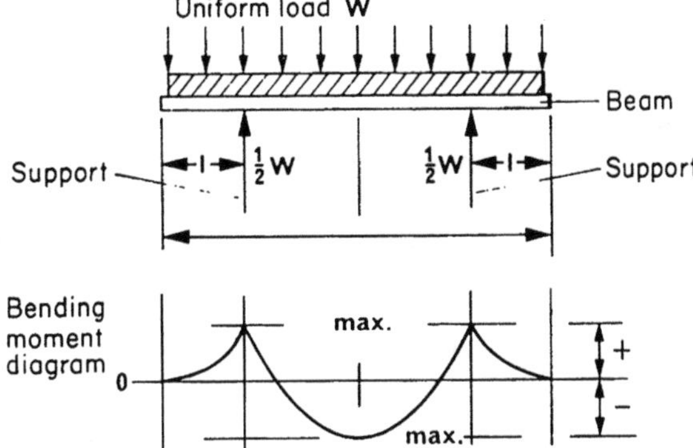

FIGURE 5.12 Bending moment in overhanging beam.

If the distances l were $0.207L$ (about $1/5L$) then the maximum bending moment at the supports would be equal to the maximum bending moment at the centre and would be $0.021WL$.

When a beam is bent under a load the stress is not regular throughout the section of the beam. To see what actually happens, take a suitably sized piece of wood and bend it slowly round your knee until it begins to break.

First, it will bend to an arc and then it will begin to crack on the outer face of the curve (the convex face). When it is practically broken it should look like the illustration in Fig. 5.13a. The outer fibres on the convex face have been pulled apart, whilst the fibres on the concave face have been squashed together. This shows that the outer face is under a tensile stress while the inner face is subjected to a compressive stress. Somewhere in between the two is a neutral axis which is not stressed at all.

In fact, the stress goes from zero at the neutral axis to a maximum in each direction at the faces.

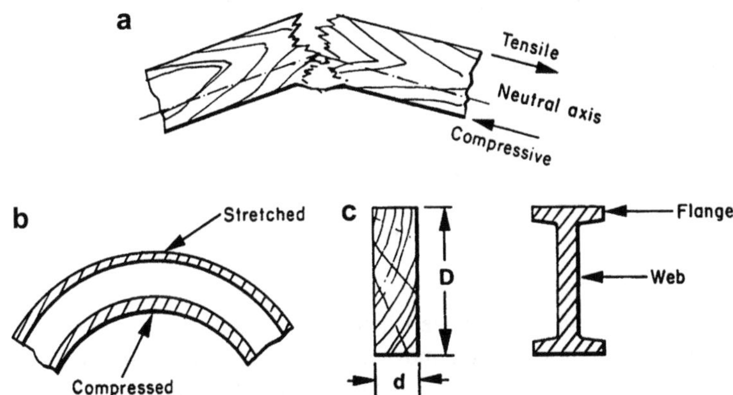

FIGURE 5.13 (a) Effect of bending wood, (b) effect of bending pipe, and (c) cross-section of wood joist and rolled steel joist.

Figure 5.13b shows the effect of bending on a piece of soft, ductile metal pipe. The tension on the outside of the bend stretches the metal and reduces the thickness of the pipe wall. On the inside of the bend the metal is compressed and the wall becomes thicker.

This effect limits the extent to which pipes of various materials and diameters may be bent (see Volume 2, Chapter 1).

Figure 5.13c shows the cross-section of a typical wood joist with the height D approximately three times the width d.

$$D = 3d \quad \text{or} \quad d = \frac{1}{3}D = \frac{D}{3}$$

To support the maximum load the joist is used, as shown, with D vertical. If it were turned through 90° so that D was horizontal, it would support only one-third of the previous load.

The final illustration shows the section through an T beam or rolled steel joist (RSJ). This form of beam is commonly used in structural steelwork since it concentrates the metal in the flanges which have to withstand the maximum stresses. For its weight this beam can support greater loads than most other sections.

PULLEYS

Pulley systems offer a method of applying a small effort to raise a much heavier load. Like levers, they do this by making the effort move much further than the load so that a small force multiplied by a large distance balances a large force multiplied by a small distance. In other words, the amount of work done by the effort equals the amount of work done on the load (assuming the system to be 100% efficient).

Figure 5.14 shows a single fixed pulley. Since the distance of the load from the pivot is equal to the distance of the effort from the pivot, both being the radius of the pulley, the amount of force needed to balance the load will be exactly equal to the load (Fig. 5.14a and b).

The distance moved by the effort will be the same as the distance moved by the load (Fig. 5.14c and d).

So,

$$E \times r = L \times r \quad \text{and} \quad \frac{d}{D} = \frac{r}{r} = 1, \text{ so that } d = D \text{ and } E = L$$

Therefore

$$\frac{L}{E} = 1$$

There is no mechanical advantage when using a single fixed pulley.

Figure 5.15 shows a single movable pulley. Movable pulleys are always combined with fixed pulleys in order to reverse the direction of the ropes. There are always equal numbers of fixed and movable pulleys in any single system using a continuous rope.

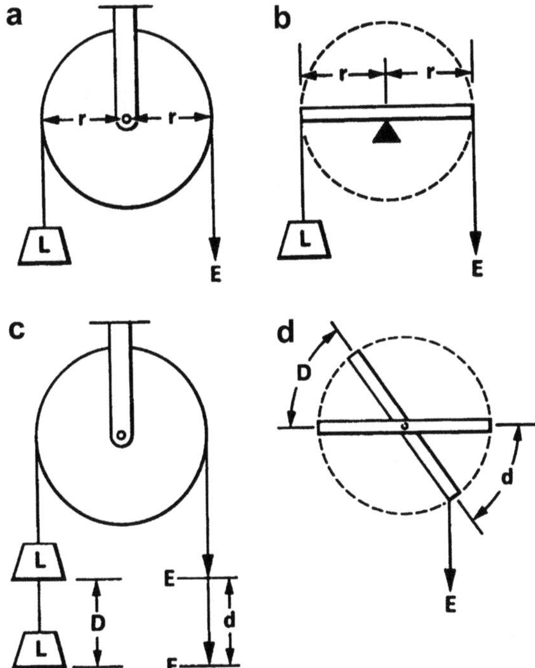

FIGURE 5.14 Single fixed pulley: (a) forces on the pulley, (b) comparison with forces on a lever, (c) ratio of movement of effort and load, and (d) comparison with movement of lever.

In Fig. 5.15a the pulleys are shown side by side so that the principle can be more clearly seen. Figure 5.15b shows the usual practical arrangement of the blocks. Figure 5.15c shows the relationship between the movement of the load, D, and the movement of the effort, d.

If the load has to be raised at a distance D then both the sections of rope a and b must be made shorter by this amount. It follows, therefore, that the section of rope c must be pulled down by a distance equal to $2D$. So the effort moves twice as far as the load. This means that the effort can support a load twice its size.

The ratio of the distance moved by the effort to the distance moved by the load is called the 'velocity ratio' and in this case

$$\text{Velocity ratio} = \frac{d}{D} = 2$$

The ratio of the load to the effort is

$$\frac{L}{E} = \frac{1}{2}$$

This is known as the 'mechanical advantage'.

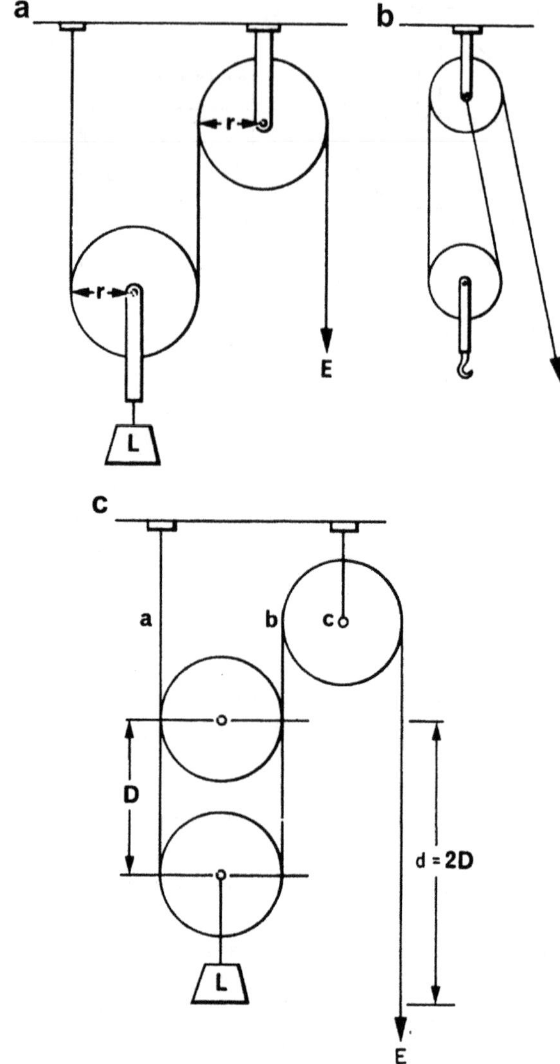

FIGURE 5.15 Single movable pulley: (a) forces on the pulley, (b) practical arrangement of pulley blocks, and (c) ratio of movement of effort and load.

For systems having any number of movable pulleys it can be shown that $E = L/2n$ and $d/D = 2n$, where n is the number of moveable pulleys.

Another easy way to find out the mechanical advantage of a pulley block is to count the number of ropes between the pulleys. And that does not include the one you are going to pull. The number of ropes equals the velocity ratio or d/D. The effort required will be:

$$\text{Effort} = \frac{\text{load}}{\text{no. of ropes}}$$

As an example look at Fig. 15b. There are two ropes between the pulleys, so the effort required will be load/2 or ½ load. And the effort will have to move twice as far, or as fast, as the load.

SLINGS

When large objects have to be lifted it is usual to employ 'slings'. Slings are ropes or chains which are attached to the object as shown in Fig. 5.16.

The safe load that a sling will carry depends not only on the thickness of the rope or chain but also on the angle between them.

This angle should be kept as small as possible and the effect of making it larger is shown in Fig. 5.17.

LIFTING AND HANDLING

The human body has adapted itself to its environment over the centuries, but it was not designed to perform great feats of strength or to withstand excessive physical labour.

With training, it is possible to develop the capacity to lift very heavy weights but an average person is well below the standard of an Olympic athlete. Many people, in the past, found themselves on the industrial scrap heap because they had so strained their bodies that they became incapable of a normal existence.

The main point to remember in lifting is to keep the spine straight and vertical. It must never be used as a lever or in the same way as the jib on a crane.

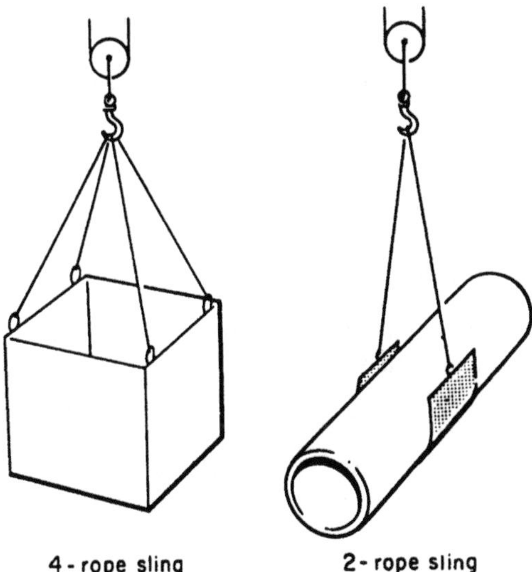

FIGURE 5.16 Types of sling.

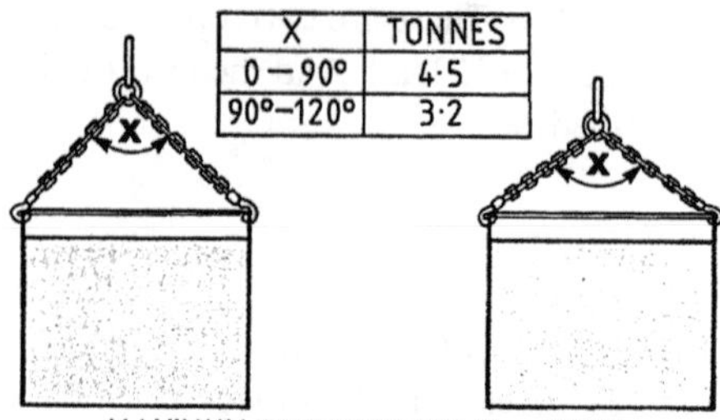

FIGURE 5.17 Maximum safe load for 10 mm chain.

Legs, arms and fingers must also never be used as levers or the strain may prove too great for them.

The method of lifting recommended is called 'kinetic' lifting, Fig. 5.18. This uses muscles best suited to the job and helps to reduce the possibility of an accident. The essential points of kinetic lifting are:

1. Get hold of the object properly, use the palms of the hands and the roots of the fingers and thumb. Do not alter your grip while carrying something.
2. Keep the spine straight and vertical. Keep the head up and the chin tucked back. Use the legs to do the lifting.
3. Keep the feet slightly apart in line with the hips and with one foot forward in the direction to be taken with the load.
4. Keep the elbows on the sides with the arms close to the body.
5. As the object is lifted, lean back so that the body counterbalances the weight of the object.

Do not persist in trying to lift an object which feels too heavy. Take precautions when dealing with objects above one-tenth of your body weight and do not attempt to lift your own body weight without help or a mechanical lifting aid.

Make sure that you understand your employer's regulations covering lifting and handling.

CENTRE OF GRAVITY

The 'centre of gravity' of an object is the point about which it can be balanced. It is the point through which passes the single, vertical force which is the resultant of all the gravity force acting on all the particles of the object.

The position of the centre of gravity of regular, flat objects can easily be found. For a square, rectangle or parallelogram it is at the point where the diagonals cross or 'intersect' (Fig. 5.19). For a circle it is at the centre (Fig. 5.20).

The centre of gravity of an irregular shape can be found by suspending it from two different points alternately and drawing lines vertically downwards

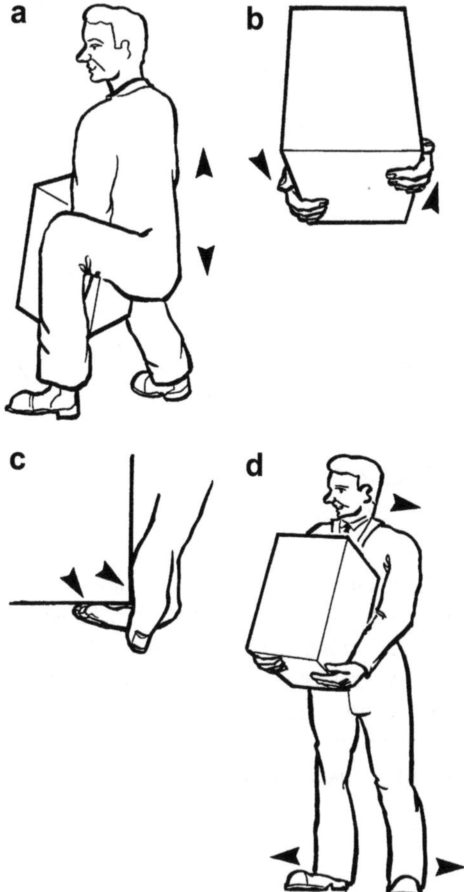

FIGURE 5.18 Lifting and handling: (a) kinetic lifting, (b) and (c) holding the load, and (d) carrying the load.

from the points each time (Fig. 5.21). Where the lines cross is the centre of gravity.

A rectangular solid has its centre at the intersection of the diagonals between opposite corners (Fig. 5.22).

Figure 5.23 shows an object being tilted. All the time when a vertical line from its centre of gravity passes through the base it will remain in,or return to, a stable position. When the line falls outside the base, the object will fall over.

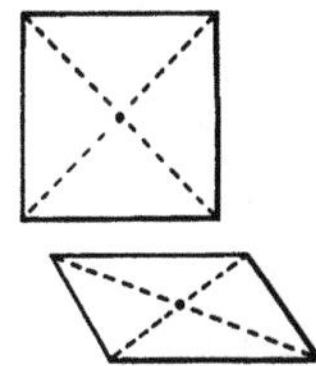

FIGURE. 5.19 Centre of gravity of a rectangle and parallelogram.

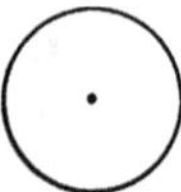

FIGURE 5.20 Centre of gravity of a circle.

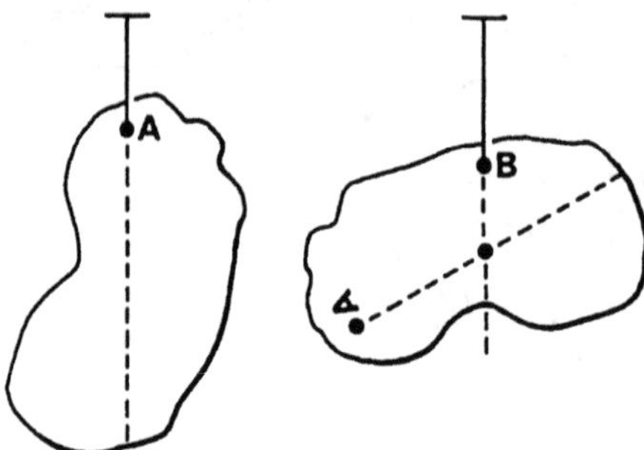

FIGURE 5.21 Centre of gravity of an irregular shape.

This is one reason why customers should not stand on the open drop-down door of a gas cooker and also why a stability bracket/chain must be fitted to most free standing domestic gas cookers. Figure 5.24 shows a cooker with its door open. The centre of gravity is still well within the base even when tins of food are placed on the door. If, however, a customer stands on the door, Fig. 5.25, the resultant centre of gravity of both customer and cooker is now outside the base and the cooker will tip over, if not prevented by some means or other.

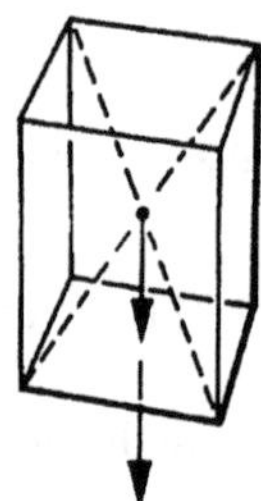

FIGURE 5.22 Centre of gravity of a rectangular solid.

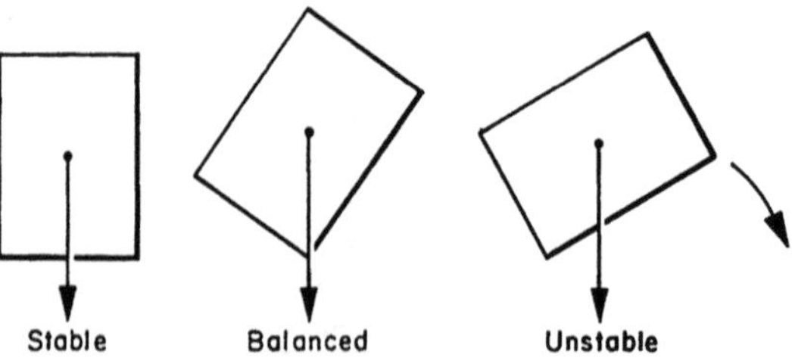

FIGURE 5.23 Tilting and balancing.

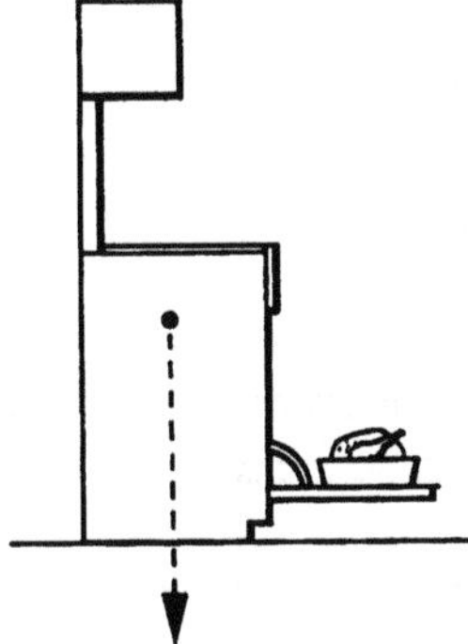

FIGURE 5.24 Centre of gravity of a gas cooker.

The method of calculating the position of the centre of gravity of a composite object is given for use if required.

First divide the object into suitable shapes and find the centre of gravity of each part (Fig. 5.26). Then draw two axes *XX* and *YY* at suitable distances from the object. Measure the distance of each centre of gravity from both the axes, a, b, c and a_1, b_1, c_1, and find the areas A, B and C of each of the parts.
Then

$$x = \frac{aA + bB + cC}{A + B + C}$$

and

$$y = \frac{a_1A + b_1B + c_1C}{A + B + C}$$

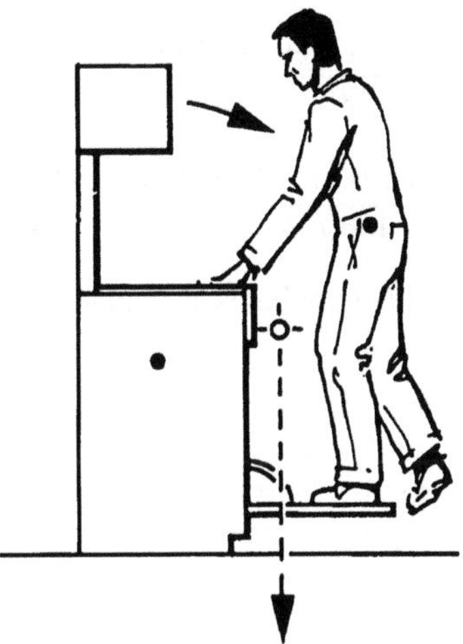

FIGURE 5.25 Centre of gravity of cooker and customer.

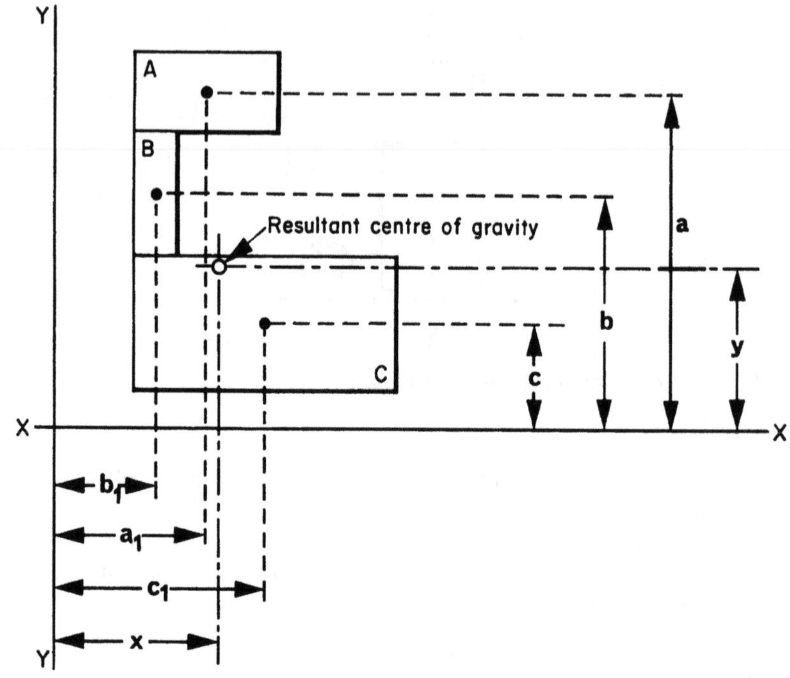

FIGURE 5.26 Centre of gravity of a composite object.

For example, if

$A = 800\ \text{mm}^2$	$a = 100$ mm	$a_1 = 55$ mm
$B = 400\ \text{mm}^2$	$b = 70$ mm	$b_1 = 30$ mm
$C = 2800\ \text{mm}^2$	$c = 30$ mm	$c = 40$ mm

then

$$x = \frac{80,000 + 28,000 + 84,000}{4000} = \frac{192,000}{4000} = 48\ \text{mm}$$

$$y = \frac{44,000 + 12,000 + 112,000}{4000} = \frac{168,000}{4000} = 42\ \text{mm}$$

where solid objects' weights may be used in place of areas.

Chapter | six

Pressure and Gas Flow

Chapter 6 is based on an original draft prepared by Mr G. Marshall

DEFINITION OF PRESSURE

'Pressure' is the same as stress, but whereas stress applies to solid objects, pressure is more concerned with fluids, that is liquids or gases. Fluids have the capacity to press themselves against the surface of the vessel that contains them (Chapter 1) and so exert a force. The intensity of this force can be measured in relation to the area of the surface. Pressure is the relationship between force and area.

$$\text{Pressure} = \frac{\text{force}}{\text{area}} = \frac{\text{newtons}}{\text{square metres}} = \frac{\text{N}}{\text{m}^2} \text{ or N/m}^2$$

The unit N/m^2 has the special name 'pascal' after the French mathematician. It is in more common use on the continent and has the symbol 'Pa'.

PRESSURE EXERTED BY A SOLID

A solid object has the ability to exert a pressure on the floor on which it stands, but this is only in a downward direction. Fluid pressure acts in all directions at the same time.

The pressure exerted by a solid object is the weight of the object divided by the area of its base. Take as an example the object shown in Fig. 6.1.

If the dimensions 25, 50 and 75 are in millimetres and the weight of the block is 5 kg, then in the different positions the pressures exerted by the base are:

(a) $$\text{Pressure} = \frac{\text{force}}{\text{area}} = \frac{5\text{ kgf}}{1250\text{ mm}^2} = \frac{5 \times 9.81\text{ N}}{0.00125\text{ m}^2} = 39.24\text{ kN/m}^2$$

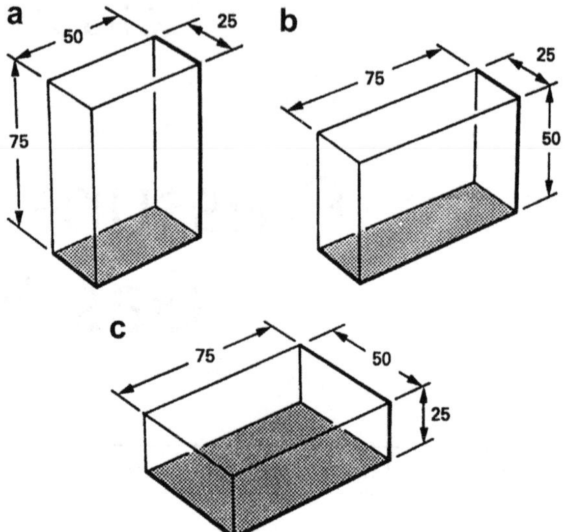

FIGURE 6.1 Pressure exerted by a rectangular block: (a) end downward, (b) side downward, and (c) face downward.

$$\text{(b)}\quad \text{Pressure} = \frac{\text{force}}{\text{area}} = \frac{5\ \text{kgf}}{1875\ \text{mm}^2} = \frac{5 \times 9.81\ \text{N}}{0.001875\ \text{m}^2} = 26.161\ \text{kN/m}^2$$

$$\text{(c)}\quad \text{Pressure} = \frac{\text{force}}{\text{area}} = \frac{5\ \text{kgf}}{3750\ \text{mm}^2} = \frac{5 \times 9.81\ \text{N}}{0.00375\ \text{m}^2} = 13.08\ \text{kN/m}^2$$

It follows that, for the same weight of object, the smaller the base area, the higher the pressure that it will exert. For example, a woman wearing stiletto heels will probably make holes in wood or vinyl flooring. The same woman in shoes with large, flat heels will not damage the floors.

PRESSURE IN FLUIDS

Figure 6.2 illustrates the difference between solids and liquids. In (a) there is a solid block which just fits into a container. The only pressure is that exerted on the base.

In (b), the container is filled with liquid. This now exerts a pressure on the sides as well as the base. In fact, the pressure will increase as the depth increases and this can be proved in a number of ways.

The vessel in Fig. 6.3 has three short outlet pipes at distances below the water level of h_1, h_2 and h_3, respectively. It is filled with water and stood at the edge of a sink so that water can pour from the three outlets. The distances d_1, d_2 and d_3 give a measure of the pressure at the different levels.

Another way of proving a relationship between depth of liquid and pressure is to refer back to Fig. 6.1.

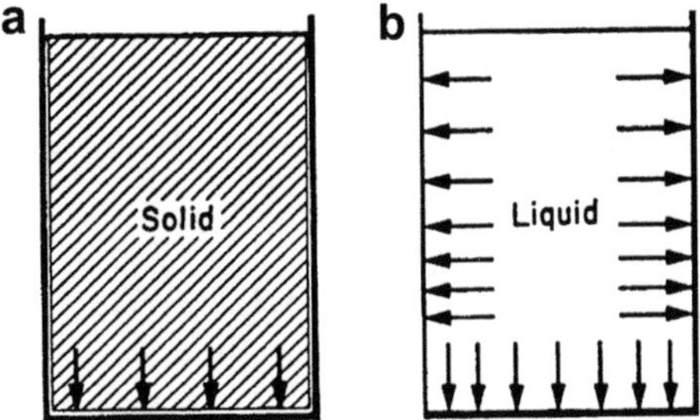

FIGURE 6.2 Difference between pressures exerted by solids and liquids.

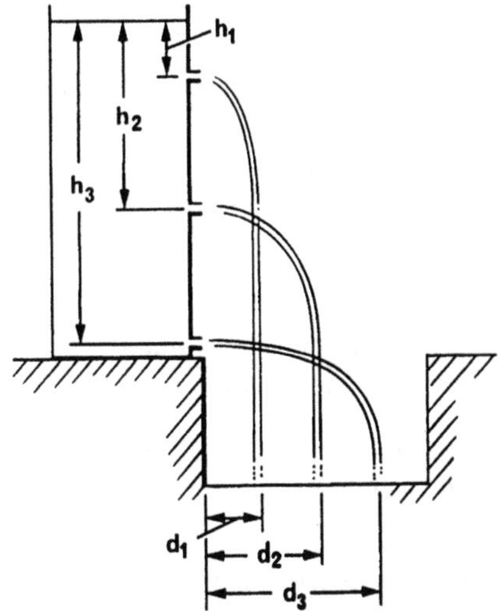

FIGURE 6.3 Relationship between pressure and depth of liquid.

If the rectangles represented tanks filled with a liquid then all three tanks would contain the same weight of liquid. If this weight was the same as the weight of the previous solid, then the pressure exerted by the liquid on the base of each tank would still be the same as before.

$$a = 39.24\ \text{kN/m}^2$$
$$b = 26.161\ \text{kN/m}^2$$
$$c = 13.08\ \text{kN/m}^2$$

Now look at the respective heights of liquid:

$$a = 75\ \text{mm}(0.075\ \text{m})$$
$$b = 50\ \text{mm}(0.050\ \text{m})$$
$$c = 25\ \text{mm}(0.025\ \text{m})$$

There is an obvious relationship

$$\text{(a)} \quad \frac{\text{pressure}}{\text{height}} = \frac{39.35 \text{ kN}}{0.075 \text{ m}} = 523.2$$

$$\text{(b)} \quad \frac{\text{pressure}}{\text{height}} = \frac{26.16 \text{ kN/m}^2}{0.05 \text{ m}} = 523.2$$

$$\text{(c)} \quad \frac{\text{pressure}}{\text{height}} = \frac{13.08 \text{ kN/m}^2}{0.025 \text{ m}} = 523.2$$

In other words, pressure is directly proportional to height. If the height is doubled, the pressure is doubled and so on. The actual pressure will depend on the density of the liquid and can be calculated, if required.

The figures $\frac{39.24}{0.075}$, $\frac{26.161}{0.05}$ and $\frac{13.08}{0.025}$ are $\frac{\text{kN/m}^2}{\text{m}} = 523.2 \text{ kN/m}^3$.

But this 523.2 kN/m^3 is the

$$\frac{\text{total weight of liquid}}{\text{total volume}}$$

or the weight of a unit volume of the liquid. This is called the ‘specific weight’ of the substance (specific weight = density × gravity).

So $$\frac{\text{pressure}}{\text{height}} = \text{specific height}$$

or $$\text{pressure} = \text{specific weight} \times \text{height}$$

In symbols this is $p = Wh$

From this you can see that the pressure at any depth in a liquid is dependent only on the height of the surface and the specific weight of the liquid. It has nothing to do with the quantity of liquid or the size and shape of the container.

Pascal proved this point by devising the apparatus shown in Fig. 6.4.

It consists of tubes of different size and shape connected to a container at the base. When filled with liquid the levels in all the tubes are found to be the same, so the pressure in each tube must also be the same at any given depth.

UNITS OF PRESSURE

Pressure can be calculated either by measuring the force exerted on a unit of area or by measuring the height of a column of liquid supported by the force. So

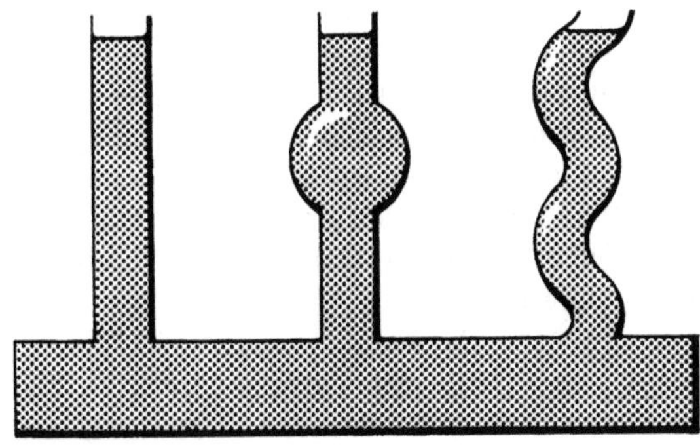

FIGURE 6.4 Pascal's apparatus.

its units are either those of force per unit area like newtons per square metre, or they are metres or millimetres height or 'head' of liquid.

There are alternative units. In some areas of activity, kilograms force per square centimetre may be used and in the gas industry pressure will be measured normally in 'bars' and 'millibars'. The symbols are 'bar' and 'mbar'.

Where liquids are used in pressure gauges, water is the most common for low pressures and mercury, which is 13.6 times as dense, for higher pressures. The gauges have scales graduated in millibars rather than in inches or millimetres which were once commonly used.

It may sometimes be necessary to convert a pressure reading from force/area units to height units or vice versa and Table 6.1 shows the comparison between them.

ATMOSPHERIC PRESSURE

The earth is surrounded by an envelope of air, held to the earth's surface by gravity. The weight of air creates a pressure on the earth's surface of about 1 bar or 101,325 N/m^2 at sea level.

Although our bodies are subjected to this pressure, they have an internal pressure which normally exactly balances atmospheric pressure so that we are unaware of it.

Table 6.1 Comparison of Units of Pressure

Height	Bars	Force/Area
1 m head of water	98 mbar	9800 N/m^2
10.2 mm of water	1 mbar	100 N/m^2
1 atm	1013.25 mbar	101.3 kN/m^2
or 760 mm of mercury	1 bar	or 101,325 N/m^2
or 10.34 m of water		

Atmospheric pressure must, however, be taken into consideration in a number of calculations involving gas pressure.

One effect of atmospheric pressure can be shown by a simple experiment (Fig. 6.5). Take a metal can which is open to the air so that the pressure inside is the same as that on the outside. Put a small quantity of water in the can, place it over a gas flame and boil the water. The steam produced will push the air out of the can.

When the can is full of steam only, seal the neck with a bung. Then place the can in a stream of cold water.

The cold water will cause the steam to condense into water in the can. The sudden reduction in volume will create a partial vacuum or 'negative' pressure and the atmospheric pressure on the outside will crush the can. This is called 'implosion' (which is, of course, the opposite of explosion).

Precautions have to be taken to prevent any negative pressures occurring in gas supplies otherwise atmospheric pressure can damage meters or similar components.

STANDARD CONDITIONS

Atmospheric pressure varies with the weather. When pressure is high the day will be fine and dry. As pressure falls the weather becomes changeable, rainy and finally stormy.

Atmospheric pressure also varies with the height above sea level at which the reading is taken. On the top of a mountain the pressure is less than the pressure on the beach.

The pressure already given, 1013 mbar or 760 mm mercury (Table 6.1), is an average figure and serves for most purposes.

The pressure exerted on a volume of gas also depends on its temperature and on whether the gas is dry or saturated with water vapour. Because of this, if there

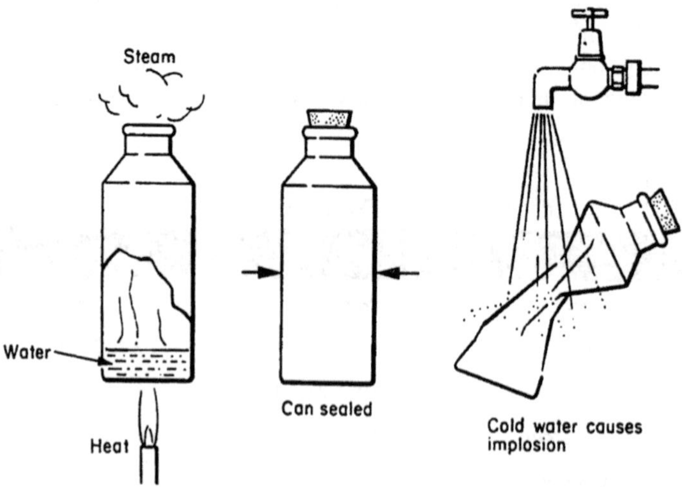

FIGURE 6.5 Experiment to show the effect of atmospheric pressure.

is a need to compare two different volumes of gas they must both be corrected to standard reference conditions (src) of temperature, pressure and moisture content.

For these, standard conditions are:

1. 15 °C,
2. 1013 mbar,
3. dry.

The average atmospheric pressure is 1013 mbar and 15 °C is the average atmospheric temperature.

A volume of gas measured under standard conditions is indicated by st in brackets following the volume abbreviation, for example,

$$1\ \mathrm{m}^3\,(\mathrm{st}).$$

Standard volumes are used in calculations of calorific value and Wobbe index.

Barometers

Atmospheric pressure is measured by a 'barometer'. The simplest form of barometer is the mercury barometer devised by Torricelli (Fig. 6.6).

This consists of a glass tube, sealed at one end, which is first filled with mercury and then inverted into a trough also containing mercury.

The mercury will begin to pour out of the tube, leaving a vacuum at the top, until the height of the column of mercury exactly balances the atmospheric pressure which is being exerted on the surface of the mercury in the trough.

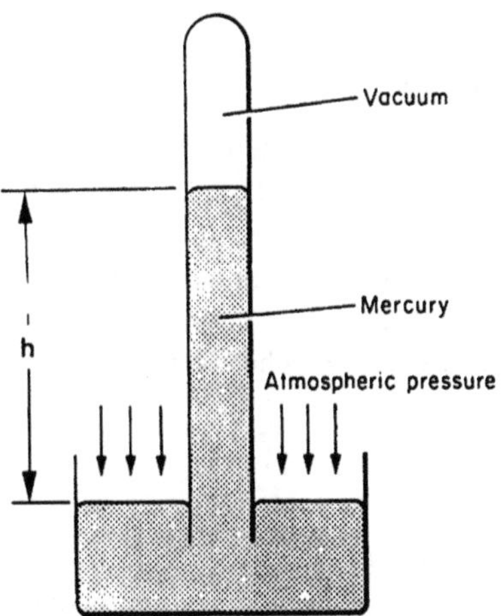

FIGURE 6.6 Mercury barometer.

The actual height of the column can be found by experiment, but it can also be calculated using the formula we discovered previously,

$$\text{pressure} = \text{specific weight} \times \text{height}, \quad p = Wh.$$

In the case of the barometer:

- atmospheric pressure = 101.3 kN/m^2
- specific weight of mercury = specific gravity of mercury × specific weight of water
- specific gravity of mercury =13.6
- specific weight of water = 9.81 kN/m^3
- specific weight of mercury = 13.6 × 9.81

Therefore $p = Wh$, so

$$h = \frac{p}{W} = \frac{101.3}{13.6 \times 9.81} \text{ m}$$

or approximately 760 mm

The reason for Torricelli using mercury becomes obvious when you calculate the length of tube which would be required if water was to be used.

$$h = \frac{101.3}{9.81} = 10.3 \text{ m}$$

The mercury barometer offers a very accurate means of measuring atmospheric pressure and there are a number of ways in which provision can be made for the scale to be set to zero and compensation made for altitude. Most domestic barometers are, however, aneroid barometers.

ANEROID BAROMETER

'Aneroid' means 'not liquid' and this type of barometer consists of a cylindrical metal box or bellows almost exhausted of air. The box has a flexible, corrugated top and bottom and is very sensitive to changes in atmospheric pressure (Fig. 6.7). Such changes cause inward or outward movements of the flexible top and bottom sections and this movement is made to rotate a pointer on a scale by means of a suitable mechanism attached to the bellows.

ABSOLUTE PRESSURE

'Absolute' pressure is the pressure from zero to that shown on a gauge. Figure 6.8 shows a container under three different conditions. In all three cases the container is subjected to atmospheric pressure on all sides.

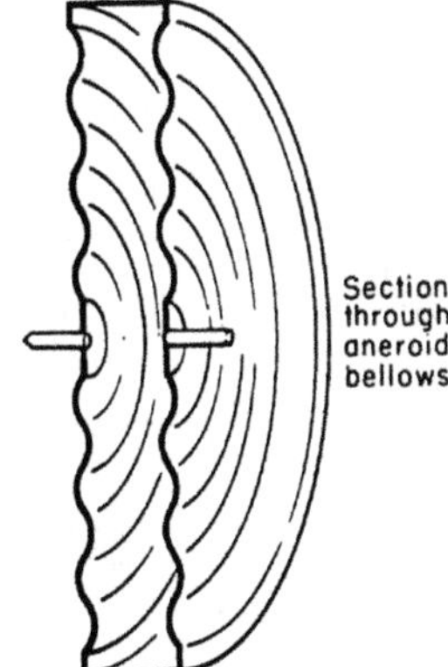

FIGURE 6.7 Aneroid bellows.

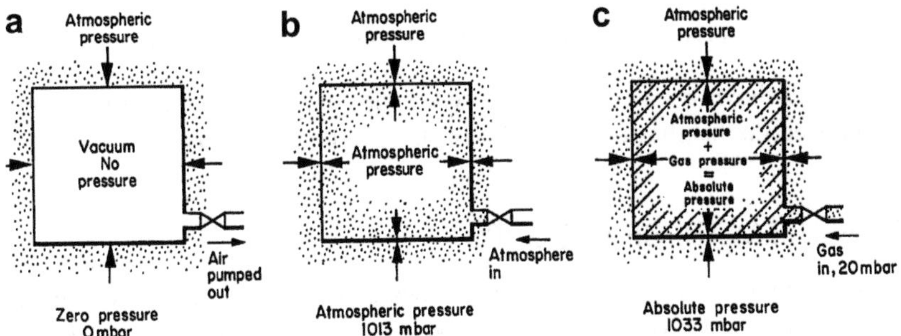

FIGURE 6.8 Absolute pressure.

1. In Fig. 6.8a the air has been pumped out and the valve closed. The pressure inside is zero.
2. In Fig. 6.8b the valve is opened and air rushes in to fill the vacuum. The pressure inside is atmospheric pressure, 1013 mbar.
3. In Fig. 6.8c the container has been connected to a gas supply at a pressure of, say, 20 mbar.

Some gas will be forced in and the pressure will now be

$$\begin{aligned} \text{atmospheric pressure} + \text{gas pressure} &= 1013 + 20 \\ &= 1033\text{mbar, absolute pressure.} \end{aligned}$$

So,

$$\text{atmospheric pressure} + \text{gas pressure} = \text{absolute pressure}$$

Any pressure gauge will have atmospheric pressure inside it before the gas is turned on, like the container in Fig. 6.8b. So it will only indicate the additional

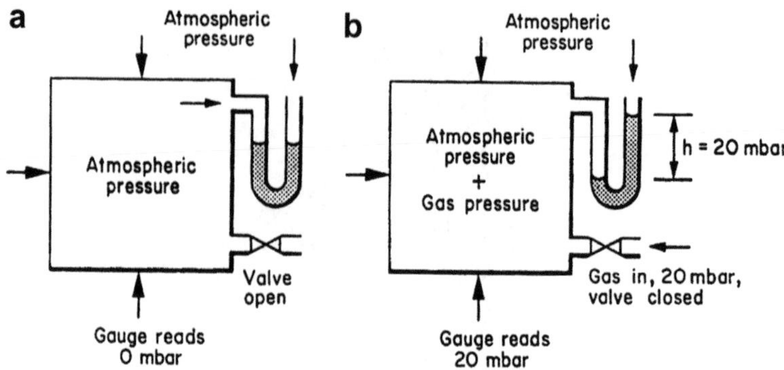

FIGURE 6.9 Pressure shown by a 'U' gauge.

pressure due to the gas. For an absolute pressure reading, atmospheric pressure must be added on:

$$\text{gauge pressure} + \text{atmospheric pressure} = \text{absolute pressure}$$

Figure 6.9 shows a water gauge attached to the container. When gas pressure is introduced the height of water indicates the gas pressure only.

MEASUREMENT OF PRESSURE

Pressure gauges, like barometers, can indicate height or force/area so they can be of liquid or dry types. The liquid types commonly use water or mercury. There are two main differences between these liquids, so far as their use in gauges is concerned.

Mercury has a specific gravity of 13.6, so it will measure pressures 13.6 times higher than the same height of water. Water adheres to the side of the gauge tube and mercury does not. Figure 6.10 shows what effect this has on the 'meniscus' or liquid level which in both cases is crescent-shaped. When reading a gauge the true value comes from the bottom of a water meniscus and the top of a mercury meniscus (Fig. 6.11).

The reason for the difference in the meniscus lies in the difference in the force of attraction between the molecules. There are two forces on the molecules.

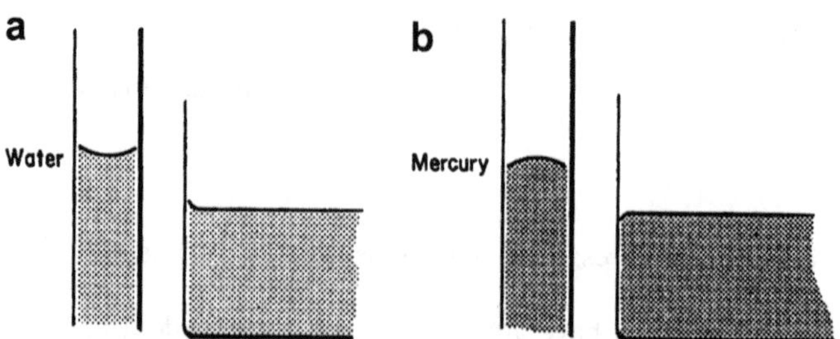

FIGURE 6.10 Effect of adhesion and cohesion.

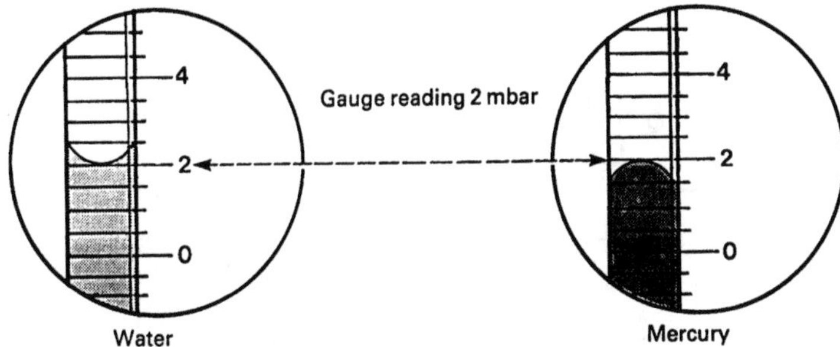

FIGURE 6.11 Reading a gauge.

1. The attraction of one molecule of the substance to another of the same substance, 'cohesion'.
2. The attraction of a molecule of the substance to a molecule of another substance, 'adhesion'.

In water, the adhesive quality is stronger than the cohesive quality. So water adheres readily to other substances. In mercury the reverse is the case. An example of the effect of adhesion and cohesion can be seen in capillary attraction.

CAPILLARITY

Capillary means 'like a hair', so a capillary tube is one with a very small bore, similar to the tiny capillaries which convey blood between the arteries and the veins.

If a very small diameter glass tube, open at both ends, is placed vertically in a glass of water, the water will be seen to rise up the tube (Fig. 6.12a). This is due to 'capillarity' or capillary attraction.

The water rises because its molecules have a greater attraction to the glass than they have to each other. With mercury (Fig. 6.12b) the reverse is true and so the level in the capillary tube is below the surface of the liquid, not above it.

There are many applications of this phenomenon. Capillary soldered joints are used on copper pipes. When melted, the solder is attracted into a narrow space between the pipe and the fitting and so seals the joint.

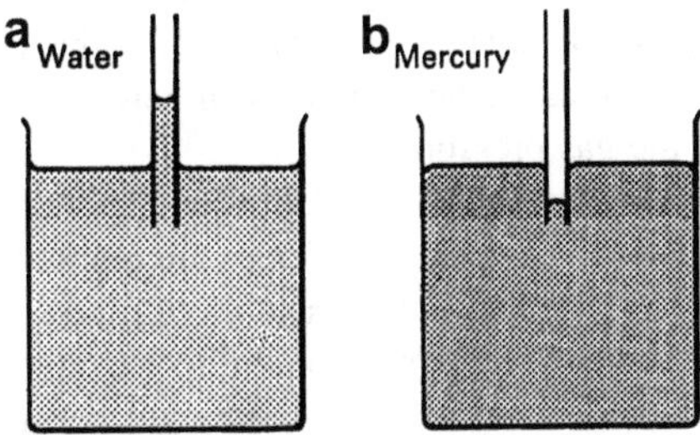

FIGURE 6.12 Effect of capillary attraction.

FIGURE 6.13 Experiment to show the effect of size of the gap on the capillary attraction.

Any substances with pores of sufficient size are able to suck up water, for example, tissues, sponges and blotting paper. Moisture rises in the roots and stems of plants by the same means.

The height up to which the water will rise depends on the diameter of the tube and becomes greater as the bore is diminished. This can be seen if two glass plates are held vertically and a small distance apart in the liquid (Fig. 6.13).

If the plates are pulled apart at one side so that they form an acute angle, the water will rise in the form of a curve, as shown, with the highest level at the point of the angle.

In the larger diameter tubes the effect of the adhesive and cohesive forces does not cause a rise or fall in the liquid but only produces a meniscus.

LIQUID GAUGES

'U' Gauge

Gauges used for measuring gas pressures are called 'manometers'. The simplest form is the 'U' tube manometer (Fig. 6.14), commonly called the 'U' gauge. This consists of a glass or transparent plastic tube usually about 300 mm long which, when containing water, will measure pressures up to 30 mbar. The tube is bent into the form of a U and has a scale fitted between the two 'limbs' or upright parts of the tube. One limb is connected to the gas supply and the other remains open to the atmosphere.

The scale is usually capable of adjustment so that the zero can be lined up with the water levels as shown in Fig. 6.14A. When gas pressure is applied, Fig. 6.14B, the water will be displaced downwards in the left limb and upwards in the right by equal amounts. The total vertical height of the column of water supported indicates the gas pressure.

There are generally two scales available, one in millibars and the other a dual scale of millibars and inches. The scales are marked off in half millibars or half inches from the zero line but each of these marks are numbered in full millibars or inches. This enables you to read direct from either limb, provided the gauge was dead level on the zero line before gas pressure was applied (see Fig. 6.15). Where the gauge has been incorrectly zeroed, it is necessary to read

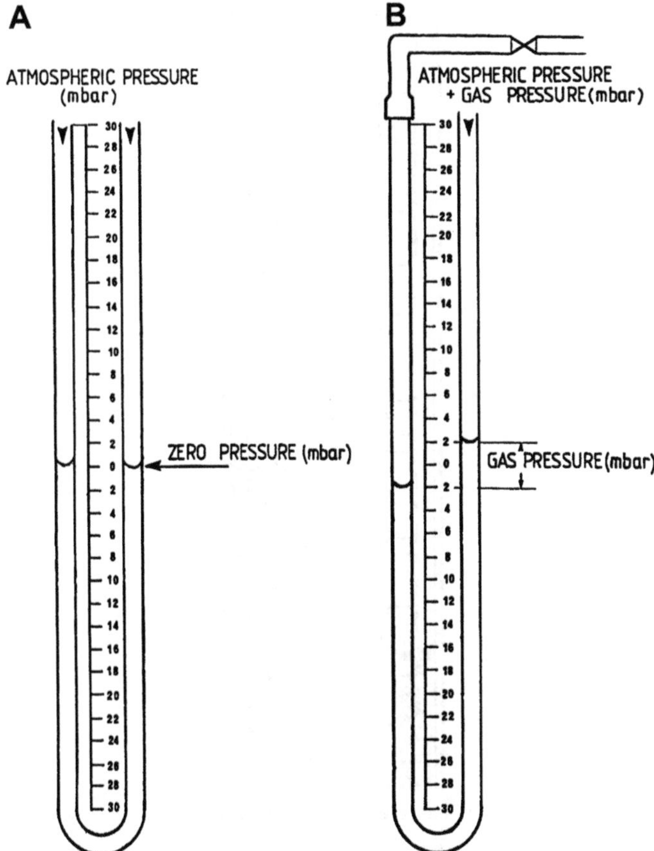

FIGURE 6.14 'U' tube pressure gauge.

the pressures indicated on both limbs, add the pressures together, then divide by two to obtain the correct pressure reading.

Single column gauge

This is really a 'U' gauge with one limb which has a much bigger diameter than the other. The gauge shown in Fig. 6.16 has a large limb, or main water container, with an area 23 times that of the small tube on the right.

When pressure is applied (Fig. 6.16b), the liquid rise in the small tube is 23 times greater than the liquid fall in the container. The scale reads only the rise of liquid and has been designed to take account of the amount by which the liquid level falls in the container.

Inclined gauge

When reading a 'U' gauge it is essential to keep the gauge vertical. What you are measuring is the vertical height of the column of liquid. If you tilt the gauge,

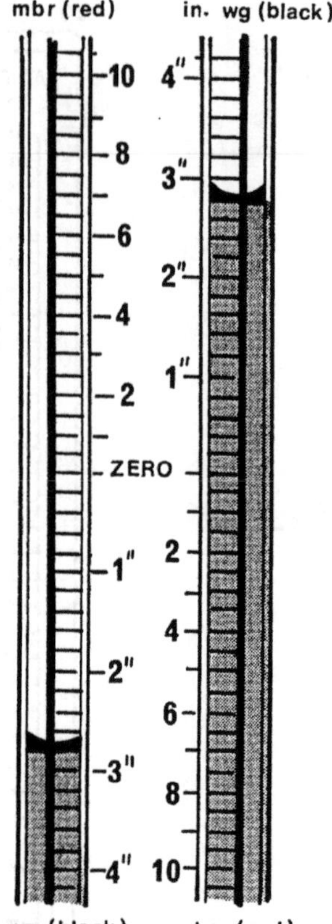

FIGURE 6.15 Dual scale 'U' gauge scales.

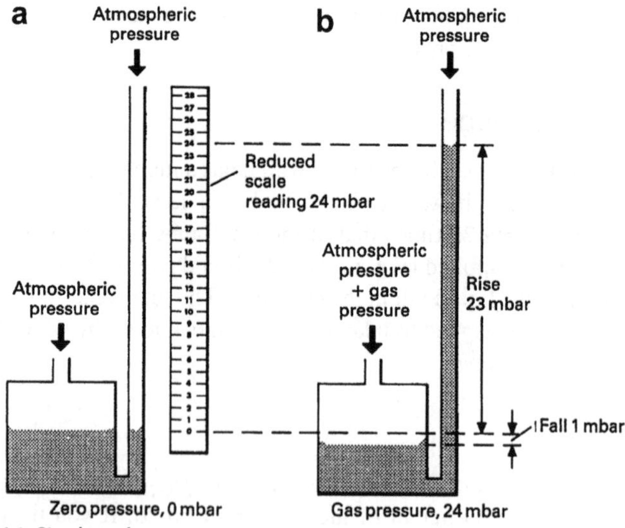

FIGURE 6.16 Single column gauge.

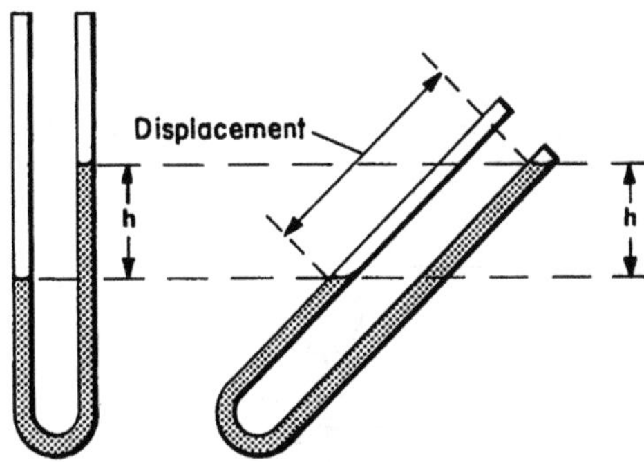

FIGURE 6.17 Tilting the 'U' gauge.

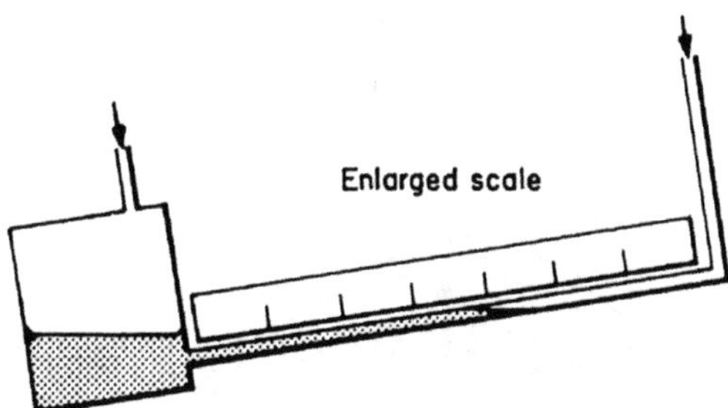

FIGURE 6.18 Inclined gauge.

Fig. 6.17, the water column becomes longer and shows a false reading on the scale. What you are measuring should still be the vertical height, *h*, between the levels.

The 'inclined gauge' in Fig. 6.18 is a single column gauge and makes use of the tilting effect to magnify the liquid movement and enable very small pressures to be measured accurately. The gauge is mainly used for measuring the difference in pressure between two points.

DRY GAUGES

Aneroid gauge

The aneroid gauge in Fig. 6.19 is similar to the aneroid barometer previously described. It has a flexible metal bellows which, when connected to the gas supply, extend in proportion to the pressure of the gas. The movement of the bellows is transferred through a lever system to a pointer on a scale.

A needle valve, fitted on the inlet, damps down any sudden pressure change and prevents the pointer from 'oscillating' or moving rapidly to and fro across the scale.

This type of gauge is used to measure pressures up to about 300 mbar.

Bourden tube gauge

This type of gauge is used normally for pressures greater than 1 atm. It consists of a tube which is elliptical in cross-section, bent in the form of a C and sealed at one end (Fig. 6.20).

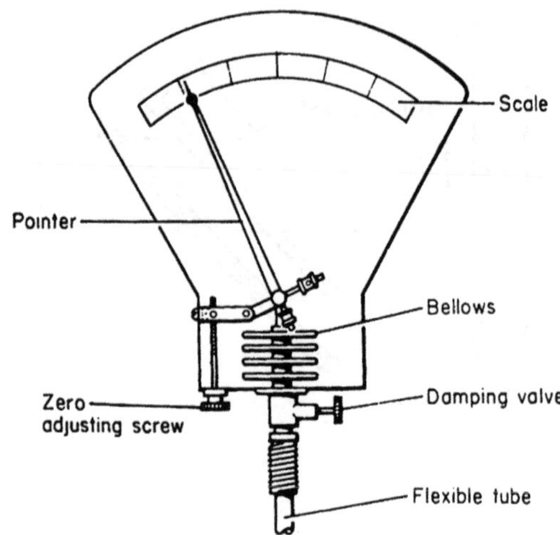

FIGURE 6.19 Aneroid gauge.

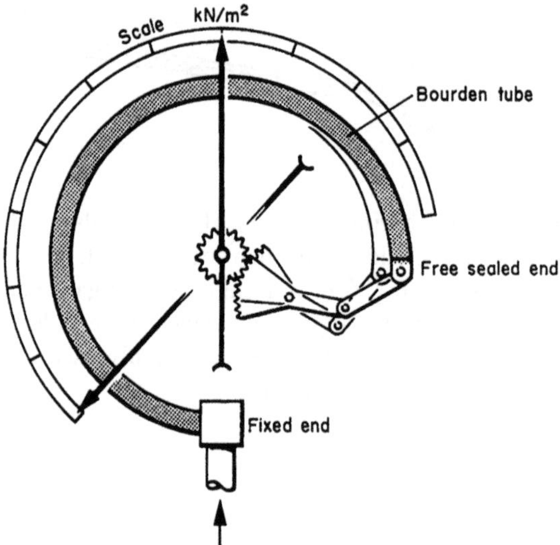

FIGURE 6.20 Bourden tube gauge.

When the fixed end is connected to a gas (or steam) supply, the pressure inside the tube tends to make the tube's section widen out and become more like a circle. As the section of the tube distends, the tube itself straightens slightly causing the free end to move outwards and upwards. This movement is transferred through a linkage to a central pinion carrying the pointer.

Pressure recorders

Whilst a pressure gauge will give an indication of the prevailing pressure at any time of the day, it is necessary to monitor the pressures at which gas is being

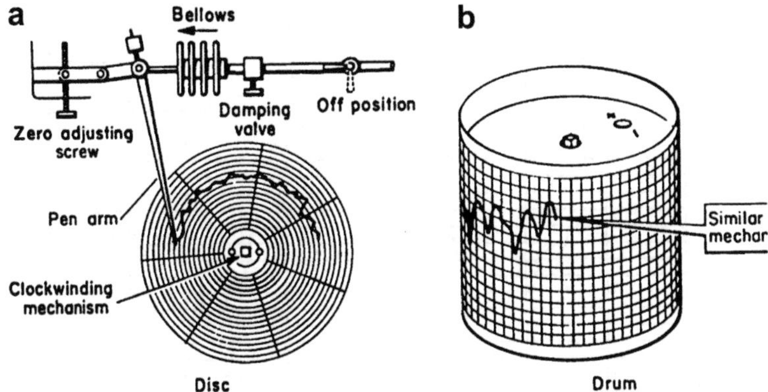

FIGURE 6.21 Recorder charts.

supplied to the district all the time. To do this, 'pressure recorders' are used. They are also invaluable for investigating reports of inadequate pressures.

A recorder is simply a gauge, with a revolving chart, a scale and a pen instead of a pointer. The chart is rotated by a clockwork mechanism and may either be in the form of a flat, circular, paper disc or a rectangular sheet of paper wound round a drum like a frill on a cake (Fig. 6.21).

Recorders, like gauges, may be operated by liquids, aneroids or bourden tubes, but the type in common use on the district is an aneroid. The usual duration of the record on the chart is one week. When a chart is completed it is removed and returned to the depot where the graph can be interpreted.

If the recorder has been fitted to investigate a problem of low pressures, the chart may be compared with those taken daily throughout the distribution system and with any other recordings from the pressure survey. Most problems of low pressure become evident at times of peak demand.

Figure 6.22 shows one day on a typical recording. Pressures are mainly between 19 and 20 mbar except at times A, B, C and D.

These are the usual peak demand times in domestic areas during the summer.

The load begins to come at around 4.30 am when the early workers are getting up, using water heaters and having breakfast. This reaches a peak between 7.00 and 8.00 am. Points B, C and D show other meal times.

The graph will change in the winter months when fires and central heating would be on all day. A heavy demand from local industry would also produce a corresponding drop in pressure on the chart.

PRESSURE IN WATER HEATING SYSTEMS

Hydrostatic pressure

'Hydrostatic pressure' is simply the pressure exerted in the system by the weight of the water. An example was given earlier in the chapter. Now, for a more practical application, Fig. 6.23 shows a domestic water heating system

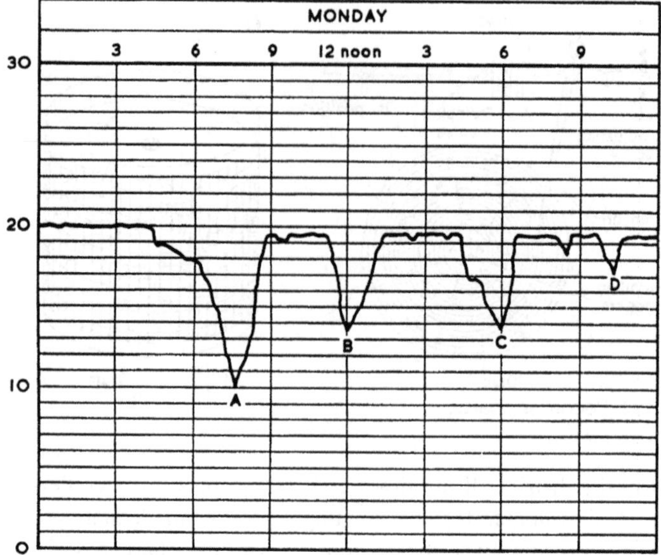

FIGURE 6.22 Record of pressure variations over one day.

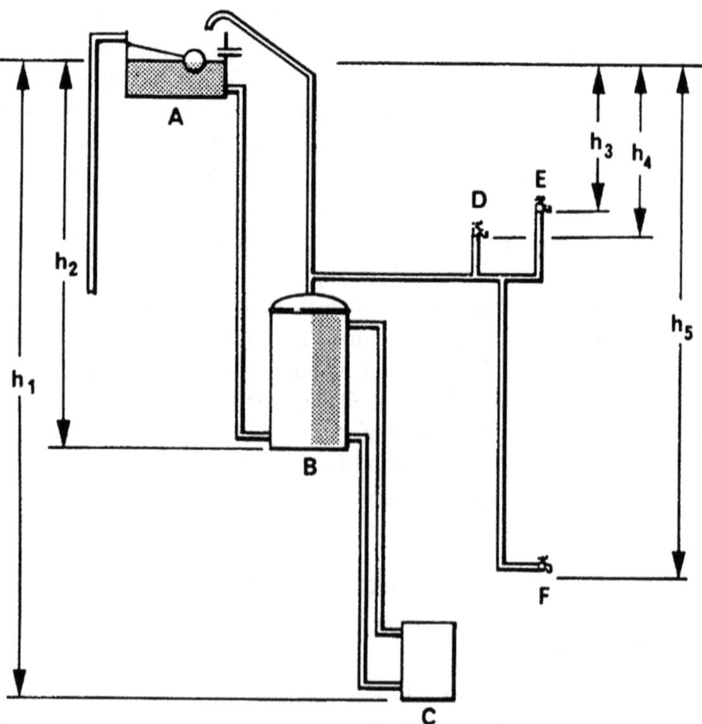

FIGURE 6.23 Pressures in a domestic water heating system.

Table 6.2 Pressures of Different Heads of Water

Head	Head (m)	Force/area (kN/m^2)	Equivalent in (mbar)
h_1	5	49 (50)*	490
h_2	245	24.5 (25)	245
h_3	1.5	14.7 (15)	147
h_4	2	19.6 (20)	196
h_5	4	39.2 (40)	392

*Figures in brackets are approximate.

consisting of a cistern A supplying cold water to a cylinder B, heated by a boiler C and with bath, basin and sink draw off taps at D, E and F.

Typical pressures in an average house are shown in Table 6.2. From the table the following points can be seen.

The pressure at the sink tap, F, is more than twice the pressure at the basin tap, E. So you would expect to get a much faster flow of water at the sink (depending on the pipes and the size of taps).

If the area of the base of the cylinder was 0.8 m^2, then the force on the base would be:

$$25\ \mathrm{kN/m^2}\ (\text{pressure}) \times 0.8\ \mathrm{m^2}\ (\text{area})$$
$$= 20\ \mathrm{kN}\ \text{or}\ 20,000\ \mathrm{N}\ (\text{about 2 tonnes})$$

If the area of the base of the boiler was 0.5 m^2, then the force on the base would be:

$$50\ \mathrm{kN/m^2} \times 0.5\ \mathrm{m^2} = 25\ \mathrm{kN}$$

This shows why makers of boilers and water heaters always state a maximum head of water up to which their appliances may be fitted. If subjected to higher pressures, the appliances could be damaged.

Circulating pressure

Water 'circulates' or goes round and round in water heating or wet central heating systems. The pressure which causes this circulation is naturally called 'circulating pressure'.

The circulating pressure in a system depends on two factors:

1. The difference in temperature between the hot water flowing from the boiler and the cold water returning to it.
2. The height of the columns of hot and cold water above the level of the boiler.

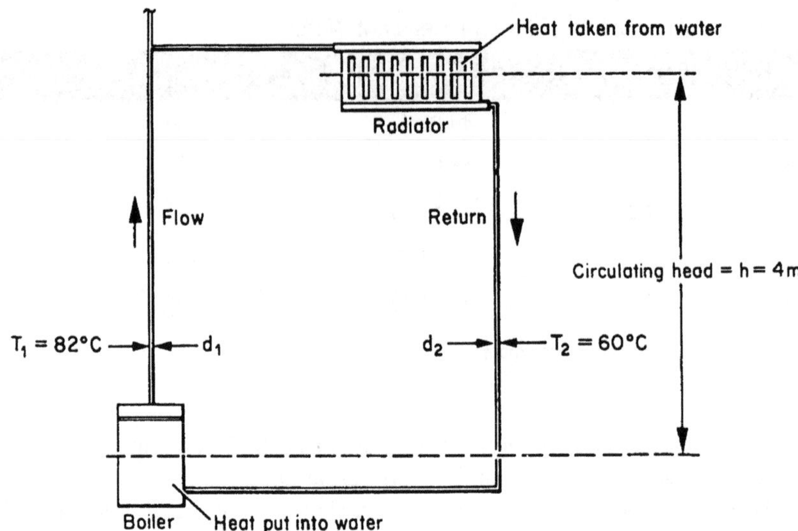

FIGURE 6.24 Circulating pressure. On a water pressure system the cylinder would be in the position occupied by the radiator. T_1 would be 60 °C and T_2 would be about 10 °C when heating began.

The example in Fig. 6.24 shows part of a simple water circulation from a boiler to a radiator and back again.

Generally, central heating systems now rely on pumps to provide the main circulating pressure so that smaller bore pipes may be used. But gravity can be used and is still the motive force in many water heating systems (Figure 6.23 shows the boiler connected to the cylinder by flow and return pipes, and the radiator in Fig. 6.24 could be replaced by a cylinder. The principle is still the same although temperatures are lower.)

All fluids expand when heated and water is no exception. When a fluid expands, the same weight of fluid occupies a bigger volume. So its density (weight/volume) is reduced.

Put simply, a given volume of hot water weighs less than the same volume of cold water.

Tables are available giving the density of water at various temperatures and, if the flow and return temperatures and the circulation height are known, it is possible to calculate the circulating pressure in a system operating on gravity circulation.

Circulating pressure is proportional to the temperature difference between flow and return water and the circulating height.

Circulating pressure will increase if:

1. the temperature difference increases,
2. the circulation height is increased.

The circulating pressure in Fig. 6.24 is approximately 5 mbar. It can be obtained from the formula:

$$p = h\frac{(d_2 - d_1)}{d_2 + d_1} \times 196.1$$

where p = circulating pressure in millibars,
h = circulation height in metres,
d_1 = density of water in flow in kg/m^3,
d_2 = density of water in return in kg/m^3,
and 196.1 is a constant.

$$\begin{aligned} \text{From tables, } d_1 \text{ at } 82°\text{C} &= 970.40 \text{ kg/m}^3 \\ d_2 \text{ at } 60°\text{C} &= 983.21 \text{ kg/m}^3 \end{aligned}$$

Therefore

$$p = 4 \times \frac{983.21 - 970.40}{983.21 + 970.40} \times 196.1 = 5.14 \text{ mbar (about 5 mbar)}$$

Other tables are available which give a reading of circulating pressure per metre height directly from a temperature difference.

THE EFFECT OF ALTITUDE ON PRESSURE

From previous sections in this chapter you already know that:

- Atmospheric pressure is about 1013 mbar at sea level.
- Atmospheric pressure becomes less as altitude increases and the weight of the column of air is reduced.
- Pressure at a depth in a liquid is equal to the depth multiplied by the specific weight of the liquid.

$$p = h \times W$$

The specific weight of a substance is the density × 9.81 kg/m^3.

So $p = h \times d \times 9.81$.

Using this information, the effect of an increase in the altitude of a gas supply can be seen.

Take a rise in altitude of 100 m. At 100 m above sea level the atmospheric pressure will be less by an amount equivalent to the pressure caused by a column of 100 m high air.

If the density of air is 1.248 kg/m^3, then the pressure exerted by the air column is p = 100 m × 1.248 kg/m^3 × 9.81 = 1224 N/m^2 or = 12.24 mbar.

So atmospheric pressure 100 m above sea level is 1013 − 12.24 = 1000.76 mbar.

But what is happening to the gas? It is also subject to a reduction in pressure, but equivalent to a column of 100 m high gas.

Gas has a specific gravity of about 0.5. So its density is only half that of air – therefore the pressure of the 100 m of gas will be half that of 100 m of air. Let us work it out to make it certain.

$$p = 100 \text{ m} \times 1.248 \text{ kg/m}^3 \times 0.5 \times 9.81 = 612.14 \text{ N/m}^2 \text{ or } = 6.12 \text{ mbar}$$

Thus, the gas pressure will be reduced by 6.12 mbar when the atmospheric pressure is reduced by 12.24 mbar.

The consequence of this is an apparent increase in gas pressure of 6.12 mbar for every 100 m increase in altitude.

Figure 6.25 shows the calculation diagrammatically. From this a formula can be developed to give the difference in pressure for any increase in altitude with any specific gravity of gas.

The difference for 100 m is due to the difference between

$$100 \text{ m of air} = 12.24 \text{ mbar and}$$

$$100 \text{ m of gas} = (12.24 \times \text{specific gravity}) \text{ mbar}$$

or $$I = 12.24 - 12.24S \text{ mbar}$$

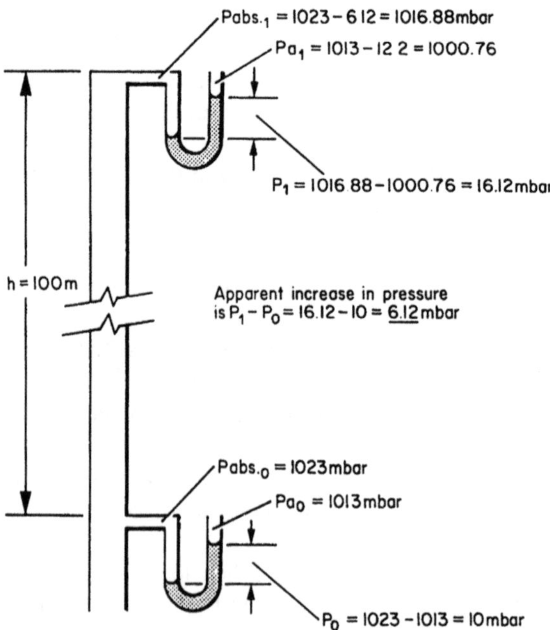

FIGURE 6.25 Calculation of the effect of altitude on pressure.

where I is the pressure difference and S is the specific gravity of the gas. Simplifying

$$I = 12.24(1 - S)/100\ \text{m}$$

for any height, h, the difference is

$$\begin{aligned} I &= 12.24(1 - S) \times \frac{h}{100}\ \text{mbar} \\ &= 0.12h(1 - S)\ \text{mbar} \end{aligned}$$

	Symbols	
Pressures	Ground level	100 m height
Gauge pressure	$P0$	$P1$
Atmos. pressure	Pa_0	Pa_1
Abs. pressure	$Pabs._o$	$Pabs._1$

If the gas being considered is heavier than air, then the difference will be a decrease in pressure, not an increase.

HOW IS GAS DELIVERED?

Gas is delivered to the seven reception points (called beach terminals) by gas producers operating Offshore Facilities from over 100 fields beneath the sea around the British Isles. In addition a newly commissioned terminal at the Isle of Grain allows Liquefied Natural Gas (LNG) to be delivered to the terminal by sea. After treatment, which includes checking the quality meets the safety requirements and measuring the calorific value (the amount of energy contained in the gas), it is transported through 275,000 km of iron, steel and polyethylene mains pipeline (see Fig. 6.26).

ORIGINS OF GAS PRESSURE

The National Transmission System (NTS) is the high pressure part of National Grid's transmission system and it consists of more than 6600 km of top quality welded steel pipeline operating at pressures of up to 85 bar (85 times normal atmospheric pressure, over 1250 psi). The gas is pushed through the system using 26 strategically placed compressor stations. From over 140 offtake points, the NTS supplies gas to 40 power stations, a small number of large industrial consumers and twelve Local Distribution Zones (LDZs) that contain pipes operating at lower pressure which eventually supply the consumer.

Gas arrives in the customer's meter at a pressure of 21 mbar.

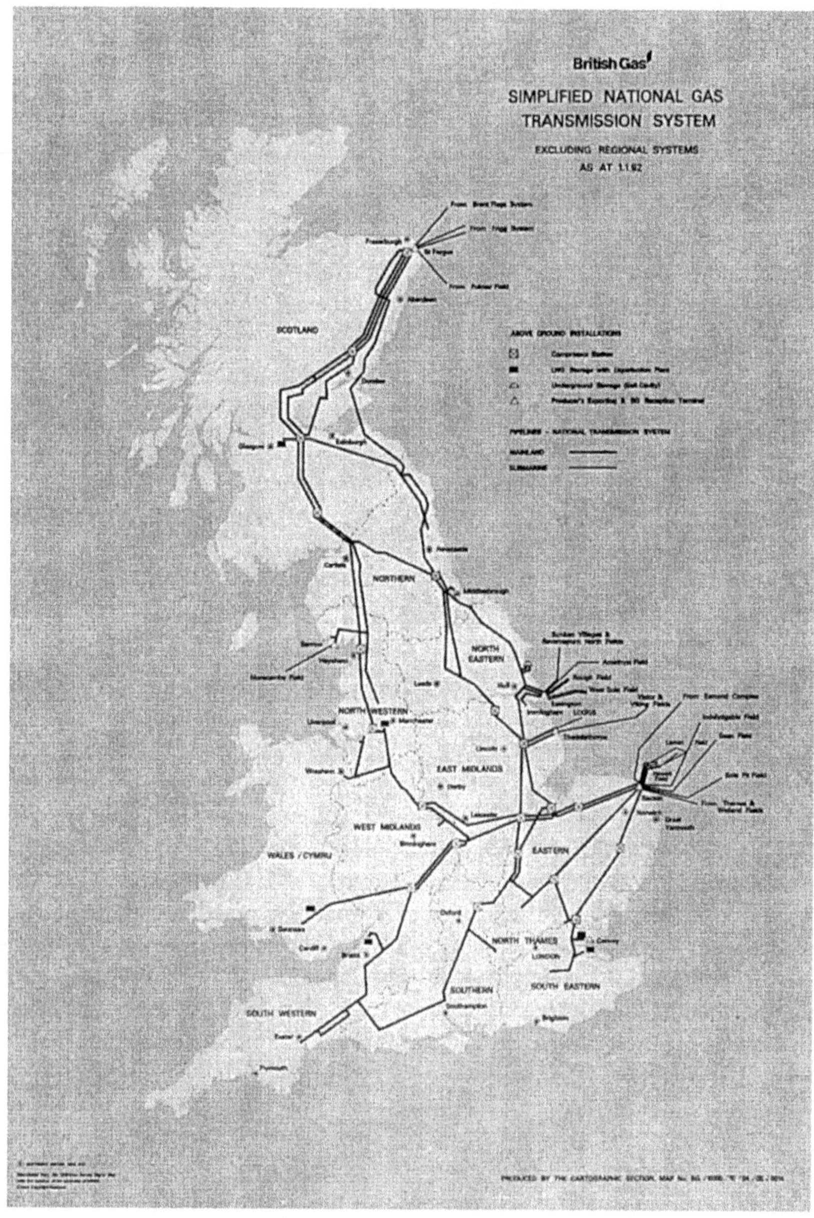

FIGURE 6.26 British Gas, Simplified national gas transmission system.

The pressure in the service pipe leading in from the road is about 30–50 mbar, and pressures in the gas main could be anything from 30 mbar to 7 bar and are classified as follows:

- low pressure: up to and including 75 mbar,
- medium pressure: from 75 mbar to 2 bar,

- intermediate pressure: from 2 to 7 bar,
- high pressure: above 7 bar.

The internal installation in normal domestic premises is at a pressure of about 20 mbar and the appliances themselves operate at pressures from about 5 to 20 mbar depending on the type of burner being used.

FLOW OF GAS IN PIPES

In Chapter 4 you saw how the potential energy of the gas at the injector was changed into kinetic energy in the burner venturi. As the velocity of the gas in the venturi throat increased, so the pressure became less. This principle holds good for any fluid flowing in a pipe.

When gas is standing still in a pipe the pressure throughout the whole length of the pipe is the same.

As soon as gas begins to flow the pressure falls progressively along the pipe as energy is changed and some is lost in overcoming the friction of the pipe walls. Figure 6.27 illustrates this point.

The diagram shows a horizontal pipe with uniform bore (i.e. having the same diameter all the way along).

The pipe is drawn to scale so that A to B, B to C and C to D are equal and each is equivalent to 8 m.

There are four pressure gauges attached at points A, B, C and D and gas can flow, in the direction of the arrow, when the valve E is opened.

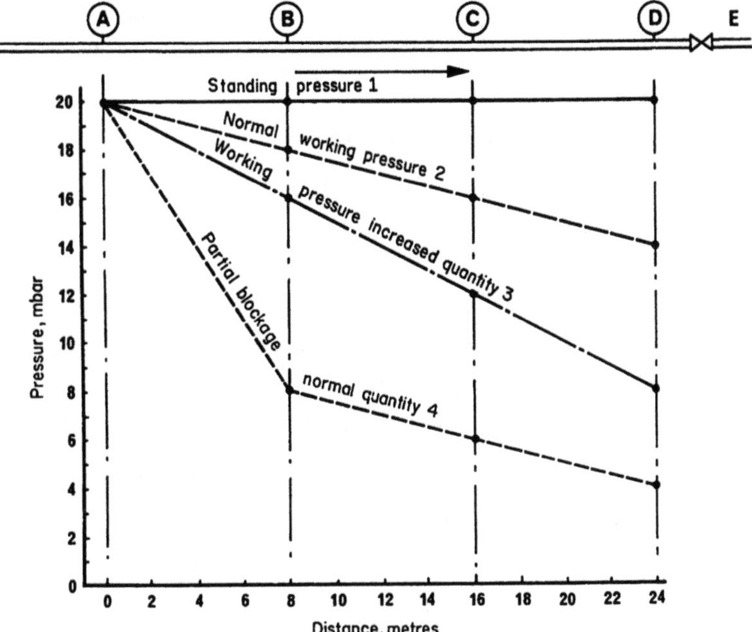

FIGURE 6.27 Flow of gas in pipes.

The graphs drawn below show clearly what happens in the installation under four different sets of conditions.

1. If the pressure at A is 20 mbar, then, with the valve E turned off the gas will stand still in the pipe and the pressures at B, C and D will also be 20 mbar.
   This is known as the 'standing pressure'. The graph of the standing pressure has been drawn at the top of the chart, 1. It shows that the pressure is the same at any point in the pipe.
2. Now the valve is turned on and a normal quantity of gas is allowed to flow through. In this case the flow causes a loss in pressure between A and B of 2 mbar. If A is maintained at 20 mbar then B will show 18 mbar.
   But B to C is just the same as A to B in size and length and the same quantity of gas is flowing. So the pressure loss in B to C will also be 2 mbar. And the same is true for C to D. So gauge C will read 16 mbar and gauge D 14 mbar.
   This is the pressure in the installation when gas is flowing and it is called the 'working pressure'. The normal working pressure is shown as a broken line by graph 2.
3. If the quantity of gas flowing is now increased then the pressure loss will also increase. The graph 3 shown is for an increase of 40% in the quantity. Pressure loss has doubled.
4. Graph 4 shows the readings obtained for a normal quantity of gas flowing but with an obstruction in the pipe between A and B. Any partial blockage effectively reduces the diameter of the pipe and this has the greatest effect on the pressure loss. After B the pressure loss is back to the normal 2 mbar in 8 m or ¼ mbar/m. The actual loss shown is due to a decrease in diameter by 30%. It caused a pressure loss of four times the normal.

FACTORS AFFECTING PRESSURE LOSS

Figure 6.27 has given examples of the effect of:

1. length of pipe,
2. quantity of gas flowing,
3. diameter of pipe.

In addition to these factors there are two more not previously mentioned:

4. specific gravity of the gas,
5. friction between the gas and the pipe wall.

The way in which these factors affect pressure loss when gas flows is as follows:

1. Pressure loss (symbol h) is directly proportional to

$$\left.\begin{array}{l}\text{length, } l\\ \text{specific gravity, } s\\ \text{friction, } \zeta\end{array}\right\}\quad \text{written as}\quad \begin{array}{l}h \propto l\\ h \propto s\\ h \propto \zeta\end{array}$$

So if any of these quantities were doubled, then the pressure loss would also be doubled.

2. Pressure loss, h, is directly proportional to the square of the quantity flowing:

$$h \propto Q^2$$

If the quantity is doubled, the pressure loss increases by 4 (2^2) times.

3. Pressure loss is inversely proportional to the fifth power of the diameter:

$$h \propto \frac{1}{d^5}$$

If the diameter was halved, the pressure loss would increase 32 (2^5) times.

Figure 6.28 shows the effect of reducing pipe diameters and Fig. 6.29 helps to show why the effect is so drastic. Whilst the diameter and the circumference have been reduced to half of the original, the actual area has been reduced by four times, to only a quarter of the original.

Collecting the factors together in one expression and ignoring friction, which only has a small effect in internal installations,

$$h \propto \frac{Q^2 sl}{d^5}$$

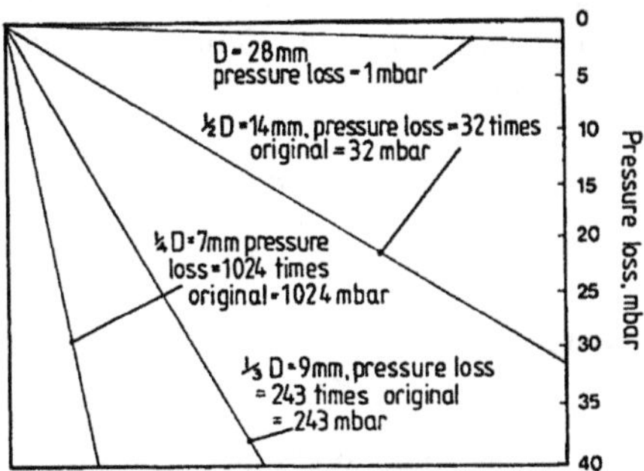

FIGURE 6.28 Effect of reducing pipe diameter.

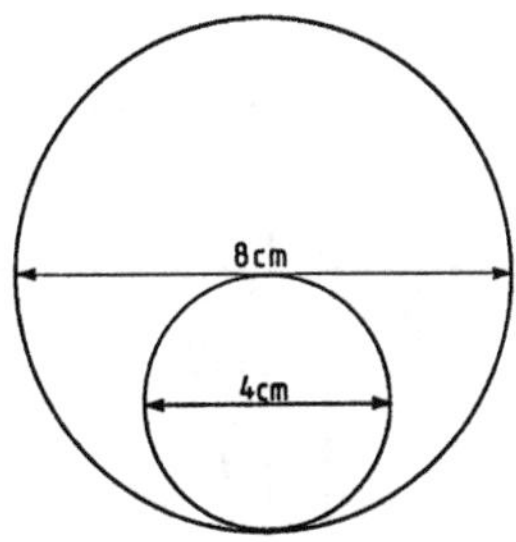

FIGURE 6.29 Comparison of pipe areas.

Transposing this to give Q, which is often what we need to know,

$$Q \propto \frac{\sqrt{hd^5}}{sl}$$

By using a constant to allow for friction and to convert the units into compatible figures we have a formula:

$$Q = 0.0071\,\frac{\sqrt{hd^5}}{sl}$$

where l = length in metres,
h = pressure loss in millibars,
s = specific gravity,
d = diameter in millimetres,
Q = quantity (or gas rate) in cubic metres per hour.

This is 'Poles formula' and it is effective for calculating pipe sizes in low pressure systems. At high pressure the compressibility of the gas must be taken into account and other formulae are used.

FLOW THROUGH AN ORIFICE

The flow of gas through the orifice of an injector was dealt with in Chapter 4, section on Injectors. The formula used to calculate the orifice diameter has been derived from the basic formula:

$$V = \sqrt{2gh}$$

where V = velocity,
h = static pressure,
g = gravity.

This link between velocity and static pressure enables the rate of flow of gas, or any fluid, to be determined by simply measuring the difference in static pressure on either side of an orifice placed in the pipeline (Fig. 6.30). The quantity of gas

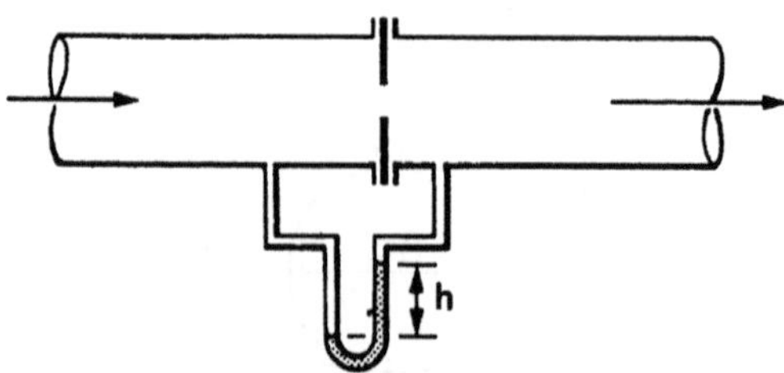

FIGURE 6.30 Flow through an orifice plate.

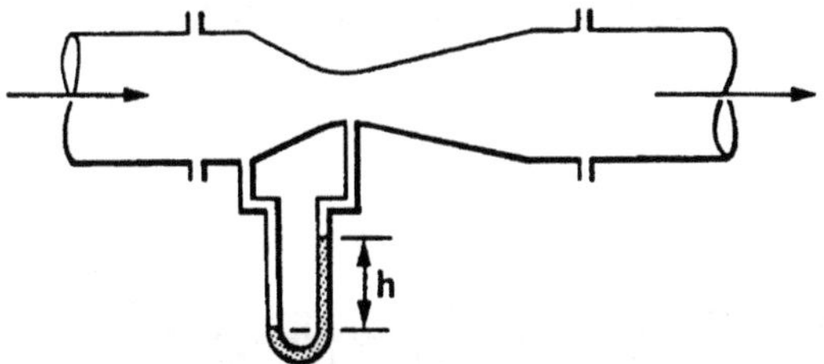

FIGURE 6.31 Flow through a venturi.

flowing may be calculated when the velocity of the gas and the area of the orifice are known.

$$\text{Quantity (m}^3\text{/h)} = \text{velocity (m/h)} \times \text{area (m}^2\text{)}$$

Figure 6.31 shows a venturi used instead of a simple orifice plate. As you know, the venturi has the advantage of having a very much lower overall pressure loss because a considerable amount of static pressure is regained as the gas stream slows down after passing through the throat.

Both the orifice plate and the venturi are forms of meter since they, in effect, measure the quantity of gas flowing through a pipe. Because this measurement is not by means of a positive displacement of actual volumes, meters of this type are called 'inferential meters' (inferential meters are dealt with in detail in Chapter 8).

MEASUREMENT OF APPLIANCE GAS RATES

For most domestic appliances it is usual to take a reading of the pressure at which gas is supplied to the burner or injector and accept this as one indication that the appliance gas rate is correct (Chapter 4, section on Fault Diagnosis).

Where it is necessary to be absolutely certain that the correct volume is being used, then the rate at which gas is passing through the domestic gas meter can be calculated from a reading of the meter test dial or on a metric meter, the test drum. The meter offers some resistance to the passage of the gas and this resistance varies with the position of the bellows and valves which are moving inside. As a result the test dial finger or test drum moves slightly faster at some parts of its revolution than at other parts. To be certain of an accurate reading the finger or drum should be allowed to rotate through at least one complete revolution.

The test dial on an index measuring cubic feet is located on the right-hand side of the index (Fig. 6.32a).

The test drum of a meter measuring cubic metres is the numeral on the extreme right of the index and indicates cubic decimetres.

$$\text{Gas rate per hour} = \frac{\text{Test dial/drum capacity per revolution} \times 3600}{\text{Time in seconds for one revolution of test dial/drum}}$$

Test dials usually have a capacity of 1 ft^3, test drums of 10 dm^3.

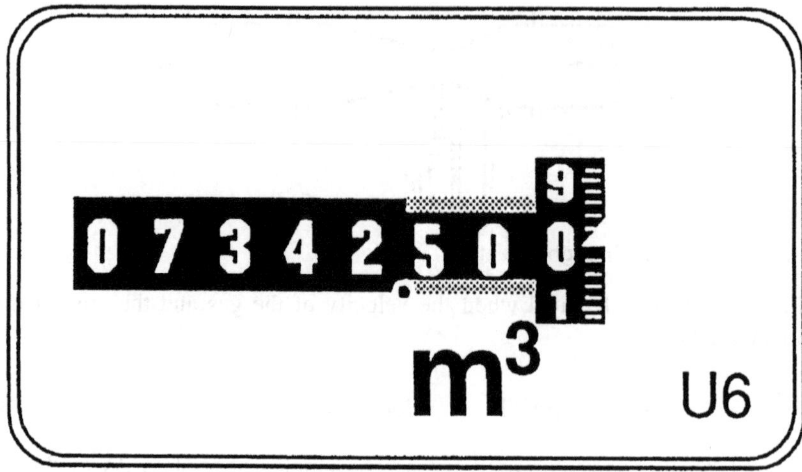

FIGURE 6.32a Meter index (cubic feet).

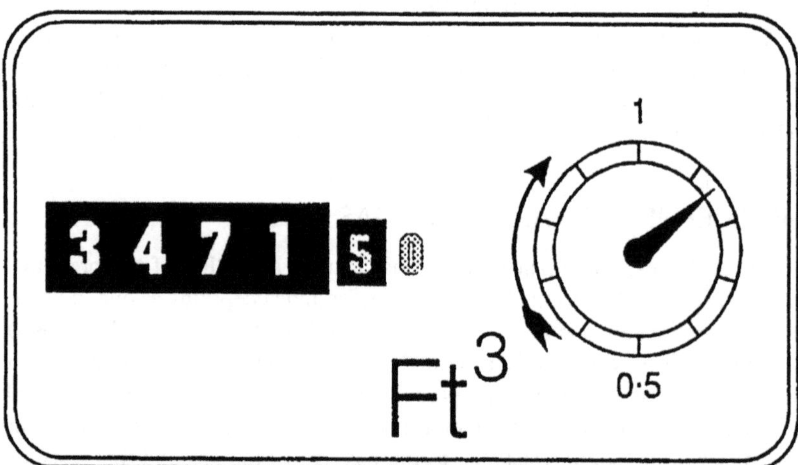

FIGURE 6.32b Meter index (cubic metres).

Examples

1. A meter test dial has a capacity of 1 ft^3. If the time taken for one revolution was 1 min 30 s, what is the gas rate per hour?

$$\text{Time taken} = 1 \text{ min } 30 \text{ s} = 90 \text{ s}$$

$$\begin{aligned}\text{Gas rate} &= \frac{1 \times 3600}{90} \text{ ft}^3/\text{h} \\ &= 40 \text{ ft}^3/\text{h}\end{aligned}$$

2. A meter test drum has a capacity of 10 dm^3. If the time for one revolution was 36 s, what is the gas rate per hour?

$$\text{Time taken} = 36 \text{ s}$$

$$\text{Test drum capacity} = 10 \text{ dm}^3 = 0.01 \text{ m}^3$$

$$\text{Gas rate} = \frac{0.01 \times 3600}{36} \text{ m}^3/\text{h} = 1 \text{ m}^3/\text{h}$$

ESTIMATING PIPE SIZES

When deciding on the correct size of pipes for an internal installation, it is recommended practice to allow a pressure loss of 1 mbar between the outlet of the meter and each of the points for the appliance.

The size of pipe selected should be capable of carrying the maximum gas rate of the appliances to be connected, taking into consideration the maximum demand likely at any one time. When the requirement is borderline between two pipe sizes, then using the larger size would allow for additional appliances to be added later.

Pipe fittings which are used to branch one pipe into another are called 'tees' and those which give a change of direction are called 'elbows'. Both tees and elbows offer considerable resistance to the flow of gas and, for pipes up to 25 mm (steel) or 28 mm (copper), a tee or elbow is equivalent to 0.5 m (2 ft) of pipe.

An example of a typical internal installation for a bungalow is shown in Fig. 6.33. The lengths of pipe and the gas rates of the appliances are given. The pipes have been sized, using Table 6.3 which has been taken from British

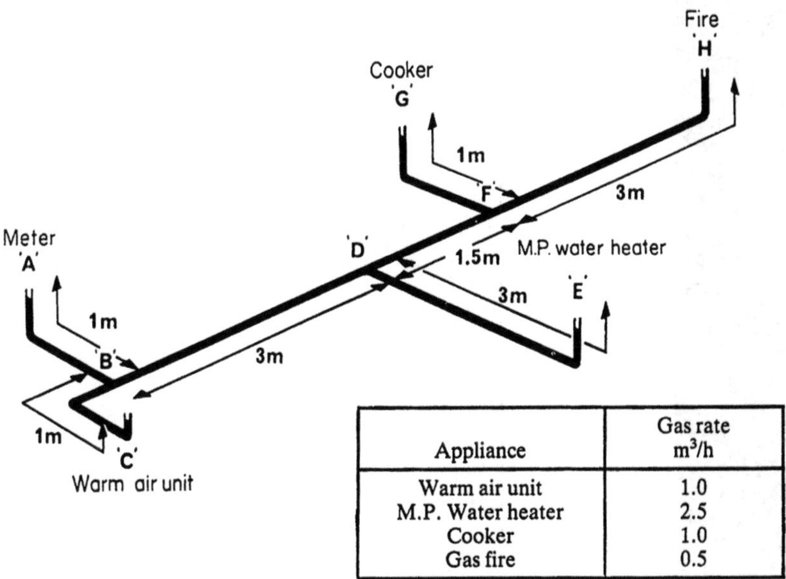

Appliance	Gas rate m^3/h
Warm air unit	1.0
M.P. Water heater	2.5
Cooker	1.0
Gas fire	0.5

FIGURE 6.33 Internal installation pipework.

Table 6.3 Discharge in a Straight Horizontal Copper Tube with 1.0 mbar Differential Pressure Between the Ends, for Gas of Relative Density 0.6 (air = 1).* Piping in accordance to BS EN 1057 1996

Nominal pipe size				Length of pipe							
Steel medium	Copper	Copper Corrugated Stainless	PE	Discharge (m^3/h)							
BS 1387	BS EN 1057 (OD)	BS 7838 (OD)	BS 7281 (OD)	3	6	9	12	15	20	25	30
6	8	–	–	0.29	0.14	0.09	0.07	0.05	–	–	–
8	–	–	–	0.8	0.53	0.49	0.36	0.29	0.22	0.17	0.14
–	10	10	–	0.86	0.57	0.50	0.37	0.30	0.22	0.18	0.15
10	–	–	–	2.1	1.4	1.1	0.93	0.81	0.70	0.69	0.57
–	12	12	–	1.5	1.0	0.85	0.82	0.69	0.52	0.41	0.34
–	15	15	–	2.9	1.9	1.5	1.3	1.1	0.95	0.92	0.88
15	–	–	–	4.3	2.9	2.3	2.0	1.7	1.5	1.4	1.3
–	–	–	20	4.0	2.7	2.1	1.8	1.6	1.3	1.2	1.05
20	–	–	25	9.7	6.6	5.3	4.5	3.9	3.3	2.9	2.6
–	22	22	–	8.7	5.8	4.6	3.9	3.4	2.9	2.5	2.3
25	–	–	32	18	12	10	8.5	7.5	6.3	5.6	5.0
	28	28	–	18	12	9.4	8.0	7.0	5.9	5.2	4.7
–	–	32	–	29	20	15	13	12	10	8.5	7.6
32	35	–	–	32	22	17	15	13	11	9.5	8.5

When using this table to estimate the gas flow in pipework of a known length, this length should be increased by 0.5 m for each elbow or tee fitted, and by 0.3 m for each 90° bend fitted.

Table 6.4 Pipe Lengths and Gas Rates for the Installation in Fig. 6.33

Pipe Section	Gas Rate (m^3/h)	Pipe (m)	Length Fitting*		Total (m)	Pipe Size** (mm)
			Type	m		
A–B	5	1	elbow	0.5	2	22
	tee	0.5				
B–C	1	1	2 elbows	1.0	2	12
B–D	4	3	–	–	3	22
D–E	0.5	3	tee	0.5	4	22
			elbow	0.5		
D–F	1.5	1.5	–	–	1.5	12
F–G	1	1	tee	0.5	2	12
			elbow	0.5		
F–H	0.5	3	elbow	0.5	0.5	10

*Most appliances have horizontal connections, so in practice an additional elbow would be required on each point.
**Copper tube in accordance with BS EN 1057 1996.

Standard 6891: 2005 – British Standard Specification for Installation of low pressure gas pipework of up to 35 mm (RI $1^1/_4$) in domestic premises (2nd family gas). The results are listed in Table 6.4.

When carrying out an exercise of this nature consideration has to be given to the permissible pressure loss in each section of the installation. For example, the pressure loss between A and H should not exceed 1 mbar.

But A to H is made up of four sections of pipe, A–B, B–D, D–F and F–H. Each section carries a different gas rate and must be sized separately.

If A to H is to have a pressure loss of not more than 1 mbar, then the pressure losses in each of the four sections should be about $^1/_4$ mbar. So A–B, B–D, D–F and F–H each must be sized to a pressure loss of about $^1/_4$ mbar.

The table of discharges only allows for pressure losses of 1 mbar. However, pressure loss is proportional to length, so, if you select a pipe size for a length which is four times longer than you need, the pressure loss on the actual length will be $^1/_4$ mbar.

As an example, take section D–F in Fig. 6.33.

Section D–F:

- has a length of 1.5 m,
- is to carry a gas rate of 1.5 m^3/h,
- should have a pressure loss of ¼ bar maximum.

But a pressure loss of $^{1}/_{4}$ mbar in a length of 1.5 m equals $(4 \times {}^{1}/_{4}) = 1$ mbar in $(4 \times 1.5) = 6$ m.

Look at the column in Table 6.3 under 6 m, for a discharge of 1.5 m^3/h. You will find:

$$12\text{ mm} = 1.0\text{ m}^3/\text{h}$$

$$15\text{ mm} = 1.9\text{ m}^3/\text{h}$$

The first size, 12 mm, will give slightly less than is required. It would do the job but with a slightly higher pressure loss.

The larger size, 15 mm, will carry the 1.5 m^3/h with little pressure loss and could allow for appliances to be added to the installation at some future date if required. This is the size to be used.

Chapter | seven

Control of Pressure

Chapter 7 is based on an original draft prepared by Mr P.M. Mulholland and updated by Route One Ltd. (2005)

NEED FOR PRESSURE CONTROL

The previous chapter showed how the pressure on a district varies with changes in the demand for gas. When very few appliances are in use pressures are at their highest. As more and more gas flows, the loss of pressure is greater and district pressures are reduced.

But, if customers' appliances are to work satisfactorily, there are limits below which the pressure must not be allowed to fall. And gas suppliers are required by statute to maintain a safe minimum pressure at all times, see Table 7.11.

Table 7.1 Health and Safety Gas Safety (Management) Regulations (GS(M)R) 1996

Standard of Pressure

With effect from 31st October 1996 the following regulations must be complied with:

The minimum pressure in a main, or in a service pipe having an internal diameter of 50 mm or more shall not be less than that specified in the following table:

Range of Wobbe Number (MJ/m^3)	Type of Gas	Minimum Pressure (mbar)
Not exceeding 31	LPG/air	5
Exceeding 31 but not exceeding 40	Natural gas (L)	10
Exceeding 40 but not exceeding 72	Natural gas (H) SNG	12.5
Exceeding 72	LPG	20

So there is a need to control pressures on the district to maintain adequate supplies.

Of course, a pressure which is too high is just as bad as one which is too low. Excessive gas rates on appliances can affect the combustion and may present hazards. What is needed is for each appliance to be supplied with gas at the pressure for which it was designed.

So pressures must be controlled within fairly close limits. And, once set, they must be kept constant.

Appliances can be affected not only by changing district pressures but also by changes in pressure within the internal installation itself. Excessive variations may be caused by:

- inadequately sized pipework;
- installation of additional or larger appliances to existing pipework;
- defective gas meter;
- partial stoppages or obstructions in the pipes;
- any combination of these factors and district pressure fluctuations.

The need for pressure control is then clear. It is required:

1. throughout the district, depending on the form of the distribution system;
2. at the inlet of the customer's meter;
3. at the appliance when the gas rate is critical or when the pressure loss in the internal installation varies with the number of appliances in use.

METHODS OF PRESSURE CONTROL

Pressure can be controlled by means of any valve or variable restrictor. But, because conditions are continually changing, what is required is a valve which will adjust automatically so that it will always provide an adequate gas supply at a constant outlet pressure. It must do this irrespective of changes in the inlet pressure or in the rate of flow of gas. A device which does this is called a 'regulator' or pressure regulator'.

There are many different types of regulators ranging from those which fit into large 900 mm diameter pipelines down to the smaller regulators fitted on older small domestic gas appliances (such as refrigerators, circulators and gas fires). With the introduction of Natural Gas many of the smaller appliance regulators were removed. On most of these modern small domestic appliances, the pressure is regulated by the meter regulator. Older large domestic gas appliances such as Central Heating boiler appliances retained there regulators, the modern domestic Central Heating boilers have the regulator incorporated into a multi-functional valve (covered in Chapter 11). The service engineer deals with all domestic regulators whilst district and some larger industrial regulators are the responsibility of specialist personnel.

Figure 7.1 shows where different types of regulator may be used in a distribution system. Figure 7.2 shows the type of regulator with which most people are familiar, a 'meter regulator' fitted to the inlet of a domestic meter.

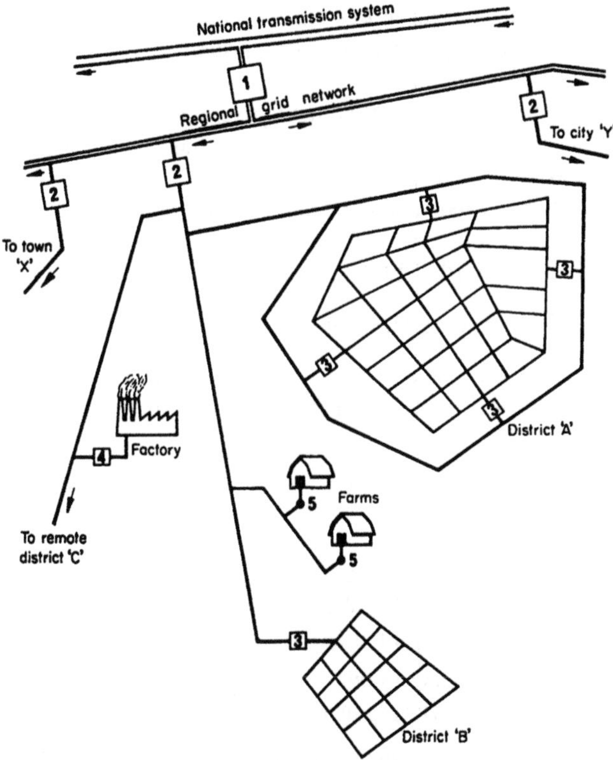

FIGURE 7.1 Distribution systems and regulators.

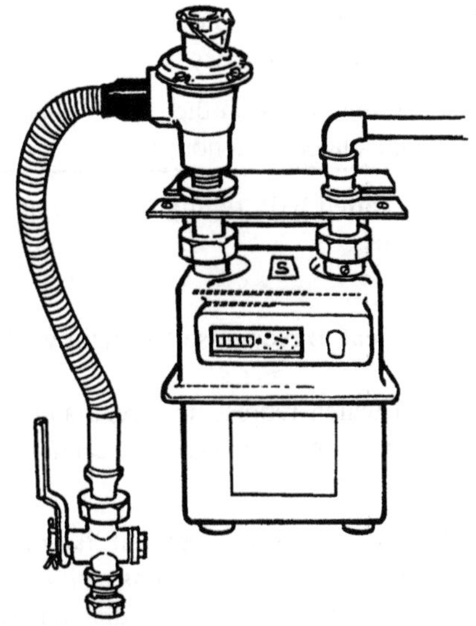

FIGURE 7.2 Meter regulator installation.

The regulators dealt with in this chapter are all 'constant pressure regulators'. That means that they deliver gas at a constant outlet pressure over a range of inlet pressures and gas rates.

It is possible to have 'constant volume governors' which deliver a fixed gas rate. These can only be used domestically on appliances which have one burner and are always used at a full-on rate. They are generally found only on instantaneous water heaters and are dealt with in Volume 2, Chapter 8.

CONSTANT PRESSURE REGULATORS

Operation of simple regulator

In order to understand how a regulator works, it is necessary to know a little about its construction. This will be explained in detail later on.

		Pressure	
Number	Regulators	Inlet	Outlet
1	Regional offtake, above-ground installation (AGI)	42–70 bar	21–35 bar
	Multi-stream, multi-regulators HP regulators		
2	City gate stations HP to IP regulators	21 to 35 bar	2–7 bar
3	District regulators IP to LP regulators	2–7 bar	30–50 mbar
4	IP to MP regulators on the service to a remote factory	2–7 bar	1.4 bar
5	HP* to LP regulators on the services to isolated farm houses	2–7 bar	20 mbar

*Although these are called 'HP' regulators, they usually operate on intermediate pressures.

For the moment it is sufficient to recognise the main working parts which are shown in Fig. 7.3.

This is a line diagram which represents a regulator cut in half down the middle and with one half removed. It is called a 'sectional elevation'.

The essential parts are:

1. Flexible diaphragm made of some material like rubber. This is usually circular and has the valve stem fixed to its centre.
2. Valve stem or spindle. This is fixed to the diaphragm at one end and the valve at the other. As the diaphragm moves, the valve will move.

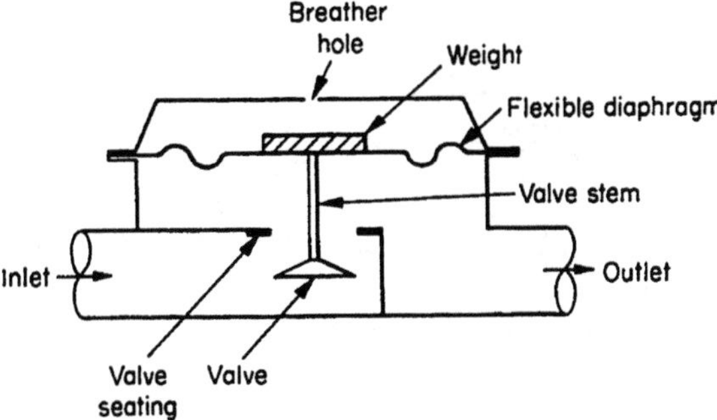

FIGURE 7.3 Simple constant pressure regulator.

3. Valve, made of metal, or hard plastic-type material, cone-shaped. Figure 7.4 shows a view of the valve and seating from below.
4. Valve seating. Circular orifice with machined seating on its edge (like the valve seating in an engine cylinder head).
5. Weight. This sets the regulator to the required outlet pressure. It rests on a plate on the diaphragm.

Breather hole. Vents the upper compartment to atmosphere so allowing the diaphragm to move up and down.

The forces which are acting on the valve and diaphragm are illustrated in Fig. 7.5.

Ignoring the weight of the moving parts (that is the valve, diaphragm and valve stem) and also any pressures acting on the valve itself, the main forces on the diaphragm are:

$$\begin{aligned}\text{Downward force} &= \text{Weight or the force exerted by the mass of the weight}\\ &= W\end{aligned}$$

$$\begin{aligned}\text{Upward force} &= \text{Outlet pressure} \times \text{the effective area of the diaphragm}\\ &= P_o A\end{aligned}$$

When these two forces balance, the valve is held suspended below the seating in a state of equilibrium. The amount of valve opening will be that which will

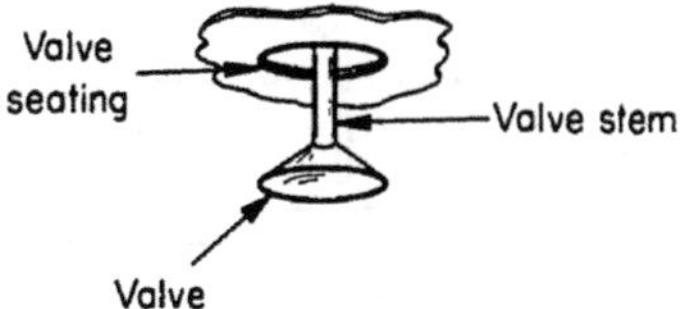

FIGURE 7.4 Regulator valve and seating.

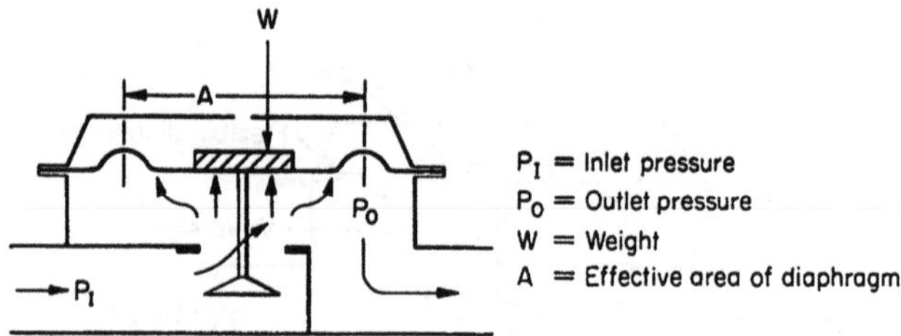

FIGURE 7.5 Forces acting on the diaphragm.

permit enough gas to pass through to maintain the outlet pressure at a value which will just maintain the balance,

$$P_O A = W \qquad \text{or} \qquad P_O = \frac{W}{A}$$

Figure 7.6 shows how changes in the conditions affect the amount of valve opening.

At Fig. 7.6a the valve is only partly open, supplying a small amount of gas to, perhaps, one burner of the appliance at an outlet pressure of P_o millibars.

If a second burner is turned on, gas will be taken from under the diaphragm more quickly than it is being supplied through the valve seating. P_o will be reduced, the balance will be upset and the downward force W will be greater than the upward force $P_O A$. So the weight will push the valve down until the opening is big enough to supply gas to both burners and again bring the outlet pressure back up to $P_O = W/A$. The valve will now be in the position shown in Fig. 7.6b.

Turning on another burner could result in the valve opening to the position shown at Fig. 7.6c. So the position of the valve depends on the rate of flow of gas required.

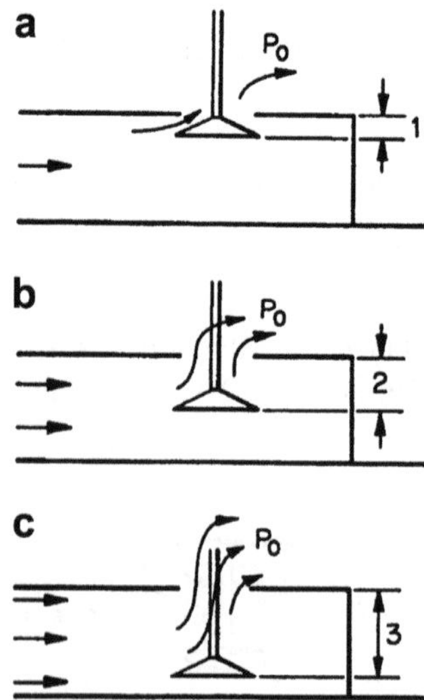

FIGURE 7.6 Effect of gas rate on valve position: (a) one burner in use, (b) two burners in use, and (c) three burners in use.

It also depends on the inlet pressure P_I.

If the valve was in the position shown at Fig. 7.6b, and if inlet pressure was increased it would tend to push the gas more quickly past the valve and outlet pressure would increase also. Once again the balance would be upset, but the stronger force would now be the upward one, $P_o A$, and the valve would close slightly as in Fig. 7.6a, so offering a greater resistance to the higher pressure.

The reverse of this would happen if the inlet pressure fell. The valve would open, as in Fig. 7.6c, to maintain a constant outlet pressure.

The regulator can operate satisfactorily only when the inlet pressure is at least slightly above that required on the outlet.

Since $W = P_o A$ for a simple constant pressure regulator, it is possible to calculate the weight required to give any particular outlet pressure on a known size of regulator.

For example, a regulator has an effective diaphragm diameter of 71.5 mm and is required to give an outlet pressure of 20 mbar. What weight is required?

$$\text{Effective area} = \frac{\pi d^2}{4} = \frac{3.14 \times 0.0715^2}{4}\,\text{m}^2$$
$$= 0.004\ \text{m}^2$$

$$\text{Pressure required} = 20\ \text{mbar} = 2000\ \text{N/m}^2$$

$$W = \text{pressure} \times \text{area}$$
$$= 2000\ \text{N/m}^2 \times 0.004\ \text{m}^2$$
$$= 8\ \text{N} = \frac{8}{9.81}\ \text{kg}$$
$$= 0.815\ \text{kg or } 815\ \text{g}$$

In small regulators the weight of the moving parts is likely to be only a few grams and may be disregarded. In very large regulators the weight may be considerable and must be subtracted from the loading weight since it provides an additional downward force.

$$W = P_o A - W_m \text{ where } W_m = \text{weight of moving parts}$$

NEED FOR COMPENSATION

Since the inlet pressure on a regulator must always be higher than the outlet pressure, the upward and downward forces on the valve are unequal. As shown in Fig. 7.7, inlet pressure pushes up under the valve and outlet pressure pushes down on top of the valve. The result is an additional upward force which tends to close the valve and reduce the outlet pressure.

On most domestic appliance regulators, the valves are small and the difference between inlet and outlet pressures is not great. So the extra pressure loss can be

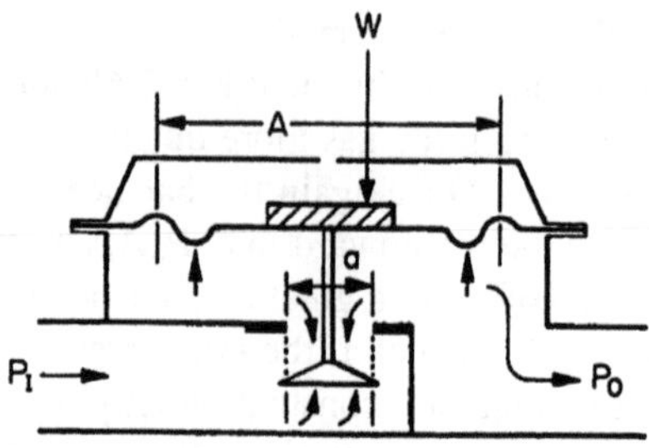

FIGURE 7.7 Forces acting on a single valve.

tolerated. In a regulator having a valve of 15 mm diameter and with a difference of 10 mbar between inlet and outlet pressure the additional upward force is 18 g.

The actual forces on the regulator are:

$$\text{Downward} = W + P_o a$$

$$\text{Upward} = P_o A + P_I a$$

In equilibrium these are equal.

$$W + P_o a = P_o A + P_I a$$

Therefore

$$W = P_o A + P_I a - P_o a = P_o A + a(P_I - P_o)$$

The additional upward force $a(P_I - P_o)$ needs to be compensated for on larger regulators and those operating at higher inlet pressures.

COMPENSATED CONSTANT PRESSURE REGULATORS

A simple method of compensating for unequal pressures on the valve is to have two valves, both attached to the same spindle, Fig. 7.8.

Now the inlet pressure P_I pushes up on the top valve and down on the bottom valve. So the resulting force is zero.

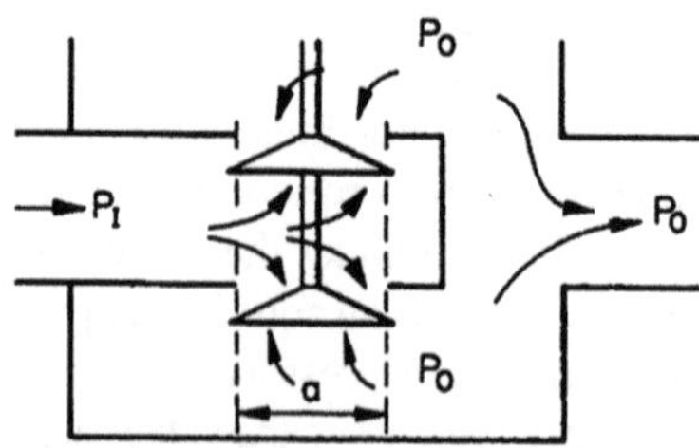

FIGURE 7.8 Forces acting on two valves.

And the outlet pressure P_o pushes down on the top valve and up on the bottom valve. So the resulting force is again zero.

$$\begin{aligned} \text{upward force} &= P_I a + P_o a \\ \text{downward force} &= P_I a + P_o a \\ \text{resultant} &= (P_I a + P_o a) - (P_I a + P_o a) = 0 \end{aligned}$$

This is a good way of compensating for forces on the valve and it is used on some large regulators. In small regulators it is difficult to adjust the valves to close at exactly the same time.

Instead of a second valve, an additional diaphragm is used (Fig. 7.9).

The small, auxiliary or secondary diaphragm, D, takes the place of the top valve in Fig. 7.8. Like the valve, it is attached to the valve spindle. It has the same effective area as the valve and so is subject to the same forces. Once again the upward and downward forces due to the inlet and outlet pressures are equal and opposite and so total zero.

The outlet pressure, P_o, reaches the underside of the main diaphragm through an orifice P. This may simply be a drilling in the cast body of the regulator or, in some large sizes, it may be in the form of a pipe. Known as the 'impulse pipe' it connects with the regulator outlet and prevents sudden surges of pressure causing rapid movement of the valve.

This is a popular regulator in the medium-sized range for use on low inlet pressures. The constraining factor is the auxiliary diaphragm which cannot be made too small or it would not have sufficient flexibility. The diaphragm is also subjected to inlet pressure and will not withstand pressures in the higher ranges. Meter regulators and service regulators are of this type.

A simple method of compensation, used on small appliance regulators, is shown in Fig. 7.10.

The diaphragm and the valve have the same effective area, like the auxiliary diaphragm and the valve in the previous type in Fig. 7.9.

In this regulator, however, the outlet pressure is exerted only on the underside of the valve, which replaces the main diaphragm as the controlling area. The space above the single diaphragm is open to atmosphere.

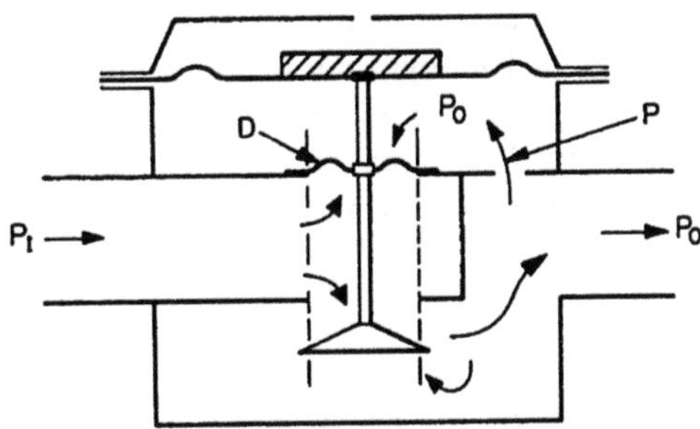

FIGURE 7.9 Compensated regulator with auxiliary diaphragm.

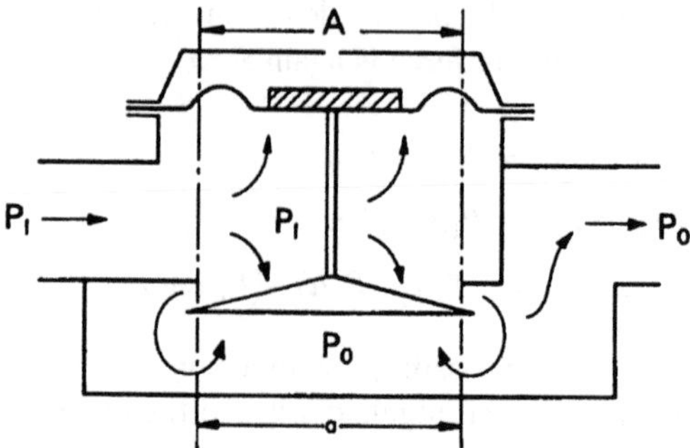

FIGURE 7.10 Single valve compensated regulator.

REGULATOR CONSTRUCTION

Development, over the years, has led to a steady reduction in the size and weight of both service and appliance regulators. The main body of the regulator which was originally cast iron or hot-stamped brass is now generally made of die-cast light alloy, based on aluminium or zinc.

The diaphragms have, in the past, been made of leather and are now generally produced from a synthetic rubber material which is impervious to natural gas. They are usually sandwiched between two metal plates or 'pans' which keep the centre of the diaphragm rigid and provide a platform for the loading weights.

Valves may be brass on small regulators and rubber covered on the slightly larger ones. Brass is used for valve seatings where these are not just machined out of the casing. The valve spindles may be stainless or cadmium plated steel.

Figure 7.11 shows the constructional details of a simple regulator with a single diaphragm and Fig. 7.12 shows details of a compensated regulator with auxiliary diaphragm.

The regulator in Fig. 7.12 has its loading in the form of a spring, rather than by weights.

REGULATOR LOADING

Regulators may be loaded in a number of ways. Some of the old work (or 'station') regulators were loaded by water in a tank on top of the bell. District regulators may be loaded by gas pressure above the diaphragm.

The most common forms of loading are by weights or by a spring. Adding weights or compressing the spring increases the outlet pressure.

Weight loading

Weights are simple to apply and can give an accurate setting. They were frequently used on appliance regulators preset by the manufacturers to the

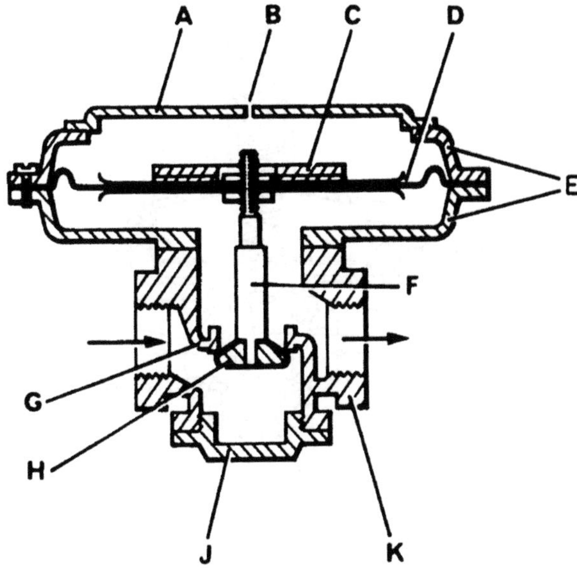

A : *Top cover plate*
B : *Breather hole*
C : *Lead weights*
D : *Diaphragm and pans*
E : *Diaphragm casings*
F : *Valve stem*
G : *Valve seating*
H : *Valve, rubber-covered*
J : *Cap*
K : *Body*

FIGURE 7.11 Construction of single diaphragm regulator.

required pressure by a single weight, usually of lead. Weight loading has two disadvantages:

1. The regulator must be fitted in a horizontal position and must be at level.
2. The weight possesses inertia which causes the valve to overshoot.

Spring loading

A spring which has been compressed possesses potential energy and so is capable of exerting a force. This force may be used to load a regulator. The advantage of spring loading is that it enables a regulator to be fitted in any position, even completely vertically and, for this reason, it is popular with appliance designers. There is a disadvantage, however. As the regulator valve opens, the diaphragm moves down and the spring is allowed to extend. This reduces the amount of compression and so reduces the loading force slightly. Spring-loaded regulators are, therefore, less accurate over the full range of gas rates and tend to be used on appliances with fixed gas consumptions.

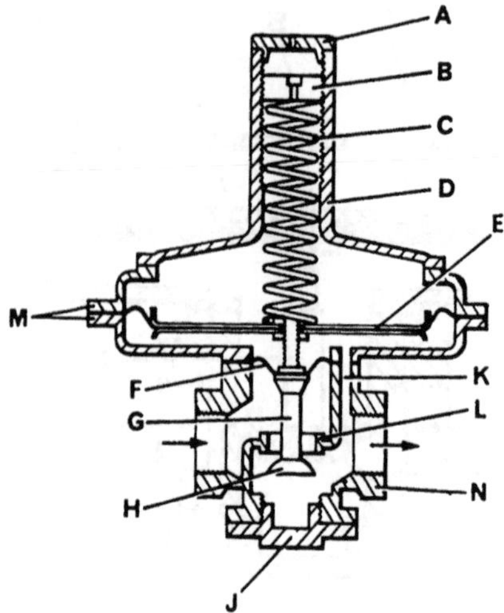

A : Cover cap with breather hole
B : Spring pressure adjuster
C : Spring
D : Spring housing
E : Diaphragm assembly
F : Auxiliary diaphragm
G : Valve stem
H : Valve
J : Alternative outlet
K : Impulse pipe
L : Valve seating
M : Diaphragm casings
N : Body

FIGURE 7.12 Construction of double diaphragm regulator.

HIGH-TO-LOW PRESSURE REGULATORS

The supply of gas to customers usually comes from district mains operating at low pressures (up to 75 mbar). Controlling these pressures calls for the types of regulators already described.

Occasions do arise, however, when the supply must come from an intermediate or high pressure main. This is because:

- The mains' pressure has been increased to meet higher gas demand.
- The customer's premises are in a position isolated from low pressure distribution mains but adjacent to a high pressure main.
- Modern 'insertion' mains and service replacement techniques (using smaller diameter PE pipework inserted into larger redundant steel pipework). Reducing the pipe size and increasing the pressure, remove the need for expensive and time consuming excavation.

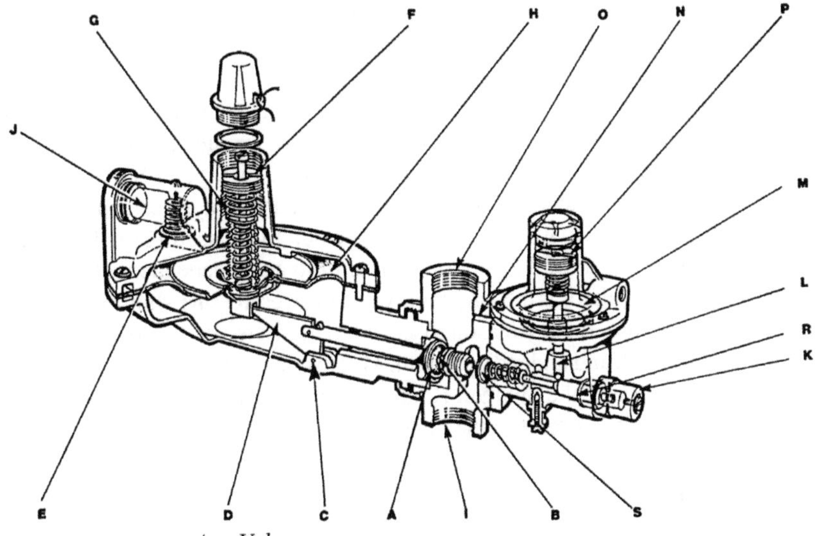

A : Valve
B : Valve seating
C : Pivot
D : Lever
E : Relief valve
F : Pressure adjustment
G : Spring
H : Diaphragm assembly
I : Inlet
J : Vent pipe
K : Manual reset
L : Diaphragm stem and arrestor
M : Slam-shut valve diaphragm
N : Impulse duct
O : Outlet
P : Pressure adjustment
R : Valve plunger
S : Shut-off valve

FIGURE 7.13 High-to-low pressure regulator.

The second case is illustrated in Fig. 7.1, where the farmhouses are shown supplied from the main feeding District B.

In both cases, the customers would be supplied through 'high-to-low pressure regulators', or 'high-pressure service regulators' as they are also called. These regulators are designed to reduce pressures from over 2 bar down to 20 mbar in a single reduction and a typical regulator is shown in cross-section in Fig. 7.13.

Gas enters through inlet, I, and passes through valve seating, B (the valve is in the open position when the regulator is at rest). The seating has only a small orifice and this restriction causes a drop in pressure. When the valve, A, closes, the area subjected to high pressure is small and the force on the valve is kept to a reasonable minimum.

The regulator is controlled by outlet pressure acting under the diaphragm, H, which is spring loaded. The diaphragm is large and operates the valve through the lever, D, which is pivoted at the bottom of the angle, at C. So the vertical movement of the end attached to the diaphragm results in a horizontal movement of the valve which, like a piston, is pushed on to the seating on the orifice.

On the opposite side of the inlet/outlet manifold and independent of the high-to-low pressure regulator is a 'slam-shut' valve or overpressure shut-off (OPSO). If the outlet pressure, for any reason, rises to too high a level, it is sensed through the impulse duct, N, and under diaphragm, M. This raises the diaphragm and stem (arrestor), L, releasing the plunger, R, and forcing the shut-off valve, S, against the orifice. The valve is reset to the open position manually at K.

LPG REGULATORS

These are similar in principle to natural gas regulators but because of the high pressures involved, LPG regulators are generally of a more robust construction. On tank supplies governing of LPG is in two stages. In the first stage pressure regulator is connected to the LPG storage tank. In the second stage regulator can be fitted in any position after the first, even on the inlet to the meter, if one is fitted. It must be designed to BS 3016, that is the under pressure shut-off (UPSO) and over pressure shut-off (OPSO) must be separate and the customer must be able to reinstate the UPSO only. Figure 7.14 shows a cross-section of a first stage regulator and Fig. 7.15 a second stage regulator incorporating UPSO and OPSO. The UPSO can be re-established by taking off the cap and lifting the spindle on the regulator.

INSTALLING REGULATORS

Appliance regulators

Appliance regulators are usually incorporated with the appliance and may, in a number of cases, be an integral part of a multi-functional control device. Figure 7.16 shows a small, spring-loaded regulator.

When a regulator has to be fitted as a replacement or on a new installation the following points should be noted.

- Before fitting, examine the regulator for any defect or damage in transit.
- Remove any plugs from inlet and outlet which have been supplied to keep out dirt.
- Ensure that the regulator is the right one for the gas rates and pressures required.
- Remove the packing used to hold the valve off its seating and prevent damage. If the packing is on top of the diaphragm, leave it in until you have finished screwing the regulator on to the supply.

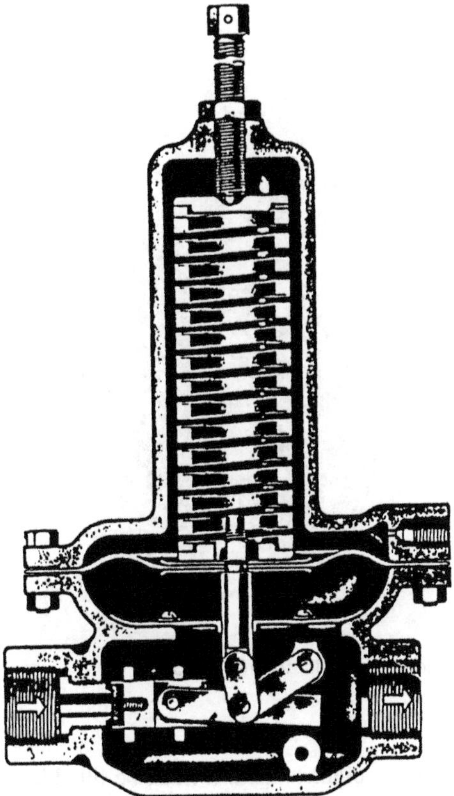

FIGURE 7.14 LPG first stage pressure regulator.

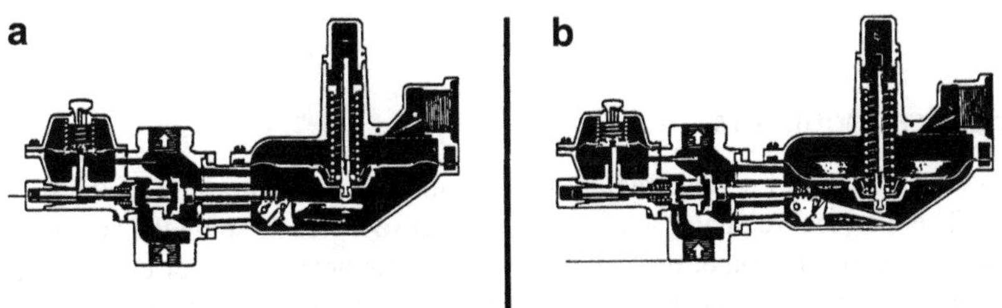

FIGURE 7.15 (a) Regulator with over-pressure and under-pressure shut-off devices in cocked (open) position and (b) regulator with the shut-off device in the tripped (closed) position.

- Make sure the regulator is the right way round.
- Locate the regulator in the coolest position, away from direct heat and excessive dirt or spillage.
- Check that a pressure test point has been provided somewhere on the outlet of the regulator so that the pressure may be checked or adjusted.

FIGURE 7.16 Spring-loaded appliance regulator.

- If the regulator is weight-loaded, it must be fitted in a horizontal position and level. It is preferable to fit all regulators in this way, although spring-loaded regulators may be fitted vertically if required.
- Ensure that the regulator is accessible for servicing.

Service regulators

The installation of service regulators is dealt with in Volume 2, Chapter 4.

The 'service' regulator was so called because it was fitted on the customer's end of the service pipe, usually between the customer's control cock and the inlet of the meter. Comparatively few service regulators were fitted until natural gas and its higher distribution pressures made it necessary to fit regulators on all premises.

The regulator now in use is the double-diaphragm type and it is commonly provided with an outlet connection which fits directly on the meter (Fig. 7.2). Now known as the 'meter regulator' it may also incorporate a dust filter on some districts where excessive rust in the mains can cause problems. Figure 7.17 shows a typical meter regulator.

The principles of installation are basically the same as those for appliance regulators. After breaking the seal to adjust the regulator, to comply with Gas Safety Regulations, it is essential that the seal is remade.

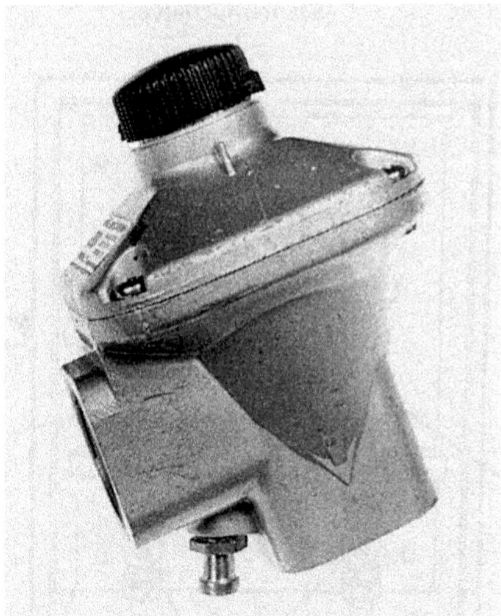

FIGURE 7.17 Meter regulator.

FIGURE 7.18 HP to LP regulator.

HP to LP regulators

Because of the high pressures involved, this type of regulator is usually fitted outside the premises in a suitable housing or an underground pit. Figure 18 shows a regulator and an example of an installation is shown in Figs 7.19a–c. The main features are:

- A valve should be provided to isolate the installation from the district mains system.

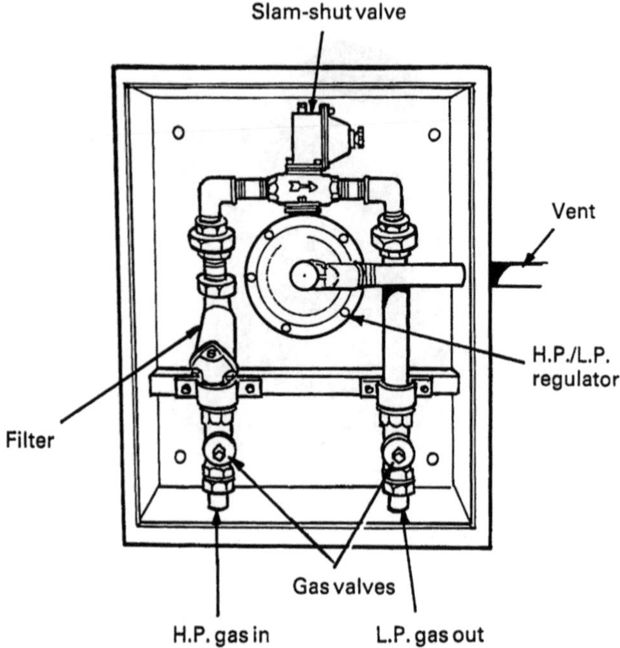

FIGURE 7.19a Natural gas installation of HP to LP regulator fitted in plastic housing.

FIGURE 7.19b MP to LP regulator.

- The valve should be situated at least 2 m from the installation and must be accessible, for instant shut-down, in the event of serious failure.
- A full-size vent pipe should be provided to discharge gas from the relief valve to a safe place.
- The end of the vent pipe should be turned down to prevent the entry of rain or snow and a gauze should be fitted to keep out insects.
- A flame trap need not be fitted if the vent pipe is less than 20 m in length.

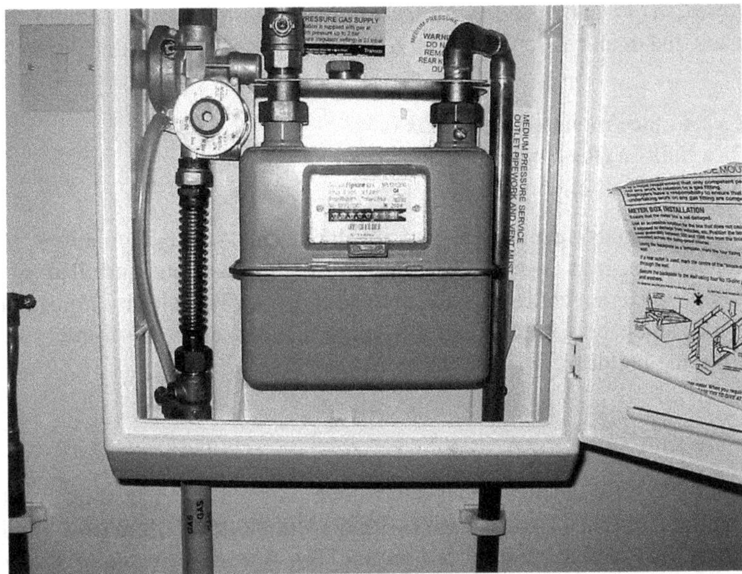

FIGURE 7.19c MP to LP meter box installation.

- The regulator must be the right size and type for the installation.
- Makers provide different sizes of inlet valve orifice and loading springs to cover inlet pressures ranging from 50 mbar to over 8 bar.

A new series of tailor-made medium-to-low pressure regulators have evolved for domestic gas installation. These compact regulators have an angled body designed to fit in to existing domestic external meter boxes. The medium-to-low pressure regulator manufacturer's installation and commissioning instructions must be followed as they may differ. (Currently there are three manufacturers Jeavons, Francel and Mesura.)

Ongoing gas mains renewal programmes have increased the number of gas service replacements from steel to plastic. Improved insertion techniques allow the gas transporters to insert smaller plastic pipe into old steel services without the need for ground excavations. Medium gas pressure can then be used through the smaller plastic pipe to achieve the required supply, a medium-to-low pressure regulator is required (this replaces the standard low pressure regulator).

Note that medium pressure is above 75 mbar but below 2 bar and low pressure is up to 75 mbar

Warning notices: There is a requirement for durable warning labels to be fitted adjacent to the meter.

Where medium pressure is used, additional safety controls are fitted to provide over pressure protection to the meter and downstream system. Gas operatives need to be aware of medium-to-low pressure installations and be able to identify the component parts which make up a typical medium-to-low pressure meter installation.

Location of medium-to-low pressure fed meter installations: Medium-to-low pressure fed meter installations need to be located outside the property, in either:

1. a surface mounted meter box or
2. a built-in (flush mounted) meter box or
3. a purpose-built housing.

Purpose-built housing must comply with all relevant statuary regulations. Meter boxes and housings need to be designed, constructed and installed so that escaping gas cannot enter the wall cavity or property. This means *all* installation pipework and cables need to exit the box on the outside of the property before entering the building.

LPG regulators

Tank supply LPG: The main features are as follows. The tank must be a minimum of 3 m (1 tonne tank), 7.5 m (2 tonne tank) from any building, boundary line or fixed source of ignition. The first stage regulator should be fitted to the tank valve, the second stage regulator may be fitted in any position after the first, up to the meter inlet, but outside the building (see Figs 7.20–7.24).

Cylinder Supply LPG: The main features are as follows. Two cylinders supply gas and two act as spares. When the service cylinders are empty the changeover valve automatically switches to the spare cylinders without interruption to the gas supply. A visual indicator tells you that the change has taken place.

Figures 7.25a and b illustrate a typical changeover valve. The actuator spring, diaphragm and inlet valves use previously described methods for

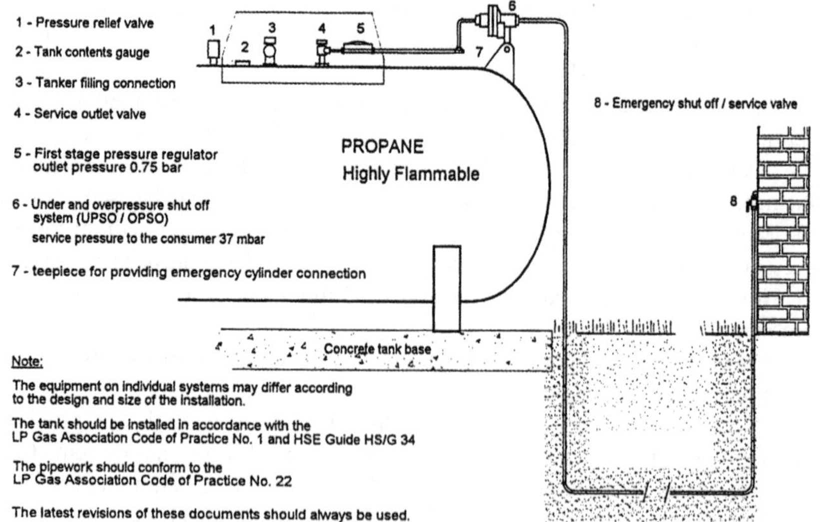

FIGURE 7.20 Installation of first and second stage regulators on a domestic LPG tank.

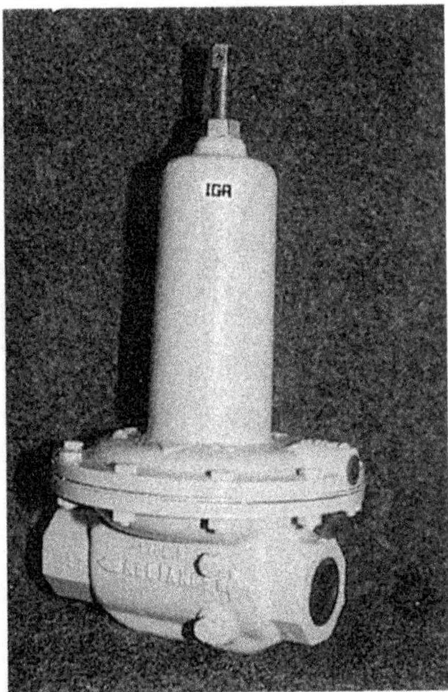

FIGURE 7.21 First stage pressure regulator.

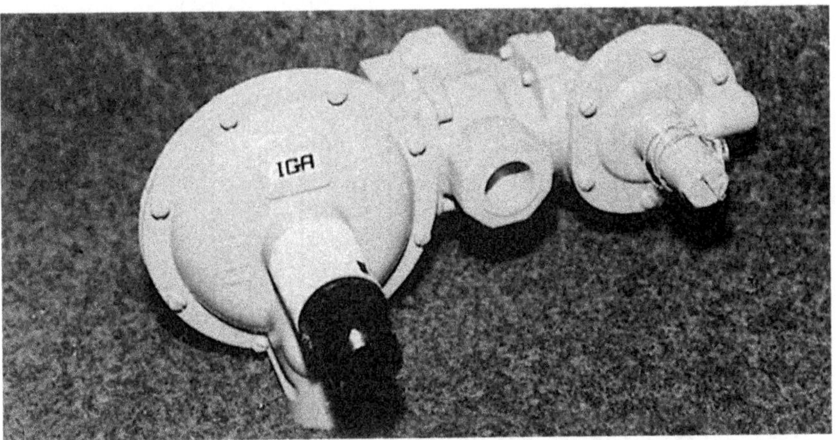

FIGURE 7.22 Second stage pressure regulator.

reducing the gas pressure. The right hand inlet valve controls the pressure when gas is being fed from the right hand cylinders and the left hand valve controls the pressure when gas is being fed from the left hand cylinders.

The spring loaded actuator has an angled base with a manually operated lever attached to it. When this lever is in the 'off' position, Fig. 7.25a, the action of the springs fitted underneath the inlet valve ensures that no gas flows. When the lever is turned through 90°, Fig. 7.25b, the action of the angled base of the actuator onto the diaphragm centre dish overcomes the spring under the right

FIGURE 7.23 Second stage regulator fitted on meter.

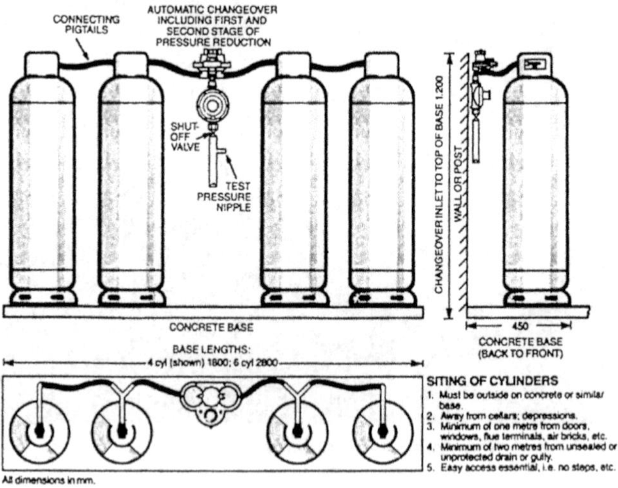

FIGURE 7.24 An example of an LPG cylinder installation.

hand inlet valve and allows gas to flow from the right hand cylinders, between the inlet valve and its seating, thus controlling the pressure.

When the right hand cylinders are exhausted the gas pressure from these cylinders will fall. The pressure of the actuator spring will cause the diaphragm to push down on the diaphragm centre dish acting against the left hand inlet valve. This actuator spring pressure will overcome the pressure of the spring under the left hand inlet valve and allow gas to flow from the left hand (reserve)

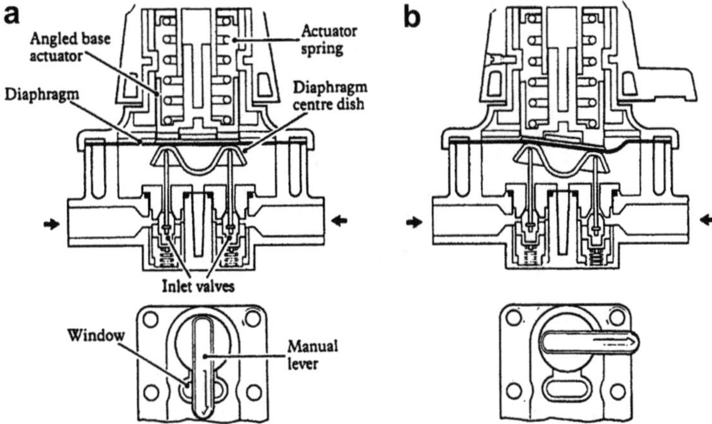

FIGURE 7.25 Automatic changeover valve.

cylinders. The pressure from these bottles will again be controlled by the gas pressure acting under the diaphragm against the pressure of the actuator spring on top of the diaphragm.

Figure 7.26 shows a single cylinder supplying a cabinet heater. The regulator, fitted to the top of the cylinder, again uses the same basic principles of high-to-low pressure regulators previously described.

Figure 7.27 shows one type of regulator used on the cylinder supply to a cabinet heater. The regulator contains a spring, diaphragm and a valve as previously illustrated. To fit the device onto the outlet of the cylinder valve it is necessary to lift a plastic locking ring against the action of a spring, push it onto

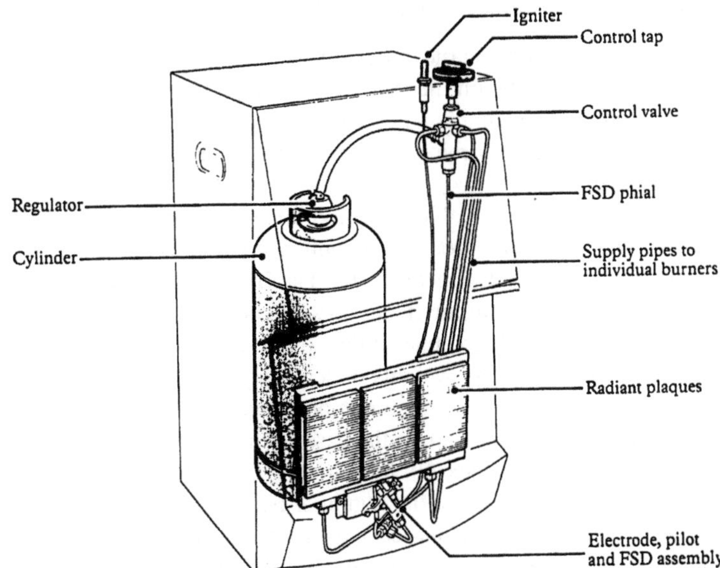

FIGURE 7.26 Typical butane cabinet heater.

FIGURE 7.27 Low pressure butane regulator.

the cylinder valve and release the locking ring. This allows the regulator's coupling to fit into a groove on the cylinder valve and make a gas-tight connection. In the event of fire the plastic locking ring will melt and release the regulator from the cylinder, thus shutting off the gas supply.

The cylinders in Fig. 7.24 would usually be stored in an outside location, they contain liquefied propane which has a low boiling (evaporating) point of −42 °C.

The cabinet heater shown in Fig. 7.26 would normally be used inside a building, the cylinder would be filled with liquefied butane which boils (evaporates) at just below 0 °C.

COMMISSIONING REGULATORS

The actual way in which regulators may be brought into operation or 'commissioned' depends on the type of regulator concerned. However, the following general procedure should be applied.

1. Fit a pressure gauge on the outlet side of the regulator at an appropriate point. This might be on the meter outlet or on an appliance burner.
2. If the regulator loading is adjustable and not preset or sealed, then remove the weights or release compression on the spring by unscrewing the adjuster.
3. Slowly open the inlet valve and purge the regulator of air at a suitable point. This might be through the appliance burner or, in the case of a meter regulator, at the meter outlet. The HP regulator should be purged at its outlet valve or at the meter control cock.
4. Turn on an appliance to create a flow of gas and increase the loading on the regulator by adding weights or screwing down the spring adjuster until the required outlet pressure is obtained.
5. Turn off the appliance and, in the case of the meter regulator or the HP regulator, check for 'lock-up' by watching the outlet pressure for a short time. If the pressure rises, the regulator needs attention.
6. The HP regulator can be purged and adjusted by allowing the air and gas to pass out at the inlet of the meter control. Only when the regulator is purged and correctly adjusted is the meter control finally opened.

SERVICING REGULATORS

All the regulators which control the distribution mains systems are subject to periodic servicing by specialist personnel. Similarly a number of appliances are given routine contractual servicing by the Service Department and any appliance regulators which are fitted are checked during the service.

Generally, small, modern regulators can operate for considerable periods without attention, so minimising the need for periodic maintenance.

When a regulator is serviced the following general procedure should be carried out.

1. Check that it is convenient to shut off the appliance or the supply.
2. In the case of a large regulator which can be isolated by valves, turn off the inlet and outlet valves and vent the regulator to atmosphere. On small regulators, turn off the appliances and any valve or cock on the regulator inlet.
3. Take off the top cover and remove the weights or loading spring.
4. Dismantle the regulator by removing the diaphragm(s) and the valve.
5. Clean all parts of the body and casings.
6. Check the diaphragms and renew if necessary. Leather diaphragms tend to dry out and require periodic oiling.
7. Clean the regulator valve. If it has a rubber seat check and renew it if necessary.
8. Examine and clean the orifice and valve seating. Check for burrs and renew if damaged or worn. Avoid the use of abrasives on valves or seatings.
9. Valve spindles which run in guides should be lightly greased with a silicone grease. Any levers or fulcrums should work freely.
10. On HP regulators, clean and examine relief valves and seatings. Check that the vent is clear.
11. Check and clean the breather hole. This acts as a damping device and its size is critical. Never enlarge the hole.
12. Reassemble the parts in reverse order.
13. Check the regulator for leakage.
14. Recommission the regulator.

REGULATOR FAULTS

Regulators are relatively simple devices and are subject to comparatively few faults. The most common faults are as follows.

Gas escaping from breather hole

This generally indicates a perforated diaphragm. It might mean that the valve spindle has become loose and gas is passing through the central hole, but this is not a common occurrence and usually the escape is due to a split or punctured diaphragm. To remedy, turn off the gas. Change the diaphragm.

Faulty regulating

Pressure too high: The valve is not shutting down onto its seating. This is probably due to dirt on the valve or seating but it could be caused by a faulty diaphragm or the valve becoming loose on its spindle.

Pressure too low: May be due to a broken or fatigued pressure spring. It could also be due to a stiff diaphragm or to rust or dust deposits in the inlet supply or under the valve itself.

Regulator not responding: A blocked breather hole will prevent the diaphragm from moving either up or down so the regulator pressure will be either high or low depending on the inlet pressure or the gas rate prevailing.

'Chattering'

This is a noisy vibration caused by rapid up and down movement of the valve and diaphragm. It is set up when the diaphragm responds quickly to a surge of pressure. The movement causes the valve to hit the seating and bounce off again, so creating another pressure surge which repeats the process. It is usually due to the breather hole having become enlarged and can be remedied by carefully knocking the hole up until the reduction in area provides adequate damping.

'Hunting'

Hunting is a condition somewhat similar to chattering where the outlet pressure fluctuates up and down, swinging above and below the pressure required. It is confined to larger installations and occurs when one regulator reacts to another. The remedy is to alter the settings or provide some means of damping the oscillations.

Regulator not passing gas

It may sometimes happen that, after a regulator has been 'locked-up' for a period, the valve may become stuck to the seating and the force of the weight or spring is insufficient to free it. This is due to sticky deposits on the valve or seating which should be cleaned with a suitable solvent.

Chapter | eight

Measurement of Gas

Chapter 8 is based on an original draft prepared by Mr A. Alexander

THE GAS LAWS

The word 'laws' in this case refers to the 'law of nature' or the physical 'laws' and not to man-made civil or criminal legislation.

You already know that the volume of a gas is affected by changes in pressure and changes in temperature and that it is necessary to use standard reference conditions (src) before one volume can be compared with another. So we must look in more detail at the physical laws which link volume, pressure and temperature before going on to deal with the actual measurement of volume.

Boyle's law

Robert Boyle carried out experiments to discover the relationship between volume and pressure. He used the apparatus shown in Fig. 8.1.

This consists of a glass tube, bent in the form of a U and with the shorter limb closed at the top. Mercury is poured in until the level of liquid is the same in each limb. At this point the pressure of the air on the surface of the liquid must be the same in each limb and is, in fact, atmospheric pressure.

As more mercury is added, the level rises in the shorter limb, so compressing the air. But the level in the longer limb rises by a much greater amount and the difference in levels indicates the pressures being applied. By taking readings of the values of P (pressure) and V (volume) their relationship can be determined.

The graph in Fig. 8.2 shows typical results. When the absolute pressure, P, is plotted against the volume V, you can see that P multiplied by V is always the same.

when	P is 20, V is 2	$20 \times 2 = 40$
	P is 10, V is 4	$10 \times 4 = 40$
	P is 5, V is 8	$5 \times 8 = 40$
	P is 2, V is 20	$2 \times 20 = 40$

So $PV = K$ where K is a constant.

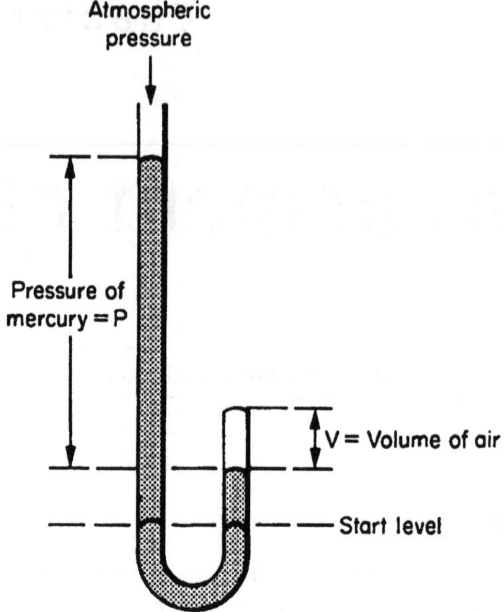

FIGURE 8.1 Boyle's Law apparatus.

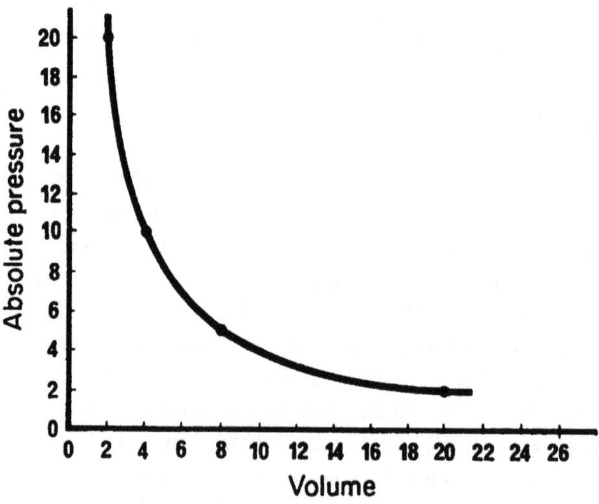

FIGURE 8.2 Graph showing variation of volume with pressure.

To put this into words, if the absolute pressure is increased four-fold, then the volume is reduced to a quarter.

Boyle's law is usually expressed as follows:

> The volume of a gas is inversely proportional to its absolute pressure, provided that the temperature remains constant.

This is written as

$$V \propto \frac{1}{P} \quad \text{or} \quad P \times V = 1 \times \mathrm{K}$$

So, again $PV = \mathrm{K}$, where K is a constant.

This means that

$$P_1V_1 = \mathrm{K}$$

where P_1 is the original pressure, V_1 is the original volume and

$$P_2V_2 = \mathrm{K}$$

where P_2 is a different pressure, V_2 is the corresponding volume.

Therefore

$$P_1V_1 = P_2V_2$$

since they are equal to the same constant.

This formula can be used to calculate the effect of changes in pressure on any volume of gas. When any three conditions are known, the fourth can be calculated.

It has been found that gases at high pressure do not behave in exactly the same way and the law is only approximately true. At ordinary operating pressures, however, the amount of error is negligible.

Charles' law

Professor Charles investigated the effect of temperature on gases. He discovered that all gases increase in volume by the same proportion when heated through the same temperature range. This proportion is 1/273 of their volume at 0 °C for each 1 °C rise in temperature above 0 °C. So you would need to increase the temperature from 0 to 273 °C to double the volume.

But −273 °C is absolute zero temperature. The temperature at which gases theoretically occupy no volume at all (Chapter 5, section on Temperature). So Charles' law may be expressed as:

> The volume of a gas is directly proportional to its absolute temperature, the pressure remaining constant.

$$V \propto T \quad \text{or} \quad \frac{V}{T} = \mathrm{K}$$

where V = volume, T = absolute temperature, K = a constant.

Therefore

$$\frac{V_1}{T_1} = \frac{V_2}{T_2}$$

for two different volumes of the same gas at different temperatures but at constant pressure. This formula enables calculations involving changes in volume and temperature to be carried out.

Because changes in conditions often involve both temperature and pressure at the same time the two laws are generally combined and most commonly expressed as:

$$\frac{P_1 V_1}{T_1} = \frac{P_2 V_2}{T_2}$$

This formula is used when comparing a volume of gas under one set of conditions with its volume under a different set of conditions. When volumes have to be compared they are corrected to src which are:

- 15 °C,
- 1013.25 mbar,
- dry.

A standard cubic metre, such as is used in defining the calorific value of the gas given by the symbol m^3(st) (Chapter 6, section on Standard Conditions).

PRACTICAL EFFECTS OF GAS LAWS

The effects of changes in temperature and pressure conditions on the gas measured in a customer's meter are shown in the following examples.

1. A customer's meter is fitted in an unheated garage or cellar and the gas then passes through the internal installation to the kitchen where it is used on the cooker. If the temperature in the garage is 4 °C, then

   $$T_1 = (273 + 4) = 277 \text{ K}$$

   If the temperature in the kitchen is 15 °C, then

   $$T_2 = (273 + 15) = 288 \text{ K}$$

   From Charles' law

   $$\frac{V_1}{T_1} = \frac{V_2}{T_2}$$

   Therefore

   $$V_2 = \frac{V_1 T_2}{T_1}$$

If $V_1 = 1\,m^3$, then

$$V_2 = \frac{1 \times 288}{277} = 1.04\,m^3$$

So $1\,m^3$ at the meter becomes $1.04\,m^3$ in the kitchen. This is an increase of 0.04 in every $1 m^3$ or

$$\frac{0.04 \times 100}{1} = 4\%$$

The customer gets $4\,m^3$ free in every $100\ m^3$ measured!

2. The pressure is reduced from 20 mbar at the meter to 6 mbar at the appliance. So the absolute pressures are:

$$P_1 = (1013 + 20) = 1033 \text{ mbar}$$

$$P_2 = (1013 + 6) = 1019 \text{ mbar}$$

From Boyle's law $P_1V_1 = P_2V_2$
Therefore

$$V_2 = \frac{P_1V_1}{P_2}$$

If $V_1 = 1\,m^3$, then

$$V_2 = \frac{1033 \times 1}{1019} = 1.014\,m^3$$

So $1\,m^3$ at the metre becomes $1.014\,m^3$ at the appliance. This is an increase of 0.014 in every $1 m^3$ or

$$\frac{0.014 \times 100}{1} = 1.4\%$$

The customer gets another $1.4\,m^3$ free in every $100\,m^3$ measured.

3. A further point which concerns the service engineer is the effect of heat on the gas passing through a burner.

   As an appliance heats up, the burner becomes hot and so raises the temperature of the gas passing through it. The gas expands in the burner so that the same heat value is contained in a larger volume. So the calorific value of the gas is, in effect, reduced. And the heat input rate to the appliance is reduced in the same proportion, often by about 10%.

Because of this it is usual to allow the appliance to heat up before checking the gas rate. Some manufacturers quote both hot and cold gas rates in their installation and servicing instructions.

MEASUREMENT METHODS

There are two basic methods of measuring gas, positive displacement or inference. Inferential meters are dealt with in more detail in Volume 3, Chapter 2, but a survey of the principles employed is useful at this stage.

INFERENTIAL METERS (INDUSTRIAL AND COMMERCIAL USE)

Two inferential meters have already been mentioned in Chapter 6, section on Flow through an Orifice. They are the orifice plate and the venturi meter. One other device which makes use of the same principle is the pitot (pronounced pee-toe) tube (Fig. 8.3). This measures the kinetic head by measuring the impact pressure A, on a tube pointing into the path of the gas flowing and also the static pressure B, on the side of the nozzle. Since the kinetic head is proportional to the velocity of gas flowing, the volume can be calculated for different pipe sizes.

Other types of inferential meter use the velocity of the gas to drive turbines, impellers or vanes.

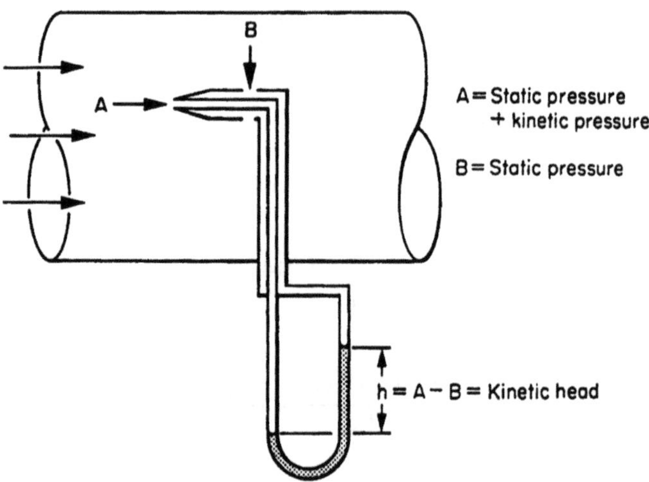

FIGURE 8.3 Pitot tube.

ROTARY DISPLACEMENT METER (ROOTS TYPE)

This meter has two impellers rotating in opposite directions inside a casing. A sectional view is shown in Fig. 8.4. The inlet is on the left and the outlet on the right. The top impeller rotates in a clockwise direction and the bottom impeller rotates anticlockwise. The four positions show the method of operation.

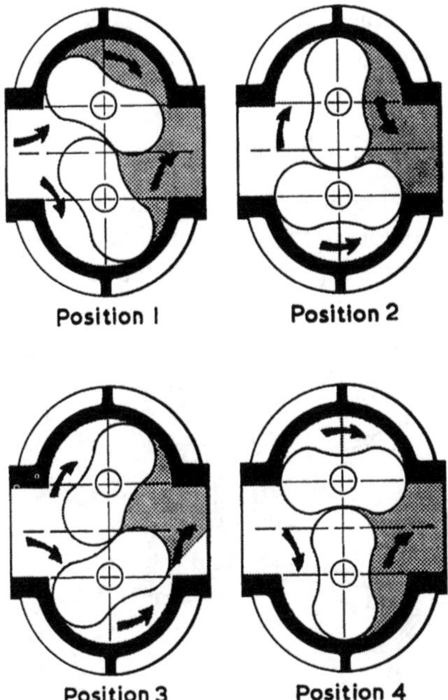

FIGURE 8.4 Rotary displacement meter cycle of operation.

Position 1

As the bottom impeller turns anticlockwise, gas enters the space between the impeller and the casing.

Position 2

At the horizontal position a measured volume of gas is contained between the impeller and the casing.

Position 3

As the impeller continues to turn the volume of gas passes to the outlet.

Position 4

The top impeller has been rotating in the opposite direction and has reached its horizontal position, so confining another equal measured volume of gas.

The process is repeated four times for each complete revolution of the impeller shafts.

Although this meter actually measures gas by displacement of volumes, it was classed as an inferential meter because it did not maintain its accuracy when passing small volumes. Since 1956, it has been stamped by the Department of Trade and Industry for use in industrial applications.

AXIAL FLOW TURBINE METER

In this meter the gas flow impinges on the specially shaped blades of a turbine and is streamlined by the contour of the casing on either side of the turbine (Fig. 8.5).

The speed of the turbine is proportional to the velocity of the gas passing and

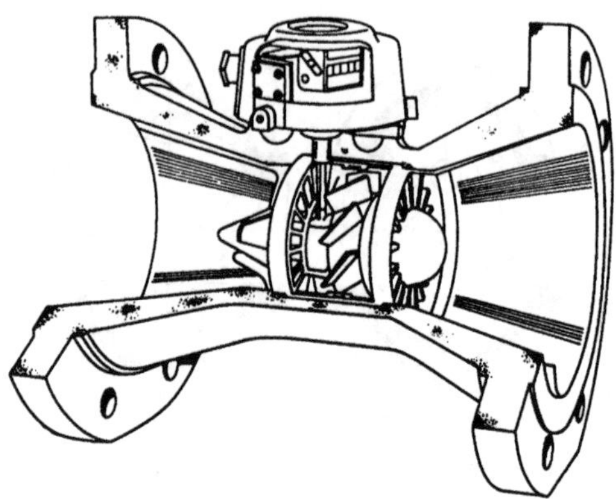

FIGURE 8.5 Turbine meter.

the area of the bore of the meter is known. So the volume passing is proportional to the speed. The revolutions of the turbine are conveyed through gearing to the index counter which is calibrated directly to read volumes of gas.

THORPE ROTARY METER

Gas passes through a series of circular ports and is directed on to the vanes of a vertical anemometer. The anemometer is rotated by the impingement of the gas on the vanes and its speed is proportional to the velocity of the gas passing through the ports. The area of the ports is known and the index, which is driven by the anemometer, converts the revolutions into volumetric units (Fig. 8.6).

POSITIVE DISPLACEMENT METERS (DOMESTIC USE)

The Gas Act 1986 requires gas sold to domestic customers to be measured by positive displacement. Originally customers were charged for the number of gas lights, used over a set period of time. When meters were introduced in the early nineteenth century, their sizes and connections were also based on the number of lights in the premises, e.g., 2, 3, 5, 10 lights, etc. Each gas light was estimated to use 6 ft^3/h (0.17 m^3) of gas. Therefore the 5 light meter was designed to pass 30 ft^3/h (0.85 m^3) and had 3/4 in. (19 mm) connections. The 10 light meter passed 60 ft^3/h (1.7 m^3) and had 1 in. (25 mm) connections. Over the years, with

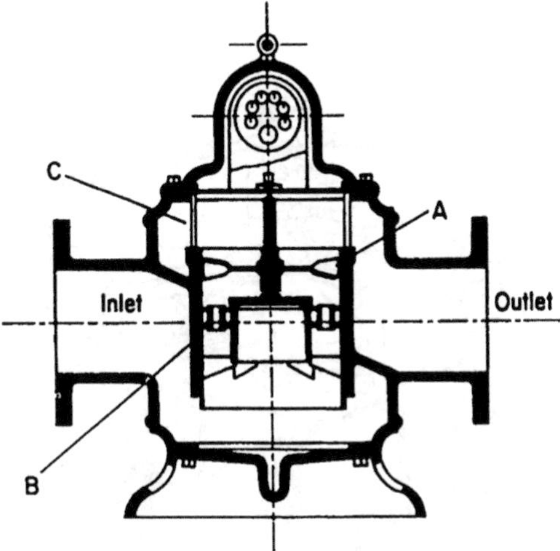

FIGURE 8.6 Rotary meter. A: anemometer vanes at 45°, B: porcelain ports, and C: removable measuring unit.

the changes in the use of gas from lighting to cooking, space heating etc. the consumption of gas by domestic, commercial and industrial customers has increased. The development of the gas meter has matched this increase by improvements in the internal mechanism, resulting in increases in capacity while decreasing the size of the meter case (see Fig. 8.7).

Note: The first meter developed was called a 'wet' meter, so named because it contained a drum, divided into measuring compartments, located in a cylindrical tank containing water. It was subsequently replaced by a meter using diaphragms; the new meters were usually referred to as 'dry' meters.

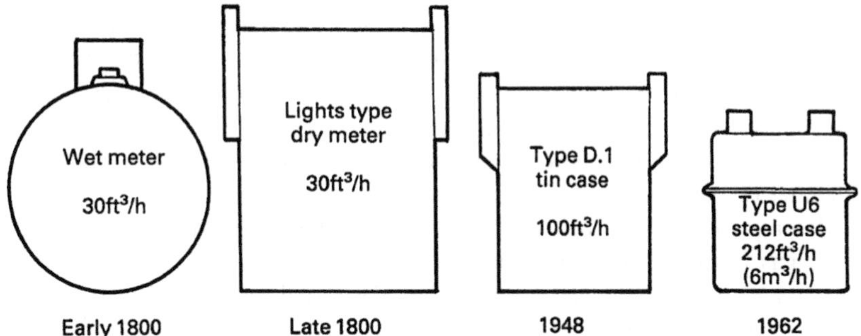

FIGURE 8.7 Comparisons of sizes and capacities of meters.

PRINCIPLES OF OPERATION

The positive displacement gas meter uses diaphragms that are alternately inflated and deflated by the pressure of the gas. The movement of the diaphragms is linked by 'flags' and 'flag-rods' to the slide valves which control the passage of the gas into and out of the four measuring compartments. Figure 8.8

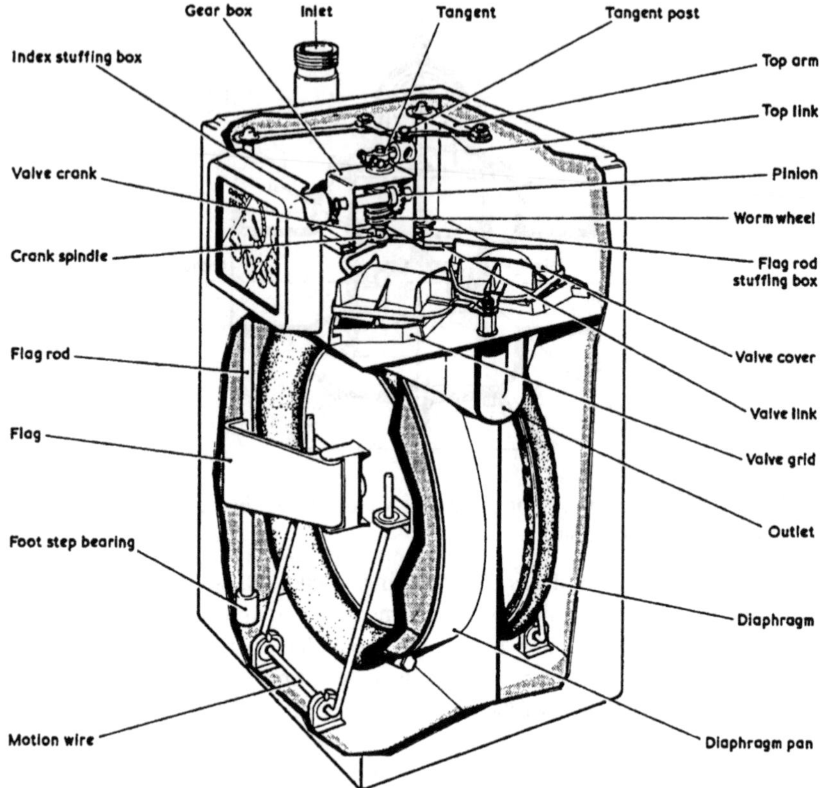

FIGURE 8.8 Construction of tin case meter.

shows the principle of a 'tin-case' meter, so called because the meter is constructed of sheets of tin-plate soldered together. The meter case is divided into three main compartments. The top section, sometimes called the 'attic', is partitioned off by the valve plate. It houses the slide valves and their seatings, together with the gearing and the meter index. The lower part is divided by a central vertical plate into front and rear compartments, each containing a leather diaphragm attached to a metal disc.

Inlet gas is fed into the top compartment through the left-hand connection. Then gas passes to the outlet through channels fixed to the underside of the valve plate. The same basic principles are described in detail after the section dealing with unit construction meters.

DEVELOPMENT OF THE DRY METER

Major design changes have included the introduction of the phenolic, self-lubricating valve and dry lubrication at all bearing points by means of acetal resin plastic bearings. Cases have been strengthened to allow meters to be fitted to rigid pipework.

The 'D' range of meters, later called the 'P' range and extended to include small industrial meters, was introduced to a specification of the Institution of Gas Engineers. The original details are shown in Table 8.1. The capacity of the meter was further increased so that the D.1. with 1¼ in. (20 lt.) connections was up-badged to 200 ft^3/h and the D.2. to 300 ft^3/h.

Table 8.1 IGE Meters

IGE Designation	Badged Capacity (ft^3/h)	Volume Per Revolution (ft^3/h)	Size of Connections
D.1	100	0.1	¾ in. (5 light) 1 in. (10 light)
D.2	200	0.2	1 in. (10 light) 1¼ in. (20 light)
D.4	400	0.4	1¼ in. (20 light) 1½ in. (50 light)

A later development is the introduction of the 'unit construction meter', shown in Fig. 8.9. Its principle of operation is identical to that of the tin-plate meter, which it is replacing, but it differs in construction (Fig. 8.10). It has a pressed steel, two-part case and the internal measuring mechanism is built as a complete unit which can be removed in one piece. This makes it easy to repair and reduces costs. It is structurally stronger and has a greater resistance to fire.

A further improvement is in the adjustment for accuracy of registration. In earlier meters this was carried out by varying the distance of the tangent post from the pivot of the tangent arm. Moving it further out allowed the diaphragms more travel, so the meter passed more gas in each revolution. Moving it in had the reverse effect (Fig. 8.8).

In recent designs the unit has a fixed-stroke tangent, and adjustment to the registration is by changing gear wheels behind the index. This can be done without dismantling the case.

The meter shown in Fig. 8.10 has its flag-arms and tangent situated below the valve plate, and the unit in Fig. 8.11 has its valve plate cut away to show the flag-arms and linkages. Some manufacturers still fit them above, as before.

The diaphragms are made of synthetic fabric called reinforced nitro-rubber. A standard version of the meter, known as the U6, is referred to in the British Standard Specification 4161: Part 3: 1989: Unit construction meters of 6 m^3 (or 212 ft^3) per hour rating. A picture of the U6 meter appears in Fig. 8.12. This meter has proved so successful that the range has been expanded to include larger models; details are given in Table 8.2 and Fig. 8.13.

Figure 8.14 shows the operational cycle of a diaphragm meter. It is easier to study the diaphragm and valve movement in one compartment first and then follow the same cycle through the second compartment.

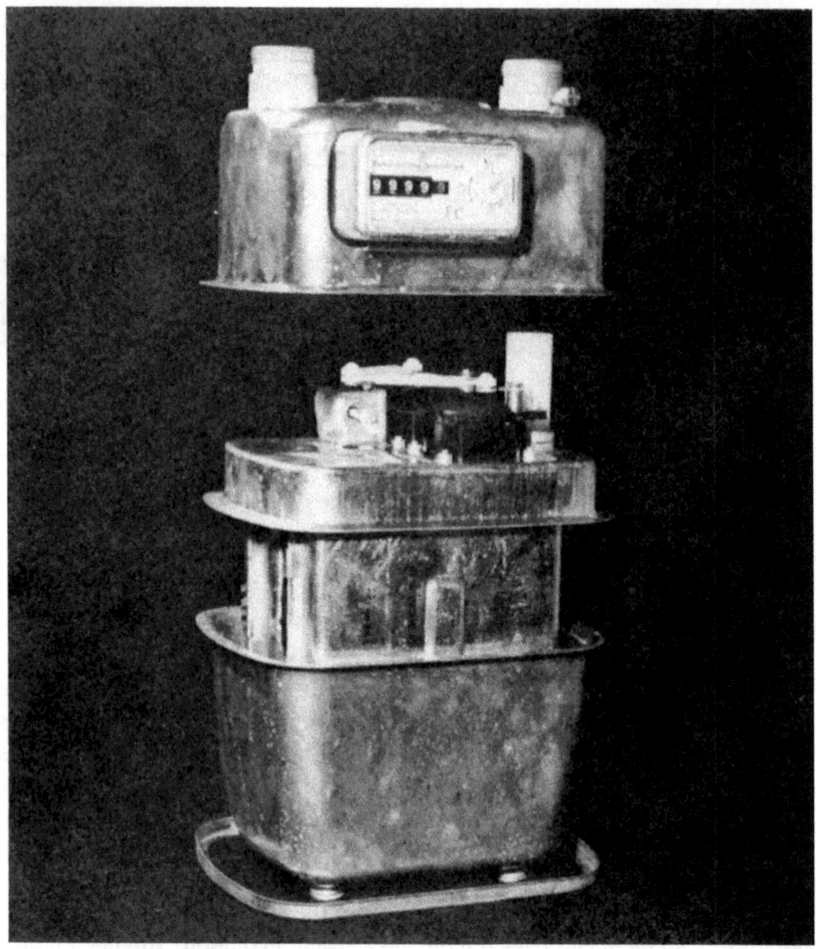

FIGURE 8.9 Unit construction meter.

The cycle in each compartment begins when the other is halfway through its travel. This ensures that there is no dead-centre in the mechanism and the meter will always restart after gas has been turned off.

THE METER INDEX

Modern meters have a digital or 'direct reading' index, similar to those shown in Figs 8.15 and 8.16, which simplifies the process of meter reading.

Reading a meter index

Figure 8.15 shows an index registering in cubic feet, and the gas consumption is indicated by the first four white numerals only, reading from left to right. They are shown surrounded by a thick black band. In the picture, the first four

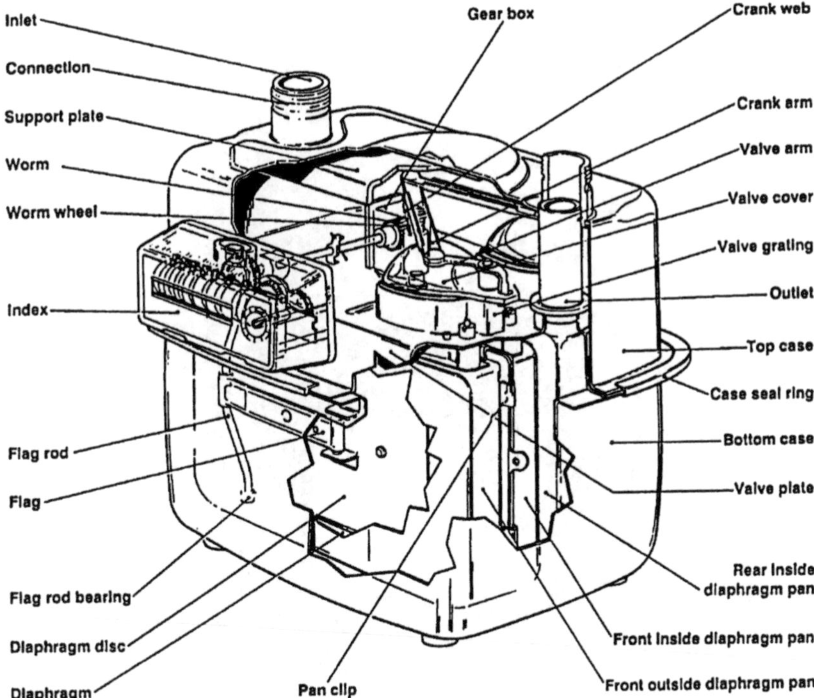

FIGURE 8.10 Construction of unit construction meter.

FIGURE 8.11 Internal construction of unit construction meter.

FIGURE 8.12 U6 meter.

numerals are followed by two more 0s, one on a dial and one on the index plate. This is because the index indicates hundreds of cubic feet and two 0s must be added after the four figures to show the correct amount of gas.

The second index in Fig. 8.16 registers cubic metres and here the first five digits give the quantity up as far as the decimal point. They are again surrounded by a thick black band. The last three digits represent the number of decimetres. The last digit is the test drum and has a further five sub-divisions between each unit division, representing 0.2 dm^3 per sub-division.

Both the indices shown indicate zero readings.

Reading an older type meter index

Older meters use the centre-pointer type index, shown in Fig. 8.18. Reading this index is a little more complicated but, again, the meter test dial and the dial showing tens of cubic feet are ignored. The lower four dials are read from left to right, and two noughts added to the answer to give the number of cubic feet.

When reading these dials look carefully at the position of the pointer and note the lower of the two figures between which it is situated. Try reading the index in Fig. 8.17 using this rule.

Take the left-hand dial. The pointer is between 1 and 2. So the lower figure is 1. On the next dial the pointer is between 3 and 4. So the lower figure is 3. The next is

Table 8.2 'U' Series Unit Construction Meters

Designation	Badged Rating		Type	Connections	
	m³/h	ft³/h		Nominal Bore	Centres
U4	4	141	Threaded (BS 746)	25 mm (1 in.)	110 mm
U6	6	212	"	"	152 mm
U10	10	353	"	32 mm (1¼ in.)	250 mm
U16	16	565	"	"	152 mm
U25	25	883	"	50 mm (2 in.)	250 mm
U40	40	1412	"	"	280 mm
U65	65	2295	Flanged (BS 4504)	65 mm (2½ in.)	335 mm
U100	100	3530	"	80 mm (3 in.)	430 mm
U160	160	5650	"	100 mm (4 in.)	"

A physically smaller meter with a capacity of 2.5 m³/h and able to operate at a higher working pressure is available for small LPG installations (Fig. 8.13).

between 8 and 9. So take the 8. On the last, the pointer is between 2 and 3. So the lower is 2. That makes the index 1382 and the amount of gas indicated is 138,200 ft³.

The difficulty arises when a pointer appears to be exactly on a number, as in Fig. 8.18. Take the left-hand dial as an example.

The pointer appears to be on 0. But is it? It could be either between 0 and 1 – in which case you read it as 0. Or it could be between 9 and 0 – in which case you read it as 9, because the 0 now means 10.

The way to check is to look at the next dial on the right. If this pointer has gone past its 0, then so has the first one.

With the index in Fig. 8.18 this still does not help. The second and also the third dials are just as ambiguous. So you have to keep going to the right. The last dial has its pointer between 0 and 9. So it has not completed its revolution and it reads 9. That means that the dial on its left has also not completed its revolution and also reads 9. And the same applies to the other two.

The index shown in Fig. 8.18 is 9999 – not zero, as you might have thought. In fact there are another 20 ft³ to go through before the index is 0000.

R5 index

New meters now being installed are fitted with a different type of index called the R5 (Fig. 8.19).

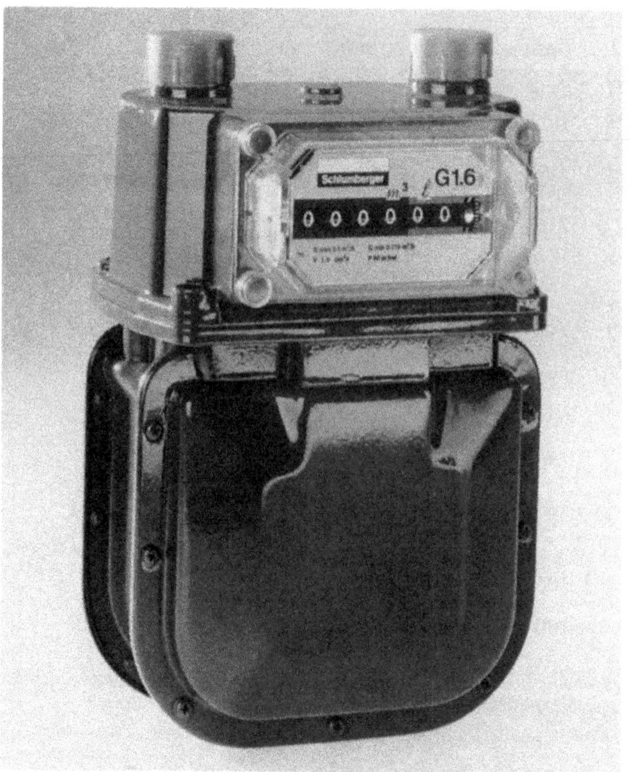

FIGURE 8.13 Meter for small LPG installation.

Whilst the meter is the current steel-cased diaphragm type, the index has an output facility incorporated, in readiness for the introduction of an Automated Reading System (AMR).

Inside the index, a magnet rotates on one of the gear wheels, this in turn activates a reed switch and produces a pulse. The pulse is transmitted to a circuit, terminating at an outlet socket located underneath the index casing. On U6 to U16 meters the index is fitted with a six pin telecom socket (Fig. 8.20), whilst on non-domestic U25 to U60 meters the socket is a four pin Fischer type (Fig. 8.21). All indexes are clearly marked R5.

The index housing has an anti-tamper rear plate and has a facility to detect tampering, by lifting or by the use of a magnetic field. The meters are supplied by the manufacturers with a security label covering the socket outlet (Fig. 8.22), to protect against damage to the outlet and safeguard against tampering.

THE SOLID STATE ULTRASONIC GAS METER

The U6 meter in use today is a development of diaphragm meters used in the 1800s. The principle of the meter is the same. It is now smaller, passes more gas, uses new materials for valves, bearings and diaphragms and has a stronger case but it is still a mechanical diaphragm type positive displacement meter. In the

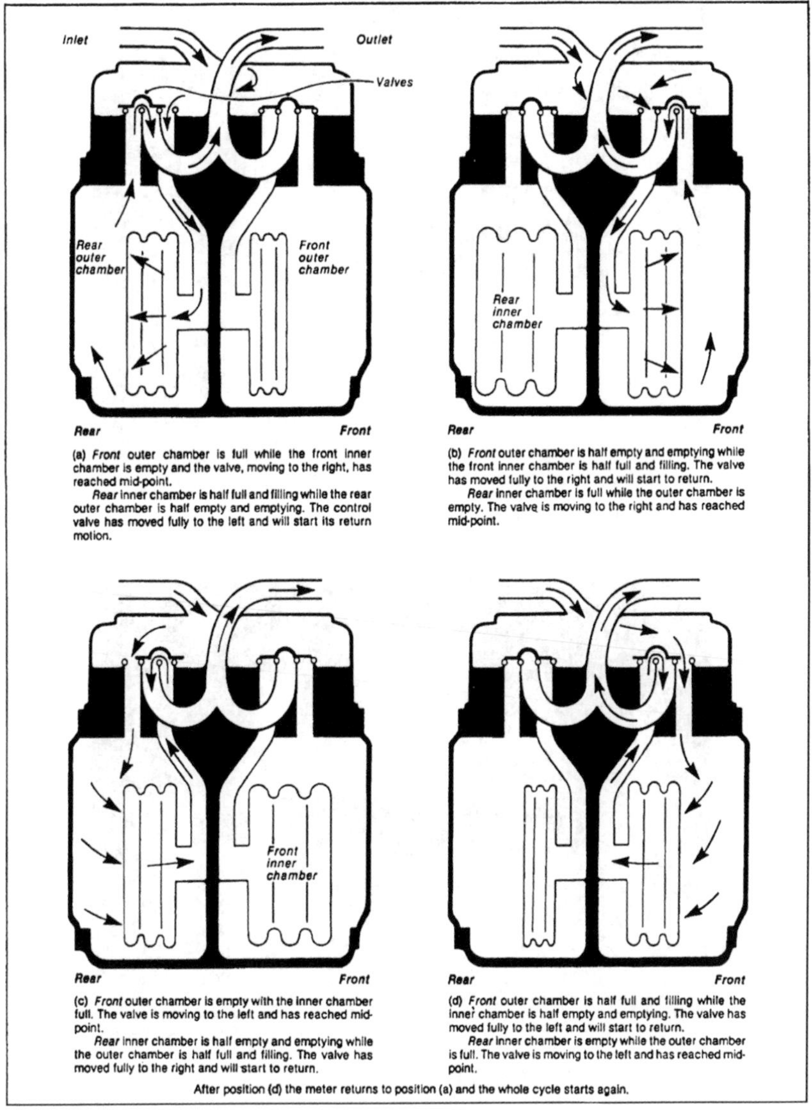

(a) *Front* outer chamber is full while the front inner chamber is empty and the valve, moving to the right, has reached mid-point.

Rear inner chamber is half full and filling while the rear outer chamber is half empty and emptying. The control valve has moved fully to the left and will start its return motion.

(b) *Front* outer chamber is half empty and emptying while the front inner chamber is half full and filling. The valve has moved fully to the right and will start to return.

Rear inner chamber is full while the outer chamber is empty. The valve is moving to the right and has reached mid-point.

(c) *Front* outer chamber is empty with the inner chamber full. The valve is moving to the left and has reached mid-point.

Rear inner chamber is half empty and emptying while the outer chamber is half full and filling. The valve has moved fully to the right and will start to return.

(d) *Front* outer chamber is half full and filling while the inner chamber is half empty and emptying. The valve has moved fully to the left and will start to return.

Rear inner chamber is empty while the outer chamber is full. The valve is moving to the left and has reached mid-point.

After position (d) the meter returns to position (a) and the whole cycle starts again.

FIGURE 8.14 The operational cycle of a typical diaphragm meter.

twenty-first century, advances in electronic technology have surpassed anything anticipated just a few years ago.

Ultrasonic technology has been used for a number of years to measure the flow of liquids. However, further research was required to measure the flow of gas. In a five-year project, British Gas plc in partnership with two manufacturers has produced the world's first electronic domestic gas meter. It is a solid state ultrasonic gas meter (Fig. 8.23) which will measure accurately regardless of the gas composition. It is designed to pass 6 m^3/h and is designated E6.

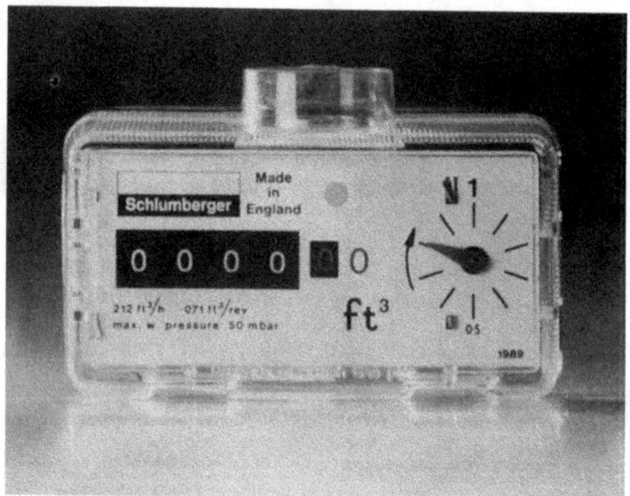

FIGURE 8.15 Index registering in cubic feet.

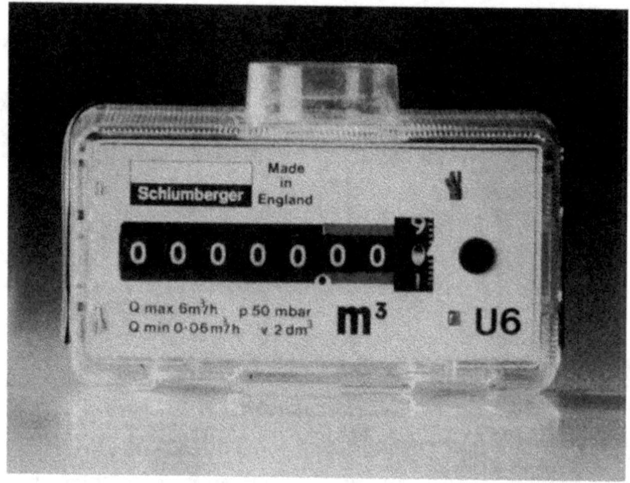

FIGURE 8.16 Index registering in cubic metres.

The Gas Act 1986 requires gas sold to domestic customers to be measured by positive displacement. The ultrasonic meter has no moving parts and is in fact an inferential meter. To accommodate this change a new Statutory Instrument has been issued – The Gas (Meters) (Amendment) Regulations 1993.

PRINCIPLES OF ULTRASONIC MEASUREMENT

Figure 8.24 shows the heart of the ultrasonic meter. It is a measuring tube fitted with two ultrasonic transducers. (A transducer is a device which will convert an electrical signal into a mechanical signal or a mechanical signal into an electrical one. It acts as both a transmitter and receiver.)

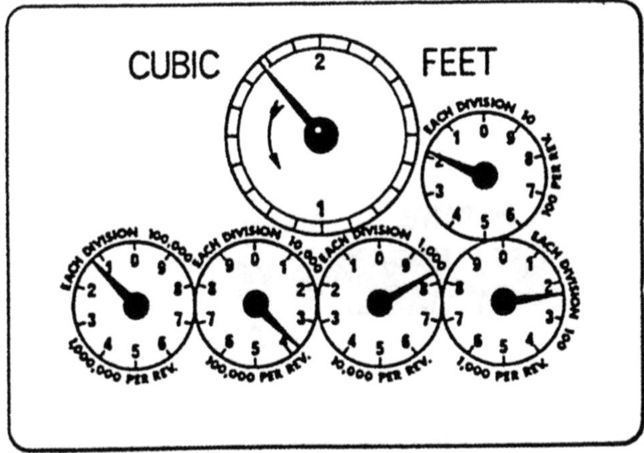

FIGURE 8.17 Dial-type meter index.

FIGURE 8.18 Older type meter index.

The transducers in the measuring tube convert electrical energy (from a battery in the meter) into sound waves. When gas flows, impulses emitted from the transducers travel in the direction of gas flow and alternatively in the opposite direction of gas flow. Their transmission times are measured, the time of flight differences calculated and the volume of gas consumption accurately displayed on an Liquid Crystal Display (LCD) readout on the meter index.

CONSTRUCTION OF THE ULTRASONIC GAS METER

The meter shown in Fig. 8.23 has a mild steel case and an injection-moulded plastic front panel. Its case dimensions are 225 mm wide, 170 mm high and 100 mm deep. The 3.6 volt 'D' cell vented battery is housed at the bottom right-hand side of the front panel.

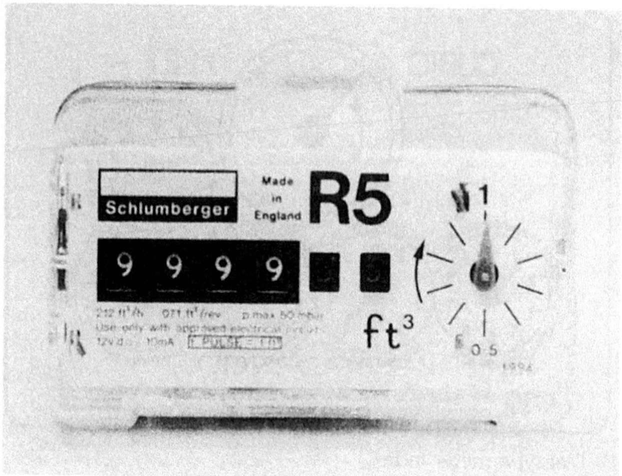

FIGURE 8.19 R5 index.

FIGURE 8.20 Six-pin socket.

The meter is of a modular construction to accommodate future developments such as integrated regulator, remote control of gas flow and remote interrogation by low power radio. The other manufacturer's meter is only about the size of a housebrick. On both meters, inlet and outlet connections are of the same dimensions and distances from each other as on the U6 meter. Figure 8.25 shows the physical differences between the three meters.

The ultrasonic meter index, Fig. 8.26, gives eight digit registration readings, five before the decimal point and three after. This is followed by a further reading which gives a diagnostic code.

Note: Meters supplied will have only one diagnostic code display, not two as shown.

FIGURE 8.21 Four-pin socket.

FIGURE 8.22 Security label across outlet socket.

Situated on the right of the index is an optical communication port. This is a glass fronted Light Emitting Diode (LED) display onto which information held in the meter is fed. Information can be transmitted into or out of the meter from this point. The optical communication port has a standard magnetic coupling which enables communication to be established between the ultrasonic meter and a handheld unit such as the Itron or Micronic already used for meter reading (Fig. 8.27).

ADVANTAGES OF THE ULTRASONIC METER

1. Unaffected by varying air moisture or temperature.
2. High resistance to dust.
3. Resists tampering.

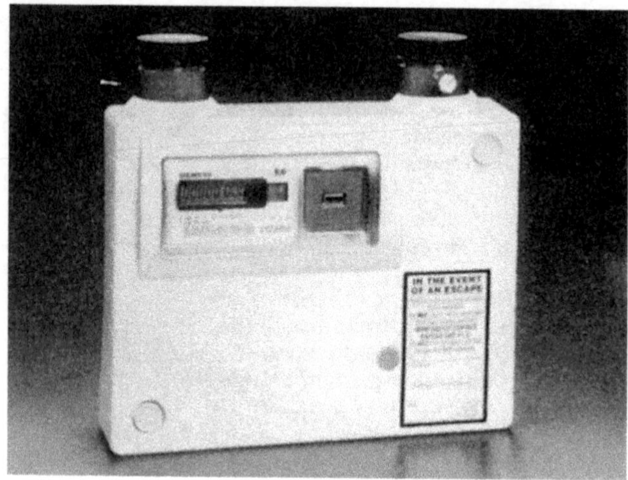

FIGURE 8.23 Solid state ultrasonic gas meter.

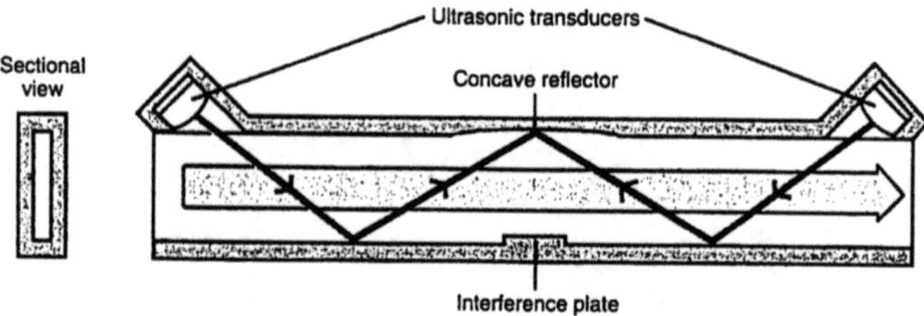

FIGURE 8.24 Cross-section of measuring tube.

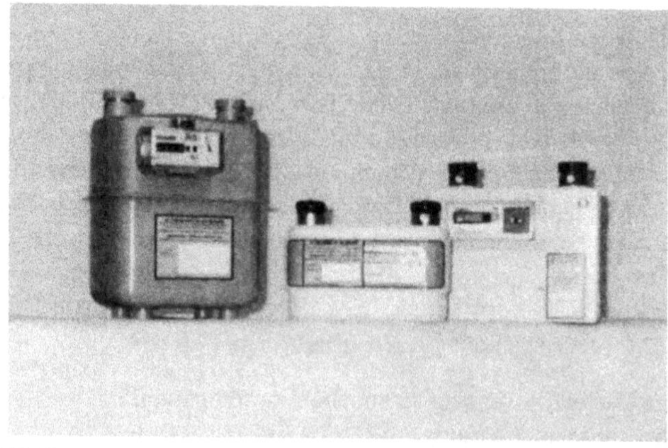

FIGURE 8.25 Comparison between U6 and new ultrasonic E6 meters.

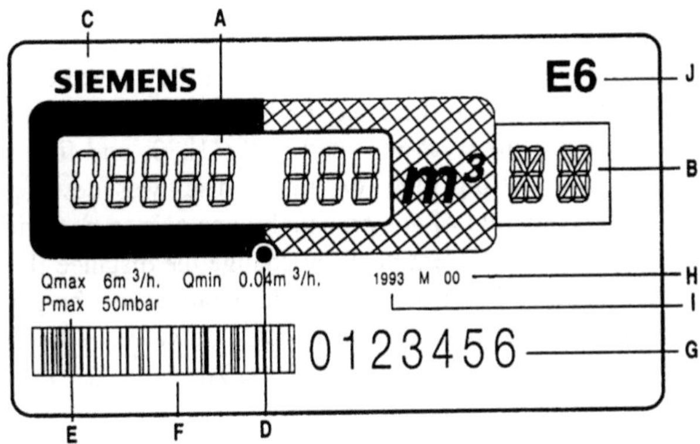

A : Meter registers
B : Diagnostic codes
C : Manufacturer
D : Decimal point indicator
E : Operational details
F : Bar code details
G : Serial number
H : Model number
I : Year of Manufacture
J : Meter type

FIGURE 8.26 Ultrasonic meter index.

FIGURE 8.27 Handheld controller.

4. Fraud detection circuitry.
5. LCD Display and Status.
6. Easily reprogrammed.
7. Can be interrogated by handheld controller.
8. Modular approach, for future use of modules.
9. Ten-year battery life.
10. More compact than U6 diaphragm meter.

The only limitation with these units is the fact that operation is via a ten-year battery.

Installation and servicing of the meter are discussed in Volume 2.

PREPAYMENT METERS

These meters are no longer in use as primary meters they have been replaced by the credit card meter, however, they may be found as secondary meters.

Prepayment attachments were invented in the 1880s and first used on wet meters. Later the coin setting plate or 'rating disc' was developed by R.T. Glover and allows the amount of gas prepaid by one coin to be varied. Rotating the disc and fixing it in a different position change the distance through which the coin can be rotated and so alter the amount charged for the gas.

Basically a prepayment mechanism consists of a valve, in the outlet of the meter, which is operated through a train of gears. The valve gearing is turned by two shafts:

(1) the coin mechanism,
(2) the meter shaft, sometimes called the 'burning-off' shaft.

Inserting a coin and turning the handle operates the coin shaft and opens the valve. Gas can then flow through the meter.

As the gas is used the meter mechanism rotates and turns the burning-off shaft, moving the prepayment gearing in the opposite direction and so closing the valve. The valve opens by an amount in proportion to the number of coins inserted.

Figure 8.28 shows the construction of a U6 prepayment meter and Fig. 8.29 shows a U6 meter with a £1 coin attachment.

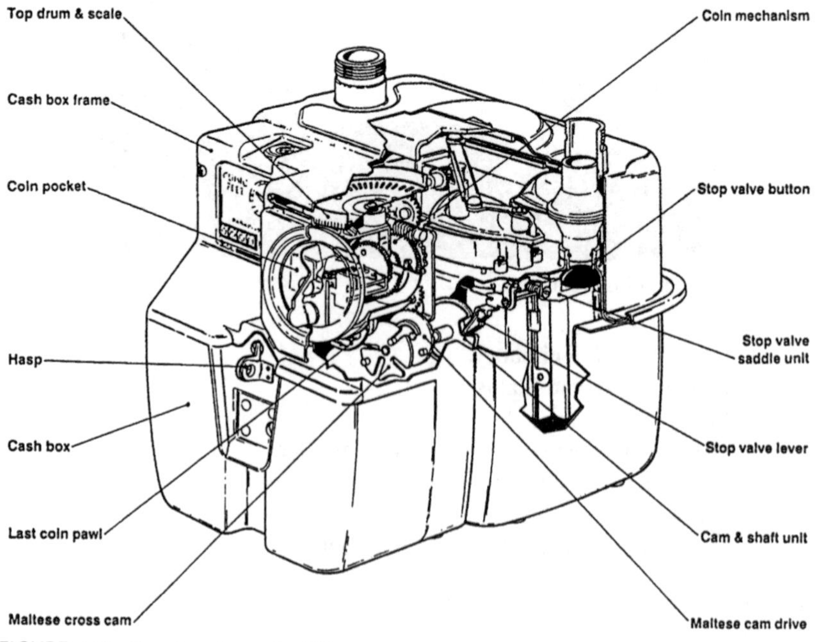

FIGURE 8.28 Construction of a U6 prepayment meter.

FIGURE 8.29 U6 meter with £1 coinplate.

Another type is the prepayment 'token' meter (Fig. 8.30). These meters operate in the same way as a standard prepayment meter, the only difference being that they take plastic key tokens rather than coins. The token, worth £1 (Fig. 8.31), is inserted into the slot as far as it will go. It compresses against a spring and must then be given a quarter turn clockwise to break off the end of the token. The meter handle can then be turned in the normal way, causing the end of the token to fall into the meter cash box. Security was the prime reason for the introduction of this meter and it is also used as a means of recovering arrears without cutting off the gas supply.

THE QUANTUM METER

Once again, thanks to the advance in electronic technology, it has been possible to develop a new type of gas meter, this time a prepayment gas meter.

The Quantum meter (Fig. 8.32) utilises 'smart card' technology. 'Smart card' is the name given to a card which resembles a credit card but which holds a computer microchip capable of storing and relaying information. The meter has been developed to replace coin prepayment and plastic token meters. It is a conventional U6 meter with an electronic section fitted onto its front.

FIGURE 8.30 U6 prepayment token meter.

FIGURE 8.31 Coinplate and plastic key token.

THE QUANTUM SYSTEM

The electronic unit on the meter has a battery compartment at its base which takes a pack of five 'D' size batteries, this supplies power to the electronic circuitry. Above the compartment is a LCD readout panel which gives clear data to both customer and service engineer. To the right of the panel are two press

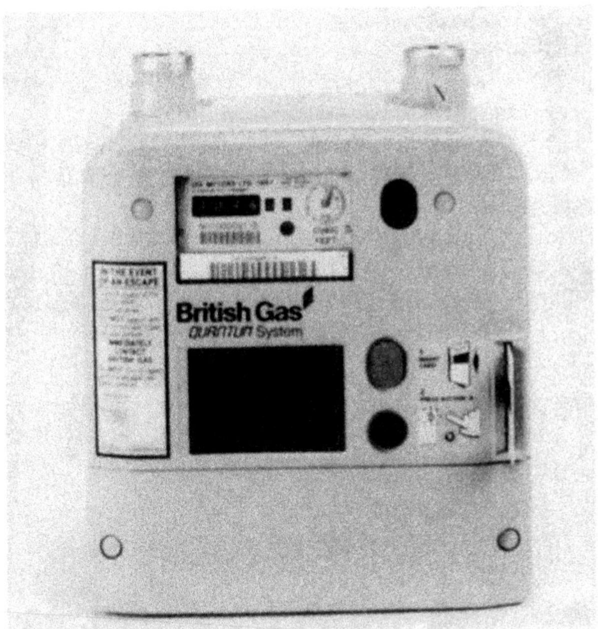

FIGURE 8.32 The Quantum meter.

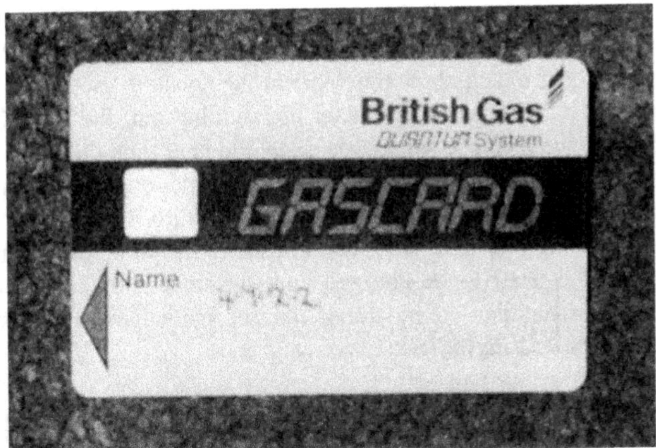

FIGURE 8.33 The Quantum system customer GASCARD.

buttons 'A' and 'B' and a card slot. The 'smart card' used by the customer is labelled 'GASCARD' (Fig. 8.33). The customers can pay for the gas they require by taking their card to a post office or other retail outlet displaying a 'GASCARD' sign and having the credit they pay for electronically inserted on their card by a charger unit (Fig. 8.34).

By inserting the card and pressing button 'A' the customer can start to use the credit available on the card. Should the credit run out, some emergency credit is

FIGURE 8.34 Quantum system charger unit.

available from the meter, until the customer can reach a 'GASCARD' point to pay for more credit. A budget option, available by pressing button 'B', allows customers to choose exactly how much credit they want to use in any particular period of time. If batteries in the meter are running out, the 'GASCARD' is programmed by the meter to send a message, the next time the card is inserted into a charger. The meter is equipped with anti-tamper devices which close the gas valve if an unauthorised attempt is made to disturb it.

When it is being installed, the Quantum meter has to be programmed with items such as the tariff to be charged and the amount of emergency credit available. To enable this to be done, the service engineer is issued with a 'SERVICE CARD', Fig. 8.35.

FIGURE 8.35 The service engineer's 'SERVICE CARD'.

METER REGULATIONS

Meters are the subject of statutory regulations. Any meter installed to measure gas which is being sold to a customer must conform to standards covered by the Gas Act and carry an official seal of approval. The seal is so placed as to prevent any unauthorised attempts to alter the registration or its adjustment.

The official body responsible for enforcing the regulations is the Gas and Oil Measurement Branch, of the Department of Trade and Industry, whose meter examiners are mostly present in the manufacturers' or repairers' premises. The examiners' job is to test the meters and safeguard the rights of the customer.

The main points to which the examiners pay particular attention are:

- accuracy of registration,
- pressure loss and mechanical pressure loss,
- noise,
- external and internal leakages.

The requirements for each of these are as follows.

Registration

The maximum gas flow for which a meter is designed is stated on the meter itself, often on a badge fixed to the case. This is called the 'badged rating' of the meter.

At rates of flow between this and 1/50th of the badged rating the accuracy of registration must be within ±2%. This means that the index may register up to 2% more than the actual amount of gas passed, or 2% less than the actual amount. Expressed in another way, the meter registration must not be more than 2% fast or 2% slow. A fast registration is in the supplier's favour, a slow registration favours the customer.

Pressure loss

The average pressure loss when air is passed through the meter at the badged rating must not exceed 2 mbar.

Mechanical pressure loss

The pressure loss when air is passed at 1% of the badged rating must not exceed 0.6 mbar.

Noise

The meter must be quiet in operation.

External leakage

The meter must not leak when subjected to a pressure of 50 mbar.

Internal leakage

The meter must register when air is passed through it at 14 dm^3/h (0.5 ft^3/h).

TEST CONDITIONS

Figure 8.36 shows a manufacturer's meter test rig. The equipment is similar to that used by the gas meter examiners. Registration and absorption tests are carried out using air at a pressure of 5 mbar.

FIGURE 8.36 Manufacturer's test rig.

Remembering Charles' law, it is important that the water in the test holder and the air in the room are at the same controlled temperature.

Although manufacturers test each meter individually, the Department of Trade and Industry examiners test only by a statistical sampling process. Should one sample from a batch fail the test, however, all the meters in the batch are tested until the quality standards are restored.

In addition the examiners can, at any time, inspect manufacturers' methods of production and testing.

METER FAULTS

The modern gas meter has been scientifically designed and, if handled carefully and installed in an approved manner, it should give years of trouble-free service.

Most faults which do occur can be rectified only by exchanging the meter. However, care taken when handling or fixing can prevent some faults from developing in the first place. Remember that a meter starts off as a very accurate measuring instrument and it is in everyone's interests that it stays that way! Some points to note are:

- Do not shake it about. Always keep it upright when handling. Otherwise the valves may be damaged and dirt can lodge between the valves and their seatings. Gas could then pass unregistered.
- Keep caps on the inlet and outlet. This keeps dirt out and prevents explosions with used meters.
- Make sure that the case does not get scratched or dented. Rust can start if the steel is exposed.
- Do not overtighten the connections. Too much force can cause leaks rather than prevent them.
- Never fit a meter in a persistently damp location or touching a damp surface. Rust causes leaks.
- If you have to wind back the prepayment mechanism, do not strain the valve. It may not open when the first coin is inserted if you do.
- Dirt on the prepayment valve can allow unpaid for gas to pass. On some meters it is possible for dirt falling down the outlet tube to settle on the valve.
- Make sure that pipes are clear of dirt or scale before connecting the meter.

The most common meter faults which may occur on the district are as follows:

1. *Leakage*

   May be due to the case rusting, damage to the case from articles stored around the meter or faults on the pipe joints or connections.

2. *Meter not registering*

   Usually due to faulty valves or diaphragms. It could also be due to internal corrosion on older meters. This is often termed 'Passing Unregistered Gas' or PUG for short.

3. *No gas*

   Meters can sometimes come to a dead stop. This is usually due to an internal fault in the mechanism, like a pin shearing.
   Prepayment meters may reach a condition when the prepaid gas is used up and it is not possible to insert any more coins because the coin box is full up. Bent or damaged coins can also jam the mechanism and have a similar effect.

Customers may also complain of 'no gas' when the prepayment valve has been strained and does not open when the first coin is inserted. These faults are usually dealt with by an emergency service.

4. *Noise*

   Older meters, which do not have dry lubrication, may squeak and groan under full flow when the oil has dried out of the bearings. The noise can be quite disconcerting and indicates that the meter should be exchanged. Domestic meters are no longer lubricated on the district.

5. *Faulty registration*

   The customer who suspects that his meter is not registering correctly may demand that it be tested. The meter will then be sent to the government testing station for examination. If it is found to be outside the permitted tolerances, the customer's account will be adjusted accordingly. If the meter is correct, the customer may be charged the cost of the testing. The registration of a meter can be checked on the district by comparing it against that of a special test meter which is fitted temporarily after the customer's meter. This is known as testing 'in situ', which means 'in its original position'. Testing in situ is used as a means of checking the accuracy of industrial meters.

GAS BILLS AND TARIFFS

Gas bills

Gas is sold on a thermal basis, the customer paying for the amount of heat energy used. Meters, however, measure the volume of gas which passes through them, so a calculation is necessary to find out how much heat energy has been consumed. As you saw in an earlier chapter, this is done by multiplying the volume of gas by its calorific value (the calorific value is stated on the back of the gas bill), thus:

$$\text{total amount of heat} = \text{volume of gas} \times \text{calorific value}$$

The units used must always be compatible. That is:

$$\text{MJ} = \text{m}^3 \times \text{MJ/m}^3$$

As explained previously, since April 1992 Gas companies have charged for gas based on the number of kilowatt-hour supplied (1 kWh = 3.6 MJ).

Therefore

$$\text{kWh} = \frac{\text{m}^3 \times \text{MJ/m}^3}{3.6}$$

At present, most gas meters register the volume of gas passed in 100s of cubic feet ($2.83\ \text{m}^3 = 100\ \text{ft}^3$).

Therefore

$$\text{kWh} = \frac{\text{ft}^3 \times 2.83 \times \text{MJ/m}^3}{3.6}$$

In 2001 a gas bill was calculated as follows:

Present meter reading:	2119 (hundreds)
Previous meter reading:	1854 (hundreds)
Volume of gas used:	265 (hundreds)

If the calorific value of the gas is 38.7 MJ/m^3, then:

$$\text{kWh} = \frac{265 \times 2.83 \times 38.7}{3.6}$$

$$\text{kWh} = 8061.9$$

Rounded down to the nearest kWh, 8061 kWh.

The cost per kilowatt-hour is known as the 'commodity charge' and for example is 1.477 pence.

The 'standing charge' which may vary in different parts of the country is, for example, 10.1 pence per day.

Value added tax has been charged at 5% in the UK since April 2001. The bill for an 83-day period would be:

Commodity charge	8061 × 1.477	=	119.06
Standing charge	83 × 10.1	=	8.38
Value added tax @ 5%		=	6.37
Total bill		=	£133.81

PREPAYMENT METERS

Gas meters available to domestic customers are either 'credit' or 'prepayment' meters. The credit meter, sometimes called an 'ordinary' meter, and the customer is presented with a gas bill at the end of each quarter. The bill is produced by a computer from the meter readings taken, so the customer pays for the gas after they have used it.

Prepayment meters are available to customers under certain circumstances and require a token or 'smartcard' which has to be inserted before gas will pass. So all the gas is paid for before it is used.

Chapter | nine

Basic Electricity

Chapter 9 is based on an original draft prepared by Mr K. Pomfrett

ELECTRICAL ENERGY

Electricity is a form of energy (Chapter 5, section on Forms of Energy), and it is produced from other forms of energy in a variety of ways.

Kinetic energy produces electricity when, in a generator or alternator, a coil is revolved in a magnetic field. Pressure on a suitable crystal (quartz, tourmaline, lead zirconate titanate) produces a spark by the 'piezo-electric' effect. Friction can cause particular materials to become charged with static electricity.

Heat energy can produce electricity when applied to a 'thermocouple' or junction of two dissimilar metals.

Light energy falling on a selenium cell or 'photo-electric' cell produces electricity.

Chemical energy in the form of dry batteries or primary or secondary cells is a common source of electricity.

All these 'sources' of electricity are in fact simply means of producing a flow of electrons within the material which forms the electric circuit.

CONDUCTORS

All substances are composed of atoms and all atoms have electrons revolving round them in a number of orbits (Chapter 2, section on Atoms). Each orbit or shell has a limit to the number of electrons it can contain. So some elements have only one electron in their outer shell. Because the single electron is much further away from the centre it has only a very small attracting force holding it to the nucleus. In a solid this electron is free to wander from one atom to the next and random movement of the free electrons takes place continually.

The materials which have this movement of free electrons are all good 'conductors' of electricity. Mostly they are metals and Fig. 9.1 shows the structure of the atoms of some good conductors. Gold and silver have one electron in their outer shell and copper and aluminium have two and three outer electrons, respectively.

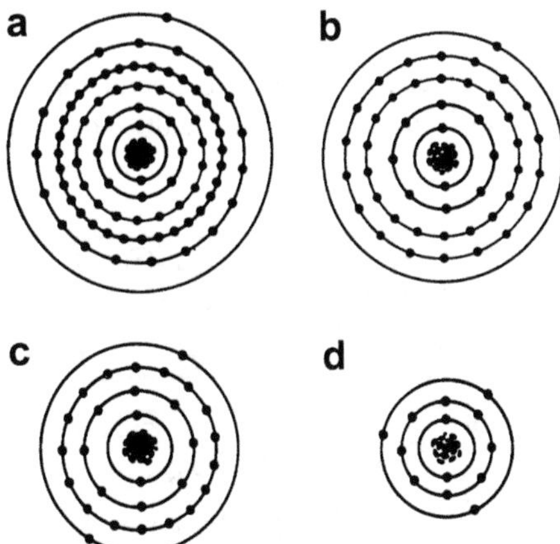

FIGURE 9.1 Structure of atoms. Representation of the electrons in their orbits: (a) gold – 2,8,18,32,18,1; (b) silver – 2,8,18,18,1; (c) copper – 2,8,17,2; and (d) aluminium – 2,8,3.

Silver is used for some applications, in spite of its high cost, whilst copper is the metal generally used as an electrical conductor. Aluminium is not as good a conductor but may be used where its low density (about one-third the weight of copper) gives it an advantage, in overhead power lines for example.

Even the best conductors offer some resistance to the flow of current but some metals, which have a high resistance, are used deliberately to reduce the amount of current flowing. Resistance wires may be made of alloys like nickel and copper ('Eureka' wire) or nickel and chromium ('Nichrome' wire) (Eureka and Nichrome are trade names of two typical resistance wires).

INSULATORS

An 'insulator' is a material made up of atoms, which have their outer shell completely full of electrons. In these atoms the force which attracts the electrons to the nucleus is very strong and there are no free electrons in the material. So an insulator cannot conduct electricity.

Insulators are non-metals, and may be solid, liquid or gas. The solids include rubber, paper, cotton and most plastics (bakelite, PVC, PE). Porcelain and mica are used where heat is present.

Liquids include oils and pure water. Since water usually contains a number of dissolved substances, ordinary tap water must be regarded as a conductor.

Gases have been used as insulators for some very specialised applications.

No insulator is completely non-conducting. Under normal conditions the amount of current passing is so small that it is difficult to measure. If, however, the electrical pressure is increased the insulator will eventually break down and a current will flow.

The point at which this happens is called the 'dielectric' strength of the material and it is expressed in volts per millimetre thickness.

ELECTRON FLOW

The movement of free electrons through a conductor is brought about by connecting the conductor in a closed circuit with a source of electrical energy (Fig. 9.2).

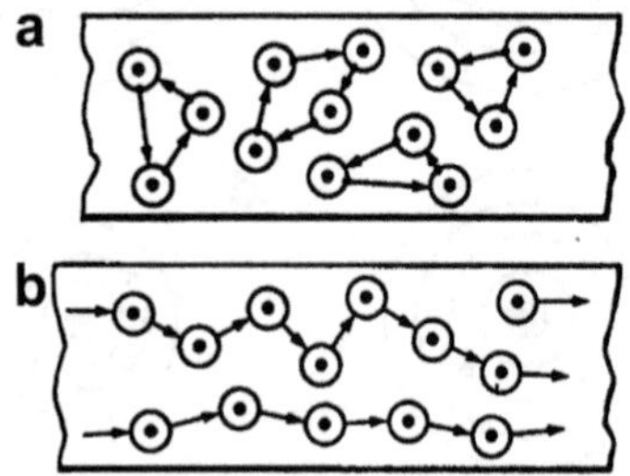

FIGURE 9.2 Movement of free electrons in a conductor: (a) no current and (b) current flowing.

Electrons obey the law of electrical charge. That is:

- like charges repel,
- unlike charges attract.

So electrons, which have a negative charge, will be repelled by another object having a negative charge and will move away. They will, however, be drawn towards an object with a positive charge. This is how a battery produces a movement of free electrons.

The 'positive' plate of a battery is so called because it has a positive charge. And the 'negative' plate has a corresponding negative charge.

When the plates are joined by a conductor, electrons will be repelled by the negative plate and attracted to the positive plate. So a flow of electrons will be set up from the negative to the positive terminals.

This is the opposite direction to the 'conventional' flow of current, which was thought to be from positive to negative. The convention still exists and, for most practical purposes, the direction of flow is not important.

IONS

A battery may be regarded as having a deficiency of electrons at its positive plate and an excess of electrons at its negative plate. So electrons will flow through any conductor joining the two plates until the excess electrons make good the deficiency and the plates no longer carry any electrical charge. The battery has become 'discharged'.

The charge is built up in the first place by chemical reactions within the cell or battery. A simple 'primary cell' can be made of a plate of copper and a plate of zinc suspended in dilute sulphuric acid, H_2SO_4.

The liquid breaks up into particles having electric charges:

$$H_2SO_4 \rightarrow H^+ + H^+ + SO_4^-$$

These charged particles are called 'ions'.

The acid consists of hydrogen ions, which have a positive charge and sulphate ions, which have two negative charges.

The positive hydrogen ions move to the copperplate or 'anode' which then has a positive charge. And the negative sulphate ions move to the zinc plate which becomes the 'cathode' and has a negative charge.

Liquids which 'ionise', or form ions, are called 'electrolytes' and are capable of carrying a flow of electrons. Ions are also formed in a gas flame during combustion. So a flame is also capable of carrying electrons and some safety devices and igniters make use of this fact.

DIRECT CURRENT

A flow of electrons through a conductor is called an 'electric current'. If the flow is in one direction only, it is called 'direct current'. (This is in contrast to 'alternating current' which will be dealt with later.) The electron is a very tiny amount of electricity, so the unit by which the quantity of electricity flowing is measured is the 'coulomb', which contains a very large number of electrons (6.26×10^{18} to be precise). The coulomb is a quantity, like m^3, whereas what is needed is a rate of flow, like m^3/h. So the unit which is commonly used is the 'ampere', often shortened to 'amp', and we commonly use the algebraic symbol I for current flow.

An ampere is a rate of current flow equal to one coulomb passing in one second. So:

$$\text{amperes} = \frac{\text{coulombs}}{\text{seconds}}$$

or

$$\text{coulombs} = \text{amperes} \times \text{seconds}$$

In symbols:

$$I = \frac{Q}{s} \quad \text{or} \quad Q = Is$$

ELECTRICAL ENERGY

The free electrons in a conductor need a charge of energy to get them all moving in the same direction in a circuit. This is called the 'electromotive force' (emf) and it comes from one of the energy sources listed at the beginning of the chapter.

The unit of emf is the 'volt' (V). An emf of one volt will give one joule of energy to each coulomb. So:

$$\text{volts} = \frac{\text{joules}}{\text{coulombs}} \quad \text{or} \quad \text{joules/coulombs}$$

When electrons pass through a conductor some energy is given up. The amount of energy is related to the degree of difficulty experienced by the electrons in passing from one atom to the next. Because of this there will be a difference in energy levels at different points in the circuit when current is flowing. This is called the 'potential difference' (pd) and it is measured in volts.

Volts are sometimes referred to as electrical pressure and they measure the electrical equivalent to gas pressure in a pipe, that is the force that moves the energy on its way.

ELECTRICAL POWER

The unit of electrical power is the 'watt' (Chapter 5, section on Power). Power is the rate of doing work and is equal to

$$\frac{\text{energy}}{\text{time}}$$

A watt is the rate of working at one joule per second. So:

$$\text{watts} = \frac{\text{joules}}{\text{seconds}}$$

Electrical power may also be derived from emf and current flow.

$$\text{emf} = \text{volts} = \frac{\text{joules}}{\text{coulombs}}$$

$$\text{current} = \text{amps} = \frac{\text{coulombs}}{\text{seconds}}$$

$$\text{power} = \text{watts} = \text{volts} \times \text{amps} = \frac{\text{joules}}{\text{seconds}} = \frac{\text{joules}}{\text{coulombs}} \times \frac{\text{coulombs}}{\text{seconds}}$$

The formula watts = volts × amps has a number of uses. One of the most common is when checking that a fuse or flex is of adequate size.

For example: An electrical appliance has a rating of 3 kW. The mains voltage is 240 V. What size of fuse is required?

$$3\ \text{kW} = 3000\ \text{watts}$$
$$\text{watts} = \text{volts} \times \text{amps}$$

So

$$\text{amps} = \frac{\text{watts}}{\text{volts}} = \frac{3000}{240} = 12.5$$

So a 13-amp fuse is the correct size to be used.

Similarly 1 kW would require:

$$\frac{1000}{240} = 4.2\ \text{amp or a 5 amp fuse}$$

and 750 watts will need 3 amps and so on.

The 'unit of electricity' on which the Electricity Supply Company's or Power Generating Company's charges are based is the kilowatt-hour (kWh). This is equivalent to electricity being used at the rate of 1 kW for a period of one hour.

For example, a 1 kW electric fire, burning for 1 h, will use 1 unit of electricity.

Similarly, a 100-watt bulb, burning for 10 h, will use 1 unit of electricity.

A 60 W bulb will burn for 1000/60 = 16.7 h to use one unit.

RESISTANCE

In some materials the electrons have more difficulty in breaking away from the atom than in others. So some materials offer more resistance to the flow of current than others.

For any particular material the total resistance depends on:

1. Length
   Resistance is directly proportional to length, twice the length, twice the resistance. More emf would be needed to make the electrons travel the longer distance.
2. Cross-sectional area
   Resistance is inversely proportional to the area. The smaller the wire, the greater the resistance. A small wire has fewer atoms to supply free electrons.
3. Temperature
   The hotter the material, the more resistance it will have. Heat causes the atoms to vibrate and slows down the movement of free electrons.

'Resistors' are devices which are used to introduce extra resistance into a circuit. This is done to control the rate of current flow to the required amount. Resistors could be used to control the speed of an electric motor.

They may be made of special wire (usually wound in a coil, for controlling large currents) or of metal film.

For small currents, resistors made from graphite, which is a form of carbon, may be used (see section on Resistors).

OHM'S LAW

The unit of resistance is 'ohm' named after the German physicist who discovered the relationship between current, resistance and emf in a circuit.

An ohm, symbol **Ω,** is the amount of resistance which will allow a current of 1 amp to pass when an emf of 1 volt is applied.

Ohm discovered that the resistance of a conductor, in ohms, is equal to the potential difference between its ends, in volts, divided by the current flowing in it, in amperes.

So Ohm's law states that:

$$\text{resistance} = \frac{\text{potential difference}}{\text{current}}$$

In symbols:

$$R = \frac{V}{I}$$

(sometimes written $R = E/I$ where $E = \text{emf}$).

This is shown in Fig. 9.3. If the resistance of the conductors which connect the resistor across the battery is negligible, then the potential difference between the ends of the resistor and the emf of the battery is the same. The diagram uses the conventional symbols for the battery and the resistor and shows conventional current flow from positive to negative.

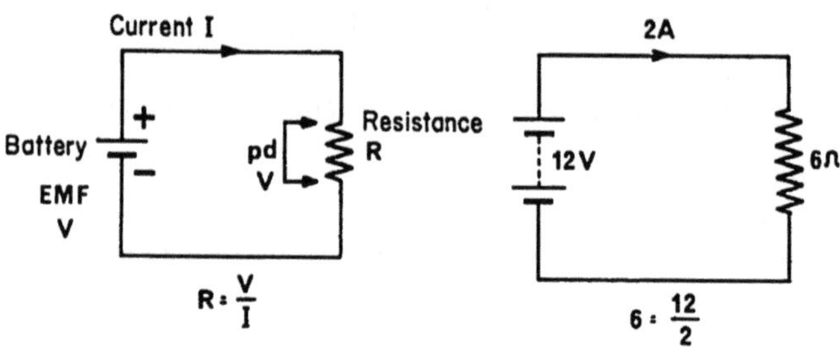

FIGURE 9.3 Ohm's law.

The two diagrams demonstrate Ohm's law, using symbols first and numbers second.

Since

$$R = \frac{V}{I}$$

$$I = \frac{V}{R} \text{ and } V = IR$$

Having now produced formulae which link watts, amps, volts and ohms it is possible to derive a set of formulae which will enable any one unknown quantity to be calculated, if required, when any two others are known.

Without going into the details of the substitutions, the formulae are listed below:

$$\text{Power (watts)}: \quad W = VI \text{ or } \frac{V^2}{R} \text{ or } I^2R$$

$$\text{Resistance (ohms)}: \quad R = \frac{V}{I} \text{ or } \frac{VI^2}{W} \text{ or } \frac{W}{I^2}$$

$$\text{Current (amps)}: \quad I = \frac{V}{R} \text{ or } \frac{W}{V} \text{ or } \frac{W}{R}$$

$$\text{Potential difference (volts)}: \quad V = IR \text{ or } \frac{W}{I} \text{ or } WR$$

Table 9.1 Comparison of Gas and Electricity

Electricity		Gas	
Quantity	Unit	Quantity	Unit
Electromotive force or potential difference	Volt	Pressure or pressure difference	Millibar
Current	Ampere	Flow rate	Cubic metre/hour
Power	Watt	Heat flow rate	Megajoule/hour or Watt
Resistance	Ohm	Pressure loss per quantity flowing*	Millibar/m^3/h

*This is not the usual way of expressing resistance to gas flow which is normally only measured as a pressure loss.

It is also possible to compare the flows of electricity and gas. The four main units and their equivalents are shown in Table 9.1.

SERIES CIRCUIT

A 'series' circuit is one where there is only one current path and all resistors are connected one after another (Fig. 9.4).

The electric charge leaving the battery passes through each resistor in turn, so the rate of current flow is the same at any point in the circuit.

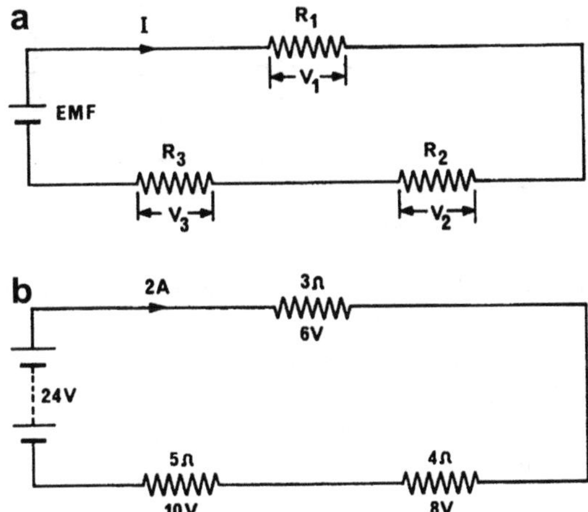

FIGURE 9.4 Resistances in series: (a) symbols; (b) numerical example.

The voltage drop across the resistors adds up until at the end of the circuit, the total drop or potential difference, is equal to the applied emf. In Fig. 9.4a the total pd is:

$$\text{Total pd} = V_1 + V_2 + V_3 = \text{emf}$$

In a similar manner the resistances add up so that the total resistance in a series circuit is the sum of the separate resistances. In Fig. 9.4a the total resistance is:

$$\text{total} \quad R = R_1 + R_2 + R_3$$

The current flowing is:

$$I = \frac{\text{total pd}}{\text{total resistance}}$$

Figure 9.4b gives a numerical example.

$$\text{Total resistance} = 3 + 4 + 5 = 12\ \Omega$$

$$\text{Total pd} = 6 + 8 + 10 = 24\ \text{V}$$

$$I = \frac{24}{12} = 2\ \text{A}$$

PARALLEL CIRCUIT

A 'parallel' circuit is one in which the resistors are connected in parallel branches, each providing a separate path through which current can flow (Fig. 9.5). This is like having several gas pipes supplying the same load. The more pipes there are, the lower will be the total resistance to the flow of gas.

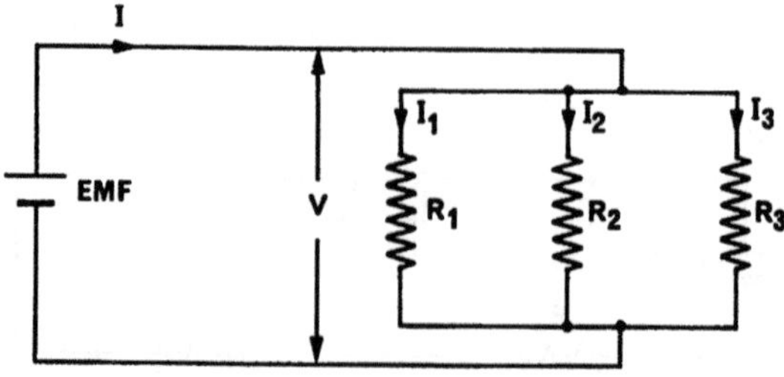

FIGURE 9.5 Resistances in parallel.

So with a parallel electric circuit, the total resistance is less than the smallest individual resistance in the circuit.

From Fig. 9.5 it is apparent that the potential difference across each resistor is the same, *V*, and that this is equal to the emf. The current divides up, some of it passing through each branch. So the total current, *I*, is equal to the sum of the currents in each of the branches.

$$I = I_1 + I_2 + I_3$$

From Ohm's law the current in each branch will be:

$$I_1 = \frac{V}{R_1}, \quad I_2 = \frac{V}{R_2}, \quad I_3 = \frac{V}{R_3}$$

But $I = V/R$ where R is the single resistance that could replace the resistors in the three branches.

So:

$$\frac{V}{R} = I_1 + I_2 + I_3 \quad \text{or} \quad \frac{V}{R} = \frac{V}{R_1} + \frac{V}{R_2} + \frac{V}{R_3}$$

Therefore, since V is common to each fraction, dividing each side of the equation by V gives:

$$\frac{1}{R} = \frac{1}{R_1} + \frac{1}{R_2} + \frac{1}{R_3}$$

This can be extended to give the equivalent resistance to any number of resistors connected in parallel.

If the resistances in parallel have the same value, then the current flowing through them will divide equally, so that the current in each branch will be equal to the total current divided by the number of branches.

$$\text{Current in one branch} = \frac{\text{total current}}{\text{number of branches}}$$

Similarly the equivalent resistance will be equal to the value of one resistance divided by the number of branches.

$$\text{Equivalent resistance} = \frac{\text{single resistance}}{\text{number of branches}}$$

For example, Fig. 9.6 shows four 100 Ω resistances in parallel. The total resistance is therefore:

$$\frac{100}{4} = 25\ \Omega$$

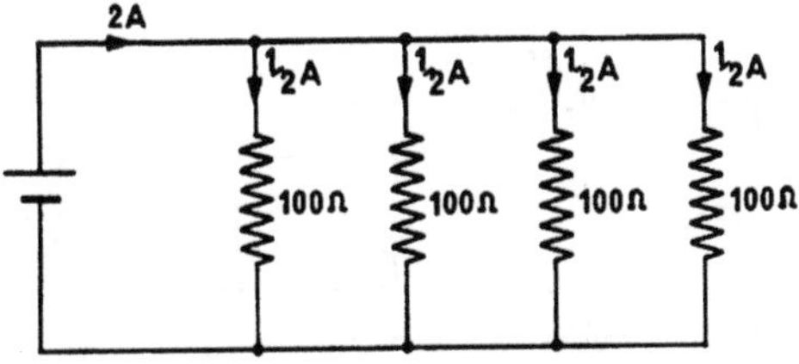

FIGURE 9.6 Equal resistances in parallel.

If the total current flowing is 2 amps, then the current in each branch would be:

$$\frac{2}{4} = \frac{1}{2}\text{ amp}$$

The total resistance of resistances with different values can only be found by using the formula:

$$\frac{1}{R} = \frac{1}{R_1} + \frac{1}{R_2} + \frac{1}{R_3} + \cdots$$

Figure 9.7 gives an example:

$$\frac{1}{R} = \frac{1}{4} + \frac{1}{8} + \frac{1}{6} + \frac{1}{12} = \frac{6+3+4+12}{24} = \frac{5}{8}$$

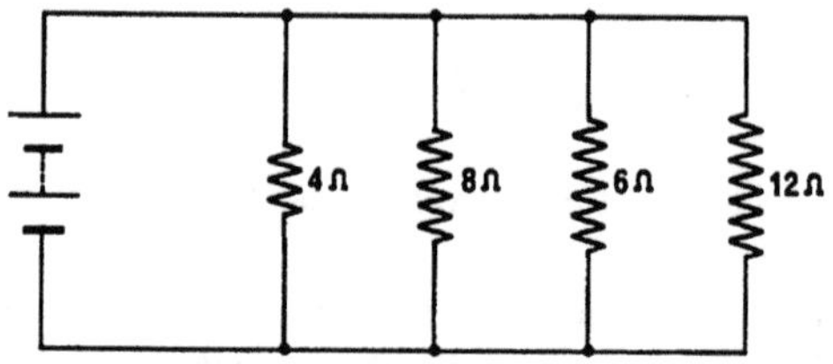

FIGURE 9.7 Unequal resistances in parallel.

Therefore

$$R = \frac{8}{5} = 1.6\ \Omega$$

COMBINED SERIES/PARALLEL CIRCUIT

Where a circuit is made up of a combination of resistors in series and in parallel, the parallel resistors must be dealt with first. Figure 9.8 shows an example. The total resistance of the three resistors in parallel (R_p) comes from:

$$\frac{1}{R_p} = \frac{1}{6} + \frac{1}{6} + \frac{1}{6} = \frac{3}{6}$$

So

$$R_p = \frac{6}{3} = 2\ \Omega$$

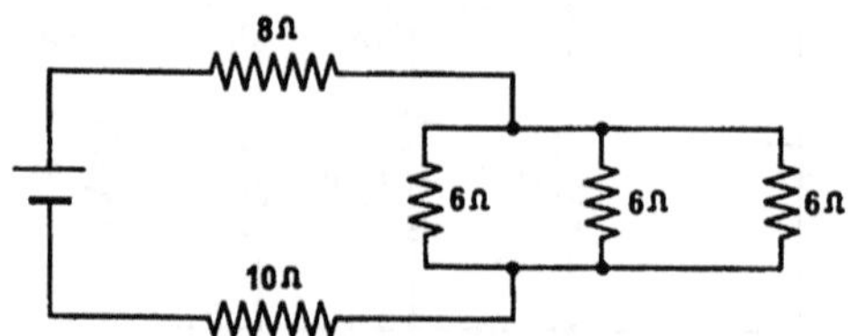

FIGURE 9.8 Combined series and parallel circuit.

The total resistance of the whole circuit is:

$$2 + 8 + 10 = 20\ \Omega$$

KIRCHHOFF'S LAWS

Kirchhoff, a German physicist, discovered two important laws which govern the behaviour of current and potential difference within a network. Although the laws have not been stated before, examples of their application have been given.

1. Current
   The total current entering a junction is equal to the sum of the currents leaving it.
   So, if a number of wires branch out from a main, the sum of the currents in the branches is equal to the current in the main wire.
2. Voltage
   In a closed circuit the algebraic sum of the products of current and resistance in each part of the circuit is equal to the total emf in the circuit.

$$\text{Total emf} = I_1R_1 + I_2R_2 + I_3R_3 + \cdots$$

MAGNETISM

A 'magnet' is an object composed of a material which is capable of attracting iron. Magnets may be found naturally in the form of rocks of lodestone, which are iron oxide or magnetic.

A magnet which is suspended will point to the North Pole, and lodestone was used by early navigators in simple forms of ships' compasses. Permanent magnets are usually made from iron, cobalt or nickel or an alloy of these metals.

Molecules of iron are themselves tiny magnets, each having a north and south pole. In an unmagnetised piece of iron they arrange themselves in closed loops as shown in Fig. 9.9a. When the iron is magnetised they form into lines, all facing the same direction, Fig. 9.9b.

Magnets obey the laws of magnetism, which are similar to those of electric charges:

- like poles repel,
- unlike poles attract.

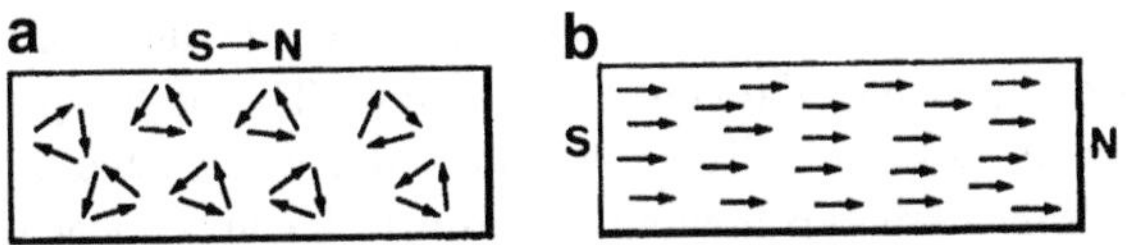

FIGURE 9.9 Magnetism: (a) unmagnetised iron and (b) magnetised iron.

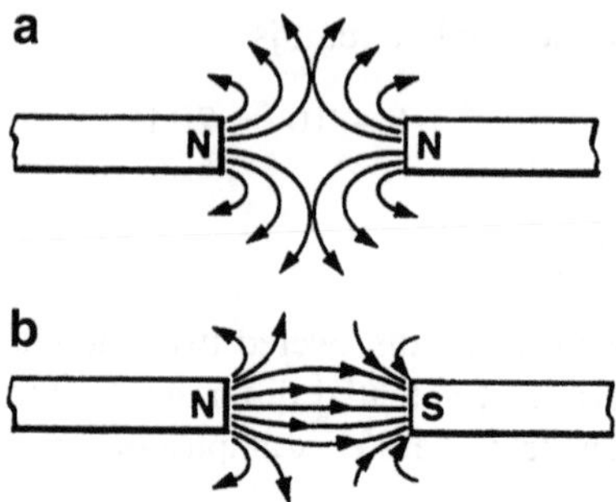

FIGURE 9.10 Forces between magnetic poles: (a) like poles and (b) unlike poles.

Forces between magnetic poles are shown in Fig. 9.10, which illustrates repulsion at Fig. 9.9a and attraction at Fig. 9.9b. The lines of magnetic force create a 'magnetic flux' or total force between the poles. These lines always form complete loops and never cross each other. Figure 9.11 shows the lines of magnetic flux which form the magnetic 'field' around a bar magnet.

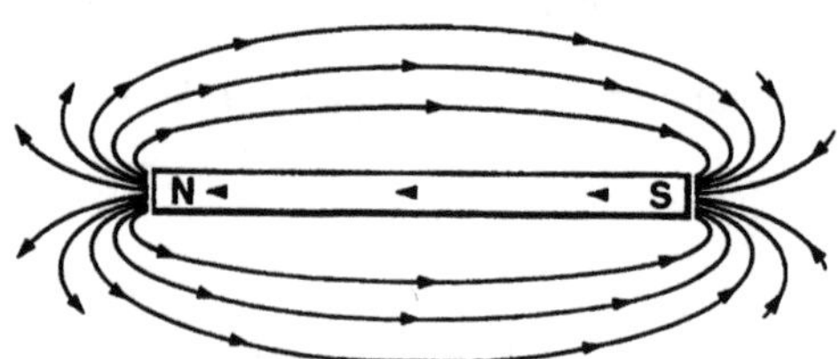

FIGURE 9.11 Magnetic field around a bar magnet.

MAGNETIC EFFECT OF A CURRENT

When an electric current is passed through a conductor, a magnetic field is created around the conductor. The direction of the field depends on the direction of the current, Fig. 9.12.

Figure 9.12a shows a vertical wire with the current flowing upwards. To determine the direction of the field flux, use the 'right-hand rule'.

If the conductor is gripped by the right hand with the thumb pointing in the direction of the current flow, then the fingers point in the direction of the field.

Figures 12b and c show the wire in cross-section with the field around it. Figure 9.12b is the wire with the current flowing upwards, indicated by ⊙, and Fig. 9.12c shows the wire with the current flowing away from you, indicated by ⊕.

The strength of the field can be increased by winding the conductor into a coil. Figure 9.13 shows how the separate fields on the turns of the conductor become concentrated into the centre of the coil. This is the principle of the electromagnet or 'solenoid'. The solenoid will only be 'energised' or made magnetic while the current is flowing. In this way, it can be made to operate like a mechanical device, like a gas valve (see section on Solenoids).

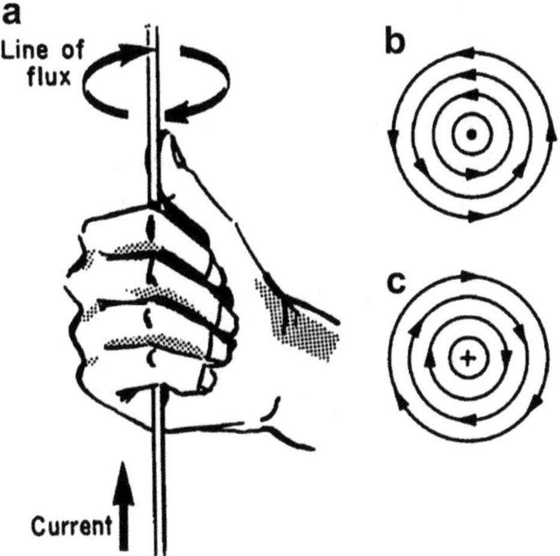

FIGURE 9.12 Magnetic effect of a current: (a) right-hand rule; (b) current flowing upwards; and (c) current flowing downwards.

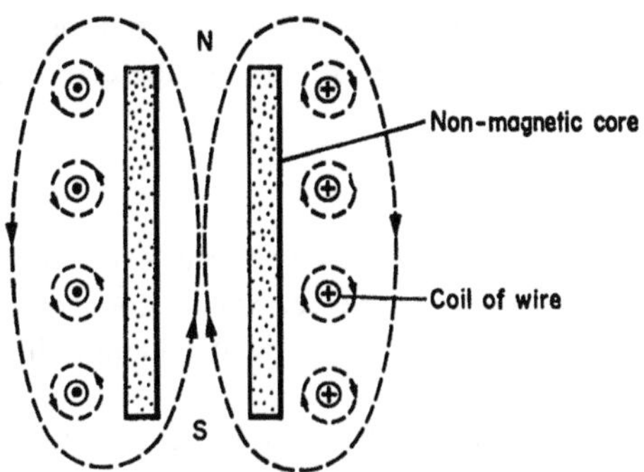

FIGURE 9.13 Magnetic field around a coil.

The polarity of the coil can be determined by using the 'right-hand rule' for solenoids.

If the coil is gripped by the right hand with the fingers pointing in the direction of the current flow in the turns, then the thumb points to the north pole of the magnetic field.

If the magnetic fields produced by a permanent magnet and current in a conductor are brought together, the resulting field produces a force which acts on the conductor (Fig. 9.14).

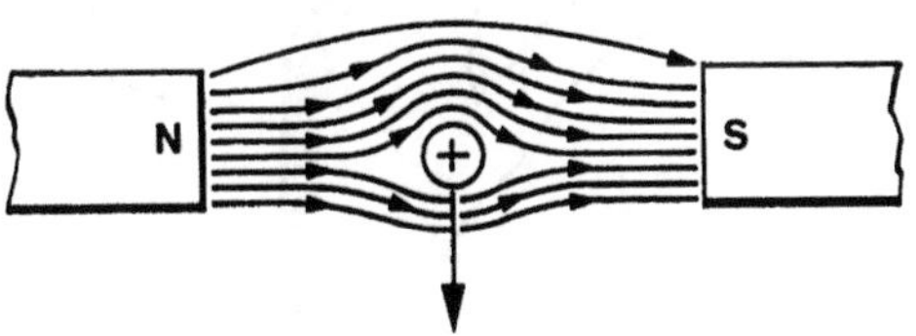

FIGURE 9.14 Force acting on a conductor in a magnetic field.

Because the lines of force do not intersect, the field becomes stronger on one side of the wire and weaker on the other. So the wire is forced away from the stronger part of the magnetic field.

This principle is used in electric motors where current is made to flow through coils mounted on a central 'armature' between the poles of a magnet (Fig. 9.15). The magnetic fields produced by current in the coils cause the armature to rotate.

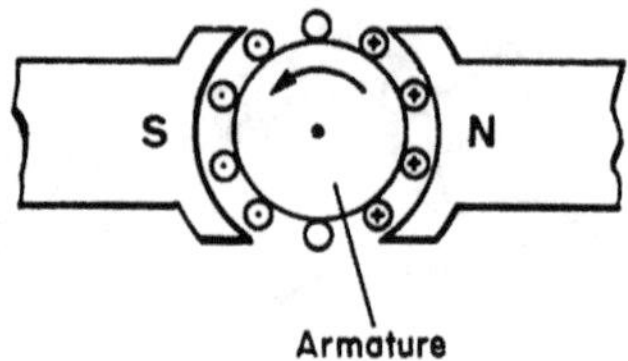

FIGURE 9.15 Principle of the electric motor.

The force exerted on the coil varies with the amount of current flowing and electrical measuring instruments make use of this principle.

In a 'moving coil' meter the coil is held between two spiral springs and mounted in jewel bearings. The coil turns against the force of the springs and a pointer attached to it measures the strength of the current flowing.

INDUCTION

Passing a current through a conductor which is situated in a magnetic field causes the conductor to move. In reverse, when a conductor is moved through a magnetic field, a current is 'induced' in the conductor. This is the basis of the electric generator.

If the direction of the conductor's movement is not changed, then the direction of the current will not change. If the conductor moves more quickly, it will pass through more lines of force in a given time and this will increase the current flow. So a larger emf will be produced.

The relationship between the directions of the magnetic flux, the current and the movement of the conductor is given by 'Fleming's right-hand rule'

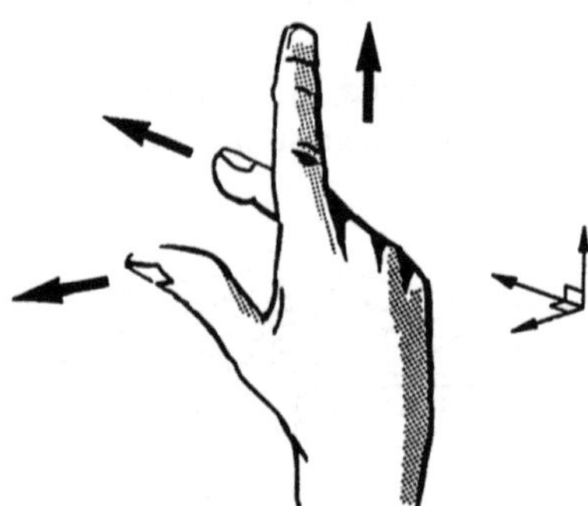

FIGURE 9.16 Fleming's right-hand rule.

(Fig. 9.16). If the thumb, forefinger and the remaining fingers are held at right angles to each other, then the direction of an induced current can be found by holding the forefinger in the direction of the magnetic field and the thumb in the direction of movement of the conductor. The remaining fingers show the direction of the induced current.

If the rule is applied to Fig. 9.17a, it shows that the induced current is moving downwards. If the conductor was moved from right to left as in Fig. 9.17b, the current would flow in the opposite direction. This is exactly what happens when a conductor is rotated in a magnetic field. The current is reversed as the conductor comes back through the field.

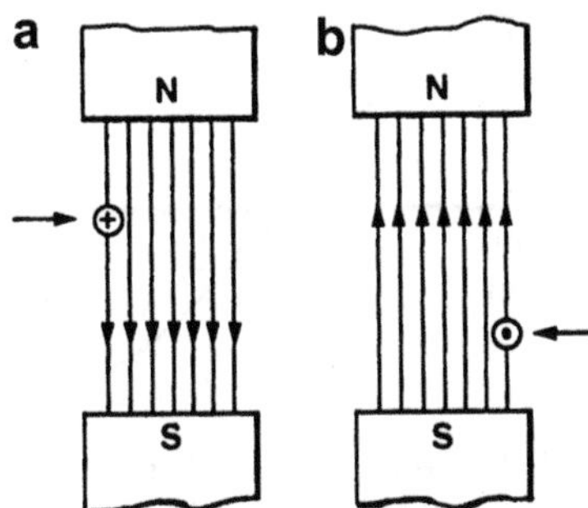

FIGURE 9.17 Direction of induced current: (a) conductor moving left-to-right, current flows down; (b) conductor moving right-to-left, current flows up.

The right-hand rule applies to generators and, in a similar way, the left-hand rule can be used to show the same three relationships with respect to motors.

ALTERNATING CURRENT

'Alternating' current is current which changes its direction of flow. Instead of passing from positive to negative all the time, like the direct current produced by a battery, alternating current (AC) changes direction very rapidly. In a normal electricity supply system, there are fifty complete cycles of change in each second.

Figure 9.18 shows how this occurs. It illustrates, in cross-section, the positions of a single coil rotating between the poles of a permanent magnet. The graph below shows how the induced emf behaves as the coil and is rotated through one complete revolution.

Starting at position (a), consider the quantity and direction of the emf induced in the coil.

(a) As the conductors move, they are parallel to the lines of force. So no lines are 'cut' and no current is induced. The emf is zero.
(b) Here the conductors have rotated through 90° and are moving at right angles to the flux. So the rate at which they cut the lines of force is at its maximum. The emf has risen to a maximum in the positive direction.

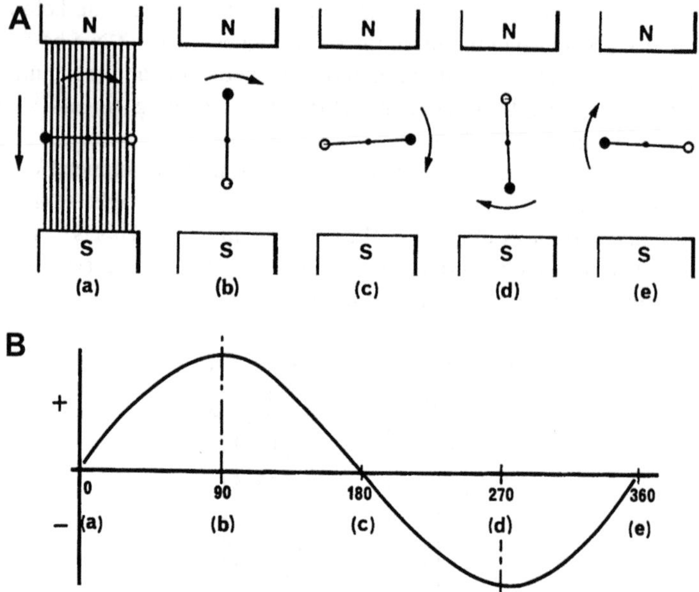

FIGURE 9.18 Production of alternating current: A, rotation of a single coil between magnetic poles; B, graph of quantity and direction of the emf.

(c) The conductors have now rotated through 180° and are no longer cutting the lines of force, emf is again zero.

(d) As rotation continues to **270**° the conductors have reached the position at which they are cutting the lines of force at their maximum rate. But this time in the opposite direction. So the induced current flows in the opposite direction around the coil and emf rises to a maximum in a negative direction.

(e) The revolution is now complete and the conductors have resumed their original position; the emf has again fallen to zero.

In one revolution the voltage reaches two maximum or 'peak' values, one positive and the other negative.

FREQUENCY

An alternating current 'wave' or graph is a 'sine' wave. This is a particular shape of graph associated with trigonometry, Fig. 9.19. It shows the length of one 'cycle', which is one rotation of the coil. The number of cycles which occur in one second is called the 'frequency' of the electricity supply. The supply frequency is measured in 'hertz' (Hz) and 1 hertz is one cycle per second. The frequency of the alternating current (AC) supply distributed by the power generating companies is 50 Hz.

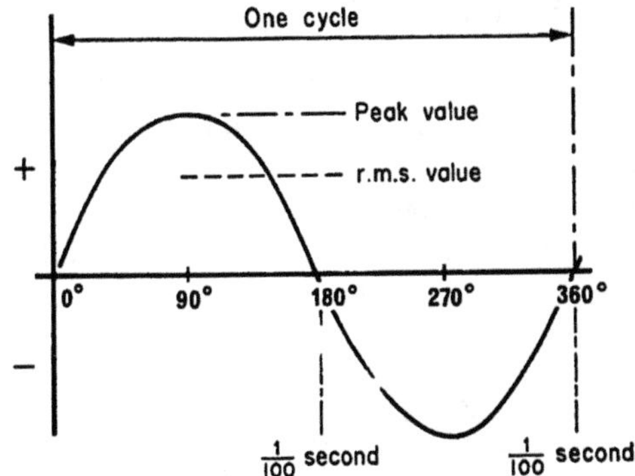

FIGURE 9.19 AC frequency.

The symbol for AC is ~, which represents the sine wave and the normal domestic supply is written as:

$$240\ \text{V} \sim 50\ \text{Hz}$$

ALTERNATING CURRENT AND VOLTAGE

The peak values of current or voltage are the maximum values reached in each cycle and may be either positive or negative in direction. But they do not give a true indication of the amount of work which the supply can do.

There is a need to find a method of averaging the values so that the same power equation $P = V \times I$ can be used for AC as well as for DC (direct current).

The value which is used is the 'root mean square' or rms value. This is obtained by plotting the graph of I^2 rather than just I, or current. Since any quantity, whether negative or positive, becomes positive when it is squared, the graph of I^2 is all above the zero line and it is easy to find its average, or 'mean', value. Because we need amps and not amps^2, it is necessary to take the square root of the mean value. Thus we have the root of the mean squared value, or rms value.

All AC voltages and currents are rms values (unless otherwise stated) and these are the values used for any calculations of power, current, voltage or resistance.

For a sine wave, the rms value is 0.707 of the peak value.

$$\text{rms} = \text{peak} \times 0.707$$

So the peak value is $1/0.707 = 1.414$ of the rms value.

$$\text{peak} = \text{rms} \times 1.414 \qquad \text{or} \qquad \text{approximately rms} \times 1\tfrac{1}{2}$$

Since the voltage of the domestic electricity supply is 240 V, and this is the rms value, the peak voltage of the supply is:

$$240 \times 1.414 = 339.4\ \text{V}$$

GENERATION OF ELECTRICITY

You have seen that electrical energy can be generated by rotating a coil within the field of a permanent magnet. As the coil revolves, the induced current changes direction.

If the ends of the coil are connected to 'slip rings' (Fig. 9.20) then an AC supply is produced. If, however, the ends of the coil are connected to a 'commutator' (Fig. 9.21), then a direct current results.

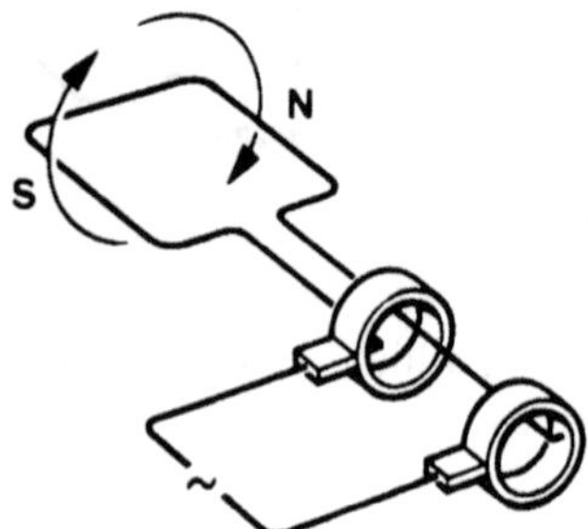

FIGURE 9.20 Slip rings producing AC.

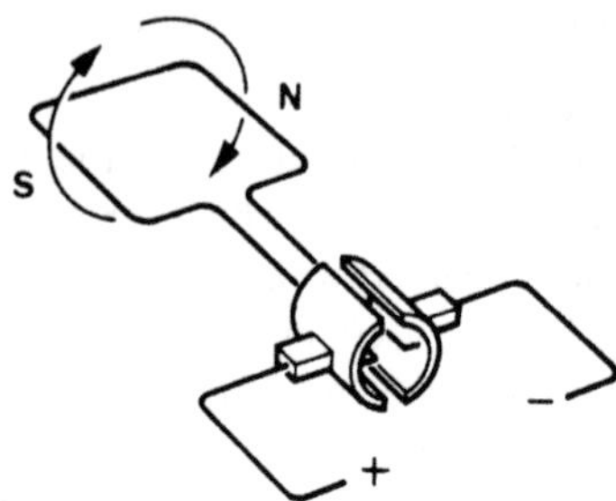

FIGURE 9.21 Commutator producing DC.

This is because the brushes are positioned on the commutator so that they always receive either positive or negative supplies.

The machine which generates AC is commonly called an 'alternator', and the one producing DC is a 'dynamo' or 'generator'.

In a power station, which produces electrical energy, the rotors of the alternators have the coils mounted on them with an angle of 120° between the coils (Fig. 9.22). This produces three sets of emf or three 'phases' as shown by the sine waves. The distance in degrees between the peak values is called the 'phase angle' and has the symbol ϕ.

The alternators at the power stations are usually driven by steam turbines, or water turbines in the case of hydroelectric schemes, although in recent years power stations using natural gas to power gas engines and gas turbines have been built (see Volume 3, Chapter 10). The voltages generated could be between 11,000 and 33,000 V (11–33 kV). The rotors revolve at a constant speed in order to maintain the frequency at a constant 50 Hz.

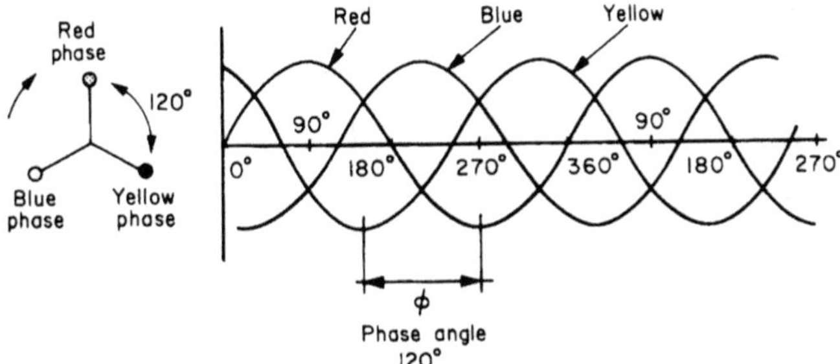

FIGURE 9.22 Production of three-phase AC.

HARMONIZED CABLE CORE COLOURS

British Standard 7671 (17th Edition) 2008 aligns the colour coding of cables across Europe.

In single-phase (domestic) the old colours were red for live and black for neutral, the new colours are brown for live and blue for neutral.

In three-phase (industrial/commercial) the old colours were: phase 1 red, phase 2 yellow, phase 3 blue and neutral black, the new colours are. Phase 1 brown, phase 2 black, phase 3 grey and neutral blue (see Fig. 9.24)

DISTRIBUTION OF ELECTRICITY

Alternating current is used because it is easy to change its voltage. The voltage can be 'stepped-up' or 'stepped-down' by means of a device called a 'transformer' (see section on Transformers). Because power (P) is equal to the product of voltage (V) and current (I), then $P = VI$ and a small current can be carried in a small cable at a high voltage. When it reaches the point of use it can be stepped down to give a higher current flow at a lower voltage. Figure 9.23 shows a typical distribution system of overhead cables with sub-stations at which the voltages are changed.

From the sub-station the supply is usually carried in underground cables in urban areas. In country districts it may continue on to the customers by overhead cables.

Connections taken from three-phase supplies may be either three-phase or single-phase (Fig. 9.24). Generally, industrial premises have three-phase so that more powerful motors can be used, and domestic premises have single phase.

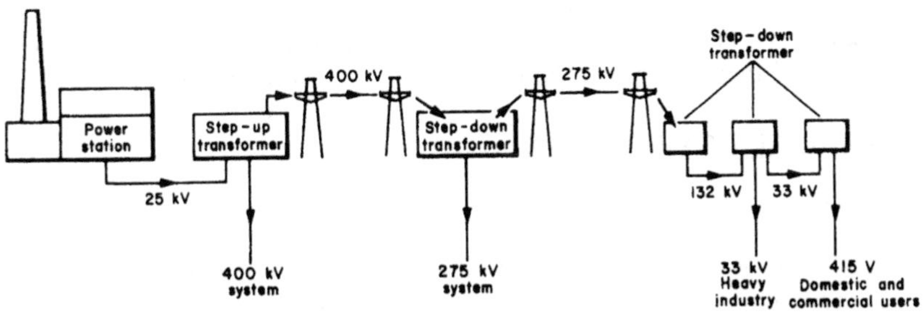

FIGURE 9.23 Electrical distribution system.

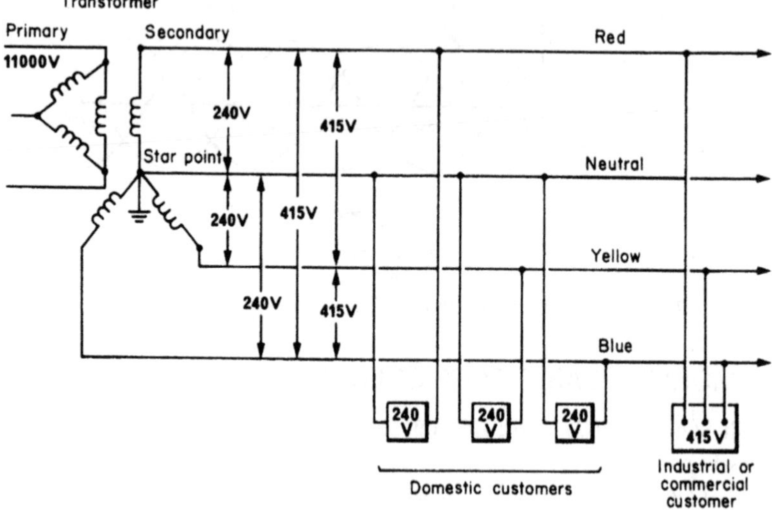

FIGURE 9.24 Single and three-phase connections to customers.

Houses are connected to the phases alternately in order to balance the loads on each phase. The voltages in a three-phase supply are:

- 240 V between any phase and the neutral,
- 415 V between any two phases.

DOMESTIC INTERNAL INSTALLATION

Figure 9.25 shows the layout of a typical domestic main connection. The supply is taken through the main fuse box and then to the meter. Both these components are sealed and can be opened only by staff of the electricity supply company or power generating company.

Current then passes through the main switch. This is a double-pole switch which isolates both the live and neutral connections. From the switch the supply goes to the distribution box where it divides up into the various house circuits. Each circuit is protected by its own fuse (see the section on Fuses and Circuit Breakers).

Power supplies in a house are commonly provided by means of a 'ring' main (Fig. 9.26). This is a method of connecting the socket outlets in a closed ring or loop. It has the advantage of feeding current to the outlets from both ends of the loop, so reducing the cable size.

Figure 9.27 shows how the flex on an appliance is connected to a component and a typical three-pin plug. The old colours are shown in brackets.

ELECTRICAL COMPONENTS

Transformers

A transformer works by induction. In this case, instead of a conductor moving through a magnetic field, the conductor remains stationary and it is the field that moves.

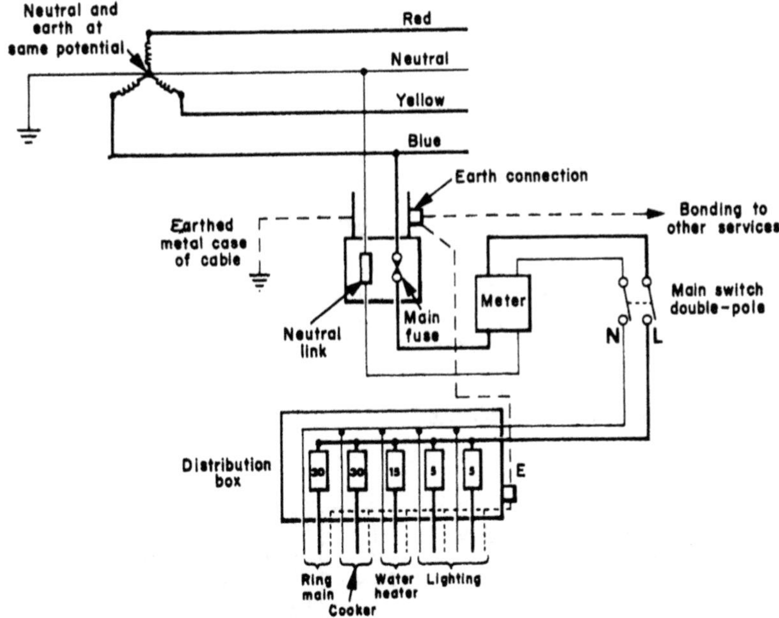

FIGURE 9.25 Domestic customer's connection.

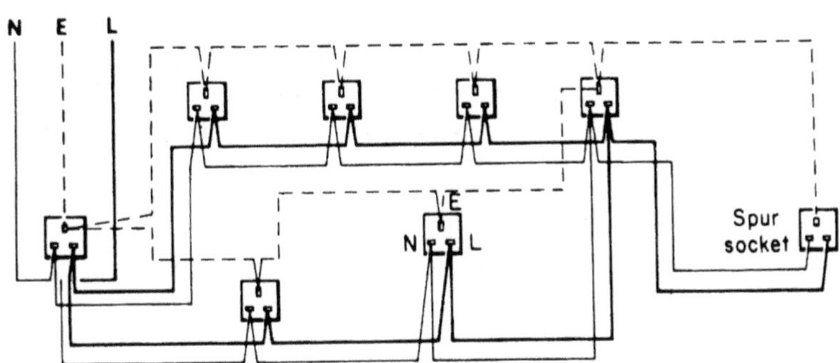

FIGURE 9.26 Ring main circuit.

When current flows in a coil it produces a magnetic field (Fig. 9.13).

When an alternating current flows in a coil it produces an alternating magnetic field. As the field grows and diminishes and changes its direction, its lines of force would be cut by the loops of any coil placed in their path. So, a coil placed in an alternating field will have an alternating current induced in it.

If two coils are placed close together and one is connected to an AC supply, then alternating current can be drawn from the other and the device is called a 'transformer'. The coil connected to the mains supply is called the 'primary' winding and the output comes from the 'secondary' winding (Fig. 9.28).

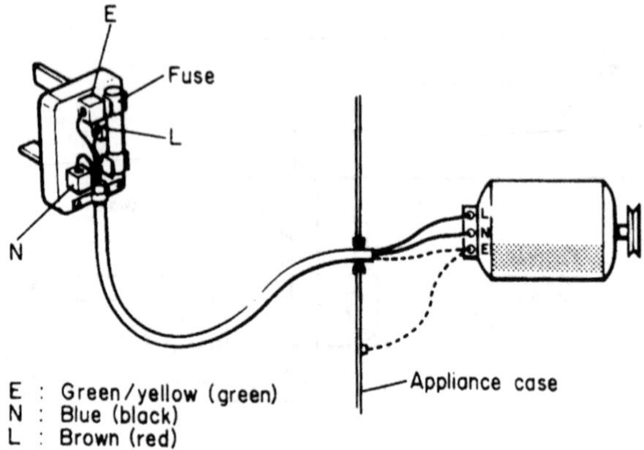

FIGURE 9.27 Connection to appliance.

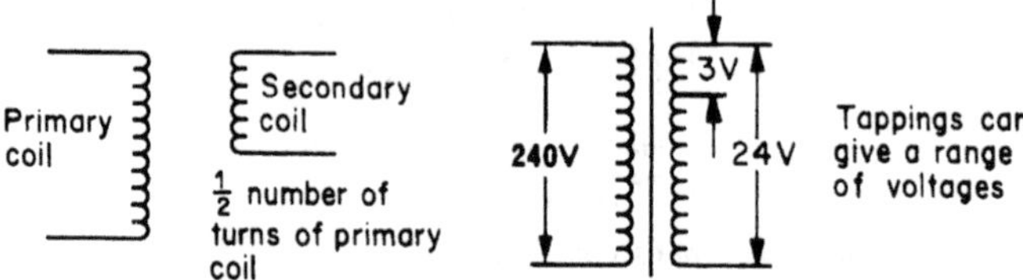

FIGURE 9.28 Transformer.

The voltage of the secondary supply depends on the primary voltage and the ratio between the number of turns in each of the coils. If the secondary coil has only half the number of turns of the primary, then its voltage will be half the primary voltage. So:

$$\frac{\text{primary voltage } V_1}{\text{secondary voltage } V_2} = \frac{\text{primary turns } N_1}{\text{secondary turns } N_2}$$

$$\frac{V_1}{V_2} = \frac{N_1}{N_2} \quad \text{or} \quad V_2 = \frac{V_1 N_2}{N_1}$$

For example: If a transformer has 2000 turns on its primary coil and 100 on the secondary, then, if the primary voltage is 240 V, the secondary voltage will be:

$$V_2 = \frac{V_1 N_2}{N_1} = \frac{240 \times 100}{2000} = 12 \text{ V}$$

The relationship between the current I and the number of turns, *N,* is:

$$N_1 I_1 = N_2 I_2 \quad \text{or} \quad \frac{I_1}{I_2} = \frac{N_2}{N_1}$$

Only a little power is lost in a transformer, so, as voltage decreases, the current will increase in the same ratio.

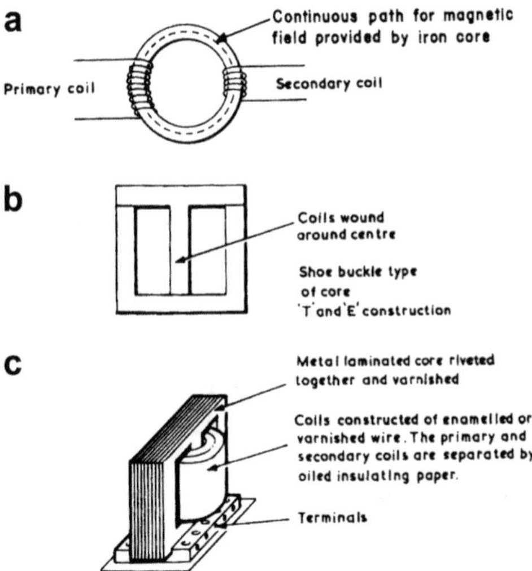

FIGURE 9.29 Construction of transformers: (a) continuous magnetic path through iron core; (b) shoe-buckle type core, T and E construction; and (c) complete transformer.

Transformers cannot work on DC supplies.

Transformers have an iron core which provides a continuous path for the magnetic field (Fig. 9.29a), and the coils are generally wound with the secondary on top of the primary around the central member of the core. The core is made up of 'laminations' or flat plates of iron rivetted together (Fig. 9.29b), and a common type of small transformer is shown in Fig. 9.29c.

Transformers are used on gas appliances to reduce mains voltage to provide a low-voltage supply for thermostats, clocks and solenoids. The common are 12 or 24 V. When spark ignition is required an ignition transformer can step up the voltage to about 10,000 V.

Some transformers have more than one secondary winding. This might be to supply a control circuit at 24 V and an igniter circuit at 3 V (Fig. 9.28). Or it could provide a series of tappings giving voltages between 200 and 100 V in order to vary the speed of a fan motor.

The iron core and a point on the secondary circuit must be connected to the earth conductor.

Capacitors

A 'capacitor' is a device which stores an electric charge. It is composed of two metal plates separated by insulating or 'dielectric' material (Fig. 9.30). The dielectric may be made from paper, mica, polystyrene, oil or ceramic materials. An air space could be used. Electrolytic capacitors are used on DC circuits. There is no electrical continuity across a capacitor.

A capacitor will not allow a DC supply to pass but, on AC, a current can flow. This is because the plates are continually charging and discharging as the emf grows to maximum and then falls to zero.

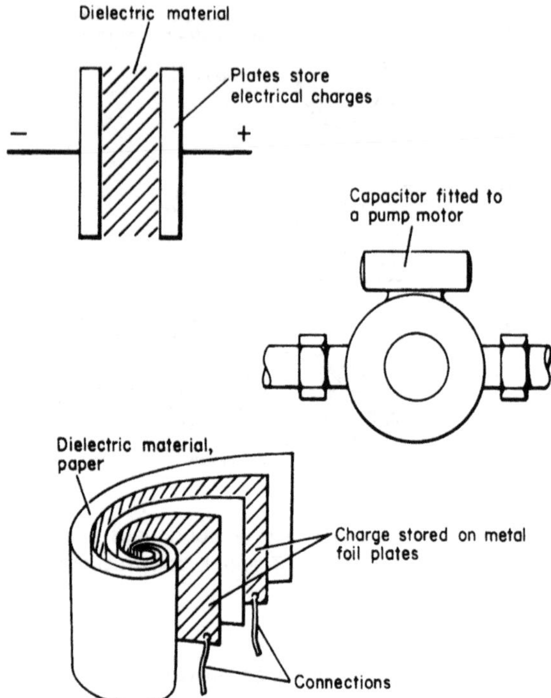

FIGURE 9.30 Capacitors, construction and location.

Capacitors, sometimes called 'condensers', have the effect of slowing down the voltage in the circuit. This is because current begins to flow before the charge builds up in the plates. So the current always 'leads' the voltage by 90°.

The ability to produce an out-of-phase voltage in a circuit is made use of in small electric motors. Motors can run effectively on single-phase AC but they need something to give them an initial impetus.

By introducing a separate coil in circuit with a capacitor the phases are 'split' and the motor becomes, in effect, a two-phase motor. On some split-phase motors the additional coil is cut out by a centrifugal switch when the motor reaches full speed (see Volume 2, Chapter 13).

The unit of capacitance is farad (F). This is the measure of the ability of a capacitor to store 1 C of electricity with a potential difference of 1 V between its plates. So a capacitor with a capacity of 1 F requires a current of 1 amp to pass for 1 s to produce a charge of 1 V, or

$$\text{farads} = \frac{\text{amps} \times \text{seconds}}{\text{volts}} \qquad \text{F} = \frac{\text{As}}{\text{V}}$$

The farad is a very large unit so, for practical purposes, the microfarad (μf) or picofarad (pf) is used.

Resistors

Resistors were introduced earlier in the chapter. They may be of various types (Fig. 9.31).

1. Wire wound
   These resistors, Fig. 9.31a, usually have a low resistance and can carry large currents. They have a high power (watts) rating.
2. Carbon
   Finely ground carbon is mixed with a binding agent and then compressed to make these resistors (Fig. 9.31b). They have a high resistance and a low current flow. The ratings are usually 2 W or less. Some high-precision resistors are made from deposits of carbon vapour on ceramic or glass formers.
3. Metal oxide or film
   These are similar to carbon resistors, but can operate at higher temperatures. Their power ratings are correspondingly higher than the same size of carbon resistor.
4. Variable resistors
   A variable resistor or 'rheostat' has a control knob, like the volume control on a radio. Turning the knob varies the length of resistance wire through which the current has to flow (Fig. 9.31c).

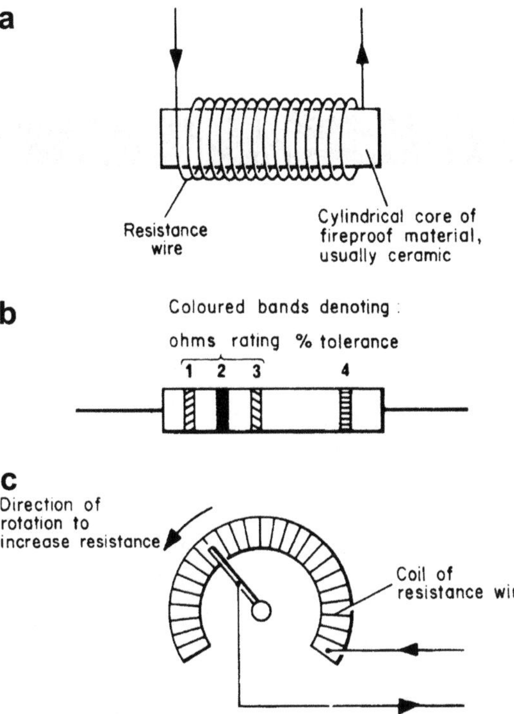

FIGURE 9.31 Resistors: (a) wire wound, (b) carbon, and (c) variable.

5. Thermistor
   Whilst most resistance wires have a higher resistivity when heated, a 'thermistor' is the opposite. When subjected to heat it allows current to pass more freely. For this reason it is used to measure temperature by being connected to an electric meter. A Multimeter has a thermistor, as part of the field kit, which measures temperatures from 0 to 120 °C (32–248 °F).

CODING OF RESISTORS AND CAPACITORS

The marking codes for resistors and capacitors are set down in BS EN 60062 2005. In the case of fixed resistors the resistance value and any tolerance in this value are indicated by coloured bands painted round the components or by a letter and digit code.

COLOUR CODING

The colour code indicates resistance values to two or three significant figures and the tolerance. The first significant figure is indicated by the band nearest to the end of the resistor. Table 9.2 details the values corresponding to colours.

As previously stated, the band indicating the first significant figure is the band nearest to the end of the resistor while the band furthest away from that end

Table 9.2 Resistors

Colour	Significant Figures	Multiplier	Tolerance (%)
Silver	–	10^{-2}	± 10
Gold	–	10^{-1}	± 5
Black	0	1	–
Brown	1	10	± 1
Red	2	10^2	± 2
Orange	3	10^3	–
Yellow	4	10^4	–
Green	5	10^5	± 0.5
Blue	6	10^6	± 0.25
Violet	7	10^7	± 0.1
Grey	8	10^8	–
White	9	10^9	–
None	–	–	± 20

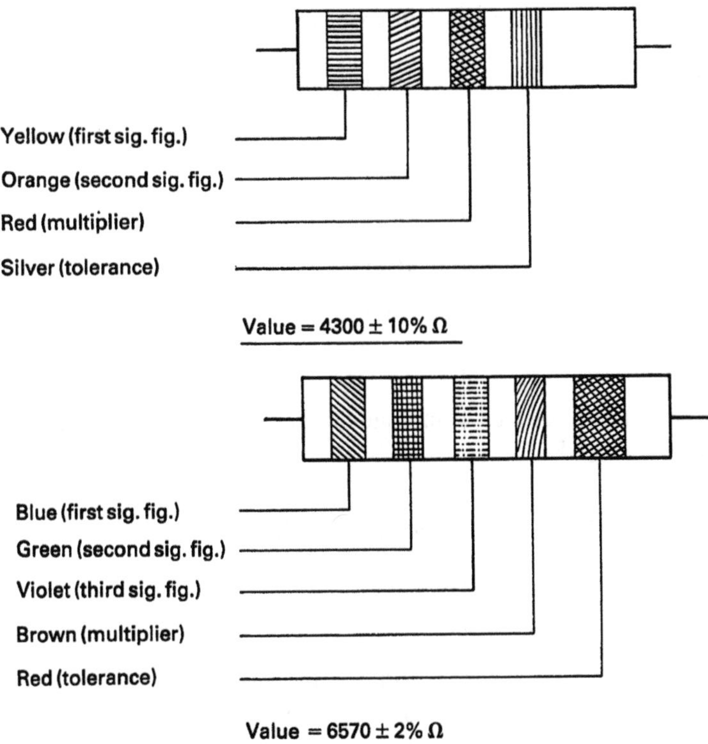

FIGURE 9.32 Examples of colour code*.

indicates the tolerance. The band next to the tolerance band indicates the multiplier to be used when calculating the value. Any further bands indicate significant figures between the first significant figure and the multiplier. Examples of colour codes are shown in Fig. 9.32.

CODED MARKING

To indicate their electrical value, resistors are often code marked with a letter and digit (number). When using a letter and digit code the value and tolerance are indicated by 3, 4 or 5 characters, consisting of 2 figures and a letter, 3 figures and a letter or 4 figures and a letter as required. The code letters replace the decimal point and are equal to the following multiples:

$$
\begin{aligned}
R &= 1 \\
K &= 10^3 \\
M &= 10^6 \\
G &= 10^9 \\
T &= 10^{12}
\end{aligned}
$$

Examples using the letter and digit code are shown below:

$$0.47\ \Omega = \text{R47}$$
$$5\ \Omega = \text{5RO}$$
$$2.7\ \Omega = \text{2R7}$$
$$75\ \Omega = \text{75R}$$
$$4\text{k}\ \Omega = \text{4KO}$$
$$15\ \text{M}\Omega = \text{15M}$$
$$125\ \Omega = \text{125R}$$

*When a fifth band is used to indicate the tolerance, it is 1½ to 2 times wider than the other bands.

$$74.6\ \Omega = \text{74R6}$$
$$67.07\ \Omega = \text{67RO7}$$
$$49.16\ \text{k}\Omega = \text{49K16}$$

A letter after the coded marking will give the tolerance.

Symmetrical tolerances are indicated by the letters in Table 9.3.

Table 9.3 Symmetrical Tolerances

Tolerance (%)	Code Letter
0.1	B
0.25	C
0.5	D
1	F
2	G
5	J
10	K
20	M
30	N

A symmetrical tolerances are indicated by the letters in Table 9.4.

A letter and digit code is also used to indicate the value and tolerance of capacitors. Capacitance is measured in farads.

Table 9.4 Asymmetrical Tolerances

Tolerance (%)	Code Letter
−10 + 30	Q
−10 + 50	T
−20 + 50	S
−20 + 80	Z

In the case of capacitors the following code is used to indicate the multiplier:

$$
\begin{aligned}
F &= 1 \\
m &= 10^{-3} \\
\mu &= 10^{-6} \\
n &= 10^{-9} \\
p &= 10^{-12}
\end{aligned}
$$

Examples of the code marking for capacitance values are shown below:

$$
\begin{aligned}
0.16 \text{ picofarads} &= \text{p16} \\
3.32 \text{ picofarads} &= \text{3p32} \\
150 \text{ picofarads} &= \text{150p} \\
1.5 \text{ nanofarads} &= \text{1n5} \\
33.2 \text{ nanofarads} &= \text{33n2} \\
4.41 \text{ nanofarads} &= \text{4n41} \\
1 \text{ microfarad} &= 1\mu 0 \\
4.80 \text{ microfarads} &= 4\mu 8 \\
415 \text{ microfarads} &= 415\mu \\
10 \text{ millifarads} &= \text{10m} \\
3.17 \text{ millifarads} &= \text{3m17} \\
4.80 \text{ millifarads} &= \text{4m8}
\end{aligned}
$$

A letter after the coded marking indicates the tolerance. The symmetrical and asymmetrical tolerances are indicated by the same letters as already given in the tables on resistance values in this section. For tolerance values below 10 pF the letters in Table 9.5 are used.

Where no tolerance has been laid down the letter A may be used.

Table 9.5 Tolerances on Capacitance Values Below 10 pF

Tolerance (pF)	Code Letter
±0.1	B
±0.25	C
±0.5	D
±1	F

SOLENOIDS

A solenoid, referred to in the section on Magnetic Effect of a Current, uses the magnetic effect of a current to operate a valve or a switch (Fig. 9.33).

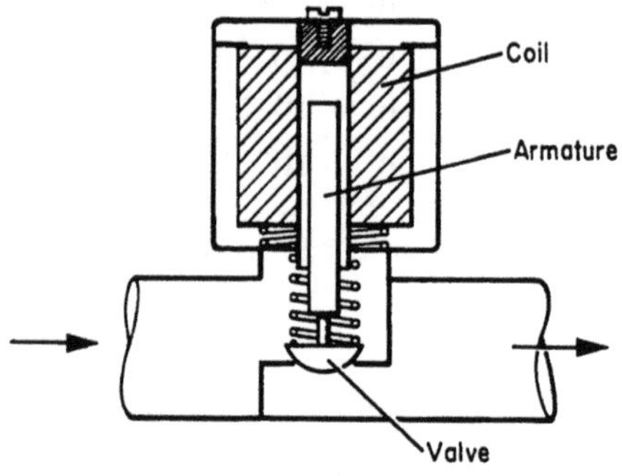

FIGURE 9.33 Solenoid valve.

The valve is attached to the armature, which is a soft iron rod free to slide in a brass tube. The tube has an iron plug at the top acting as a pole piece. The coil is wound on a bobbin which fits over the brass tube and is covered by an iron case which helps to concentrate the magnetic flux. When current flows in the coil the armature is drawn into the tube and so lifts the valve open.

Solenoids can operate on either AC or DC. On AC the reversal of the current occurs so quickly that the armature does not have time to fall. However, the reversals of flux can cause the armature to vibrate against the pole piece and produce an objectionable noise. For this reason the current in some appliances may be 'rectified', or changed to DC, so that a perfectly quiet solenoid will result.

Electrical relays (Fig. 9.34) are used for the remote control of switches and usually for switching on a mains-voltage circuit by means of a low-voltage control. They can be used to open, close or change over switches.

A relay consists of an electromagnet and a moving armature which operates the switch or switches. When the current is off, the armature is held away from the magnet by a spring. When the current flows the armature is drawn onto the coil and the switch is moved to its other position.

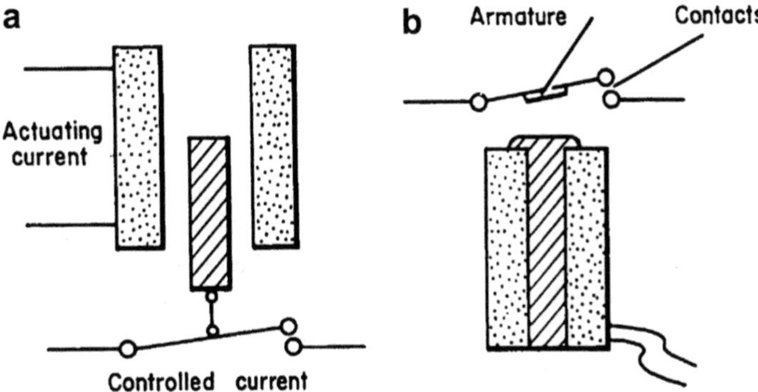

FIGURE 9.34 Electrical relays: (a) open when energised and (b) closed when energised.

FUSES AND CIRCUIT BREAKERS

To 'fuse' means to melt. And an electrical fuse is made of soft, low-melting-point metal (Fig. 9.35) so that it will 'blow' when too high a current causes it to heat up.

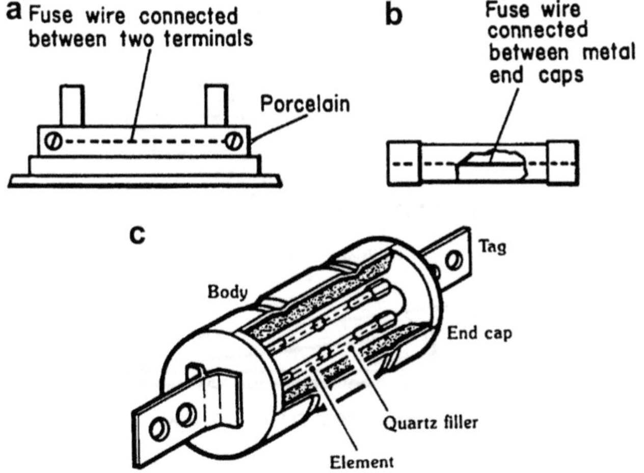

FIGURE 9.35 Types of fuses: (a) rewirable, BS 3036; (b) Cartridge, BS 1362; and (c) High rupturing capacity, BS 88.

Fuses are fitted in series with electrical equipment in order to protect the equipment from damage by excessively high currents. The size of fuse to use is one which has a rating slightly higher than the largest current required. For example, a 5-amp fuse should be used where the largest current is about 4 amp.

A fuse will blow when the circuit is overloaded or when a 'short-circuit' occurs. This is when the load is bypassed by a fault in the wiring or its insulation. Fuses can fail when the metal link ages.

When replacing fuses:

- Make sure that the supply is switched off;
- Find and remedy the fault which caused the failure;
- Replace with the correct rating.

Circuit breakers are also used to protect electrical equipment from overloading and accidental damage.

There are several different types of circuit breaker and they differ from fuses by interrupting the supply through switching rather than fusing. The advantage being a simple switching action that will restore the supply once the fault has been found and corrected.

Figure 9.36 shows three types of miniature circuit breaker (MCB). These are sometimes used in place of fuses to control the interruption of small industrial/ commercial and domestic circuits.

Miniature circuit breakers are designed to activate immediately when there is a high current surge, as in short circuit and when there are slight but continual overloads. To prevent unnecessary shutdown, however, they must be designed to

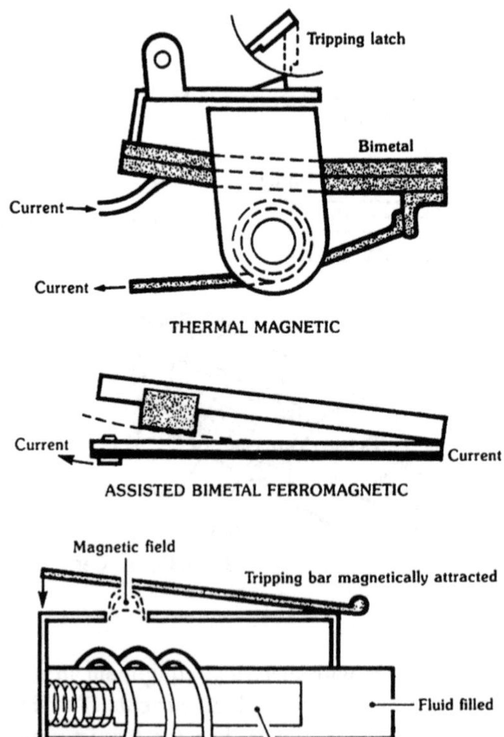

FIGURE 9.36 Miniature circuit breakers.

ignore harmless overload conditions, 'spikes' etc., that occur in an AC supply. Miniature circuit breakers have the following:

(i) Advantages
 (a) easy to observe when they have failed;
 (b) only require switching to restore the supply;
 (c) reduce nuisance shutdown by ignoring transient overloads;
 (d) tamper proof,
(ii) Disadvantages
 (a) initial cost is high;
 (b) mechanical moving parts;
 (c) need to be tested at regular intervals;
 (d) can be affected by the surrounding conditions.

RECTIFIERS

A 'rectifier' is a device for changing AC into DC. It has two plates of metallic material and so it is also called a 'diode' (two electrodes).

A rectifier has a very high resistance to current flow in one direction and a very low resistance in the other, so it resembles a non-return valve, only allowing a flow to take place in one direction. It conducts electricity only during the positive half-cycle of the AC, so voltage and current are in one direction only (Fig. 9.37). Only 'half-wave' rectification is obtained from a single diode.

Full-wave rectification can be obtained by using four diodes in what is known as a 'bridge' circuit. This takes its name from the 'Wheatstone

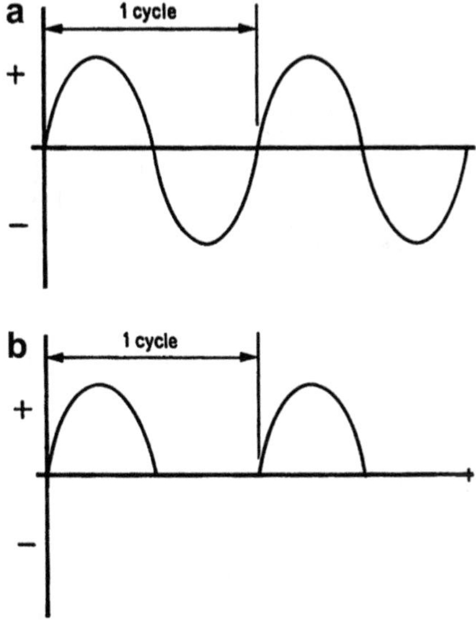

FIGURE 9.37 Rectification: (a) full AC wave and (b) rectified AC half-wave.

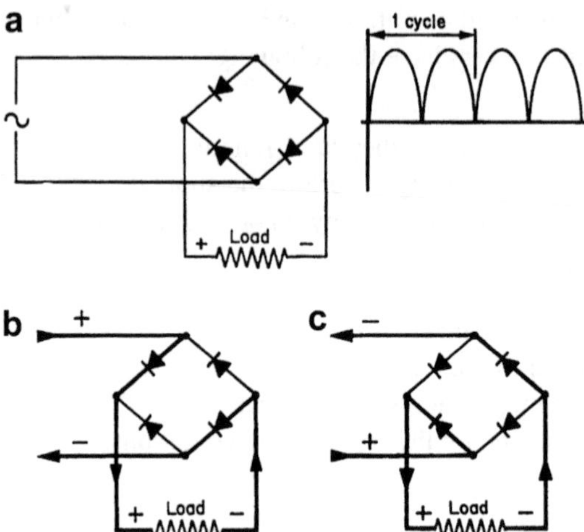

FIGURE 9.38 Bridge rectifier: (a) bridge circuit, full-wave rectification; (b) current flow; and (c) reversed current flow.

Bridge', which is a device for measuring resistance and is dealt with in Chapter 14.

The circuit of a bridge rectifier is shown in Fig. 9.38a and the direction of flow of current in each part of the bridge is shown in Fig. 9.38b and c. When the current returns from the negative side of the load it always takes the easy path towards the lower voltage.

Figure 9.39a shows a single diode. The line or band at one end indicates the outlet. Figure 9.39b is a bridge rectifier. These may be of any shape but will always have four connections indicated as shown.

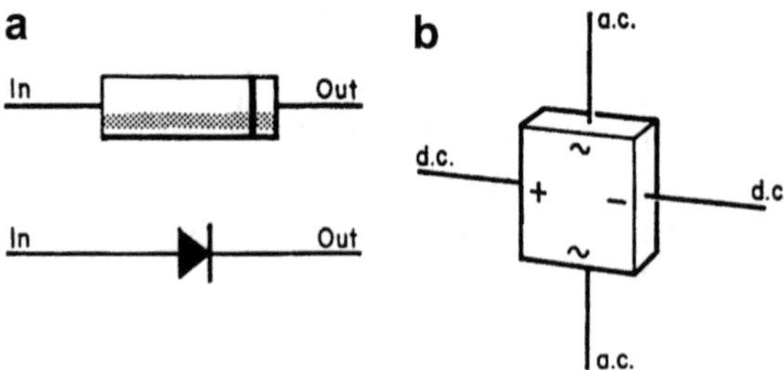

FIGURE 9.39 Rectifiers: (a) single diode and (b) bridge rectifier.

MEASURING INSTRUMENTS

The three basic electrical measurements which may need to be taken are voltage, amperage and resistance, and they are measured with voltmeter, ammeter and ohmmeter, respectively (although insulation resistance is often measured at high voltage using a 'megger').

Voltmeter

This measures the emf or the potential difference between any two points in a circuit. It must always be connected in parallel with the circuit being measured (Fig. 9.40).

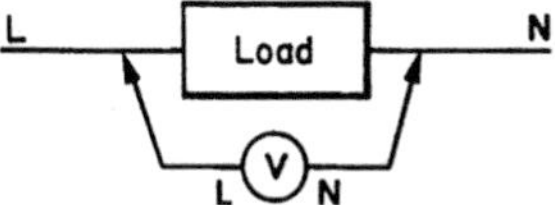

FIGURE 9.40 Method of connecting voltmeter.

Voltage measurements are useful when checking a circuit and its components since they indicate the electrical pressure at each point in the circuit.

Ammeter

An ammeter measures the current flowing and so it must be connected in series with the circuit (Fig. 9.41). It is rarely necessary to take current readings when dealing with electrical components on gas appliances.

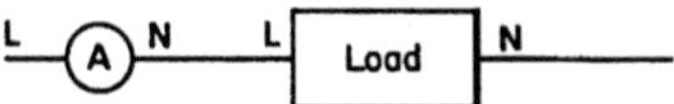

FIGURE 9.41 Method of connecting ammeter.

Ohmmeter

This measures the resistance of a circuit and should be connected in parallel with the circuit after the circuit has been isolated from the power supply (Fig. 9.42).

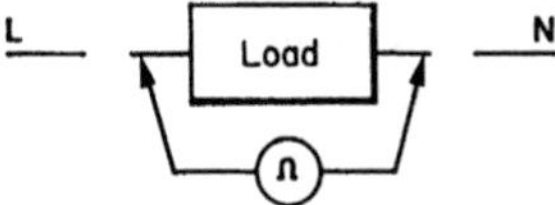

FIGURE 9.42 Method of connecting ohmmeter.

It has its own batteries which provide a power source and it would be severely damaged if connected to a live supply. The ohmmeter can also be used to check for an 'open-circuit', which shows as a reading of infinity, ∞, or 'short circuit', which would give a reading of zero.

When connecting voltmeters or ammeters to a DC supply the polarity of the meter connections must be the same as that of the supply.

Multimeters

The three instruments described may be obtained as separate devices. However, a number of multifunctional instruments are available which incorporate all

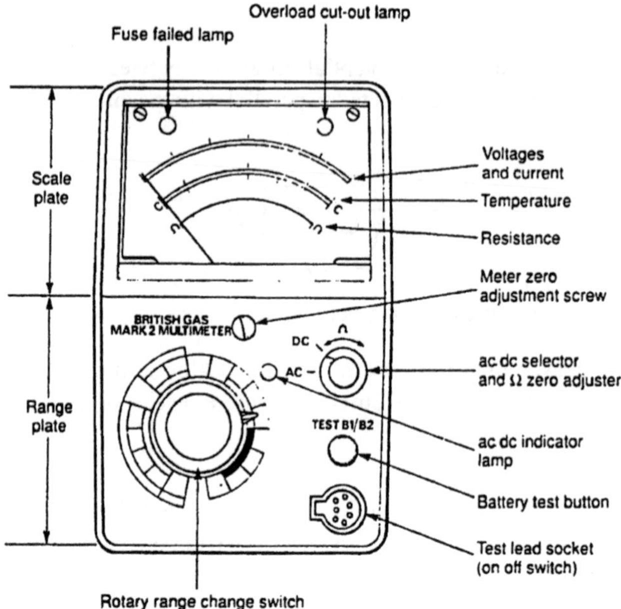

FIGURE 9.43 The British Gas Mark 2 multimeter.

three types. By means of integral switches, resistors and rectifiers these instruments can measure volts, amps and ohms over a wide range of values on AC or DC circuits. The British Gas multimeter is an example of a multifunction, multirange instrument (Fig. 9.43).

It has twenty ranges of voltage, current and resistance measurements selected by a rotary switch. The scales are:

- Red, for all voltage and current readings.
- Green, for temperature measurement.
- Black, for resistance readings.

The ranges on the rotary switch are:

- Voltage (AC or DC) 0–30 and 0–60 mV (millivolts)
  0–3, 0–30, 0–300 V
- Current (AC or DC) 0–3, 0–30, 0–300, 0–600 μA (microamps)
- Resistance 1–2 kΩ, 1000–2000 kΩ, plus continuity buzzer
- Temperature 0–60 °C, 60–120 °C

The meter includes an electronic cut-out circuit which protects against overloading by tripping the internal circuitry to prevent damage to the multimeter components. There is also a warning light that glows as long as an electrical overload is present. A check on the condition of the internal batteries can also be made by turning the rotary change switch to the appropriate position and manually depressing the rotary change switch. The internal batteries are not drained if the switch is inadvertently left in the battery test position. With special high voltage leads the meter can measure up to 600 V.

When using a multirange instrument to measure unknown voltages or current, always set the meter to the highest range first. The range can be reduced until a reasonable scale deflection is obtained.

Several types of electrical test meter can be used to diagnose and trace faults on electrical systems, controls etc. In addition to the previously described multimeter which was specially designed for British Gas plc, a DMM (digital multimeter) is being used in some of the company's regions. The multimeter is shown in Fig. 9.44.

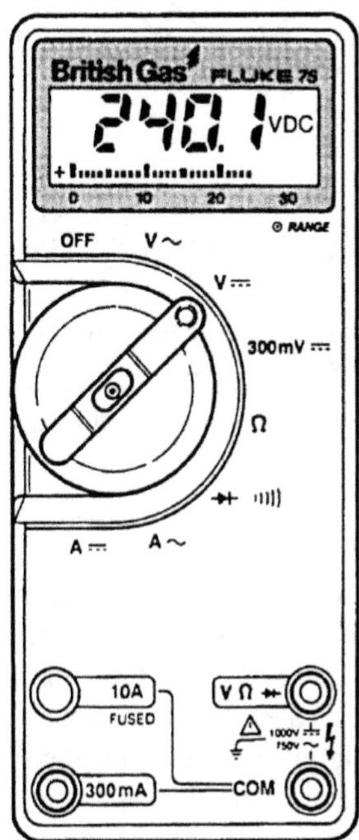

FIGURE 9.44 Fluke 75 digital multimeter.

The instrument can be used for measuring either voltage, current or resistance by operating a rotary switch. When first switched on the entire range of symbols are shown in the display area for approximately 3 s (Fig. 9.45).

This display will then clear and an audible bleep informs the user that the instrument is ready for use. The test leads must be connected to the appropriate sockets and the user must select the correct AC or DC position when measuring voltage or current.

The meter is fully autoranging therefore eliminating the need to set the meter on the highest range first when measuring unknown voltages or current. Up to 750 V AC and 1000 V DC can be measured and there are two amperage ranges, up to 10 A or up to 320 mA. The higher amperage socket has been

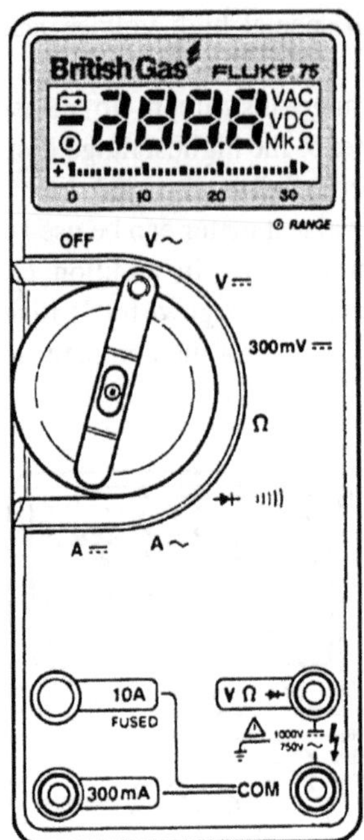

FIGURE 9.45 Automatic test display each time the Fluke 75 is switched on.

blanked off on most of the meters used by British Gas plc employees as they are never likely to measure current in this range.

The resistance displayed will be either Ω (ohms), kΩ (kilo-ohms) or MΩ (mega-ohms). If the meter is accidentally used in an overload situation, the display will indicate this by showing OL. However, if OL MΩ is displayed when checking resistance, this indicates infinite resistance (open circuit).

In addition to the digital read-out in the display window, there is a bargraph reading given below the figures. A small battery symbol appears in the top left-hand corner of the display when the meter's internal 9 V battery is becoming exhausted. There is also a facility for checking continuity audibly and the audible tone continues to sound up to a resistance of 150 Ω.

Although the resistance is autoranging, it is possible to hold the reading on a particular range by depressing the button in the centre of the rotary switch.

To return to autoranging, the same button must be pressed again. On the series 2 model of this multimeter, this same hold button provides a further function known as Automatic Touch Hold. When the hold button is depressed whilst turning the rotary switch from 'off' to any function, measurements taken are held and the reading displayed is confirmed by a 'beep'. The display reading remains held until further measurement is taken.

To exit this function the multimeter is turned off. Electrical fault diagnosis is covered in Volume 2.

ELECTRIC SHOCK

Electric shock can vary from the painful to the fatal. The 'shock' is due to the sharp contraction of the muscles in the path of the current. The severity of the damage depends on:

- the magnitude of the voltage and amperage,
- the path of the current through the body,
- the firmness of contact,
- the dryness or dampness of contact,
- the physical condition of the person.

The effects of voltages are as follows:

10–12 V — a slight tingle is experienced;
15 V — above this the pain felt increases rapidly;
20–25 V — muscles of the hand may contract so that a person holding a conductor may not be able to let go. The hand may be easily knocked away from contact. Up to 25 V may be regarded as safe.
above 25 V — danger increasing with the voltage;
60 V — lowest voltage at which death has been recorded;
above 120 V — likely to produce fatal effects. The heart muscles give rapid and irregular beats, breathing muscles may be paralysed.
200–250 V — dangerous. Alternating current is more dangerous than direct because of its high peak voltage. But DC causes more serious burns. Muscular spasms may cause other physical injuries.

The amount of current required to produce fatal results is quite small, about 0.1 A would probably be sufficient in most cases. Since the human body has a resistance of about 1500 Ω, then a current of 0.16 A will pass if the voltage is 240 V. So this is very likely to be fatal if it passes through the trunk.

Treatment for electric shock must be given immediately. The procedure is as follows:

1. Remove the casualty from contact with the live supply. Either switch it off or pull them clear using dry, insulating material, for example folded newspapers, cloth or rubber. Brooms, chairs or lengths of wood could be used to lever them clear.
2. If the casualty is unconscious or not breathing normally, begin artificial respiration at once and send for a doctor. Continue the respiration until normal breathing is restored or until death has been certified. Prompt treatment can often save life and it must be persevered with.
3. When the casualty has recovered they should be sent to hospital as there is always the risk of a relapse occurring.

Electric shocks can be avoided if sensible precautions are taken. Voltage readings may safely be taken on live supplies by using specially designed, well-insulated probes. But be particularly careful to avoid contact with any bare wires or connections. When you have to exchange components or disconnect wiring always isolate the equipment from the mains supply before you start.

Isolating is not just switching off. It means making sure that the supply cannot be turned on while you are working on the equipment. It entails removing the fuses and keeping them with you or tying up a plug or a flex so that it cannot be reconnected accidentally.

If in any doubt at all, switch off at the main. Never take chances, there is no cure for death.

EARTHING

The earth is a conductor of electricity. Connections can be made to the ground by means of buried rods or plates.

The Institution of Electrical Engineers' (IEE) Wiring Regulations define the methods necessary to give protection against electric shock and fire. These include earthing of exposed metalwork, double insulation of equipment and the use of extra low voltage (not more than 50 V AC or 120 V DC).

Earthing is the most common method and since the secondary winding of the sub-station transformer is also 'earthed', this ensures that any stray currents in the cases or metalwork attached to appliances can find their way back, through the earth, to the neutral point (Fig. 9.46). So earthing provides a means of minimising the risk of electric shock or fire which could otherwise be caused by faulty insulation or by bare wires touching a casing.

Figs 9.25 and 9.46 illustrate the most common method of ensuring that stray currents find their way back to the neutral point. Electricity Supply Companies and Power Generating Companies have, however, been introducing Protective Multiple Earthing (PME) as an additional safeguard. This is done by making a connection between the neutral and earth conductors at the customer's premises. One of the requirements for the use of a PME earth terminal is that adjacent services and metalwork should be bonded, see Fig. 9.50. This means

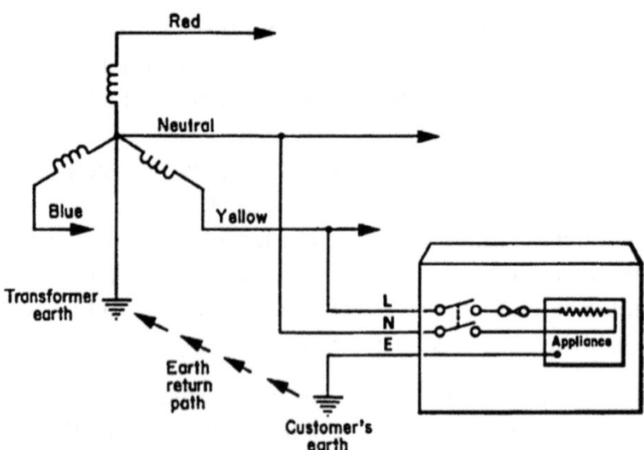

FIGURE 9.46 Earthing.

that the earth return may now carry higher currents which could cause corrosion of buried gas supply pipes. The difficulty can be overcome by fitting an insulating coupling on the gas service pipe, if the service is not already made from insulating material, that is polyethylene (PE) (see Volume 2, Chapter 4).

Protective multiple earthing systems introduce some potential dangers, and this type of system can only be installed by Electricity Boards or Power Generating Companies after government approval has been obtained.

Three types of electrical supply system are illustrated in Fig. 9.47, each showing the electrical installation and supply.

The following code is used to identify the supply.

The first letter indicates the earthing arrangement of the supply.

T = One or more points of the supply system are directly earthed.
I = The supply is either not earthed or it is earthed through a fault-limiting impedance.

The second letter indicates the installation wiring arrangements.

T = All exposed conductive metalwork is connected directly to earth.
N = All exposed conductive metalwork is connected directly to the earthed supply connector (this is often the metal casing of the supply cable).

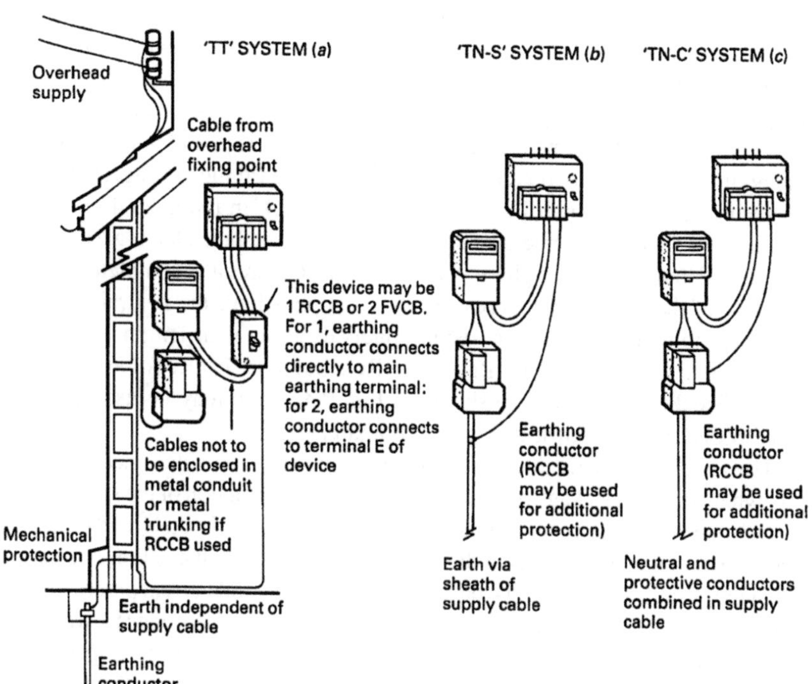

FIGURE 9.47 Typical electric supply systems.

The third and fourth letters indicate the following arrangements for the earthed supply conductor.

S = Separate neutral and earth conductors.
C = Combined neutral and earth in a single conductor.

The 'TT' system, shown in Fig. 9.47a, has one or more points of the system directly earthed, and all exposed conductive metalwork is connected directly to earth. This type of system, via two (live and neutral) overhead wires, is common in rural areas. The earth and neutral conductors are separate in this type of installation and the power supply company do not supply an earthing terminal. It is the responsibility of the consumer to provide an earth electrode (generally a metal rod hammered into the ground) for the connection of circuit protection devices. Difficulty may be experienced in obtaining an effective earth connection, and in these cases a circuit breaker should be fitted. Two types of circuit breaker likely to be found on this type of installation are described later in this section.

The 'TN-S' system, shown in Fig. 9.47b, has one or more points of the system directly earthed, all exposed conductive metalwork is connected directly to the earthed supply connector. The system has also separate neutral and earth conductors.

This type of system is usually found with underground supply cables. The power supply company connect the consumer's earthing terminal to the protective armour on the supply cable. This gives an uninterrupted metallic path back to the supplying transformer.

The 'TN-C-S' system shown in Fig. 9.47c also has one or more points of the system directly earthed. All exposed conductive metalwork is connected directly to the earthed supply connector. The neutral and earth connectors are combined in the power company's supply cable but are separate in the installation.

This type of arrangement is used on PME installations and is normally used where it is difficult to obtain other satisfactory earthing arrangements. The disadvantage of this type of system is that any break in the neutral side of the circuit will result in equipment using current beyond the break having a combined neutral and earth system at phase voltage, unless the circuit is completed through the mass of the earth. For this reason, PME supplies have the neutral conductor connected to earth electrodes at regular intervals along their length.

Residual Current Circuit Breakers (RCCBs) and Fault Voltage-operated Circuit Breakers (FVCBs) are shown in Fig. 9.47a and may be used for additional protection as shown in Fig. 9.47b and c.

The residual current circuit breaker shown in Fig. 9.48 is designed to operate only when there is a leakage of current to earth. Fast interruption of the supply, when a preset rate of leakage occurs, gives protection against electric shock and fire. A trip mechanism holds the contacts in the closed position against the action of a spring. When the circuit is free of faults, balanced phase and neutral current passes through identical coils on a transformer core. No magnetic flux is produced whilst the currents are balanced because each coil provides equal but opposite ampere-turns.

Any current leakage to earth will cause more current to flow in the phase coil than the neutral. This will cause a magnetic flux to be set up in the transformer

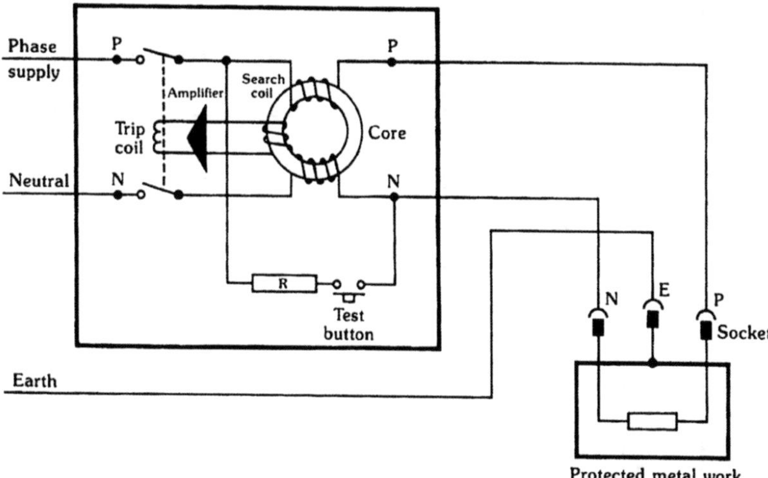

FIGURE 9.48 Residual current circuit breaker.

core and create an emf in the search coil. The current in the trip coil opens the trip mechanism and the pressure of the spring opens the switch contacts and interrupts the power supply.

For the purpose of testing the device a special circuit is provided. It is advisable to test the equipment at frequent intervals as the trip mechanism can become stiff with age.

Where an RCCB has been set to 2 mA, if there was a short-circuit to earth and this amount of current passed through a person or some other conductor, the RCCB would trip and interrupt the supply to the circuit, preventing further shock or current leakage.

When an installation has a combined neutral and earth (TN-C) there is no separate path for neutral and earth leakage currents. The phase and neutral currents remain the same even under fault conditions and an RCCB must not be used on this type of installation.

An RCCB must be fitted to a domestic installation where the power supply company do not provide an earth terminal (TT system). The device must operate on a current not exceeding 30 mA and is usually fitted between the meter and the junction box or fuse board.

A fault voltage-operated circuit breaker is illustrated in Fig. 9.49. This type of device was often used to protect 'TT' systems, but as RCCBs have developed and become more sensitive the FVCB has become less popular.

The trip coil senses any fault condition in the circuit resulting in a voltage potential between the installation metalwork and earth.

When the trip coil element is energised it activates the trip mechanism which in turn opens the double-pole switch and interrupts the power supply. This trip mechanism is usually set to cut off the supply before a fault voltage of 50 V is reached.

It is most important to remember that where a fuse has blown or a circuit breaker has tripped, then the cause of the problem must be determined and rectified before reinstating the supply.

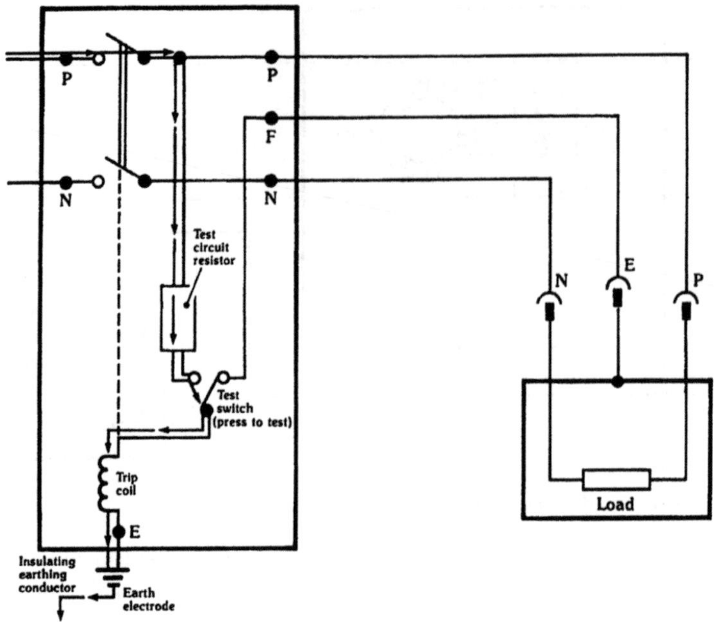

FIGURE 9.49 Fault voltage-operated earthed-leakage circuit breaker.

CROSS-BONDING

Because all three domestic services, gas, water and electricity, are brought to the premises by buried pipes or cables, they may each form an earth connection for stray currents. The strength of the current will depend on the resistance of the path provided, so the gas pipe could provide an easier path for the current than the proper electrical earth.

This could cause corrosion of the gas service. Worse still, it would mean that if someone disconnected the gas meter or part of the supply, they could get an electric shock or even cause an explosion.

To prevent this happening the three supplies are joined or 'bonded' by a conductor (Fig. 9.50). If the gas or water service pipes are of non-conducting material, the common bond should be connected to a conducting pipe. To satisfy the requirements of the Gas Safety (Installation and Use) Regulations, cross-bonding is applied to the first 600 mm (approx.) length of installation pipe from the outlet of the meter and must be attended to when installing or relocating the gas meter and associated pipework.

Electrical cross-bonding is a statutory requirement on PME systems, and the Institution of Electrical Engineers' (IEE) Wiring Regulations recommend it in other circumstances. To comply with the Gas Safety (Installation and Use) Regulations 1998, an installer who encounters a situation where cross-bonding is necessary must inform a responsible person (customer, builder etc.) that this cross-bonding should be carried out by a competent person.

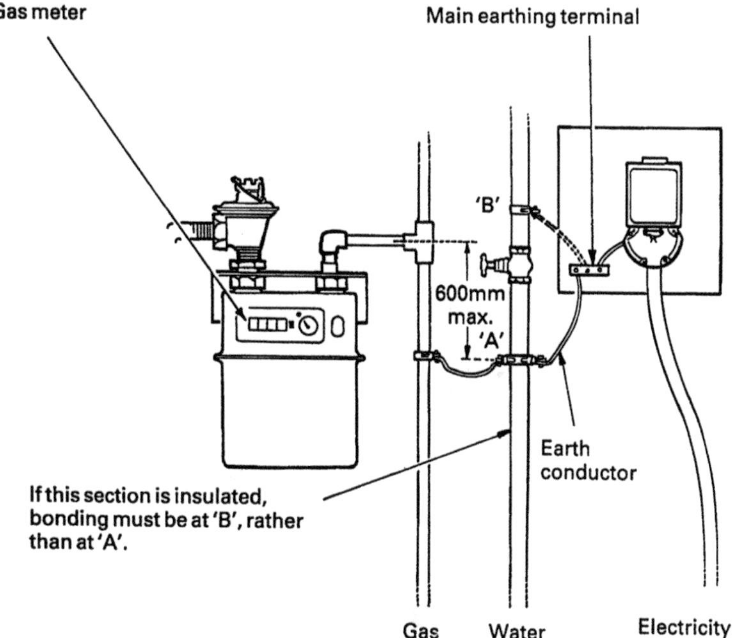

FIGURE 9.50 Electrical cross-bonding of services.

When a gas supply or a meter has to be disconnected, a temporary continuity bond must be attached before the supply is broken. This temporary continuity bond, bridging the inlet and outlet, gives any stray current a path back to earth and eliminates the risk of creating a spark, thus causing an explosion in a gaseous atmosphere, or the installer getting an electric shock. Failure to fit a temporary continuity bond in these circumstances is a breach of the Gas Safety (Installation and Use) Regulations 1998. The method of fitting and removing a temporary continuity bond is described in Volume 2, Chapter 3.

PHASE RELATIONSHIPS

In a DC circuit the voltage and the current are in phase. That is, they reach their full value at the same time.

In an AC circuit the voltage and current are rarely in phase except in the theoretical purely resistive circuit (Fig. 9.51a).

When a capacitor is introduced into the circuit the voltage is slowed down and so lags behind the current by about a quarter of a cycle or 90° (Fig. 9.51b). The property of a capacitor to oppose current is known as 'capacitive reactance'.

When AC flows along a wire it produces a changing magnetic field. As the field alternately grows and collapses, it induces an emf. This self-induced emf opposes the change of current in the wire and so causes the current to lag behind the voltage. In an inductive circuit the current lags the voltage by about a quarter

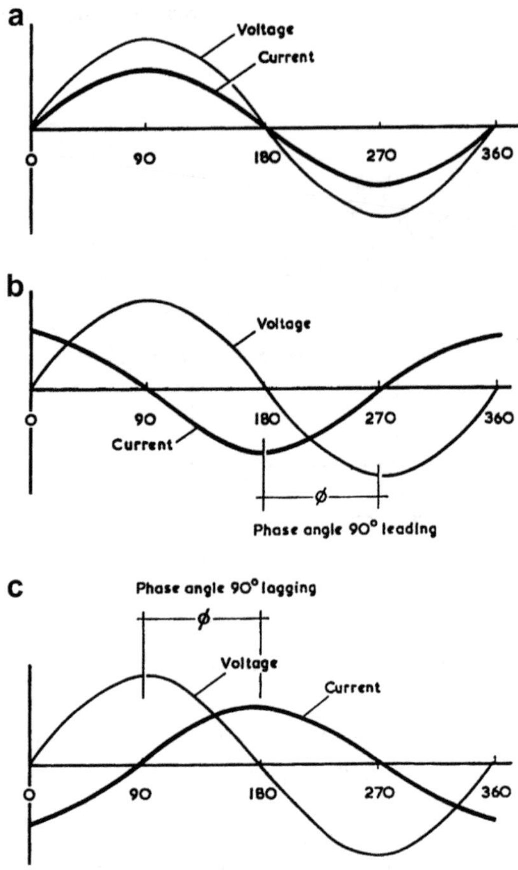

FIGURE 9.51 Phase relationships: (a) resistive circuit, voltage and current in phase; (b) capacitive circuit, current leads; and (c) inductive circuit, current lags.

of a cycle or 90° (Fig. 9.51c). Coils have much higher inductance than single wire and may be used to increase the inductance of circuits, when they are called 'inductors' or 'chokes'.

The unit of inductance is henry. When a current is changing at 1 ampere per second and the induced emf is 1 V, then the inductance is 1 henry. So:

$$\text{henry} = \frac{\text{volts}}{\text{amps per second}} \quad \text{or} \quad \frac{\text{volts} \times \text{seconds}}{\text{amps}}$$

The property of a coil to oppose current is called the 'inductive reactance'.

The total opposition to current flow in an AC circuit is a combination of resistance, capacitive reactance and inductive reactance. It is called 'impedance'. If an AC component has its resistance measured and Ohm's Law is used to calculate the rate at which current will flow, then it will be found that the actual flow is much less than that calculated.

For example, a 100 W bulb has a resistance of 40 Ω. This indicates a current flow at 240 V of:

$$I = \frac{E}{R}, \qquad I = \frac{240}{40} = 6\ \text{A}$$

But for the bulb to have a rating of 100 W, the actual current must be:

$$I = \frac{W}{V}, \qquad I = \frac{100}{240} = 0.4\ \text{A}$$

If the current is actually 0.4 A then the opposition to current flow must be:

$$R = \frac{V}{I}, \qquad R = \frac{240}{0.4} = 600\ \Omega$$

So the *impedance* of the bulb is 600 Ω, while its resistance is only 40 Ω. The difference of 560 Ω must be due to the inductive reactance of the coil and the increase in its resistance when the coil heats up.

POWER FACTOR

In a DC circuit both voltage and current are always at a maximum at the same time and power is measured by multiplying them together.

$$P = V \times I \text{ or watts} = \text{volts} \times \text{amps}$$

In an AC circuit, voltage and current may not be at a maximum at the same time, so the apparent power, volts × amps, called 'voltamperes', is likely to be greater than the true power in the circuit. The rating of equipment on an AC supply is normally quoted in voltamperes rather than watts because of this difference.

The ratio $\dfrac{\text{true power}}{\text{apparent power}}$ is called the power factor so,

$$\text{power factor} = \frac{\text{true power}}{\text{apparent power}} = \frac{\text{power}}{V \times I}$$

or

$$\text{true power} = V \times I \times \text{power factor (pf)} = \text{voltamperes} \times \text{pf}$$

In fact, the pf is the cosine of the phase angle φ between the voltage and current sine waves. So cos φ is leading in capacitive circuits and lagging in inductive circuits.

The nearer that the power factor can be made to approach 1, the greater the true power in the circuit. Inductance can be balanced with capacitance so as to bring voltage and current peaks nearer together and so raise the pf. Fitting a capacitor across the supply to an electric motor will have this effect.

Chapter | ten

Transfer of Heat

Chapter 10 is based on an original draft prepared by Mr E. Blandon

INTRODUCTION

The Gas Industry is very much concerned with the transfer of heat. Its principal task is to take the heat energy from the gas and transfer it to people, food, air, water or industrial products. It follows, therefore, that an understanding of the methods of heat transfer is required if this heat is to be effectively and efficiently transferred. Heat may be transferred by three principal processes:

1. conduction,
2. convection,
3. radiation.

Transfer can also be brought about by condensation or evaporation, that is by a change of state in a substance. When steam condenses into water, for example, it gives up its latent heat (Chapter 5).

In any practical situation the exchange of heat usually involves more than one of the transfer processes.

Heat, like water, always tries to reach a common level. A piece of pipe which has been heated will 'cool down'. That is, it will give up its heat to things in contact with it until they all reach a common temperature.

Any object at a high temperature will transfer its heat to its surroundings if they are at a lower temperature. Heat transfer always takes place in this one direction, from hot to cold. The only way to keep anything hot is to put heat into it at a faster rate than heat is being lost from it. This could be by adding more heat or by insulating the object to reduce the rate of heat loss.

CONDUCTION

In conduction, heat is passed from one molecule to the next one.

When a substance is heated the molecules gain energy and, in fluids, they move about more quickly. In a solid the molecules are like spectators at a packed football match. They cannot move about but they can jump up and down and turn around and bump into their neighbours. So, in a solid, heated molecules vibrate. And because they bump into the adjoining molecules, they pass on the vibrations and so the heat.

In some ways the conduction of heat is a bit like the conduction of electricity, and materials which are good conductors of the one are usually good conductors of the other. Copper is a good conductor of both. Similarly, non-conductors are called 'insulators' or 'insulating material' and, as with electricity, one of the best insulators is air.

If you hold one end of a piece of copper pipe about 300 mm long while soldering a fitting on to the other end with a propane lamp, you will soon learn about conduction! In a very short time the end in your hand will be too hot to hold. You will need a pair of pliers or a piece of rag to insulate you from the heat which has been transferred through the copper by conduction.

The capacity of a material to conduct heat is called its 'thermal conductivity', symbol k. This is a measure of the amount of heat energy which can be conducted in 1 s through an area of 1 m^2 across a length of 1 m for 1 °C difference in temperature between the two ends or 'faces' (Fig. 10.1).

$$k = \frac{\text{Heat flow} \times \text{thickness}}{\text{area} \times \text{temperature difference}} = \frac{\text{watts} \times \text{metres}}{\text{metres} \times \text{square} \times {}^\circ\text{C}} = \frac{\text{Wm}}{\text{m}^2{}^\circ\text{C}}$$

Table 10.1 shows the thermal conductivities of common materials.

The figures shown in Table 10.1 are the results of various experiments and are only approximate. Nevertheless, they serve to show the considerable difference in conductivity between the insulators and the gases on the one hand and the solid materials and particularly the metals on the other. Copper is obviously the best and CO_2 is the worst of those shown.

Insulators generally have cellular, granular or matted thread construction. These forms of structure break up a solid path for heat flow and trap small pockets of still air which offer considerable resistance to conduction.

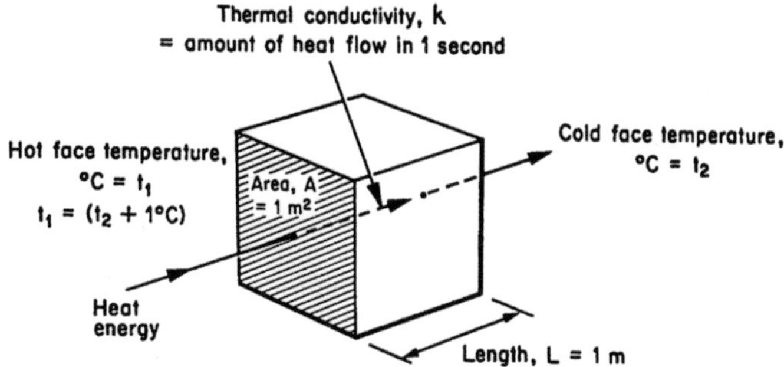

FIGURE 10.1 Thermal conductivity.

Table 10.1 Thermal Conductivities

Material	W/m °C
Metals (at 18 °C)	
Copper	384.2
Brass	104.6
Aluminium	209.2
Steel	48.1
Cast Iron	45.6
Lead	34.7
Building Materials	
Brick	1.15
Concrete	1.44
Plaster	0.58
Glass	1.05
Deal boards	0.12
Fluids (at 0 °C)	
Methane	0.029
Hydrogen	0.16
Carbon dioxide	0.014
Steam	0.015
Air	0.022
Water	0.054
Oil	0.18
Mercury	8.37
Insulating Materials	
Slag wool	0.042
Aluminium foil	0.042
Granulated cork/bitumen slab	0.15
Glass silk mats	0.040
Mineral wool slab	0.034
Fibre board	0.059
Vermiculite	0.067
Firebrick	0.61

The amount of heat which will be conducted through a section of a particular material in a given time, Q, is:

- proportional to the temperature difference between the two ends, or faces, $(t_1 - t_2)$, where t_1 is the temperature of the hot face and t_2 is the temperature of the cooler face;
- proportional to the cross-sectional area, A;
- inversely proportional to the length, or thickness, L.

So,

$$\text{heat flow} \propto \frac{\text{area} \times \text{temperature difference}}{\text{length}} \quad \text{or} \quad \frac{A(t_1 - t_2)}{L}$$

For any material,

$$\text{heat flow } (Q) = \frac{\text{thermal conductivity } (k) \times A(t_1 - t_2)}{L}$$

or

$$Q = \frac{kA(t_1 - t_2)}{L} \text{ watts}$$

where

k is thermal conductivity in W/m °C,
A is area in square metres,
$(t_1 - t_2)$ is temperature difference in degree Celsius,
L is length, or thickness, in metres,
Q is heat flow rate in watts.

This formula will enable calculations to be made to find the heat transmitted through a single piece of material, if required.

Heat is often conducted through several layers of different materials in a composite wall in, for example, a building or a furnace. The method of dealing with this is given, for use if required, as follows.

HEAT LOSS THROUGH COMPOSITE WALL

'Thermal conductance' is different from thermal conductivity.

The thermal conductance of a section of material is the amount of heat energy which will pass in 1 s through an area of 1 m^2 when the temperature difference is 1 °C, or:

$$\text{thermal conductance } (h) = \frac{k}{L} \text{ watts/m}^2\ {}^\circ\text{C}$$

where k = thermal conductivity, L = thickness.

The amount of heat passing through a composite wall of different materials having thermal conductivities of k_1, k_2, k_3, ... and corresponding thicknesses of L_1, L_2, L_3, ... is:

$$\text{Heat flow, watts} = hA(t_1 - t_2)$$

where h, for a composite wall, is obtained

$$\frac{1}{h} = \frac{L_1}{k_1} + \frac{L_2}{k_2} + \frac{L_3}{k_3} + \cdots$$

or

$$\frac{1}{h} = \frac{1}{h_1} + \frac{1}{h_2} + \frac{1}{h_3} + \cdots$$

where h_1, h_2, h_3, ... are the thermal conductances of the composite layers.

All the formulae given are for heat conduction under steady-state conditions. That is, when the materials have heated up and the temperature gradient through the layers is constant. The temperature gradients in the different layers will depend on the thermal conductivities of the various materials and their thicknesses (Fig. 10.2).

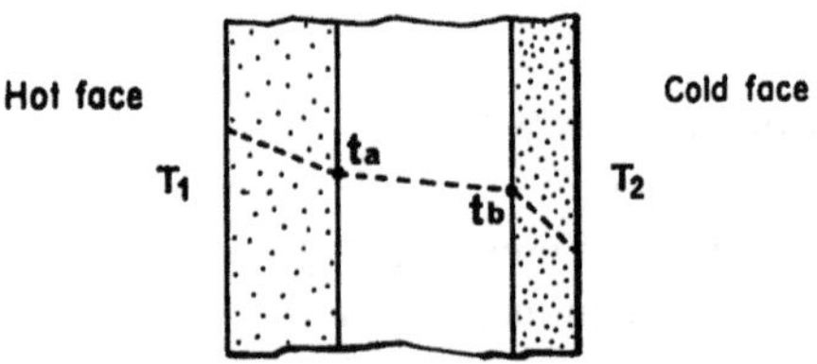

FIGURE 10.2 Temperature gradients in a composite wall.

HEATING BY CONDUCTION

Heating by conduction occurs when an object is in contact with a heat source. For example, a cooking pot, in contact with a flame or a solid hotplate, will conduct heat through the food inside. The solid hotplate itself is a means of conducting heat from the flame below to the pan above.

Wherever a fluid has to be heated the heat must be transferred to it through the walls of its container by conduction. Usually this heating is carried out in a specially designed section of the appliance called the 'heat exchanger'.

Conduction is often a major means of heat loss. It conveys heat away from those things that we are trying to keep hot. Houses lose heat through their walls, ovens lose heat through the sides and the door panels. Pan supports on a cooker hot plate conduct heat away from the pans and the flames.

Liquids and gases can conduct heat but, as Table 10.1 shows, they are not good conductors. As the molecules are heated they move about more quickly and so do not pass the heat directly to the nearest adjoining molecule. The increase in energy causes the fluid to expand so that the spaces between the molecules become even greater and a continuous transfer of vibrations is not possible.

CONVECTION

Convection is the form of heat transfer which takes place in liquids and gases due to movement of the heated fluid. When the convection is caused solely by the heating, it is called 'free' or 'natural' convection. If the fluid is circulated by mechanical means that is by the pump or a fan, it is called 'forced convection'. Forced convection gives a considerably increased rate of heat transfer.

When a fluid is being heated, it expands. So the density of the heated particles becomes less than that of the unheated portion. The heavier, colder fluid displaces the relatively lighter, heated part, pushing it upwards, away from the heat source. As the less dense fluid moves upwards, its place is taken by colder fluid which, in its turn, is heated and moves upwards. In this way a 'convection current' is set up, transferring heat throughout the whole mass of the fluid (Fig. 10.3a).

The circulation will continue while there is a temperature difference between different parts of the fluid. The rate of circulation will depend on the difference in temperature between the hot and the cold fluid and also on the height of the circulating columns of fluid.

Figure 10.3b shows part of a water heating system. It consists of a boiler and a storage cylinder connected by pipes. When the boiler is operating it sets up a convection current which eventually heats up all of the water in the same way that the fluid is heated in the pan at Fig. 10.3a.

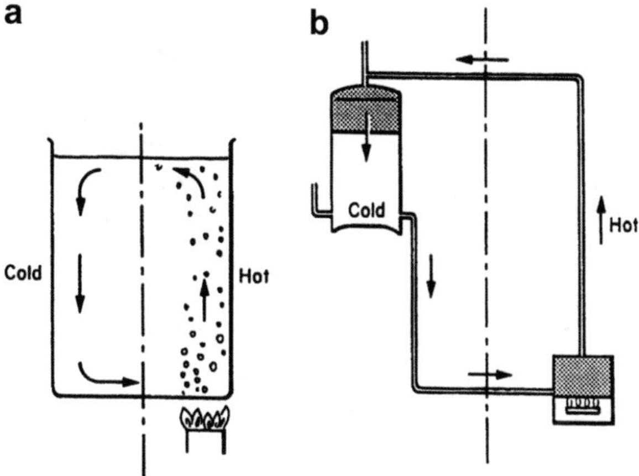

FIGURE 10.3 Convection currents in liquid: (a) in an open vessel and (b) in a closed water heating circuit.

The circulating pressure, or head, can be determined if the densities of the hot and cold water and the height of the circulation in the system are known (see Chapter 6, section on Circulating Pressure).

Convection currents set up in heated air may be used to carry heat throughout a room. The air in a room loses heat to any surfaces which are at a lower temperature. These are the walls, floor, ceiling and windows. The window and the outside walls cause the greater heat loss.

Window glass has a thickness of only about 2 mm and air in contact with it is rapidly cooled. As it becomes colder and denser the air falls downwards over the window and its place is taken by warmer air which has previously risen up to the ceiling. This movement of cold air creates a draught which travels outwards across the floor.

The effect of putting a heater in the room is shown in Fig. 10.4. In Fig. 10.4a the heater is positioned at the opposite end of the room to the window. The effect

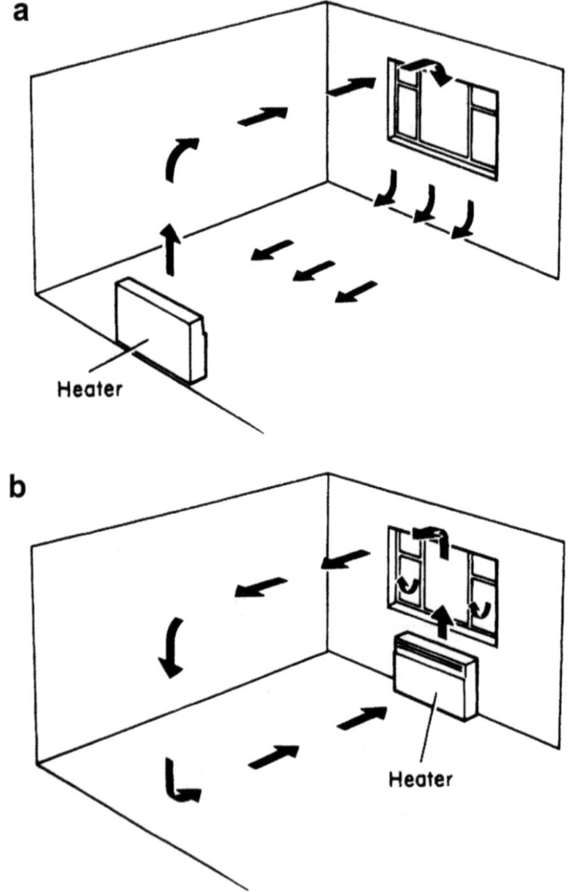

FIGURE 10.4 Convection currents in a room: (a) heater at opposite end to window – air cooled by the window becomes a cold draught across the floor; (b) heater under window – cold air is warmed and carried up by convection current.

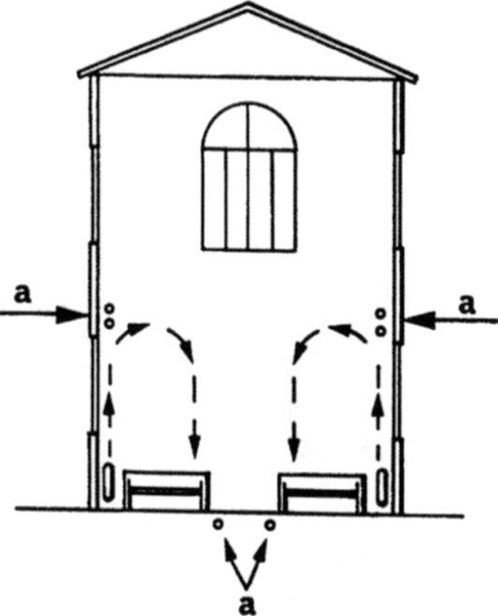

FIGURE 10.5 Convection currents in a high building.

is to set up a convection current which still allows the window to produce a cold draught. Fitting the heater under the window heats up the cold air and disperses it, so counteracting the draught.

As a general rule, the best position for a convector heater is in the place with the highest heat loss.

If convection currents are set up in a high building such as a church, the air will reach a height of about 3 m before becoming cooled so that it begins to fall (Fig. 10.5). The cold draught in the middle of the building may be counteracted in a number of ways. One method would be to provide additional heating circuits at the points marked 'a'. In this type of building, other methods of heating may be more effective.

STRATIFICATION

'Stratification', or the formation of layers, occurs in fluids when heated. This comes about in the first place because the fluid is heated by convection. So hot fluid is made to rise and forms a hot, lighter layer resting on top of a colder denser layer. Figure 10.6 shows a hot water storage cylinder containing water at 60 °C lying on top of cold water at 15 °C.

Having produced this condition by a convection current, it will remain stable for a considerable time if nothing is done to disturb it. This is because water, like most fluids, is a very poor conductor of heat. So there is little conduction of heat between the two layers and no convection is set up because the hot layer is already on top of the cold layer.

This phenomenon is made use of in water heating because it allows small quantities of hot water to be drawn off long before the whole storage has been heated. The rate of circulation is adjusted so that the circulator or boiler delivers hot water at about 60 °C. When hot water is required cold water is introduced into the bottom of the cylinder, so pushing the hot water upwards and out of the taps.

Stratification also occurs in air and a room heated by natural or free convection will have a warm layer of air just below the ceiling.

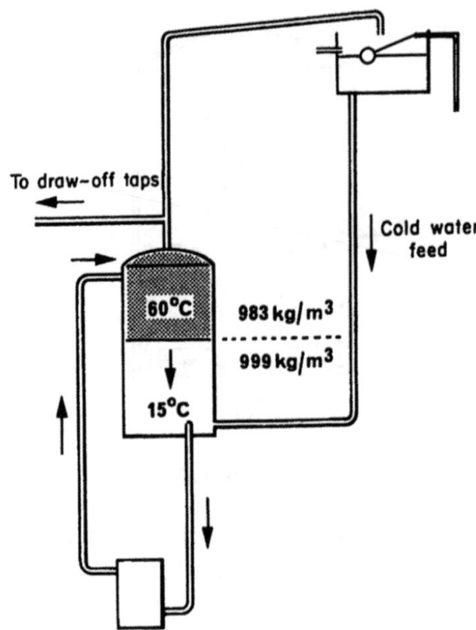

FIGURE 10.6 Stratification in a cylinder.

Between 0 and 4 °C water reverses its behaviour; it expands when cooled and contracts when heated. Convection takes place in the opposite direction, the colder water rising until it freezes and the ice forms on top of the water. Because of this reversed stratification deep ponds and rivers rarely freeze solid and fish continue to live under the ice.

CONVECTED HEAT FLOW

The rate of heat flow from a surface transmitted by free convection depends on:

- the area of the heated surface;
- the temperature of the surface;
- the temperature of the surrounding air (the 'ambient' temperature);
- the aspect of the surface, that is whether it is horizontal facing upwards or downwards, or vertical.

The amount of heat can be calculated, if required, using the following formula:

$$Q = CA(T_1 - T_2)^{1.25}$$

where Q = heat flow in watts,

A = area of surface, m^2,
T_1 = temperature of the surface, Kelvin,
T_2 = temperature of ambient air, Kelvin,

C = coefficient of convection for the following aspects of the surface:
vertical, $C = 1.9$;
horizontal facing downwards, $C = 1.3$;
horizontal facing upwards, $C = 2.5$.

These values for C are for freely exposed surfaces, that is not complete floors or ceilings.

The calculation also assumes draught-free conditions. If there is some air movement, the amount of heat transfer increases considerably. With air movement of 0.5 m/s an increase of 35% could be obtained.

RADIATION

Radiation is the transmission of heat by means of electromagnetic waves. The waves which carry rays of heat are the same as those which transmit cosmic rays, X rays, light, and radio and television signals. They all travel at the same speed which is 186,282 miles per second or 2.9979×10^8 m/s. The difference between them is only in the length of the wave form. Figure 10.7 shows the different 'wavelengths' of the various rays.

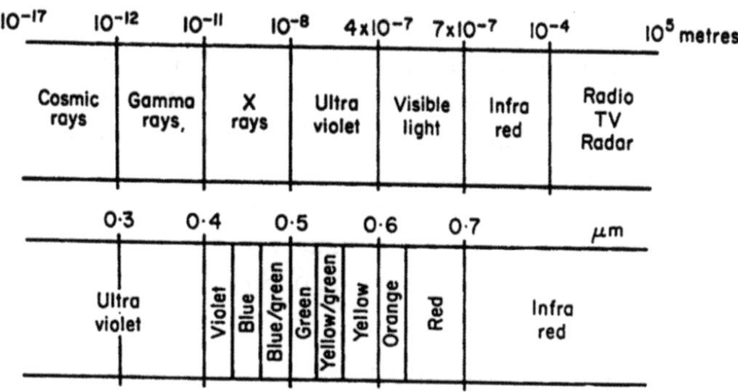

FIGURE 10.7 Electromagnetic wavelengths.

The sun is a good example of a source of radiant energy. Its rays come to us through ninety million miles of space and a few miles of air. Solar energy can be used to provide heat for our homes and our water systems. It can warm the seas and the land and ripen our crops. It provides light for our days and its ultra-violet rays produce our holiday suntans.

Radiant heat passes through the air without heating it. Like light it travels in straight lines, so it only heats those objects directly in its path. Radiant heat gives an immediate feeling of warmth but a whole room will only become warm when the radiant heat has heated up the surroundings and the furniture, and these have heated the air by convection. Figure 10.8 shows the areas being heated by a gas fire and convection currents beginning to circulate.

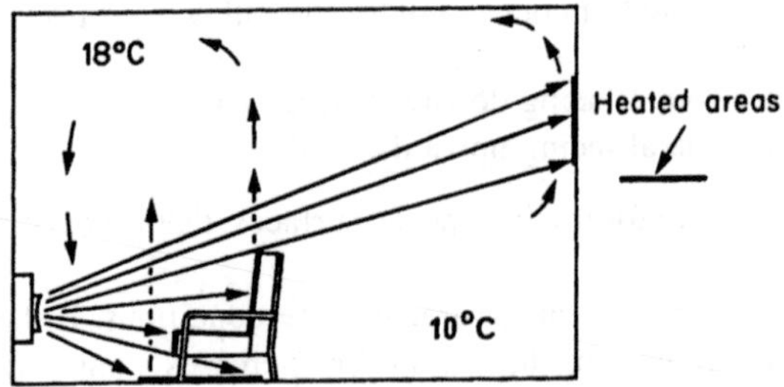

FIGURE 10.8 Room heated by radiation from a gas fire.

For convenience, radiant heating appliances are divided into three categories:

1. high temperature radiation,
2. medium temperature radiation,
3. low temperature radiation.

High temperature radiation

In this category are the gas fires or 'room heaters' which have incandescent refractory radiants. These appliances are often designed to have, in addition to their radiant output, an output of convected heat in order to heat the room to a comfortable temperature more quickly (Volume 2, Chapter 7).

Overhead radiant heaters are used in tall buildings such as factories or warehouses where heat is needed only in those places where people are actually working. They can be used in ice rinks or in the open air at stadiums to warm the spectators (see Volume 3, Chapter 8).

Grilling or spit roasting is cooking by high temperature radiation.

In all these appliances the heat source can be kept small because its temperature is high, probably in the region of 800–1350 °C.

Medium temperature radiation

Panels at a temperature of about 345 °C can be used for space heating. They emit in the infra-red band and are widely used in many industrial drying processes (Volume 3, Chapter 3).

Low temperature radiation

Because their temperature is below 82 °C, these appliances require a greater surface area to give the same heat output as those at higher temperatures. The most familiar is probably the hot-water radiator fitted on a central heating circuit. All radiators give some heat in the form of convection and the type in which the waterways were made in the form of small columns was principally a convector heater.

Heating a floor, wall or ceiling to a temperature of about 27 °C provides a low-temperature heat source for a room. It will also provide a source of convected heat, depending on its aspect.

RADIANT HEAT FLOW

The sun is an obvious source of radiation. What is not so obvious is that all objects which are at temperatures above absolute zero (−273 °C or 0 K) are giving off radiant heat. At the same time all objects or 'bodies' are receiving radiant heat from those which are at higher temperatures than themselves.

The rate at which heat is absorbed, or emitted, by a body depends on its absolute temperature and also on its capacity for emitting (or receiving) radiation. This is called its 'emissivity'. Materials or surfaces which do not reflect the waves, or allow them to pass straight through, are the best absorbers of heat and also the best emitters. They have the highest emissivities. A matt black finish is the best, and a 'perfect' emitter, which would theoretically have an emissivity of 1, is called a 'black body'. Table 10.2 gives the emissivities of various materials.

Table 10.2 Emissivities of Common Materials

Material or Surface	Emissivity (at 50 °C)
Aluminium	
Polished	0.04
Anodised	0.72
Carbon black	0.96
Cast iron	0.80
Chromium plate	0.10
Copper	
Polished	0.03
Oxidised	0.86
Stainless steel, polished	0.15
Galvanised steel	0.25
Paint	
Aluminium	0.50
Gloss white	0.95
Matt black	0.96

Joseph Stefan, an Austrian scientist, discovered that the rate (Q) at which a black body emitted heat was proportional to the fourth power of its absolute temperature (T), or

$$Q \propto T^4$$

From this was derived the Stefan–Boltzmann Law which is:

$$Q \propto AET^4$$

where Q = heat units radiated in unit time (one second)
A = area of surface (projected area),
E = emissivity (black body = 1),
T = absolute temperature of the body.

If it is required to calculate the actual amount of heat flow, from a surface to its surroundings, the formula is:

$$Q = CAE(T_1^4 - T_2^4)$$

where C is Stefan's constant, which is 5.67×10^{-8} W/m^{2}K^4

Q = heat flow, watts,
A = area of the radiating surace, m^2,
E = emissivity of surface,
T_1 = temperature of surface, kelvin,
T_2 = mean temperature of surroundings, kelvin.

INVERSE SQUARE LAW

Because radiation travels in straight lines it obeys the 'law of inverse squares'. This means simply that as heat travels further from its source it is spread over a larger area. The actual area increases as the square of the distance. Therefore the heat flow, per unit area, is proportional to the square of the distance from the source (Fig. 10.9).

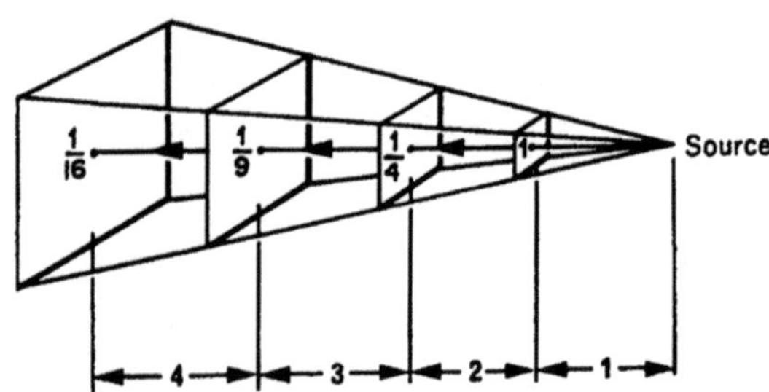

FIGURE 10.9 Inverse square law.

WAVELENGTH

The radiation from a body at a particular temperature is within a band of wavelengths. As the temperature increases the band moves to shorter wavelengths until part of the band has moved into the visible section of the spectrum. The body emits light as well as heat.

Most heat is transmitted within the infra-red waveband, but there can be an overlap at each end of the band. At the shorter wavelengths some materials can become red-hot, by contact with a flame.

At the larger wavelengths there is 'micro-wave' or 'radio-wave' cooking using waves from 1 to 300 mm or from almost the infra-red up to the shorter radio waves.

HEAT EXCHANGERS

The 'heat exchangers' with which we are concerned are constructed to transmit heat from one fluid to another. This heat exchange may be from:

- gas to water,
- gas to air,
- water to water,
- water to air.

Because they are both fluids, the two participants in each exchange must be kept apart, usually by the dividing wall of a container or by confining one fluid within a pipe. And the dividing medium must be long enough or large enough to allow the required amount of heat to be transferred.

A common problem with all fluid heat exchangers is associated with the tendency of the fluids to adhere to the dividing wall. This results in the formation of a static layer of fluid on each side of the wall. The only way heat can pass through the layers is by conduction and fluids are not very good conductors.

To overcome this, the static layers or 'films' are kept to a minimum thickness by creating turbulence in the fluids. This can be done by repeatedly changing the direction of movement of the fluid, by, for example, successive bends in the pipe which carries it. Or, in the case of air or gas, by blowing it through the heat exchanger with a fan. That is, by using forced convection.

So there are five stages in the transfer of heat from one fluid to another (Fig. 10.10).

1. Convection from hot fluid to its static film.
2. Conduction through static film.

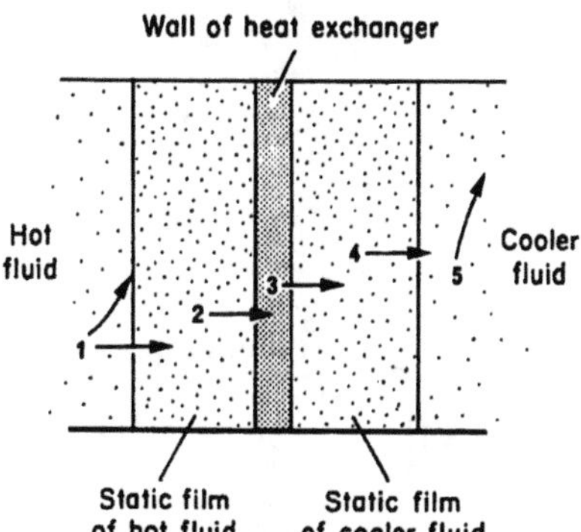

FIGURE 10.10 Heat transfer from a heat exchanger.

3. Conduction through wall of heat exchanger.
4. Conduction through static film of cold fluid.
5. Convection from static film to cold fluid.

TYPES OF HEAT EXCHANGER

Gas to waters

Under this heading are grouped the heat exchangers for instantaneous and storage water heaters and central heating boilers. They frequently use 'fins' to provide a large area in the path of the hot gases from the flames. The fins conduct heat into the pipes or annulus carrying the water. The water vapour in the products of combustion will condense on a cold surface. So, if continued condensation is to be avoided, the heat exchanger must be able to heat up fairly quickly.

These heat exchangers are made from copper, stainless steel and, in the case of some boilers, cast iron or welded plates. Fins are often in the form of a bank of flat plates and, in cast iron, conical 'pips' or rods may also be formed.

Figure 10.11 shows a section through the heat exchanger of an instantaneous heater. Both hot and cold water pipes are attached to the 'skirt' of the heat exchanger and prevent it from becoming overheated. Other finned types are illustrated. The heat exchangers illustrated in Fig. 10.11 are typical of types fitted in modern water heaters and high-efficiency boilers. Generally they take out 5–80% of the heat available from the flue gases and transfer it to the water. The remaining 20–25% of the heat is used to operate the flue and passes to the atmosphere.

Modern boilers have been developed, using mechanical means to remove the flue products that can transfer over 95% of the heat available to the water. In this type of boiler, shown in Fig. 10.12, primary and secondary heat exchangers are used. When working at its highest efficiency the water vapour in the flue gases condenses and is drained away. More details of condensing boilers are given in Volume 2.

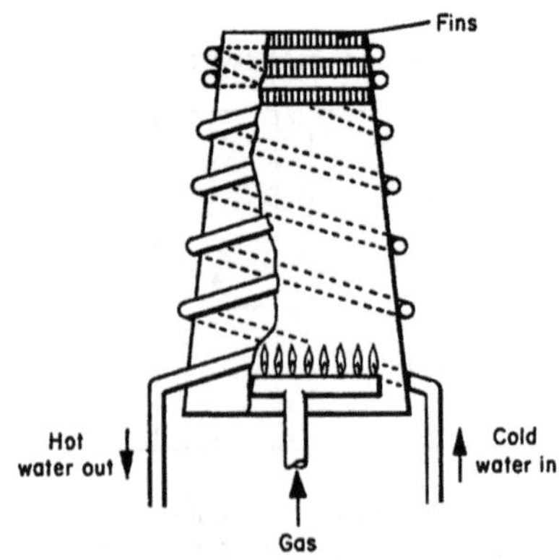

FIGURE 10.11 Heat exchanger, instantaneous water heater.

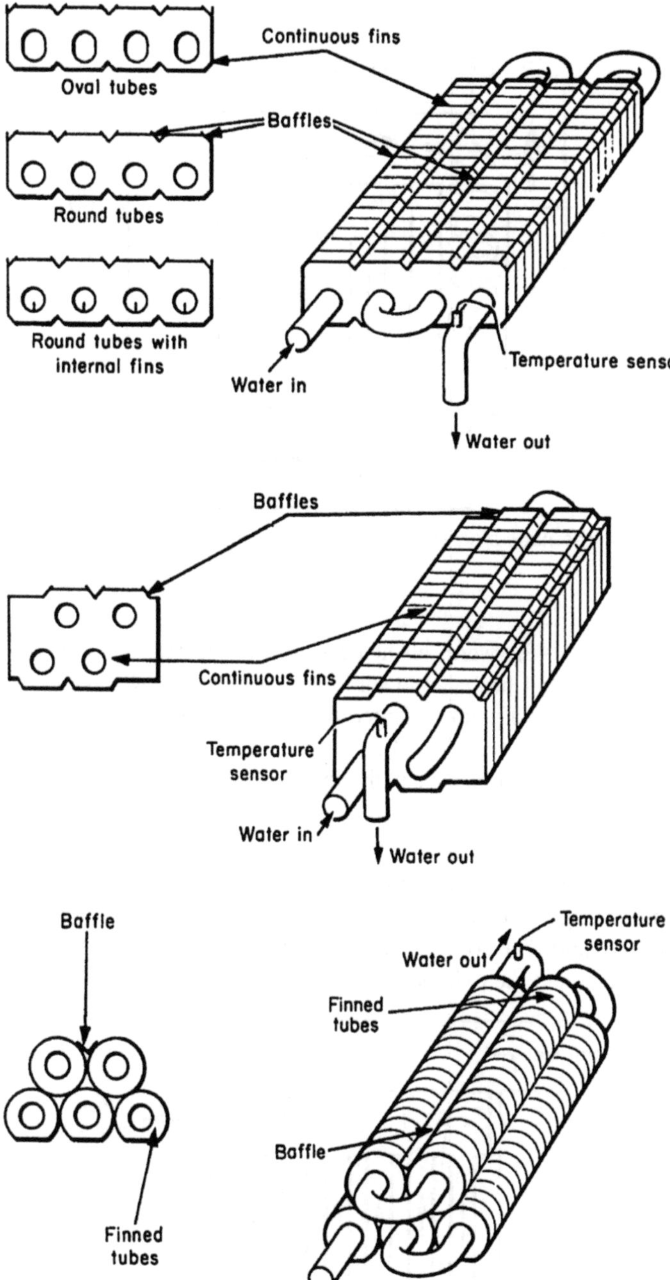

FIGURE 10.11 Cont'd. Other finned types of heat exchangers.

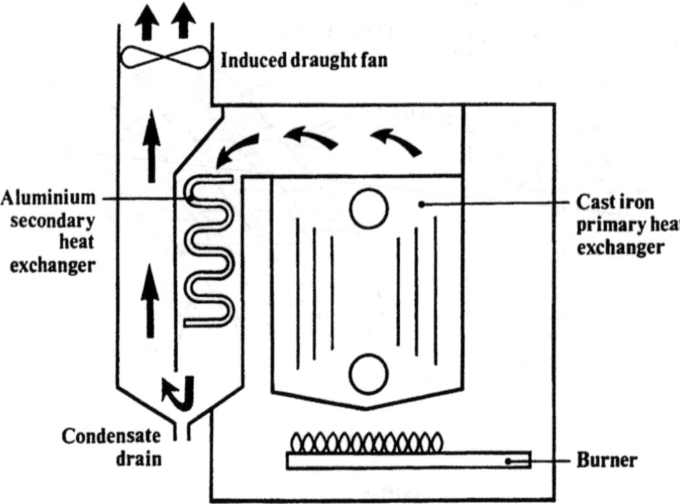

FIGURE 10.12 Schematic diagram of a condensing boiler.

Gas to air

In this category are the warm air heaters, unit air heaters and convector room heaters. The warm air heaters and some unit air heaters use forced convection, blowing the air around plain or finned tubes through which the hot gases pass towards the flue outlet (Fig. 10.13).

Convector room heaters have plain heat exchangers with natural convection.

Water to water

These are the 'calorifiers' and 'immersion tubes' used to heat domestic water in a hot water storage cylinder by means of hot water circulated from a boiler or central heating unit. The hot circulating water and the water being heated are not in direct contact and the cylinder is called an 'indirect cylinder'.

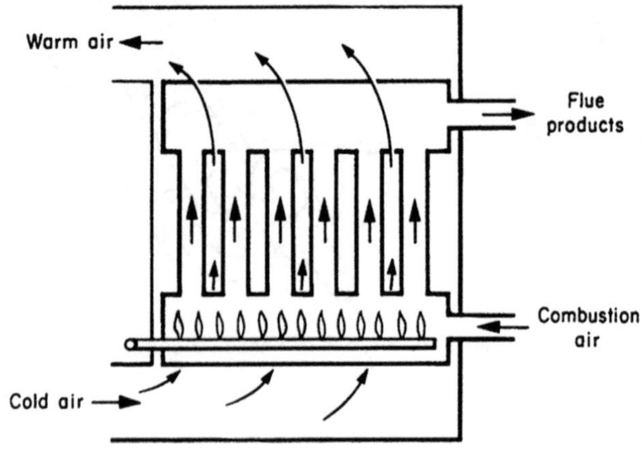

FIGURE 10.13 Heat exchanger, warm air unit.

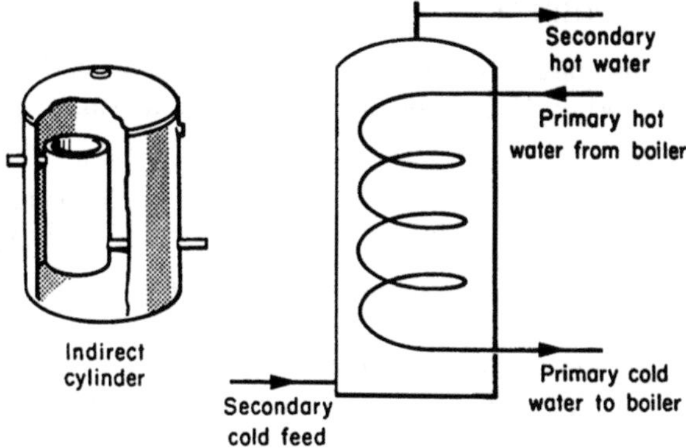

FIGURE 10.14 Calorifiers.

FIGURE 10.15 Immersion calorifier (Wednesbury Ltd).

Calorifiers usually take the form of hollow cylinders or coils of pipe. The hollow cylinders may be corrugated to have a larger surface area (Fig. 10.14).

Now that the water in the heating circuit is often pumped it is possible to insert immersion tubes in the domestic cylinder in place of the electric immersion heater (Fig. 10.15). Some tubes may have small fins formed on them.

Water to air

Examples of this form are the convectors, fitted on central heating circuits, and the 'radiators'. (A car radiator is another example.) The convectors usually have tubes containing the hot, circulating water, passing through banks of fins. A fan forces air through the fins and into the room. The hot-water radiator has a large flat surface in order to transmit more of its heat by radiation.

HEAT LOSS FROM BUILDINGS

Reference has already been made to the fact that rooms lose heat to the outside air. This loss is due to:

1. heat loss by ventilation, that is by the warmed air in the room being changed at least once or twice an hour;
2. heat loss by conduction, through walls, floor, ceiling and windows.

Heat loss by ventilation

To raise the temperature of 1 m^3 of air, through 1 °C in 1 s will require approximately 1340 W. So, the amount of heat required to make good loss of air by ventilation is:

$$\text{Heat required (W)} = \text{volume of room } (m^3) \times \frac{\text{air change/hour}}{3600} \times \text{temperature rise } (°C) \times 1340$$

For example: A room is 4 m × 5 m and 3 m high.
Air change is 2 per hour. Inside air temperature is 21 °C.
Outside air temperature is 0 °C.

So, volume of air to be heated per second

$$= (4 \times 5 \times 3) \times \frac{2}{3600} = \frac{120}{3600}\,m^3$$

$$\text{Temperature rise} = 21 - 0 = 21\ °C$$

$$\text{Heat required} = \frac{120}{3600} \times 21 \times 1340\ W = 938\ W \text{ or } 0.938\ kW$$

Heat loss by conduction

As you have seen, it is possible to calculate the amount of heat which can be conducted through the various layers of material in a structure. In dealing with buildings the factor used is the 'thermal transmittance' coefficient or '*U*' value. This takes into consideration the thickness of the structure and the resistance to heat flow caused by the air film on the surfaces of the structure.

The *U* value is the amount of heat which will flow in 1 s through a particular piece of the structure for every 1 °C temperature difference. So,

$$U = W/m^2\ °C$$

Tables of *U* values are available and these take into consideration the aspect or orientation and the degree of exposure of the building. An example is given in Table 10.3.

The amount of heat lost through a section of the structure, such as a wall, is obtained from:

$$\underset{(W)}{\text{heat loss}} = \underset{(m^2)}{\text{area}} \times \underset{(°C)}{\text{temperature rise}} \times \underset{(W/m^2\ °C)}{\text{U value}}$$

For example:
A room has an exposed wall 5 m × 3 m. It is a 280-mm brick wall with an unventilated cavity and it faces east. The *U* value is 1.8 W/m^2 °C.

Table 10.3 *U* Values for Typical Building Structures

Construction		U value (W/m^2 °C)		
		Sheltered	Normal	Severe
Windows	Single glazing wood frame	3.8	4.3	5.0
	metal frame	5.0	5.6	6.7
	Double glazing wood frame	2.3	2.5	2.7
	metal frame	3.0	3.2	3.5
Floors	Solid, ground floor, 3 m × 3 m			
	4 exposed edges		1.47	
	2 exposed edges at 90°		1.07	
	Suspended wood ground floor			
	3 m × 3 m		1.05	
	bare or lino tiles		1.05	
	carpet or parquet		0.99	
Roofs	Pitched roof, tiles on battens, felt and rafters. Plasterboard ceiling	1.4	1.5	1.6
	As above with boarding on rafters	1.3	1.3	1.3
	As above with 100 mm glass fibre insulation	0.33	0.34	0.35
Walls	Brick. solid, plastered, 220 mm	2.0	2.1	2.2
	Cavity, plastered, 260 mm	1.4	1.5	1.6
	375 mm	1.2	1.2	1.2
	Concrete block. Solid with tile and plaster, 150 mm	0.95	0.97	1.0
	Cavity, plastered, 75 mm outside, 100 mm inside	0.82	0.84	0.86

The temperature difference is from −1 °C outside to 18 °C inside = 18 − (−l) = 19 °C.

Area of the wall is 5 × 3 = 15 m^2. So, heat loss = 15 × 19 × 1.8 = 513 W or 0.513 kW.

The total heat loss calculation for a complete heating installation should also take into consideration heat gains from occupants and other appliances and sources.

INSULATION

Some of the properties of insulating materials have already been dealt with. The main methods of insulating dwellings are by:

- weather-stripping doors and windows to prevent excessive air changes;
- insulating the roof-space with fibreglass or exfoliated mica;
- double-glazing the windows;
- insulating cavity walls with rock-wool or other appropriate materials.

The effect of these methods on an average semi-detached house would result in a reduction of the design heat loss from 12 kW to only 4 kW.

Moves towards energy conservation have resulted in development work on low-energy housing which in turn influences the levels of insulation defined in Building Regulations and other specifications. From 1 April 2005, revised Approved Document L1 of the Building Regulations sets down revised guidance for the efficiency of hot water central heating gas and oil boilers installed in new and existing dwellings. From that date condensing boilers with a Seasonal Efficiencies of Domestic Boilers in the UK (SEDBUK) efficiency in band A or B shall be installed, unless there are exceptional circumstances that make this impractical or too costly.

Figure 10.16 shows the method which should be used effectively to insulate a domestic dwelling.

From 14 December 2007, properties marketed for sale in England and Wales will require a Home Information Pack, which includes a home energy rating. From October 2008 properties in the rental sector came into scope.

The certification of buildings is carried out in an independent manner by qualified and/or accredited experts. (A Domestic energy Assessor.)

One of the biggest contributors to global warming is carbon dioxide. The way we use energy in buildings causes emissions of carbon. The energy we use for heating, lighting and power in homes produces over a quarter of the UK's carbon dioxide emissions and other buildings produce a further one-sixth.

The average household causes about 6 tonnes of carbon dioxide every year. Adopting the recommendations in the energy reports can reduce emissions and protect the environment. You could reduce emissions even more by switching to renewable energy sources. In addition there are many simple every day measures that will save money, improve comfort and reduce the impact on the environment, such as: check that your heating system thermostat is not set too high (in a home, 21 °C in the living room is suggested) and use the timer to ensure you only heat the building when necessary.

Make sure your hot water is not too hot – a cylinder thermostat need not normally be higher than 60 °C. Turn off lights when not needed and do not leave appliances on standby. Remember not to leave chargers (e.g. for mobile phones) turned on when you are not using them.

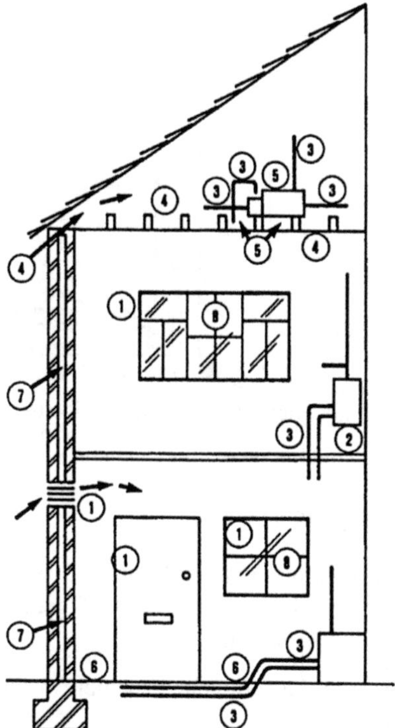

1: Weatherstrip windows and titemal doors *but DO NOT* block any purpose* mode ventilation such al air bricks.

2: Fil an 160mm thick mintrol fibrt jacket, to *BS* 5o75, onto the hot water storage cylinder.

3: Insulate any water carrying pipes not in the heated part of the dwelling.. Install Energy efficient condensing boiler

4: Lay 300mm thickness of mineral fibre between roof joists but do not block ventilation into roof space.

5: Insulate cold water cistern and feed and expansion cistern but *NOT* their bases nor the area between the joists beneath them.

6: Lay continuous floor covering, preferably carpet, after sealing any gaps between skirting and floor boards.

7: Have cavity walls professionally filled with appropriate materials.

8: Consider fitting double glazing to windows.

FIGURE 10.16 How to insulate a domestic dwelling effectively.

Two indicators are displayed on the Energy Certificate:

1. Energy Efficiency Rating (SAP* rating) and
2. Environmental Impact Rating (Carbon dioxide (CO_2) emissions), see Fig. 10.17.

The certificate will also: (a) suggests improvements that might be carried out, (b) show how they would improve the home's rating improvements explained in financial detail.

---

*Standard Assessment Process (SAP) takes account of: heating system, fuel type and control systems, DHW system & controls, ventilation and infiltration (air permeability), thermal insulation and any thermal gain, energy used for lighting.

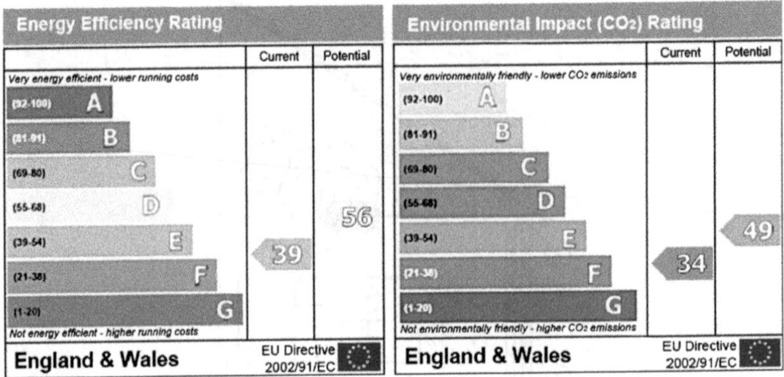

FIGURE 10.17 Environmental Energy/Impact Rating (Pictures courtesy of the Energy Saving Trust, EST).

COMFORT CONDITIONS

These were discussed in some detail in Chapter 5 and it is now necessary to examine the effects of the different methods of heating on human comfort. Basically we are comfortable when:

- we are losing heat at the same rate at which we produce it, about 0.5 MJ/h or 0.14 kW;
- we are as cool as is compatible with comfort;
- there is adequate, but not excessive air movement, between 0.1 and 0.2 m/s;
- the relative humidity does not exceed 70%;
- the air is changed about 1½ times per hour to keep it fresh and relatively germ-free;
- our heads are not more than 2.8 °C hotter than our feet. The floor to ceiling gradient should not exceed 4.5–5.6 °C;
- heads are not subjected to excessive radiant heat;
- the surrounding walls are at least as warm as the air.

Whilst the quickest way to heat a room is to fill it with warm air, that is to heat it by convection, it is obvious from the criteria for comfort that it is better to have some radiant heat source, if possible.

With convected heat only there is likely to be a fairly high floor to ceiling gradient. The walls are likely to be at a lower temperature than the air and a temperature of about 18–21 °C is required for comfort.

With radiant heat the air temperature can be as low as 15–18 °C without producing discomfort. The surroundings may become as warm as the air and a direct gain of radiant heat source will produce cheerful warmth in a healthy atmosphere. The radiant heat needs to be at low level and cold draughts should be prevented by weather-stripping and insulation.

The range of gas space heating and central heating appliances available makes it possible to select the type, or combination, of heating methods appropriate to any particular situation.

Chapter | eleven

Gas Controls

Chapter 11 is based on an original draft prepared by Mr R.S. Pryor

INTRODUCTION

One of the main advantages of gas, as a fuel, is its ease of control. It can be turned up or down, on or off, quickly and simply, manually or automatically. A wide range of devices has been developed and those used in domestic appliances are dealt with in this chapter. Controls may be used for a variety of purposes including:

- to provide safe cut-off in the event of hazards due to pressure failure or fire;
- to give protection from the results of flame failure;
- to control for time, temperature or pressure;
- to provide safe ignition.

Pressure control has been dealt with in Chapter 7 and although there is brief reference to some heating and electrical controls in this chapter, these will be covered in greater detail in Volume 2.

COCKS, TAPS AND VALVES

The simplest form of manual control of gas flow is by means of a cock, tap or valve. These three terms are often confused but strictly they have slightly different meanings when applied to gas.

A 'cock' is a device fitted to a gas supply usually controlling the flow of gas by the rotation of a drilled or slotted, tapered plug.

A 'tap' is a form of cock fitted as an integral part of an appliance. It may have a plug or a disc to control gas flow.

A 'valve' controls gas flow through an orifice by movement of a 'gate', flap or disc. It is usually larger than a cock and generally used on service pipes, mains and industrial supplies. It can withstand higher pressures than the ordinary plug cock.

A simple cock is shown in Fig. 11.1a. It is made of brass and consists of a tapered plug, through which a hole has been made, held in a tapered barrel by

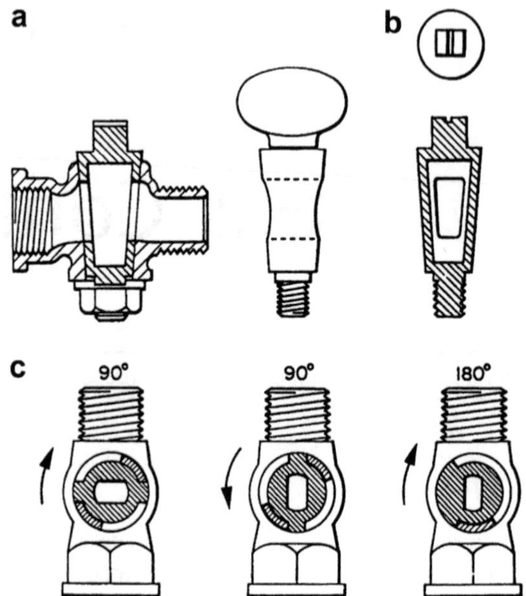

FIGURE 11.1 Plug cock: (a) section through cock; (b) fan or square head; (c) niting washers.

means of a washer and nut. Rotating the plug to bring the hole in line with a gas way through the body of the cock allows the gas to pass.

The plug is turned by means of a thumb-piece, called a 'fan' or by having a square top on which a lever can be fitted (Fig. 11.1b). The fan is in line with the drilling in the plug, so the tap is turned on when the fan is in line with the gas supply. The square head on the plug has a groove cut across it which also indicates the direction of the drilling and the lever should be fitted so that, like the fan, it lines up with the supply when the cock is on.

An arrangement is usually employed to make sure that the plug stops in the full-on and full-off positions. This is called the 'niting'. On the cock in Fig. 11.1, the washer which adjusts the tightness of the plug in the body is also a 'niting washer'. It has a tab which engages with stops on the cock body and limits the plug movement. Niting of 90° is common, Fig. 11.1c, although the larger cocks with loose levers have 180° niting.

Figure 11.2 shows the type of cock used on the inlet of a domestic customer's meter to control the supply of gas to the premises. This has had several different names. Often called the 'main cock' it eventually became the 'consumer's control' and now (perhaps because it never did control the consumer) it is the 'meter control' or 'meter control cock'. This type of cock, or the one shown in Fig. 11.5, is generally used as an emergency control when the meter is situated remotely from the premises and the Gas Safety (Installation and Use) Regulations 1998 require a control to be fitted.

There is a variety of cocks available for controlling gas supplies to the different appliances. These will be dealt with in Volume 2, when the installation

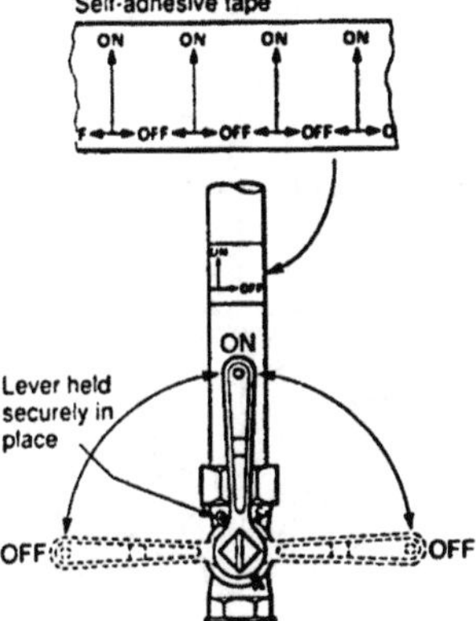

FIGURE 11.2 Emergency/meter control.

of the various appliances is discussed. Similarly, the types of taps used on appliances are many and varied and will be dealt with in detail later on. Examples of plug- and disc-type taps are shown in Fig. 11.3.

The type of valve commonly used to control large service pipes or industrial gas supplies is the gate valve shown in Fig. 11.4.

Ball valves are also extensively used for pipe sizes up to 100 mm and butterfly valves are used to control gas supplies downstream of the meter. A ball valve is shown in Fig. 11.5 and a butterfly valve is shown in Fig. 11.6.

FAULTS AND THEIR REMEDY

Faults in plug-type cocks are usually associated with the lubrication between the plug and the barrel. The grease used has a high melting point and is usually impregnated with graphite.

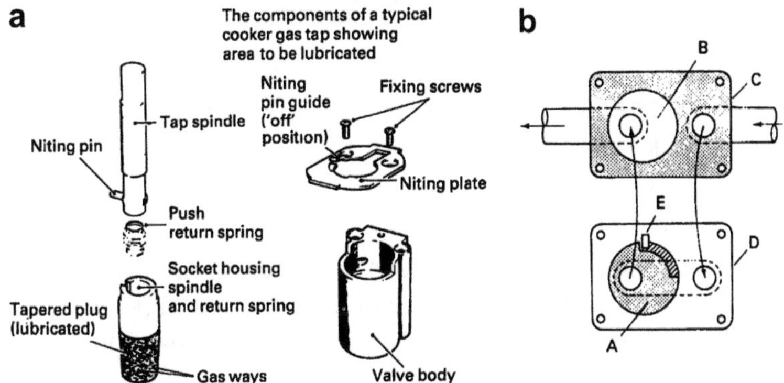

FIGURE 11.3 Appliance taps: (a) plug-type; (b) disc-type: A – disc, rotated by tap knob; B – disc seating; C – tap body with inlet and outlet connections; D – front cover shown removed; E – niting stop.

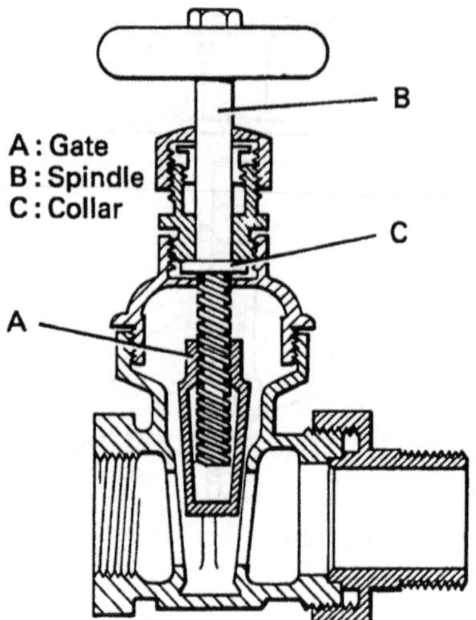

FIGURE 11.4 Gate valve.

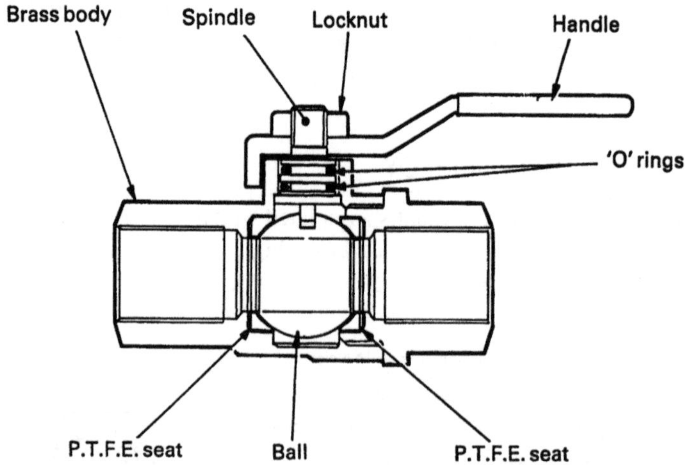

FIGURE 11.5 Ball valve.

Meter controls are very seldom turned on and off. So they may become stiff to turn and even corroded in the course of time. The special precautions to be taken when easing and greasing meter controls are covered in Volume 2, Chapter 3.

Appliance taps often have a considerable amount of use and may be subject to heat. This destroys the lubrication and they may become stiff and jerky in operation.

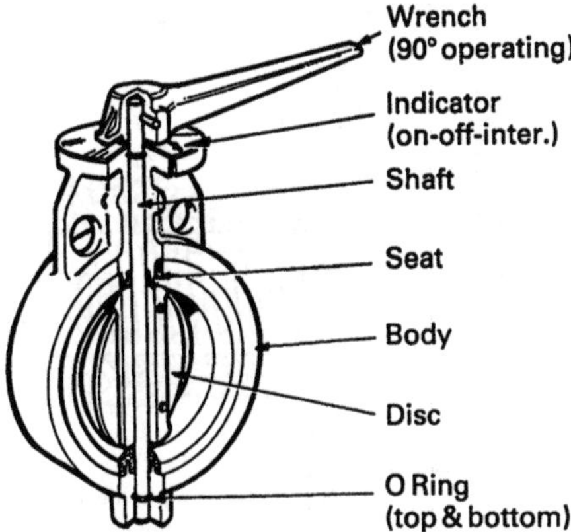

FIGURE 11.6 Butterfly valve.

To grease a plug tap turn off the gas and follow the following procedure:

1. Dismantle the tap. If this is a complex device, lay out the components on a piece of paper or cloth in the order in which they are removed.
2. Clean the plug and the barrel. Make sure gas ways are clear.
3. Grease the plug, insert it in the barrel and turn it from side to side two or three times.
4. Withdraw the plug, check that the grease is evenly distributed and that the plug is seated properly in the barrel.
5. Remove any surplus grease from the holes in the plug and the barrel.
6. Re-insert the plug and re-assemble in the reverse order.
7. Check for leakage with soap solution.

In the case of cocks with niting washers securing the plug, take care not to overtighten the nut and cause distortion of the cock body.

If a cock is leaking or letting gas through when it is turned off, the plug can be 'ground-in' to its seating in the barrel. Domestic scouring powders are quite coarse enough for small taps. Never use carborundum on any brass valve surfaces – it can become embedded in the metal.

THERMAL CUT-OFFS

The Gas Safety (Installation and Use) Regulations 1998 prohibit the fitting of meters under stairways or in sole-means-of-escape routes on new installations above two storeys. It is only allowed in single or two-storey premises if no other site is practicable, in which case the meter and its connections should be either:

- of fire resistant construction or
- housed in a special compartment with a fire resistance of not less than ½ hour or
- have a thermal cut-off device incorporated between the meter control and the inlet regulator.

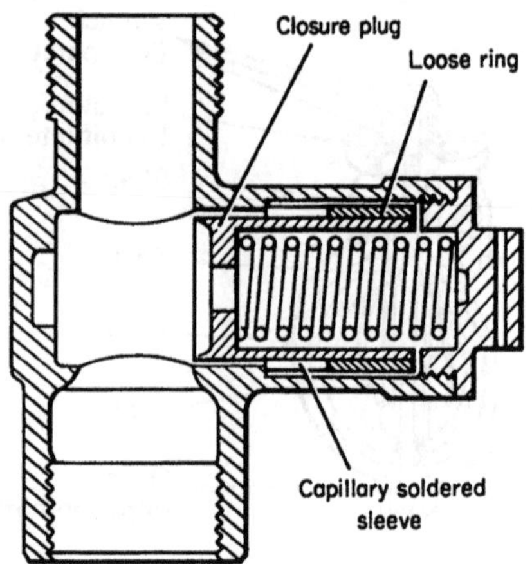

FIGURE 11.7 Thermal cut-off.

Several cut-offs have been developed and a typical device is shown in Fig. 11.7. This consists of a malleable iron tee fitting with a spring-loaded plunger fitted in the branch of the tee. The plunger is held open against the force of the spring by low-melting-point solder.

The device must be fitted immediately upstream of the meter or, if an inlet regulator is fitted, immediately upstream of the regulator. In the event of fire the solder must melt when there is an ambient temperature exceeding 95 °C. When the solder melts, the action of the spring will cause the closure plug to cut off the gas supply to the meter and its connections.

LOW PRESSURE CUT-OFFS

The supply of natural gas to an appliance may be interrupted either by a failure of the district supply (which is rare, fortunately) or by the operation of a prepayment meter. In the case of LPG the failure of supply may also be due to the tank or cylinder contents becoming exhausted or extremely cold weather conditions causing no evaporation of the liquid.

There is a possibility that an appliance has or appliances have been left turned on when the supply has failed. Someone might then put money into the meter, recharge the tank, replace the cylinder or the weather conditions might change. This could cause unburned gas to escape unless the appliances have a flame protection device. Where such a risk exists, a 'low pressure cut-off' can be fitted to protect either the whole installation or an individual appliance. A typical cut-off is shown in Fig. 11.8.

The cut-off has a body and a diaphragm similar to those of a constant pressure regulator. But the valve shuts down on top of the valve seating. Like the regulator, the space above the diaphragm is open to the atmosphere, through a vent hole. When the valve is open, the diaphragm is supported by the gas pressure and will remain open as long as the pressure is higher than that

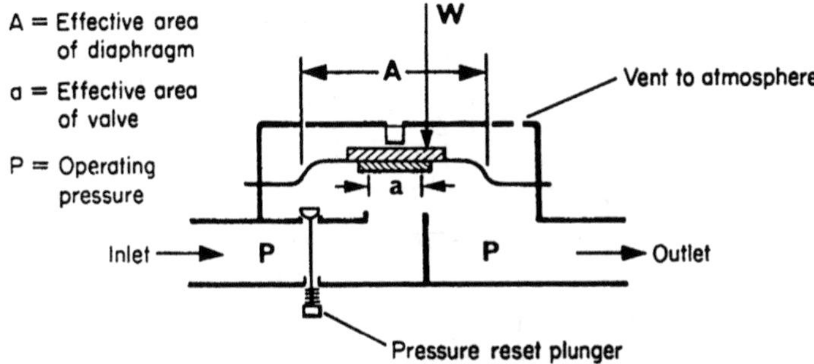

FIGURE 11.8 Low pressure cut-off with pressure reset.

predetermined by the weight acting against the area of diaphragm – usually 1.25–2.5 mbar.

When the pressure falls below this value the upward force on the diaphragm is unable to support the total load of the weight and the moving parts, and the valve closes. So failure of the gas supply causes immediate closure of the cut-off.

Once the supply is restored the gas pressure only acts on the underside of the valve, which has a very small area compared with the diaphragm. The upward force is now not sufficient to open the valve and no gas passes to the appliance.

The valve can be lifted off its seating by a number of methods. One way is to have a handle/spindle connected to the valve and manually lift it from its seating.

The cut-off shown in Fig. 11.8 uses the pressure reset method.

When the plunger is pressed in, the small, spring-loaded valve is opened and allows gas to pass through to the outlet and up, under the diaphragm. If the appliance cock has been turned off, pressure will build up and lift the valve off its seating. If the user has forgotten to turn off the appliance, gas will simply weep away and the valve will not lift.

So this device not only closes under low pressure conditions but also cannot be reset until the appliance has been turned off. The low pressure cut-off relies on gravity to close the valve when fault conditions prevail, so the control must be fitted horizontally.

RELAY VALVES

A 'relay valve' is a pressure-operated valve controlled by a diaphragm. It is used to control a gas supply by means of small, remote control devices. A diagram of a valve is shown in Fig. 11.9.

The relay valve is somewhat similar to the low pressure cut-off. But the diaphragm is now on the inlet side of the valve seating. And there is now a small orifice which allows gas to pass from under the diaphragm into the space above.

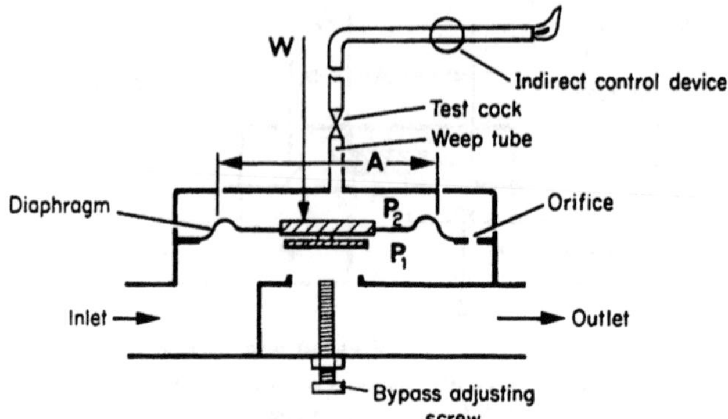

FIGURE 11.9 Relay valve.

This 'weep' of gas through the orifice is conveyed by a weep tube through the small control devices and usually back to the burner of the appliance.

When the control devices are all open, gas can flow through the orifice at full rate and pass away to the burner. So the pressure above the diaphragm (P_2) is considerably less than the inlet pressure under the diaphragm (P_1). The pressure difference lifts the diaphragm and opens the valve, turning on the appliance.

When any of the small indirect controls shuts off the weep tube, the flow through the orifice slows down and finally stops when pressure above the diaphragm (P_2) has built up to be the same as that underneath (P_1). The diaphragm is no longer supported and the valve closes, shutting off the appliance.

When the indirect control opens again, the pressure above the diaphragm is released and the differential pressures necessary to lift the diaphragm and the valve are restored.

When the valve is used as a high–low control, some form of by-pass is incorporated. The simplest form is an adjustable screw in the base which is made to prop the valve off its seating and allow a small flow of gas to pass to maintain flames on the burner. Some valves have a separate by-pass controlled by a needle valve.

Faults and remedies

The valve depends on the flow of gas through the orifice creating a pressure loss so that it will open, and on the pressure building up in the weep tube and above the diaphragm so that it will close.

This means that blockage of the weep tube will cause the valve to close and remain closed. The blockage can occur at the burner or be due to a fault in one of the small control devices.

Damage to the weep resulting in a leak of gas will cause the valve to stay open. An escape on the connections to a control device will have the same result. Similarly, blockage of the orifice will prevent gas passing into the upper part of the relay and the valve will remain open.

Care must be taken when cleaning the orifice to avoid making it larger. This would increase the rate of flow and reduce the pressure difference so that the valve would not lift. A punctured diaphragm has the same effect.

Where the weep tubes are long, more time is needed for the pressure to build up after a control has closed and the valve is sluggish in operation. This is also true of the large relay valves which have a large volume to fill above the diaphragm.

FAIL-SAFE RELAY VALVES

The 'fail-safe relay valve' was developed to overcome the possibility of the valve failing to close due to damage to the weep.

In appearance it resembles a constant pressure governor, fitted upside-down, Fig. 11.10. The operating diaphragm is in the lower part of the device and the valve and the valve seating are above. A secondary diaphragm separates the two parts and the space between the diaphragms is vented to the atmosphere. The two diaphragms and the valve are connected to the valve spindle.

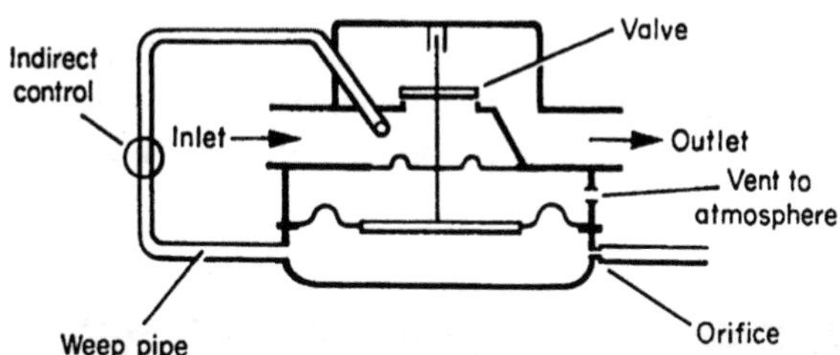

FIGURE 11.10 Fail-safe relay valve.

The weep is taken from a drilling on the inlet of the valve and asses through the indirect controls and back to the underside of the operating diaphragm. As it leaves this chamber on its way through the remainder of the weep tube to the burner, it passes through the restrictor orifice. The effect of the restriction is that, with the controls open, gas can flow into the lower chamber faster than it can flow out, so pressure builds up under the diaphragm. This pressure lifts the diaphragm and opens the valve.

As the indirect control closes, so the flow of gas into the lower chamber is reduced until the pressure can no longer support the diaphragm and the valve shuts. With the control fully closed no more gas enters the chamber and all the gas pressure weeps away.

Any leak in the weep tube from the control will now reduce the supply of gas to the lower chamber and tend to close the main valve. So the relay valve fails to a safe condition with its valve shut.

FLAME PROTECTION DEVICES

The purpose of a 'flame protection device' is to prevent unlit gas passing into the combustion chamber of an appliance. This is done by shutting off the main gas supply if:

- the pilot or the main burner fails to light when ignition is applied,
- having been lit, the pilot or the main burner becomes extinguished.

The flame protection device, also called the 'flame failure device (FFD)' or the new preferred term 'flame supervision device (FSD)', must detect the presence of a flame to maintain the supply and shut off the supply when the flame is not present. To do this, the device is usually 'actuated' or brought into action by a pilot flame. The presence of the flame may be sensed in a number of ways including:

- using heat from the flame,
- by the flame conducting and rectifying an alternating current,
- the effect of ultra-violet or infra-red rays on a photo-electric cell.

The devices described in this chapter use the methods outlined to control a gas valve situated in the supply to the main burner. All of them can, however, be made to operate electrical switches and relays. In this way they may be used in more sophisticated control systems, often giving completely automatic operation to the appliance.

BS EN 297:1994, 625:1996 and 483:2000 are issued in a number of parts, each applying to a particular domestic appliance. The standard specifies the safety requirements of appliances burning 1st, 2nd and 3rd family gases.

Table 11.1 Flame Supervision Device Shut Down Times

Appliance Type	Maximum Shut Down Time in Seconds
Boilers up to 60 kW	60 seconds
Warm air units up to 60 kW	60 seconds
Instantaneous water heaters	60 seconds
Storage water heaters 8–60 kW	60 seconds
Cooker ovens	60 seconds
Storage water heaters below 8 kW	60 seconds
Circulators	60 seconds
Cooker hotplates and grills	60 seconds
Refrigerators	60 seconds
Gas fires	60 seconds
Decorative gas fires	60 seconds

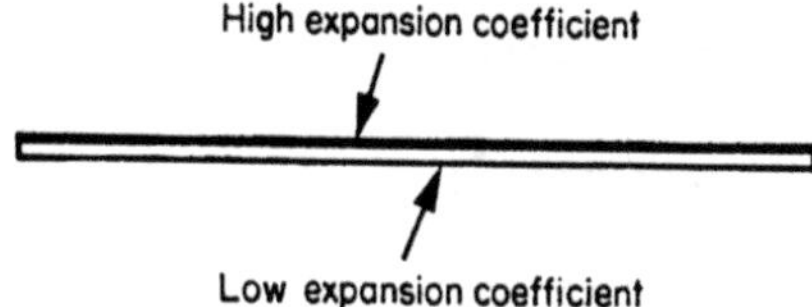

FIGURE 11.11 Bimetallic strip, cold.

Specified in these standards are maximum shut down times for flame supervision devices and Table 11.1 gives some of the times specified.

BIMETALLIC DEVICES

Metals expand when heated, but different metals expand by different amounts for the same heat input. If two small strips of two different metals were joined together to form a 'bimetal' strip (Fig. 11.11) then, if the top metal expands more than the one underneath, the strip will bend when it is heated (Fig. 11.12).

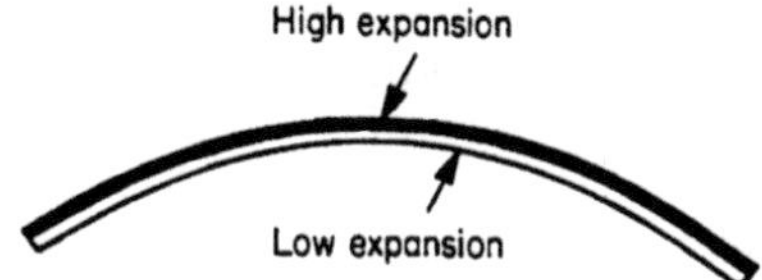

FIGURE 11.12 Bimetallic strip, hot.

The distance that the end of the strip will move, when heated, can be increased by first bending the strip into a curve (Fig. 11.13). This also shows the amount of distortion due to heating. If the strip is anchored at one end and a valve spindle is attached to the other, the valve can be made to open the gas supply to the main burner, when it has been heated by a pilot flame (Fig. 11.14). If the pilot is blown out, or becomes blocked or shortened, the strip will cool and so close the valve.

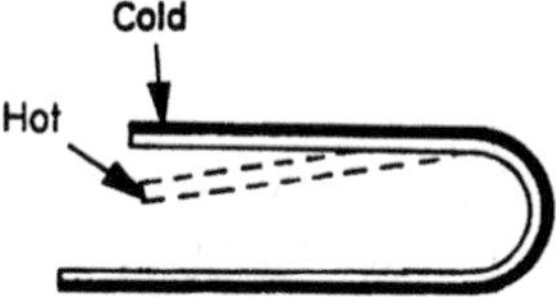

FIGURE 11.13 Curved bimetal strip, cold.

These devices are simple and inexpensive and have been used on the smaller and cheaper domestic appliances, mainly water heaters and cookers. Their disadvantages are that they do not usually shut off the pilot supply and they are slow to cool down, taking about $1^1/_2$ min to shut off. These devices are no longer used on new appliances but may be encountered on old appliances.

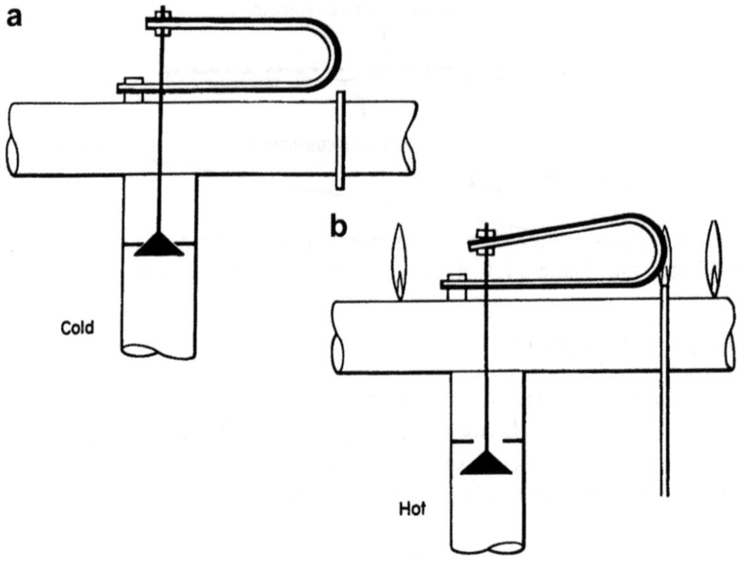

FIGURE 11.14 Bimetal flame protection device: (a) cold and (b) hot.

Faults and remedies

The most common faults are:

1. Metal fatigue in the bimetal strip. The strip becomes permanently distorted and does not respond to heating. Renew the strip.
2. Valve spindle becomes seized in the gland on the burner. This may be due to corrosion or to the spindle becoming bent or distorted. Clean and straighten the valve spindle or renew it if necessary.
3. Badly positioned or adjusted pilot. Even if the device itself is in perfect working order it will not operate satisfactorily if it is wrongly positioned in relation to the pilot flame or if the pilot does not provide sufficient heat.

VAPOUR PRESSURE DEVICES

When liquids are heated sufficiently, they turn into vapour or boil. The vapour has a much greater volume than the liquid from which it came and so causes a rapid increase in pressure which can be made to operate the valve in the vapour pressure flame protection device (Fig. 11.15).

A tube connects a sensing probe A, containing liquid, with a flexible metal diaphragm B. When the probe is heated by a separately supplied pilot flame the liquid vaporises. The pressure produced by the increase in volume distends the diaphragm B, moving the lever C, and opening the valve. So, gas can pass to the burner, where it is lit by the pilot.

If the flame fails, the probe will cool, the vapour will condense and contract the diaphragm and the spring F will close the valve on to its seating.

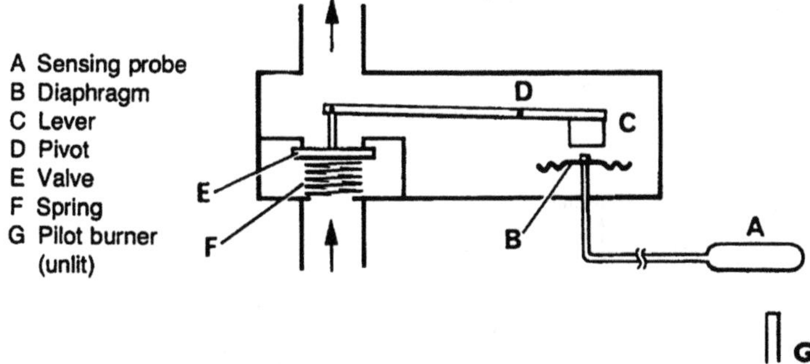

FIGURE 11.15 Vapour pressure flame protection device.

When the probe is cold, there is a gap between the diaphragm and the operating lever. This allows the liquid to expand without opening the gas valve if ambient air temperature increases. When the liquid vaporises, the rapid increase in volume closes the gap and operates the valve.

The device is used for a number of appliances including room heaters, water heaters and cooker ovens. Its advantage is that any damage resulting in loss of liquid causes the valve to fail to a safe condition. Some older devices contain mercury. Whenever these are removed they must be disposed of in a safe manner, as detailed in the code of practice for disposal of dangerous substances.

Faults and remedies

Apart from failure of the sealed system containing the liquid, which entails changing the device, the lever system and valve may require cleaning and lubricating from time to time.

The main source of complaints results from incorrect siting of the probe, relative to the pilot flame.

The vapour pressure principle is also used for other devices including appliance thermostats (see later in this chapter).

THERMO-ELECTRIC DEVICES

Thomas Seebeck, an Estonian scientist, discovered that heat can produce electrical energy when applied to a 'thermocouple'. In its simplest form a thermocouple consists of a loop of two different or 'dissimilar' metals joined together at one end with the other ends connected to an electrical meter or a magnet. When the joint or 'junction' is heated, an emf is produced in the circuit.

The voltage produced depends on the difference in the temperature between the hot and cold junctions and on the particular metals used. Like some of the other heat-operated devices, thermocouples are used for temperature measurement as well as for flame protection. The metals commonly used are

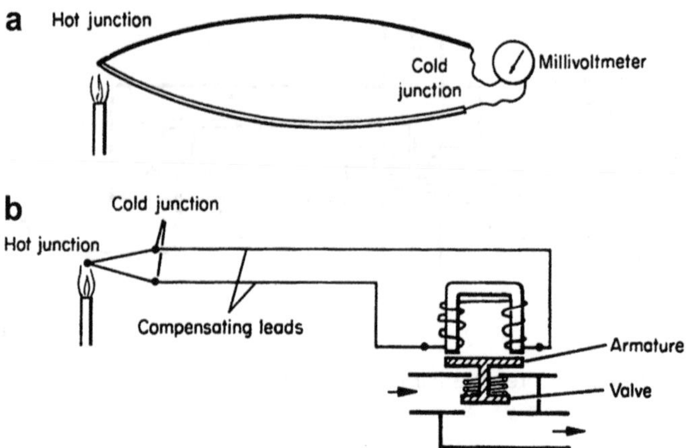

FIGURE 11.16 Principle of thermocouple: (a) connected to millivoltmeter; (b) connected to electromagnet.

copper or iron with a copper–nickel alloy, called 'constantan' or, for higher output 'chromel' with 'alumel'. Chromel is an alloy of nickel and chromium and alumel is nickel and aluminium. The output voltages are about 15–30 mV when not on load and 10–20 mV when on load. Figure 11.16 illustrates the principle of the thermocouple with an electric meter in Fig. 11.16a and an electromagnet in Fig. 11.16b. If the thermocouple is heated by a pilot flame and a plate or armature, attached to a spring-loaded valve, is lifted up to the magnet, then the armature will be held up as long as the magnet remains energised. If the pilot flame is extinguished, the thermocouple will cool quickly and cease to produce an emf. The coil will no longer be magnetised and the spring loading will close the valve.

There are a number of devices of slightly different design which operate on the principle described. These all use low-output thermocouples which make it necessary for the armature to be raised up to the magnet by hand to open the valve initially. A typical device is shown in Fig. 11.17a.

In order to establish the pilot flame, the main gas valve F must be raised off its seating. This is done by pressing in the reset button A. When this is pushed in it first allows the flow interrupter valve E to come into contact with the lower valve seating, so shutting off the outlet. Further pressure on the push button raises the spindle through the interrupter valve and lifts the main valve F off the top valve seating. This brings the armature H into contact with the pole pieces of the magnet and also allows gas to pass to the pilot connection (Fig. 11.17c).

The pilot can now be lit, heating the thermocouple and energising the electromagnet. This can take up to 60 s, during which time the push button must be held in. When the button is released, Fig. 11.17a, the spring-loaded interrupter valve and push button return to their starting positions and gas can now pass through both valve seatings to the burner.

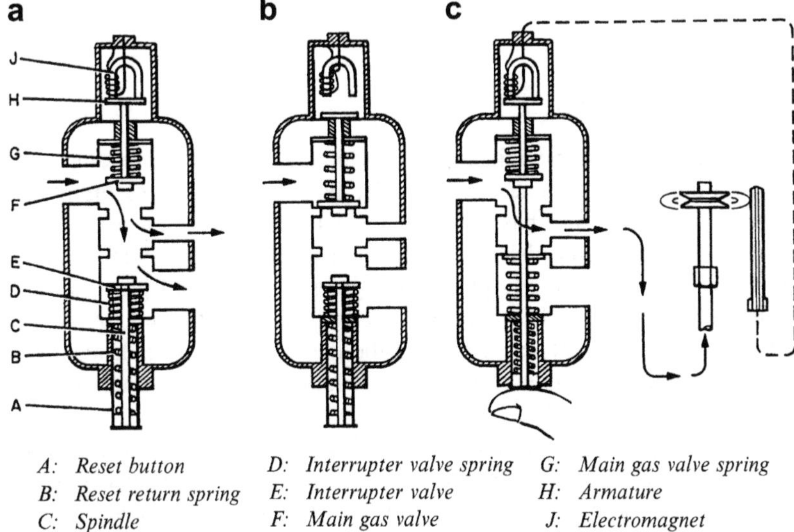

A: Reset button
B: Reset return spring
C: Spindle
D: Interrupter valve spring
E: Interrupter valve
F: Main gas valve
G: Main gas valve spring
H: Armature
J: Electromagnet

FIGURE 11.17 Thermo-electric flame protection device: (a) operating position; (b) off-position; (c) setting position.

If the gas supply is interrupted or pilot failure occurs whilst the main burner is shut down, then the thermocouple will cool, the magnet will be de-energised and the main valve will shut off in about 30 s. The setting process must be repeated and the pilot re-established before gas can pass to the main burner.

Because the thermocouple is remote from the magnet it is connected to it by means of a special electrical lead. The conductors are made of copper with the outer one in the form of a tube. The inner wire is insulated from the tube by a glass fibre sleeve and the two form a coaxial cable. At one end the conductors are brazed on to the thermocouple and at the other they are connected to the magnet housing by a union nut (Fig. 11.18).

Thermo-electric devices are in common use on a wide range of gas appliances. They are limited in size and for large appliances they may be

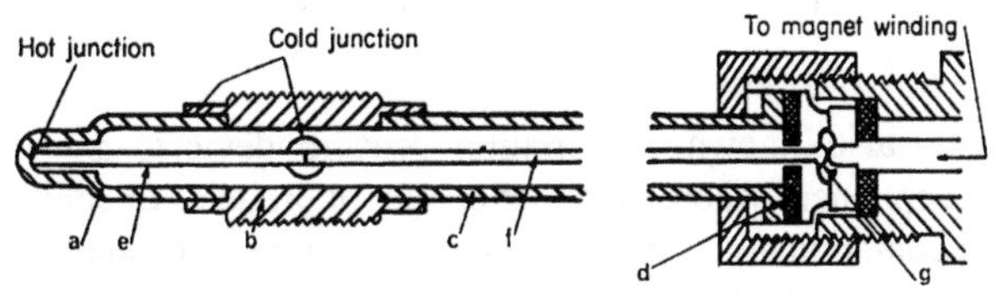

a: Tip of chrome-nickel alloy
b: Brass sleeve
c: Copper earth conductor
d: Flanged earth connection
e: Internal wire of copper-nickel alloy
f: Internal conductor wire
g: Connection of electromagnet

FIGURE 11.18 Thermocouple and lead connections.

fitted on the weep tube of a relay valve. In this case a separate pilot supply is required.

Faults and remedies

Pilot flame: This must be adjusted to provide the required temperature at the hot junction. A slack flame can be easily blown off the tip of the thermocouple by draughts and has a lower flame temperature than a flame with greater aeration. Ensure that the burner and its primary air ports are clean. The flame must be correctly positioned, generally it should play on the tip or the top 12 mm of the thermocouple, to produce the maximum emf. Although it is necessary to ensure that the thermocouple tip is heated to the required temperature, overheating can reduce the life expectancy of the thermocouple.

Thermocouple lead: This must not be kinked or bent too sharply otherwise the conductors may be short-circuited. The contacts must be clean and tight but over-tightening the union nut can distort the insulating washer and may cause a short-circuit. About a quarter of a turn beyond hand-tight is normally sufficient.

Thermocouple: The tip should be clean and undamaged. Although it must be heated to a high temperature, it should be kept below red heat if it is to have a reasonable life. The output can be measured by a millivoltmeter and, if an interrupter is used, the performance can be checked both on and off load.

Electromagnet: The armature and magnet are housed in a sealed unit which must be exchanged if a failure occurs. This is a rare occurrence. Valves have been known to close as a result of vibration, as with a sharp blow, which can dislodge the armature from the magnet. In most domestic situations the device is located in a position where this is unlikely to occur. A main advantage of this type of flame protection device is that most faults which can occur will make the valve fail to safety. And when the valve closes it shuts off the pilot as well as the main gas supply. Whilst the ordinary thermocouples do not generate sufficient emf to attract the armature to the magnet poles, high output thermocouples are available with sufficient energy to operate electrical relays.

OXYGEN DEPLETION SYSTEMS (ATMOSPHERIC SENSING DEVICES)

These devices, also known as vitiation sensing devices, are installed on domestic gas appliances where there is a danger of products of combustion from the appliance contaminating the air in the room in which the appliance is fitted. Generally, these are flueless space heaters and water heaters or open-flued appliances such as gas fires and central heating boilers. This follows a ruling from the European Gas Directive that from 1 January 1996 all appliances connected to a flue for the dispersal of products of combustion must be so constructed that in abnormal draught conditions there is no release of products of combustion in a dangerous quantity into the room concerned. The Department of Trade and Industry has granted a derogation period for appliances sold in the UK only. This is under review and does not apply to appliances fitted in bedrooms, as specified in Regulation 30(3) (b) of the Gas Safety (Installation and Use) Regulations 1998.

Two devices have been developed to enable appliances to conform to the new directive. These are:

1. Oxygen depletion system (ODS) also known as oxy pilot, anti-vitiation device or atmospheric sensing device (ADS).
2. Thermal backflow prevention device also known as a TTB stat (TTB comes from an unpronounceable Dutch term 'Themishe terugslag beveiliging').

OXYGEN DEPLETION SYSTEMS

These devices are normally found on flueless LPG cabinet heaters, flueless water heaters and gas fires. Some manufacturers have adapted them to protect central heating boilers, particularly back boiler units, from downdraught causing products of combustion to be emitted into the room.

Under normal circumstances domestic gas appliances draw in more air than is required for combustion. This ensures that there is plenty of air for the main burner and pilot burner, the excess air dilutes the products of combustion and passes up the flue, or in the case of flueless appliances, into the room. This air comes from the room in which the appliance is installed and providing that the air contains sufficient oxygen then the appliance will operate satisfactorily. If the appliance fails to clear the products of combustion via the flue because of a blockage or reverse flow then the air in the room will become contaminated or 'vitiated' reducing the oxygen content in the air and possibly producing carbon monoxide (CO) putting inhabitants at risk. The Oxygen Depletion System is designed to shut down the gas supply to the burner if the level of CO exceeds 200 parts per million concentration in the air of the room in which the appliance is fitted.

Operation

The device relies on a precision pilot holding open a thermo-electric valve via the thermocouple (see earlier in this chapter). The positioning of the pilot flame in relation to the thermocouple is critical. For this reason, if the thermocouple or pilot becomes faulty, they must be exchanged as a unit.

Under normal circumstances with the air containing normal oxygen levels, the pilot plays on the tip of the thermocouple generating sufficient heat to hold open the main gas valve (Fig. 11.19). As the air loses its oxygen content the pilot flame becomes unstable as it searches for air reducing the heat to the thermocouple tip (Fig. 11.20). If the oxygen levels reduce even more then the pilot

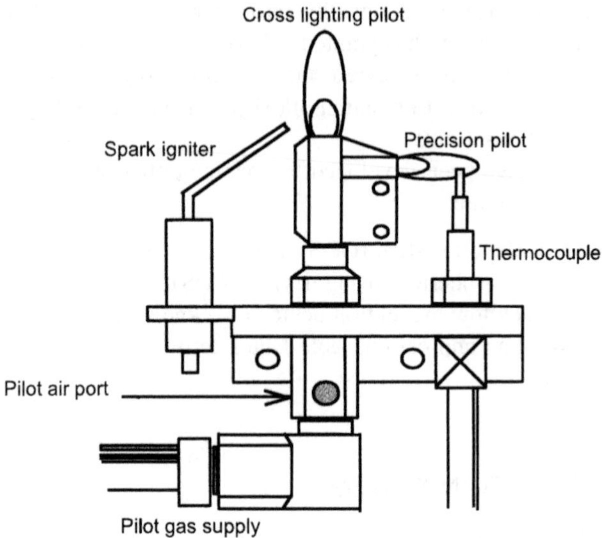

FIGURE 11.19 Air supply satisfactory; precision pilot heats thermocouple tip.

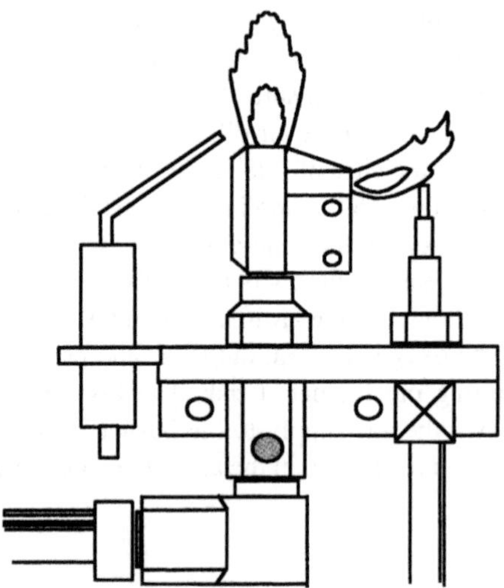

FIGURE 11.20 As the air becomes contaminated the oxygen level falls, the precision pilot lifts away, cooling the thermocouple. If this situation persists the thermoelectric valve will shut off the gas.

almost becomes extinguished, the thermocouple cools rapidly and the main gas valve closes stopping the flow of gas to the burner (Fig. 11.21).

Servicing

Because of the critical nature of this device, it is essential that the correct servicing procedure is used, manufacturer's instructions must be followed and

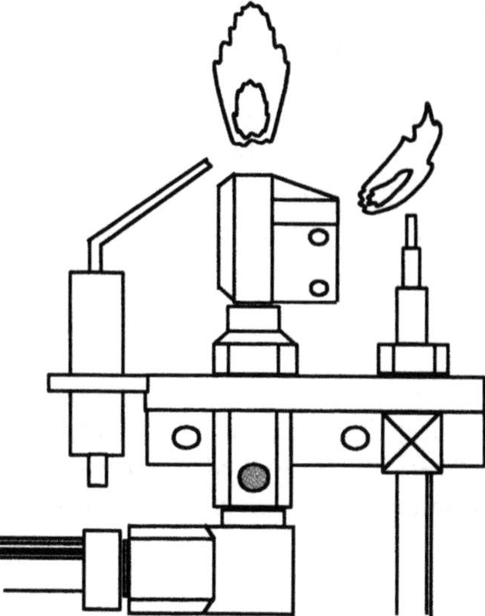

FIGURE 11.21 As the air becomes further contaminated the precision pilot hunts for air and completely lifts away from thermocouple tip. The thermo-electric valve will shut off the gas.

all safety checks must be carried out if any work is carried out on the appliance or there appears to be a problem. The main checks to be applied are:

- the ODS is complete and not damaged,
- the unit is securely mounted in its recommended position,
- the aeration port is clean and free from obstruction,
- the flame picture is not distorted and the sensor flame and cross-lighting flame are correctly positioned.

If the device has operated and shut down the burner, then the flue system must be examined to investigate the cause and any remedial work carried out.

THERMAL BACKFLOW PREVENTION DEVICE (TTB)

This device can be installed on any open-flued appliance but in particular it is used on central heating boilers. Its purpose is to shut down the gas supply should downdraught occur, so preventing products of combustion entering the room. After shutdown under fault conditions it is a requirement that manual intervention must be used to re-start the system. For this reason the device is normally connected as a thermocouple interrupter if a thermo-electric flame failure system is in use or in the lock out circuit on a full sequence controlled appliance.

Operation

The device itself will be either a klixon type or a liquid expansion type thermostat similar to an electrical overheat thermostat. It is positioned in the skirt of the appliance downdraught diverter and set to operate at either 65 or 95 °C dependent on type and position. Under normal draught conditions the skirt of the downdraught diverter has cool air from the room flowing through it keeping the device cool and therefore allowing the burner to operate. If reverse flow occurs because of downdraught or a blockage in the flue, the hot products of combustion will pass over the device, raise its temperature and, if it persists, will lead to shut down of the burner as the device reaches its set temperature (see Figs 11.22 and 11.23).

Servicing

The device must be tested to manufacturer's instructions and checked for damage, position and security. If installed as a thermocouple interrupter, check the tightness and cleanliness of the electrical connections to prevent nuisance

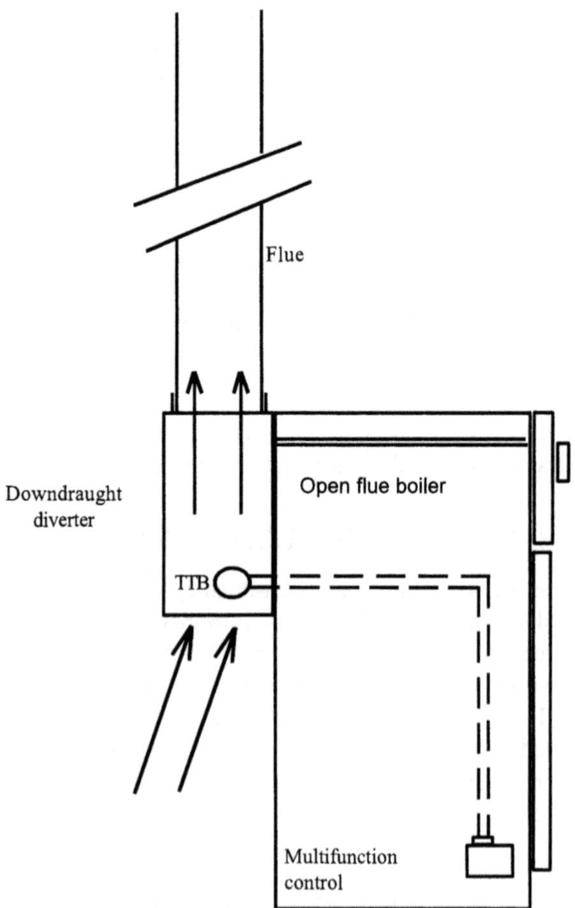

FIGURE 11.22 Normal flue conditions, airflow keeps the TTB cool – system remains on.

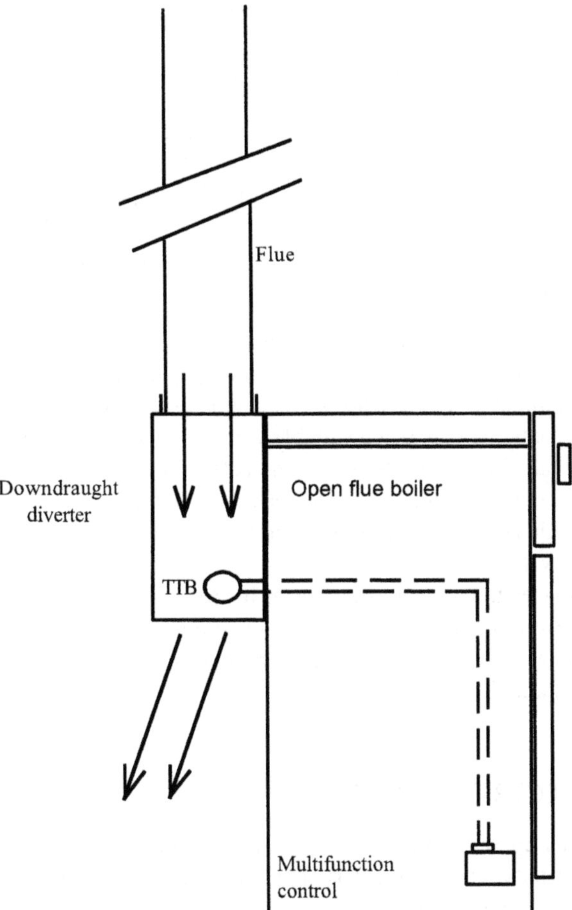

FIGURE 11.23 Reverse flow conditions or downdraught, the hot products of combustion heat the TTB and close the system down.

shutdowns. If the device has operated and shut down the burner, then the flue system must be examined to investigate the cause and any remedial work carried out.

Originally, the electronic flame protection devices were designed for use on industrial or commercial equipment. However, the development of semi-conductors and printed circuitry has reduced the size and the cost of electronic systems and they are increasingly being applied to domestic appliances.

As you know (Chapter 9) the chemical reaction taking place in a flame produces ions. These electrically charged particles can be made to pass between two electrodes so a flow of electrons is set up and the flame conducts a current (Fig. 11.24). In the early systems the current was amplified and made to operate an electrical relay which controlled solenoid valves in the pilot and main gas supplies. The disadvantage of the flame conduction device was that any short

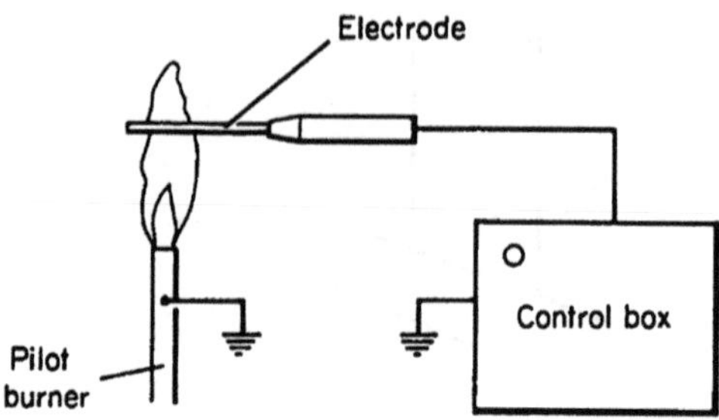

FIGURE 11.24 Flame rectification device.

circuit within the system had the same effect as the presence of a flame, so the device failed to danger and is therefore no longer used.

Flame rectification devices do not have this disadvantage. They make use of the fact that the flame can, when suitable electrodes are used, rectify the AC and produce a small DC output. So the flame acts as a conductor for the AC input and, at the same time, rectifies it to DC. The DC output is made to operate a relay, through a suitable electronic circuit. This in turn controls the electrically operated gas valves.

These devices have the advantages of instant response and immediate shut-down in the event of flame failure. They are commonly used in conjunction with full-sequence automatic control of gas burners which is dealt with in more detail in Volumes 2 and 3.

PHOTO-ELECTRIC DEVICES

These devices operate on the principle that some metallic substances will not conduct electricity when in the dark, but allow a current to pass when light falls on them. Their electrical resistance decreases when they are subjected to radiation. Such a substance is selenium, which is used in a photo-electric cell, often called a 'magic-eye'. This can be made to open doors, detect smoke and operate burglar alarms. It can also switch on the lights when the level of illumination falls to a selected value.

The devices used to protect gas flames are sensitive to either ultraviolet or infra-red rays and not sensitive to light. The infra-red detector can be made to accept only the pulsating rays from the flames and not the steady output from the combustion chamber itself. Because it may be affected by radiations from the combustion chamber which have been modulated by air currents, the infra-red detector is generally used in lower temperature combustion chambers, for example boilers. The cell makes use of lead or cadmium sulphide.

THERMOSTATS

A thermostat is a device designed to maintain a steady, predetermined temperature of an object or an environment. It must sense the prevailing temperature and control the flow of gas to the burner of the appliance so that the selected temperature is reached and maintained.

Some thermostats give a gradual change of gas rate or 'modulating' control whilst others are designed to produce a snap-action on-off operation.

Thermostat settings may be either:

- fixed by the manufacturer or
- adjusted by the service engineer or
- selected by the user.

The principles used in gas appliance thermostats are those of:

- differential expansion of metals,
- liquid expansion,
- vapour pressure.

As with the flame protection devices, these methods can be used to actuate valves in the gas supply to the burner. They may also be made to operate electric switches in more complex control systems.

ROD-TYPE BIMETAL THERMOSTATS

The operating element, or 'temperature-sensitive' element, in this device consists of a rod of one metal fixed inside a tube of another metal. The rod is made of nickel–steel alloy like 'invar', which has 36% nickel, or the 42% nickel in steel used in cooker and similar thermostats controlling temperatures up to 300 °C.

Tubes are usually made of brass, which has a coefficient of linear expansion roughly twelve times that of the nickel–steel. The two coefficients are:

$$\begin{aligned} \text{invar} &= 1.5 \times 10^{-6} \text{ per } ^\circ\text{C} \\ \text{brass } (65/35) &= 19.0 \times 10^{-6} \text{ per } ^\circ\text{C} \end{aligned}$$

Figure 11.25 shows the arrangement of the rod in the tube. The two metals are usually brazed together at one end so that expansion of the tube, when heated, will draw the rod into the open end (Fig. 11.26). As the temperature increases, the rod moves further into the tube, its own expansion having a negligible effect on the differential expansion.

If a spindle with a valve attached is now held on the end of the rod by a spring, the movement caused by the expansion will close it on to its seating and so control the heat output of the burner (Fig. 11.27). The greater the distance

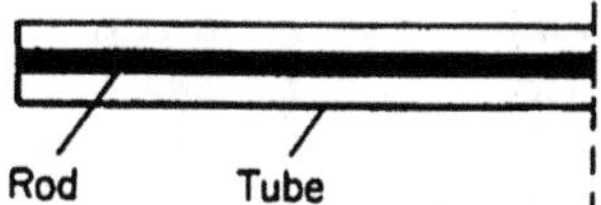

FIGURE 11.25 Bimetal rod-type thermostat, rod and tube, cold.

between the valve and its seating when the thermostat is cold, the higher will be the temperature which is reached before shut-down.

Figure 11.28 shows the type of thermostat used in a gas cooker oven. Turning the control knob F rotates the adjusting screw and positions the valve A to give the required oven temperature. In normal operation the valve will partly close until the quantity of gas passing to the burner is just the right amount to

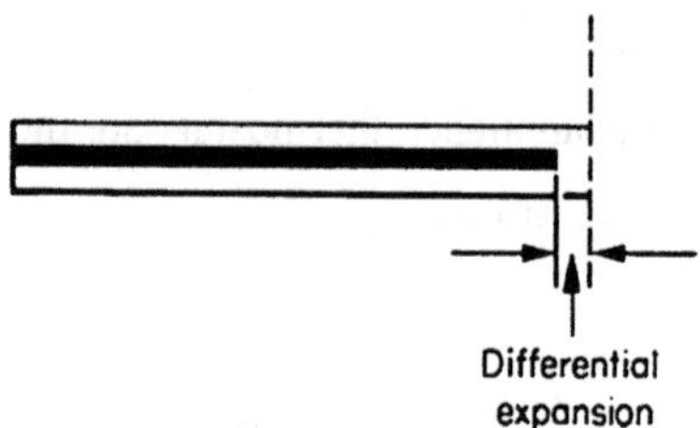

FIGURE 11.26 Rod-type thermostat, hot.

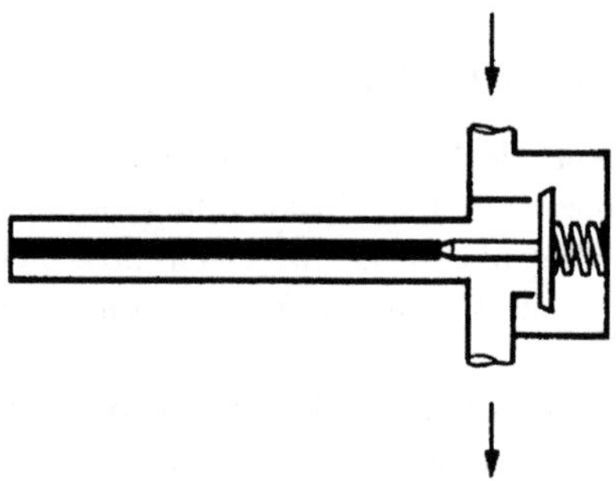

FIGURE 11.27 Simple rod-type thermostat.

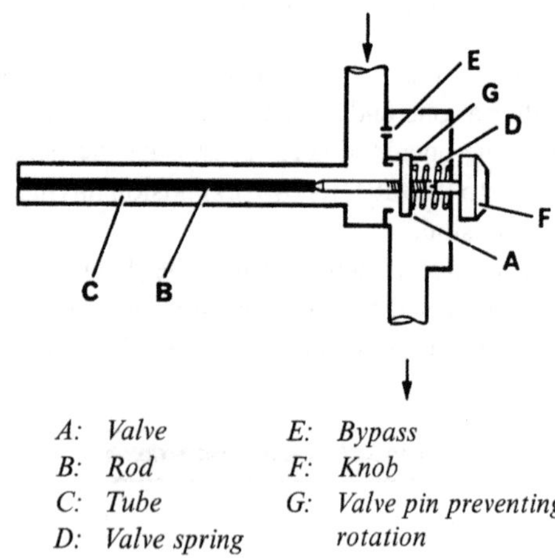

A: Valve | *E: Bypass*
B: Rod | *F: Knob*
C: Tube | *G: Valve pin preventing rotation*
D: Valve spring

FIGURE 11.28 Cooker oven thermostat.

maintain the oven at the selected temperature. So, at cooking temperatures the thermostat valve does not close completely.

It is possible, however, for the user to turn the knob to a low setting after the thermostat has been heated up on a high one. This shuts the valve completely off and would extinguish the burner if it were not for the bypass, E. This maintains a stable flame on the burner but is insufficient to affect the lower cooking temperatures. Usually the bypass will give a temperature of about 70 °C in the oven, which is useful for keeping food warm.

Similar thermostats may be used to control the temperature of water, either by inserting the tube into a storage vessel or by fitting it in the path of the circulating water. Figure 11.29 shows a water heater thermostat in which the valve is pivoted on one edge of its seating. This magnifies the movement of the rod and gives the valve a quicker response than the conventional arrangement.

The thermostats described have carried the full gas rate of the burner and are called 'direct-acting' thermostats. When large amounts of gas have to be controlled, relay valves are used with small 'indirect-acting' thermostats. These do not have a bypass and this is now provided by the relay valve.

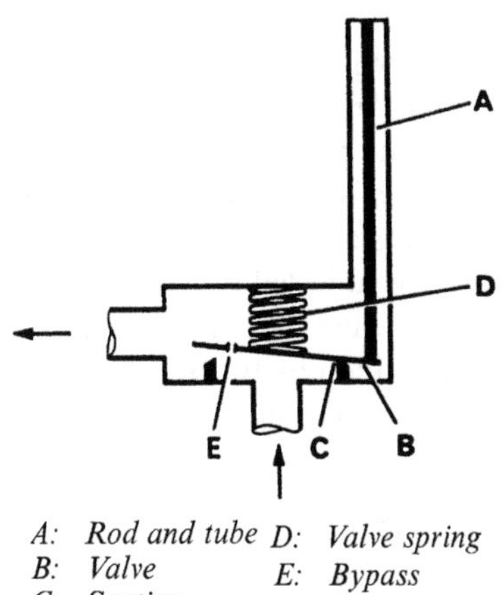

A: Rod and tube D: Valve spring
B: Valve E: Bypass
C: Seating

FIGURE 11.29 Lever-type, quick-acting thermostat.

Faults and remedies

Rod and tube: The tube must be kept clean and free from burnt-on-grease (or scale, in the case of water heaters). Both will act as insulators and affect the operation. Bending or distorting the rod or tube will also affect the operation and the thermostat will require renewal.

Valve and bypass: These may be checked by heating up the thermostat on a high setting and then turning the knob to a low setting. There should then be a small, stable flame on the burner. If the flame has gone out, the bypass is choked. When cleaning it take care not to enlarge the gasway.

If the flames have not been reduced, then probably the valve has not shut down properly. This might be due to dirt on the valve or seating or to some mechanical fault. Most thermostats can be dismantled for cleaning without altering the calibration.

In the case of a thermostat with an adjustable bypass, this fault might be due to faulty adjustment giving a very high bypass rate.

Calibration: Most thermostats have some facility for adjusting the calibration. Usually this involves altering the position of the setting knob in relation to the valve spindle. The problem, in the case of cooker ovens, is not in making the adjustment, but in accurately measuring the temperature at the centre of the oven.

BIMETAL STRIP THERMOSTATS

Bimetal strips may be used to detect changes in temperature. The bending movement which results from heating can then be used to operate a small gas valve or an electric switch. In order to get a large movement a long strip is used, but this is wound into a flat spiral or a helical coil so that the thermostat may be reasonably compact.

An example of a small, indirect, water heater thermostat is shown in Fig. 11.30. The coil of bimetal is housed in a tube which is located in the hot water storage cylinder. The twisting movement of the coil A, as the water is heated, turns the central spindle D and closes the flap valve E on to its seating. This shuts off the weep and the relay valve then shuts off the gas to the burner. When the water cools the strip winds up again and the flap opens, the weep is opened and the relay valve restores the gas supply.

This thermostat cannot be adjusted by the user. Removing the cover plate allows the service engineer to alter the temperature by rotating the fixed end of the helix which is held in position by two screws at C.

Bimetal air thermostats are described in Volume 2, Chapter 12.

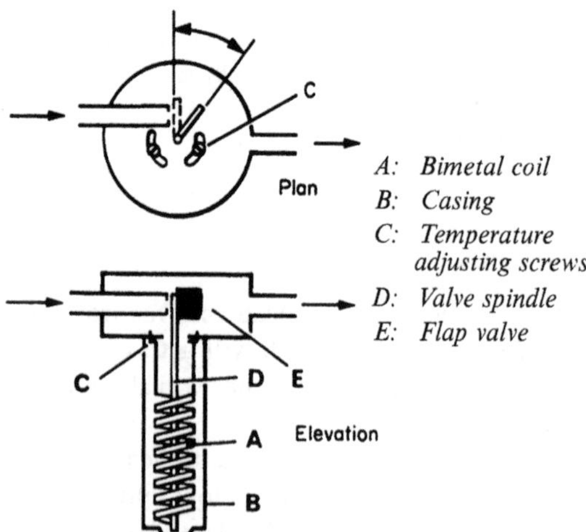

FIGURE 11.30 Bimetal strip thermostat.

Faults and remedies

Bimetal thermostats operate at much lower temperatures than bimetal cut-offs and give very little trouble. The strip is usually protected within a case and not likely to be damaged. Strips which have been distorted by interference can sometimes be formed back to their original shape. But generally a distorted strip means an inaccurate thermostat and it is better renewed.

LIQUID EXPANSION THERMOSTATS

Where the thermostat valve is housed remotely from the point at which the temperature is to be sensed, use is often made of devices where the controlling element is operated by the expansion of a liquid. The temperature is sensed by a phial containing the liquid, connected by a capillary tube to a flexible bellows or a diaphragm which operates the valve.

The capillary tube, which is a very small copper tube with thick walls and a very tiny bore, can be bent to very small radii and enables the phial to be easily positioned in any suitable location. Figure 11.31 shows a typical thermostat.

The valve A is spring loaded and operated by lever B, which also serves to multiply the small movement produced by the bellows D. The lever is pivoted on the adjusting screw at C. Turning the range setting knob G moves the fulcrum and alters the distance of the valve from its seating, so setting the temperature required.

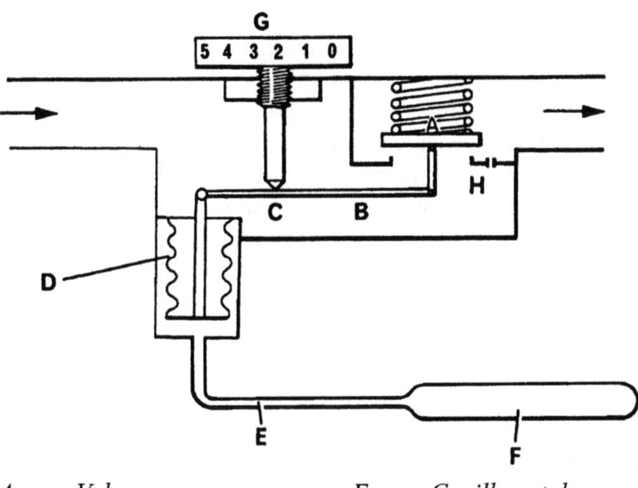

A:	*Valve*	*E:*	*Capillary tube*
B:	*Lever*	*F:*	*Phial*
C:	*Pivot on adjusting screw*	*G:*	*Range setting knob*
D:	*Bellows*	*H:*	*Bypass*

FIGURE 11.31 Liquid expansion thermostat.

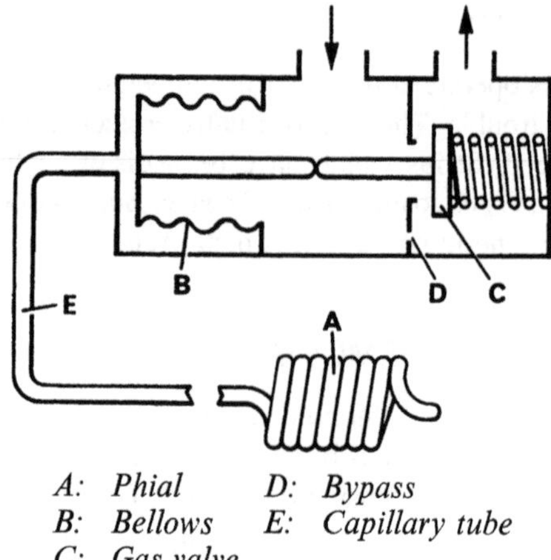

A: Phial *D: Bypass*
B: Bellows *E: Capillary tube*
C: Gas valve

FIGURE 11.32 Refrigerator thermostat.

The bellows is designed to hold as small an amount of liquid as possible. This is to keep the liquid in the phial F and not allow it to be affected by overheating in the bellows if the valve becomes hotter than the phial.

This type of thermostat may be found on gas cookers for both oven and hotplate thermostats. It is also used on room heaters and on some central heating units and other heating appliances. A modified type is used as a refrigerator thermostat (Fig. 11.32).

In this thermostat, the liquid contracts as it is cooled and closes the gas valve instead of opening it. The end of the capillary tube E is wound into a helical coil to form a phial A. This is located in a socket in the evaporator (the coolest part) in the refrigerator cabinet. If the temperature in the cabinet rises, due to opening the cabinet door and putting in warm food, the bellows B expands and opens the gas valve C against the pressure of the spring. The gas speeds up the refrigeration cycle so cooling the cabinet again. The liquid now contracts the bellows and the valve shuts down. A bypass is provided at D. An arrangement would be incorporated to allow the temperature to be varied. This has the effect of altering the pressure of the spring so that the force required to close the valve can be made greater if cooler temperatures are required.

Faults and remedies

Like other types of thermostat, these can be checked by heating up on a high setting and turning down to a low one. Calibration may get out of adjustment and need resetting. Damage to any part of the liquid system will require its renewal. Always make sure first that the phial is properly located in its right position. On some room heaters the phial or the capillary tube may be displaced or distorted by the user when cleaning round the appliance.

The phial on this thermostat should contain the bulk of the liquid in order to act as the sensing element for the temperature. However, the capillary tube and the bellows or diaphragm also contain liquid. If the tube or bellows are subject to high temperatures then liquid is expanded further and the valve may shut off before the required temperature is reached. So local overheating of the device can cause faulty operation.

VAPOUR PRESSURE THERMOSTATS

These are similar to the liquid expansion types but the system is now only partly filled with liquid. As temperature increases, the expansion of the liquids is accommodated in the free space within the system. When the temperature reaches the boiling point of the particular liquid used, the liquid vaporises with a rapid increase in volume. The pressure produced by this increase in volume expands the bellows and closes the gas valve.

With these thermostats the movement of the valve takes place over a relatively small temperature range and they are used principally for room heaters and boilers.

The faults which are liable to occur and their remedies are similar to those of the liquid expansion types.

POWER OPERATED VALVES

When electrical devices are used to control gas appliances the valves in the gas supply may be actuated by various means including:

- electric heating coils acting on metal rods or bimetal strips,
- electromagnets or solenoids,
- electric motors,
- electro-hydraulic devices.

HEAT MOTORS

A ‘heat motor’ makes use of the heating effect of an electric current to operate a small gas valve. Heat from a coil of resistance wire is used to bend a bimetal strip or disc, or to expand a rod or tube.

Figure 11.33 shows a heat motor incorporating a bimetal strip A. The strip is heated by the coil B. An additional heater E is provided for use when the device is used on lower voltages. As the strip is heated it bends, bringing the arm C towards the seating D. The valve itself is in the form of a diaphragm G, which is quickly shut by the pressure of the gas behind it when it nears the seating.

When the current is switched off, the strip cools and tries to straighten out. But it is opposed by the gas pressure holding the valve on to the seating. So forces build up in the strip until it can overcome the pressure and the valve snaps open. The valve can be made to be ‘normally open’ or ‘normally closed’ depending on which side of the operating arm the bimetal strip is positioned.

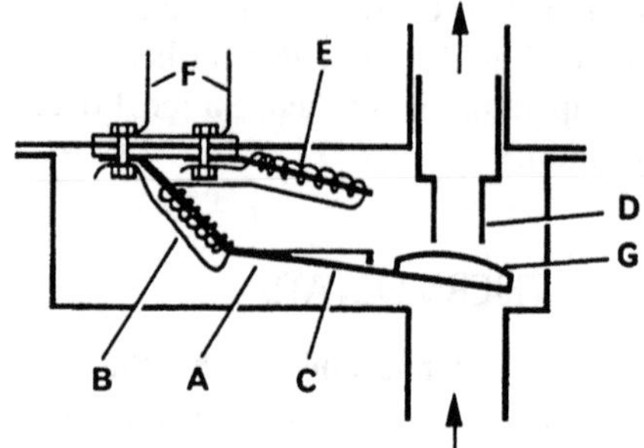

A: Bimetal strip
B: Heating coil
C: Valve arm
D: Seating
E: Additional heater
F: Electrical connections
G: Diaphragm valve

FIGURE 11.33 Heat motor.

Heat motors are used on gas cookers and on central heating systems where automatic control is required. They may be incorporated into multifunctional controls instead of AC solenoid valves.

SOLENOID VALVES

These were described in Chapter 9. They can be operated from an AC mains supply and in this case the early types had a tendency to be noisy. The rapid polarity changes, whilst not detracting from the overall magnetic effect, can cause the valve to hum and at times this could become a loud 'chatter'. One way of overcoming the problem was to rectify the control with a DC input.

Solenoids are used mainly in a range of sizes from 6 to 25 mm connections. The smaller sizes are used to control pilots or weep lines. They may be operated at voltages from 12 to 250 V AC and are also incorporated into multifunctional controls.

For large appliances a more positive shut-off is required and slower opening is an advantage. So motorised valves are generally used instead.

MOTORISED VALVES

As its name implies, a 'motorised' valve is one which is actuated by an electric motor. There are a number of ways in which this can be done.

A simple form of valve has a small motor attached to the top of the plug controlling the fluid flow. When current is switched on, the motor runs, turning the plug through a quarter of a revolution, or 90°, to the 'on' position. At this point the motor changes over a switch contact and breaks the circuit. So the valve remains in its new position until current is again switched on to the motor

FIGURE 11.34 Motorised valves.

by a control circuit. When this happens the motor turns the plug another 90° and switches itself off. So the valve is shut off again.

The valve continues to rotate 90° each time the motor runs, stopping alternatively at the 'on' and 'off' positions.

The low voltage version of this valve is often fitted with two stator windings wired so that the motor can be reversed. This valve is opened or closed by supplying current to the appropriate winding.

In some motorised valves the action of opening the valve compresses a spring which closes the valve when current is switched off.

Valves are used for a variety of purposes. Some which give on-off control have a simple plug, or a rotary vane, called a 'butterfly valve'. Others with more complex plugs can be used as two-way valves to change over the circulation of water in central heating or water heating systems. These are known as 'zoning valves' (see Volume 2). Typical valves are shown in Fig. 11.34.

ELECTRO-HYDRAULIC VALVES

This device uses hydraulic power to open the valve and derives its oil pressure from a pump driven by a small electric motor. The valve has a soft seating and is closed by a powerful spring. A typical example is shown in Fig. 11.35.

When current is switched on, the electromagnetic relief valve H closes and the motor A starts. Oil is pumped from the reservoir B through the check valve C into the space above the diaphragm D. The oil pressure forces the diaphragm down, opening the valve E against the spring F.

When the valve is fully open, the diaphragm trips the limit switch G and stops the motor. The relief valve remains energised and the oil pressure keeps the valve open.

Switching off the current de-energises the relief valve which opens, allowing the oil under pressure to escape back into the reservoir as the spring presses the valve up on to its seating.

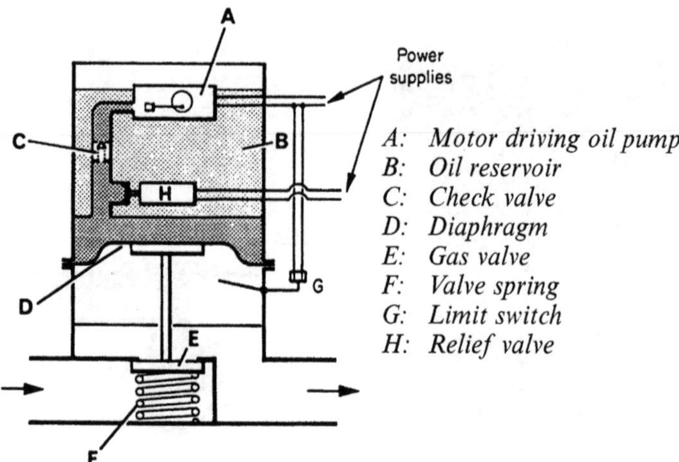

FIGURE 11.35 Electro-hydraulic valve.

When current is switched on, the valve opens progressively, taking about 10 s to reach its full-on position. Shut off is very quick and the use of soft-seated valves ensures complete closure. If the power supply fails or there is damage to the oil reservoir, the valve fails to safety.

MULTIFUNCTIONAL CONTROLS

A 'multifunctional' control is a composite control which incorporates all the control devices required by a particular appliance. It consists of a chassis fitted in the gas supply to the appliance to which the individual control units are attached. The chassis can remain in the gas supply whilst the various units are removed for servicing or renewal. The multifunctional control may include:

- main control cock,
- constant pressure regulator,
- flame protection device,
- solenoid valve.

The control might also incorporate an ignition device or a thermostat.

A diagrammatic sketch showing the principal components of a multifunctional control is shown in Fig. 11.36. This type might be fitted to a central heating boiler or warm air unit.

When the appliance is off, gas entering the inlet connection would be stopped by the safety shut-off valve A. This is part of the flame protection device which includes the thermo-electric valve unit B, to which a thermocouple is connected at C.

To put the appliance into operation the knob D must be turned to the 'pilot' position. This turns the disc-type cock so that the main gas way is still closed, but the knob can now be pushed in against the spring loading. Depressing the

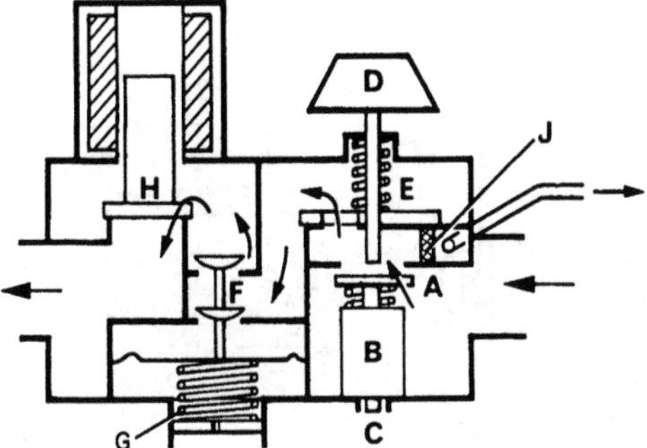

A: Safety shut-off valve
B: Thermo-electric valve unit
C: Thermocouple connection
D: Knob
E: Disc-type cock
F: Regulator
G: Regulator spring
H: Solenoid valve
J: Pilot filter

FIGURE 11.36 Multifunctional control.

knob lifts the shut-off valve from its seating and allows gas to pass through the filter J to the pilot burner. It also brings the armature into contact with the electromagnet in the thermo-electric device.

The pilot can now be lit and the knob must be held down until the thermocouple energises the magnet so that it can retain the shut-off valve in the open position. This takes about 20 s.

Then the knob can be released and turned to the 'on' position, so allowing gas to pass through the drilling in the disc. A latch is provided to prevent the user accidentally turning the knob off and then on again immediately. This would extinguish the pilot and then allow main gas to pass to the burner, without it being lit. The latch prevents the cock being turned back to 'on' until the thermocouple has cooled and the shut-off valve has closed.

Gas can now pass to the regulator F where a constant pressure is maintained by the spring loading G. From the regulator, gas is finally controlled by the solenoid valve H.

This main control valve might be a conventional solenoid or a heat motor. On a central heating boiler it would usually be controlled by the time switch and the boiler thermostat. It could operate at either mains or low voltage.

Many different types of multifunctional controls are produced but the main characteristics are similar. As an example, a more complex control is shown in Fig. 11.37. It has push-button operation and the small solenoid uses a servo mechanism to operate the main valve. This has the advantage that the valve fails to safety if the supply is interrupted or the diaphragm fails. A strong spring ensures complete closure.

Gas entering the inlet first passes through the flame protection device (Fig. 11.38). The shut-off valve and the armature are mounted at opposite ends of a double, pivoted lever mechanism which also carries a safety latch.

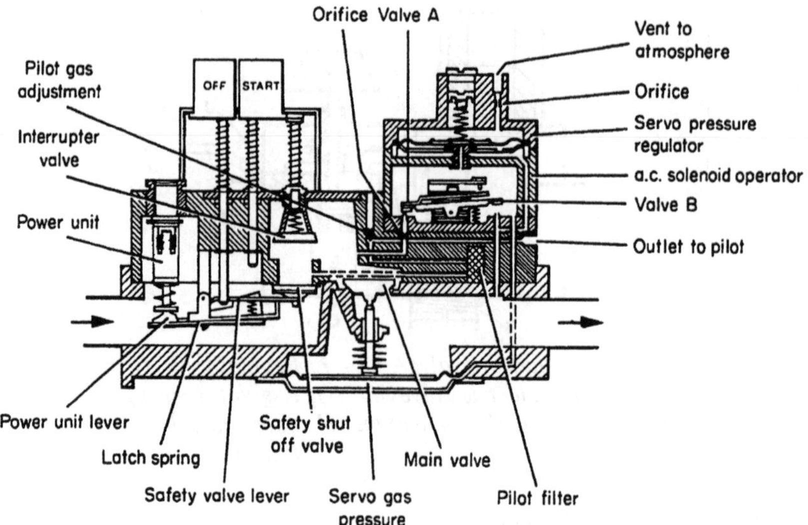

FIGURE 11.37 Multifunctional control with servo regulator in 'off position'.

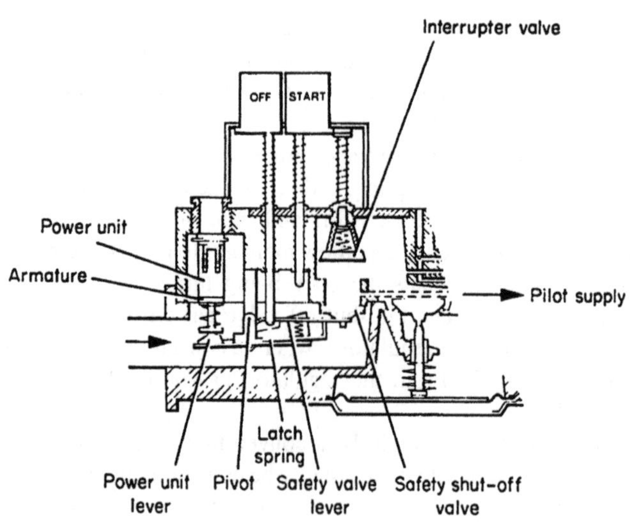

FIGURE 11.38 Flame protection device.

Depressing the 'start' button operates two plungers (Fig. 11.39). The first closes the flow interrupter valve whilst the second opens the safety shut-off valve. It also engages with the safety latch and pivots the power unit lever up to bring the armature on to the magnet. Opening the shut-off valve allows gas to the pilot which energises the magnet by heating the thermocouple. Releasing

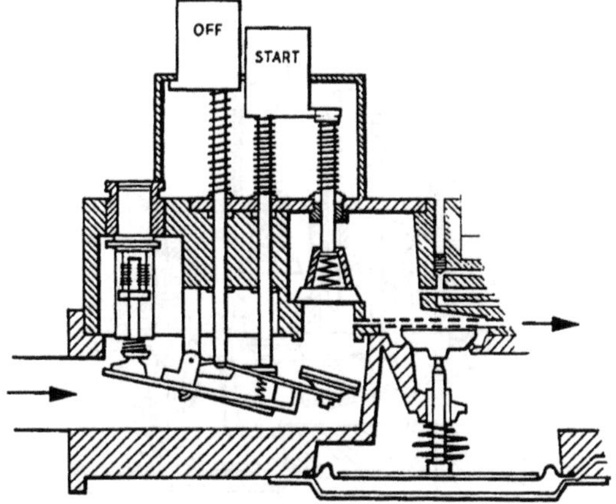

FIGURE 11.39 Push buttons in 'start' position.

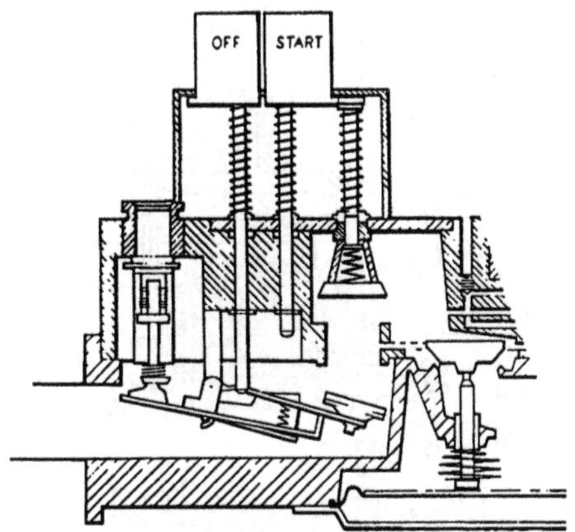

FIGURE 11.40 Push buttons in 'on' position.

the 'start' button after 30 s lifts the interrupter valve and allows gas to the main control (Fig. 11.40).

To shut down the appliance, the 'off' button is depressed, disengaging the latch and allowing the shut-off valve to close (Fig. 11.41). The valve cannot be re-opened until the thermocouple has cooled and the power unit lever re-engages the latch. This prevents the user from releasing gas into the appliance when the pilot has been extinguished. The re-lighting sequence may be carried out again when the levers are in the positions as shown in Fig. 11.42.

The main control valve is actuated by a diaphragm which is controlled by the servo pressure regulator. This is operated in the first place by the AC solenoid.

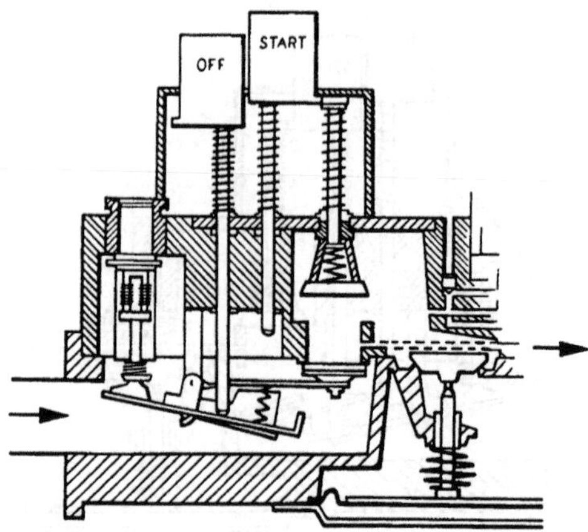

FIGURE 11.41 Push buttons in 'off' position.

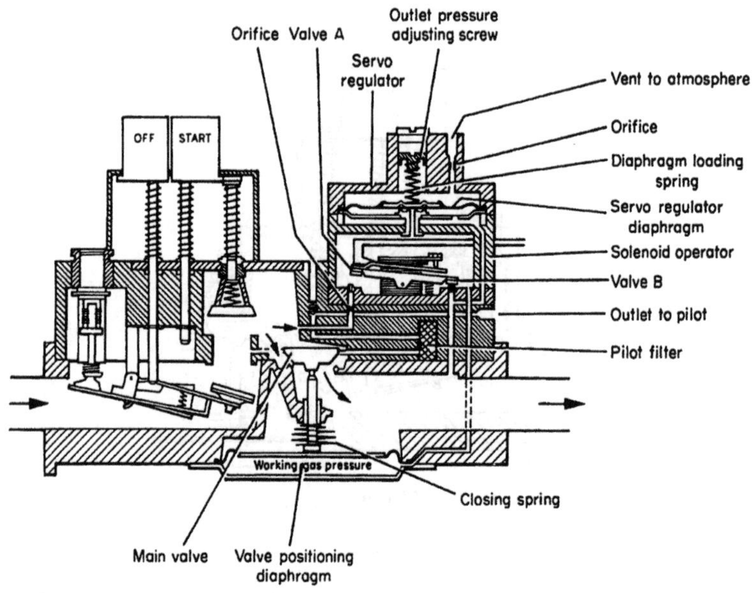

FIGURE 11.42 Servo-regulated valve in operating position.

The solenoid pivots a lever which has a small valve at each end. The sequence of operations is as follows:

- Current switched on. Solenoid energised. Valve A open. Valve B closed.
- Current switched off. Solenoid de-energised. Valve B open. Valve A closed.

When valve A is open (Fig. 11.42) gas enters through the orifice and passes into the regulating chamber. It exerts a pressure on the underside of the servo regulator diaphragm, lifting it against the spring loading, and it is also communicated to the underside of the valve positioning diaphragm so lifting the main valve off its seating.

The outlet gas is also conveyed, through drillings, to the underside of the servo regulator diaphragm. In this way the pressure exerted on the valve positioning diaphragm is regulated to position the main valve and maintain a constant outlet pressure.

When current is switched off, the solenoid trips the lever, closing valve A and opening valve B (Fig. 11.37). This shuts off the gas supply to the regulating chamber and allows the gas it contains to weep away quickly to the outlet. So servo gas pressure falls to zero and the spring returns the main valve to its seating.

Sophisticated multifunction controls are used to give full sequence control and/or modulation of the gas supply to an appliance.

A full sequence control does not require a permanent pilot. When the heating system (external control) calls for heat, the device electronically tests the integrity of the control system before generating an ignition spark and allowing gas to an intermittent pilot or the appliance burner. The pilot or main burner flame is then monitored, generally by a flame rectification device – described earlier in the chapter. In the case of the intermittent pilot, once the pilot flame has been proved, gas is then allowed to flow to the main burner where it is ignited by the pilot flame.

DIRECT SPARK IGNITION (DSI) SYSTEMS: PRINCIPLES OF OPERATION

Figure 11.43 shows a direct spark ignition system and below it is a description of the principle of operation. Figure 11.44 shows an intermittent pilot system with a description of its operation using a Honeywell VR4700/VR7700 gas control shown in Fig. 11.45.

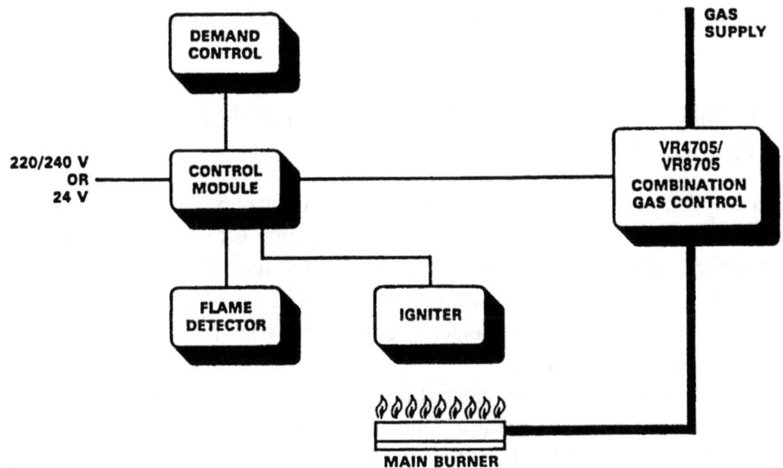

FIGURE 11.43 Direct spark ignition (DSI) system (Honeywell).

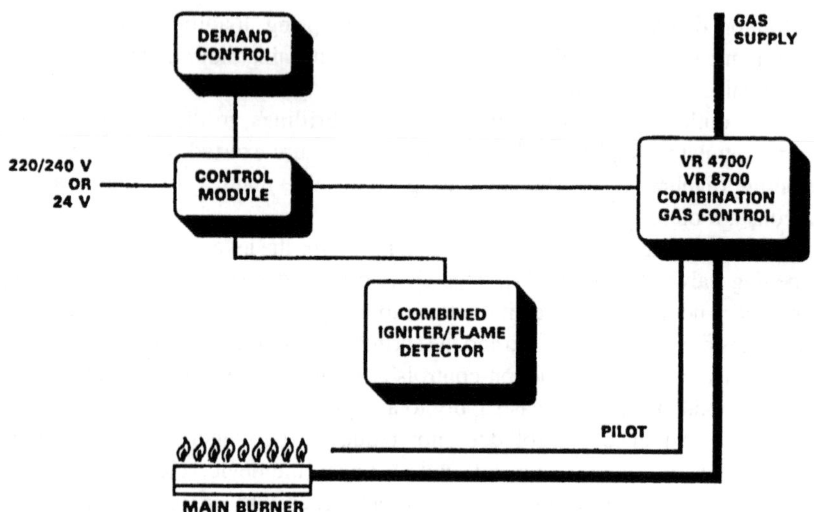

FIGURE 11.44 Intermittent pilot (IP) system (Honeywell).

On a *call-for-heat* from the demand control the control module performs a safe-start check. If a flame (or flame simulating) condition is found the start sequence is inhibited. If everything is in order, the control module triggers the igniter and energises the on–off electric operators which control main gas flow to the burner.

If the burner gas is not ignited during the 'trial-for-ignition' period, ignition stops, the operator de-energises and the control module locks out.

When the burner is ignited during the trial for ignition period the flame is detected by the flame sensor and the burner continues to function until the demand controller stops *calling-for-heat.* If flame is lost during normal operation then one further start sequence is attempted. If flame is not established the operator is de-energised and the control locks out. When the control module locks out it must be reset manually or electrically before another attempt to start the burner is made.

INTERMITTENT PILOT (IP) SYSTEMS: PRINCIPLES OF OPERATION

On a *call-for-heat* from the demand control the control module performs a safe-start check. If a flame (or flame simulating) condition is found the start sequence is inhibited. If everything is in order the control module triggers the igniter/sensor and energises the operator that controls the gas to the pilot.

If the pilot flame is not detected during the trial for ignition period the control module will lock out.

When the pilot flame is detected by the igniter/sensor, the controller energises the on–off electric operator controlling the main gas valve. Gas is fed to the burner and ignited by the pilot. The burner continues to function until the demand controller stops *calling-for-heat.*

When the control module locks out it must be reset manually or electrically before another attempt to start the burner is made.

MODULATION

Modulation of the gas supply can be achieved by using the water temperature in the appliance heat exchanger, to control the voltage transmitted to the solenoid valve in the multifunction control. The water temperature is sensed by a thermistor in the heat exchanger. The thermistor then transmits a signal to a potentiometer (adjustable resistor) in the appliance printed circuit board (PCB). The potentiometer controls the voltage to the solenoid coil. This in turn determines the position of the gas valve in relation to the valve seat and the volume of gas passing to the main burner.

This type of control is quite common on combination appliances giving domestic hot water and central heating, see Volume 2, Chapter 10. In the domestic hot water mode the appliance requires its maximum gas rate; however, when set in the central heating mode the modulating gas valve comes into its own.

Figure 11.46 shows a modulating valve fitted onto a multifunction control.

Faults and remedies

The faults which may occur on power operated control devices are dealt with in Volume 2, Chapters 12 and 13.

The faults which may occur in the regulator sub-unit of a multifunctional control are, in essence, the same as those already described in Chapter 7, and may be rectified in the ways already suggested.

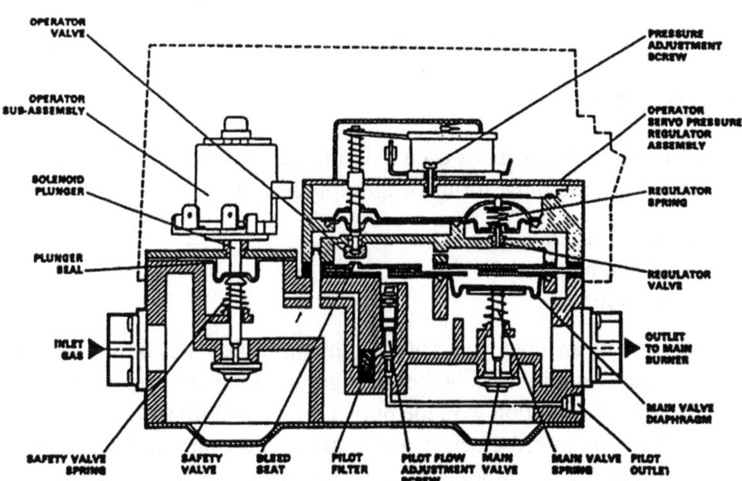

FIGURE 11.45 Schematic cross-section through VR4700/VR8700 (Honeywell).

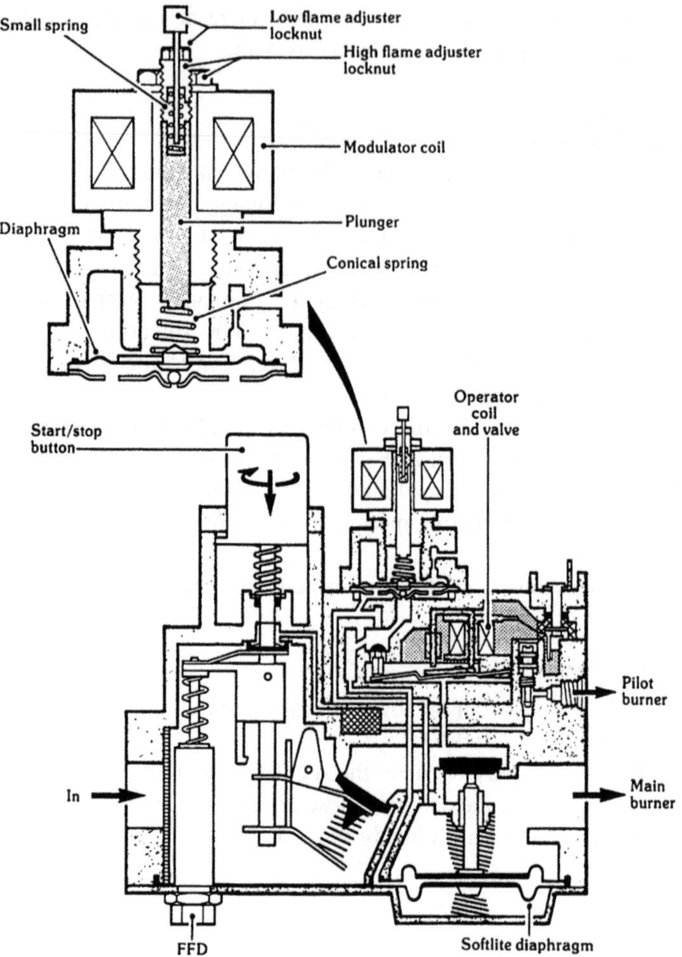

FIGURE 11.46 Multifunction control, fitted with 'modulating' pressure regulator.

IGNITION DEVICES

An introduction to the methods used to ignite gas was given in Chapter 2. These dealt briefly with the three forms of ignition device:

1. permanent pilots,
2. filament igniters,
3. spark ignition.

Igniters are usually an integral part of the appliance. They may be designed to light a particular pilot burner, main burner or a group of burners and can be operated separately or in conjunction with the burner tap or multifunctional control.

Some filament, piezo-electric and battery-operated spark generators are produced as separate, hand-held devices. They may be carried about and used to ignite any gas burner.

PERMANENT PILOTS

These have the advantages of simplicity, cheapness and independence of any other source of energy.

However, in the interests of economy and energy conservation, the pilot jet is no longer used on new appliances, however, many are still in use on older appliances. So the jet is very small and can be subject to blockage by dust, products of corrosion and condensation. It may easily be affected by draughts. High standards of manufacture must be maintained to produce jets which will give stable flames and avoid sooting. The aeration ports must be prevented from linting. Pilots must be sited to light the main burner smoothly and non-explosively. They may also operate a flame supervision device.

Where the pilot cannot be located immediately adjacent to the burner, or where a number of burners are to be lit from one pilot, as on the older type of cooker hotplate, 'flash tubes' were used. A flash tube carries an ignition flame (or 'flash') from the pilot to the burner. A typical layout of flash tubes on a cooker hotplate is shown in Fig. 11.47.

Each single flash tube is positioned between the pilot flame and the burner, as shown in Fig. 11.48 so that the drilling A is in line with the centre of the tube B. When the burner tap is turned on, an air/gas mixture flows into the burner head and some of the mixture passes through A into the tube and onto the pilot C. As it enters the tube it entrains more air at D. This increases the aeration of the mixture and so increases its flame speed.

As it emerges from the tube at the pilot end, the mixture is ignited. Because the flame speed is high and the speed of the mixture in the tube is relatively low,

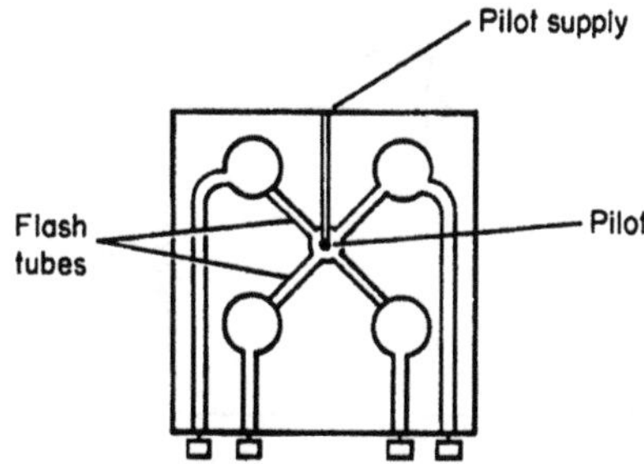

FIGURE 11.47 Flash tubes on cooker hotplate.

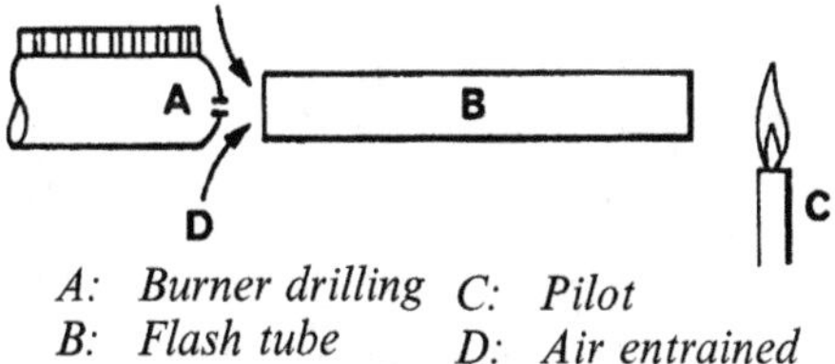

FIGURE 11.48 Arrangement of single flash tube.

the flame flashes back down the tube and lights at A, so igniting the gas at the burner.

Flash tubes must be kept clean and unobstructed. They must be accurately positioned in line with the burner drillings and the pilot. The primary aeration of the burner must be correctly adjusted. If it is inadequate the ignition flame will simply burn on the pilot end of the flash tube. The maximum length of flash tube which will work satisfactorily is 200 mm. For reasons of economy and energy conservation, previously mentioned, this method of automatic ignition on cooker hotplates has been discontinued.

FILAMENT IGNITERS

This type of ignition system was popular for igniting manufactured gases with a high percentage of free hydrogen. Most modern appliances use spark ignition systems and these are described later in this chapter. Some filament systems are however still in use.

The igniter consists of an electrically heated filament adjacent to a small gas jet. The filament is a small coil of thin resistance wire, usually of platinum, because of its catalytic properties. The jet may be a pilot or a special ignition flame which is only present during the ignition period. The power for heating the filament may come from dry batteries or mains electricity and a potential of 3 V is required. This may be obtained from two 1.5 V batteries or from a step-down transformer. The transformer has the advantage of consistent power output and avoids the need for battery replacement.

Typical wiring circuits are shown in Figs 11.49 and 11.50.

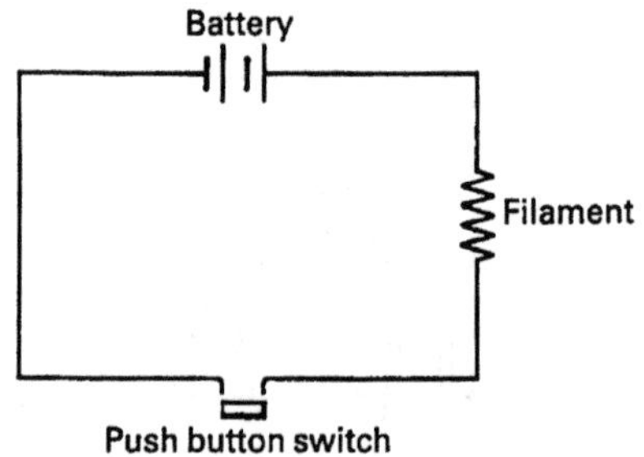

FIGURE 11.49 Battery-operated filament igniter.

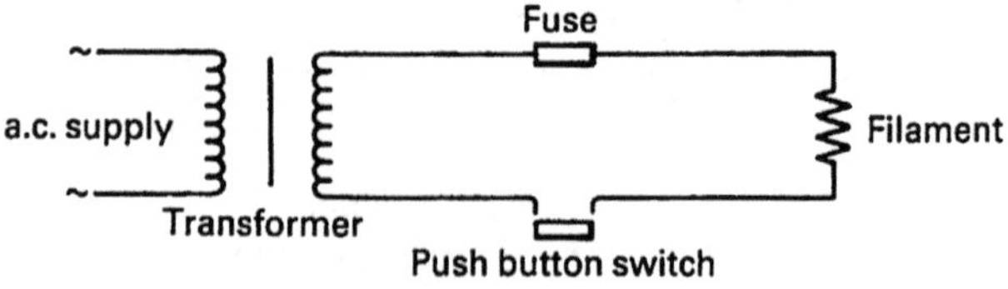

FIGURE 11.50 Mains-operated filament igniter.

An igniter head which incorporates a lighting jet is illustrated in Fig. 11.51. The protective shield which surrounds the filament must allow air into the jet so that it may form a combustible mixture with the gas. Failure of an igniter to light up can be due to lack of air at the filament as well as the more obvious reasons. These include damage to the filament, blockage of the jet or gas supply and electrical failure in the source or the circuit.

The gas and electrical supplies to the igniter may be switched on together by a combined switch and valve. A typical arrangement is shown in Fig. 11.51 with the contacts and the gas valve mounted on the same, spring-loaded spindle.

On some of the older models of gas fires a common arrangement was for the tap to be allowed to turn beyond the full-on position to an 'ignition position'. The extra rotation allows gas to the igniter jet and turns a cam which closes a switch, so supplying current to the filament (Fig. 11.52). When ignition takes place, the main burner is lit immediately by the igniter flame.

When the tap handle is released, spring loading returns the tap to its full-on position, shutting off the power and the gas supplies to the igniter.

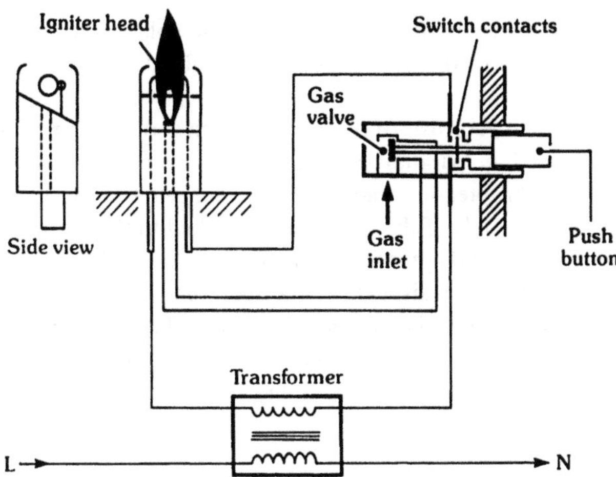

FIGURE 11.51 Igniter circuit.

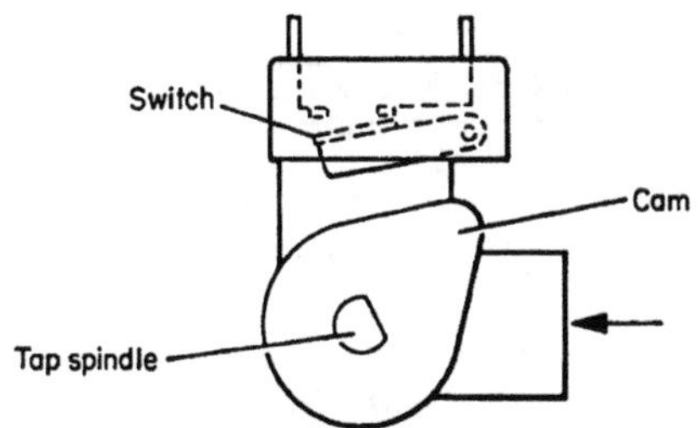

FIGURE 11.52 Cam operated igniter switch on tap spindle.

Glowcoil igniters have a heavier filament and are heated by mains electricity, through a step-down transformer. They operate at voltages from 2.5 V but require a heavier current than a catalytic filament.

SPARK IGNITION

Gas may be ignited by a high-voltage spark, using an emf of 5000–15,000 V. These voltages may be obtained from:

- piezo-electric generator,
- mains transformer,
- electronic pulse system.

Piezo-electric igniters

Piezo-electric ignition systems are commonly used on gas fires and for lighting permanent pilots on water heaters and boilers. Some cookers have also used this method of ignition.

The lead zirconate–titanate crystals used on these igniters are exposed to powerful electric fields during manufacture. These polarise the material so that, when stressed or deformed, it produces an emf of about 6000 V between the two ends of the crystal.

The igniter usually consists of two crystals each about 12 mm long and 6 mm diameter connected in parallel (Fig. 11.53). The crystals are shown at A, separated by a metal pressure pad which is connected to the spark electrode supply D. The other ends of the crystals are earthed to the casing.

Depressing the lever B applies pressure to the crystals. So the voltage builds up until it overcomes the resistance of the spark gap and sparks are produced. More sparks are produced when the lever is released and the stress is removed. The lever could be operated in a number of ways. In this case a cam C, which would be attached to the tap spindle, as shown.

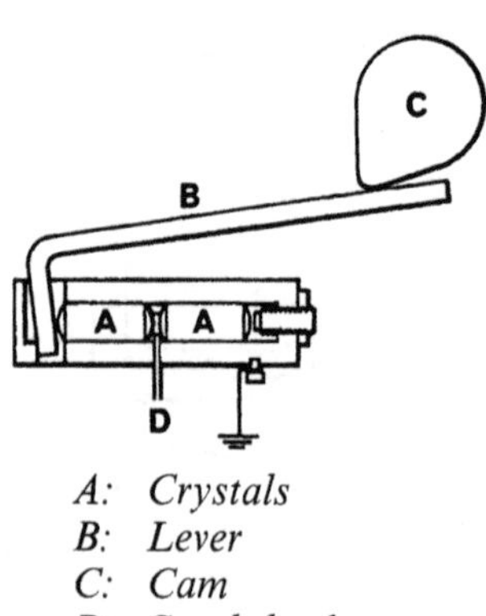

FIGURE 11.53 Cam operated piezo-electric igniter.

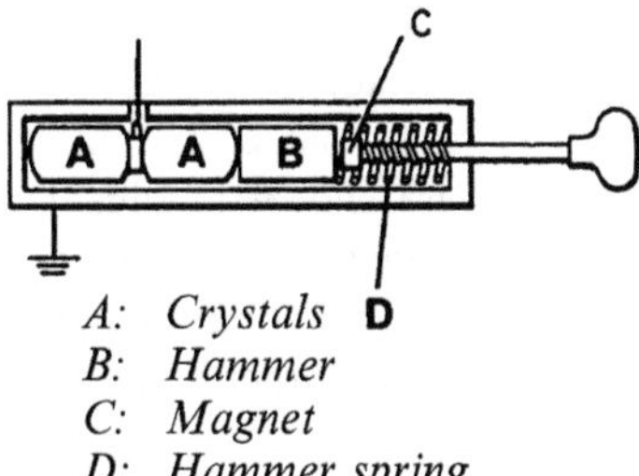

FIGURE 11.54 Impact type piezo-electric igniter.

It is also possible to apply pressure to the crystals by impact (Fig. 11.54). This device has a similar arrangement of the crystals A. It also has a hammer B, which is attracted to the magnet C on the end of the knob spindle.

When the knob is pulled, magnetic force takes the hammer along with the knob assembly against the pressure of the spring D. When the spring is fully compressed a further pull overcomes the magnetic force and the hammer is returned smartly by the spring to deliver a blow to the crystals. This causes sufficient stress to produce a spark at the electrode. This method has the advantage of applying a consistent load to the crystal since the hammer and the spring always produce the same amount of stress, irrespective of any action on the part of the user.

Mains transformer

An ignition transformer is simply an ordinary step-up transformer usually with separate input and output windings. The electrode is connected between the two ends of the secondary winding or between one end and an earth return (Fig. 11.55a).

On the higher voltage arrangement (Fig. 11.55b) the secondary winding is earthed from a centre tapping. This is to keep the potential between the two output connection and earth down to half the total output voltage, (2 output connections at 5000 V = 10000 V).

Electronic pulse/coil ignition

Various methods have been employed to produce high-voltage electric sparks. With the development of solid-state devices it has become possible to produce

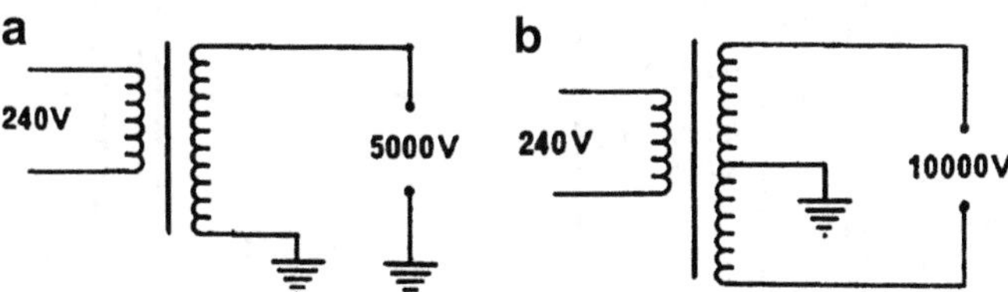

FIGURE 11.55 Mains ignition transformer: (a) electrode connected to one end of secondary winding; (b) electrode connected across both ends of secondary winding.

sparks at speeds up to eight sparks per second at several electrodes simultaneously, from a single, small generator. Although the voltage of the spark may be 15,000 V, the actual danger from contact with the electrodes is negligible.

Examples of these devices are shown in Figs 11.56 and 11.57. A more sophisticated unit is also available that will sense when the flame is alight and at that point stop the spark. If the flame is extinguished during use, the spark probe senses the loss of flame between the burner and the probe, and starts to spark

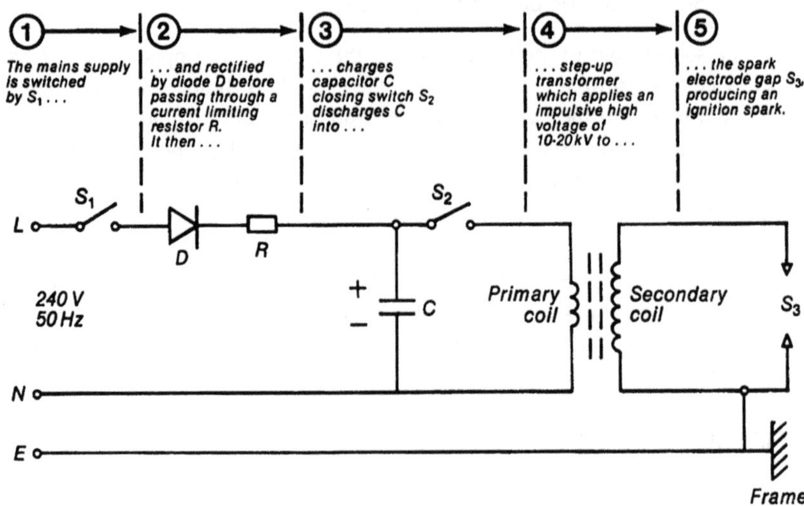

FIGURE 11.56 Basic concept of a mains powered coil ignition system.

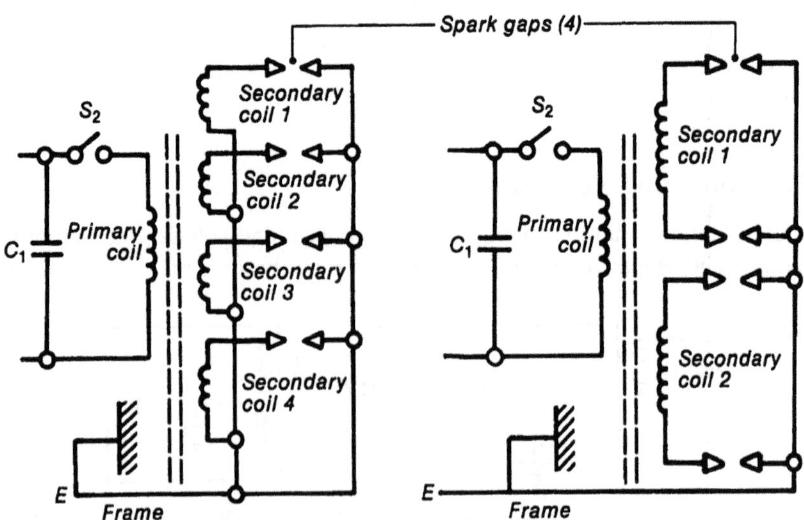

FIGURE 11.57 Multi-outlet ignition.

again until ignition occurs. These re-ignition devices are found on cookers and appliances with full sequence controls.

Faults and remedies

In all the spark igniters the position of the electrode is critical. Not only must the gap be correct but also the spark must occur in a place where the air/gas mixture is well within the flammability limits. The speed of the mixture must allow the flame produced to light back on to the flame port.

The insulation must be very good to prevent leakage at the high voltages produced.

On piezo-electric crystals, dampness on the crystals' assembly may result in the current shorting to earth with consequent loss of spark at the electrodes. When faults occur within the spark generators themselves, the entire unit should be renewed.

CONTROL SYSTEMS

The way in which control devices are arranged into a complete system depends on the type of appliance and the way in which it is required to operate. On some appliances a measure of manual control may be satisfactory. On others, like central heating appliances, fully automatic operation is essential.

There are many possible combinations of controls, and the examples given are merely typical of the methods commonly used. Figures 11.58 and 11.59 show systems which have been superseded by electrical controls but which may still be in use.

PRESSURE OPERATED SYSTEMS

Figure 11.58 shows a simple, traditional system of gas pressure operated controls for a smaller central heating boiler. On the main gas supply to the burner F the sequence of the direct controls is:

1. Main gas control cock A,
2. Regulator B,
3. Flame protection device C (thermo-electric),

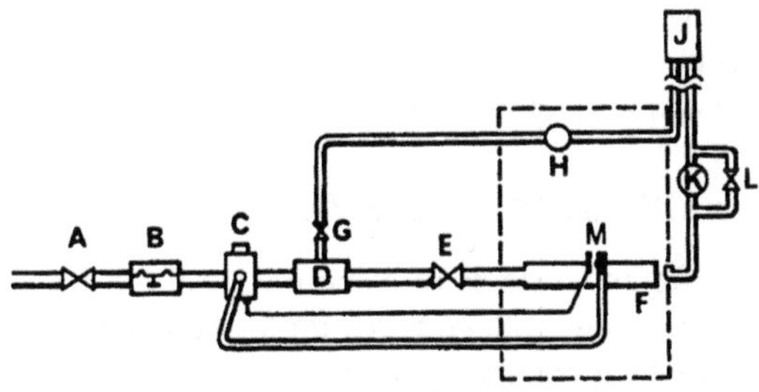

FIGURE 11.58 Pressure operated control system for central heating boiler.

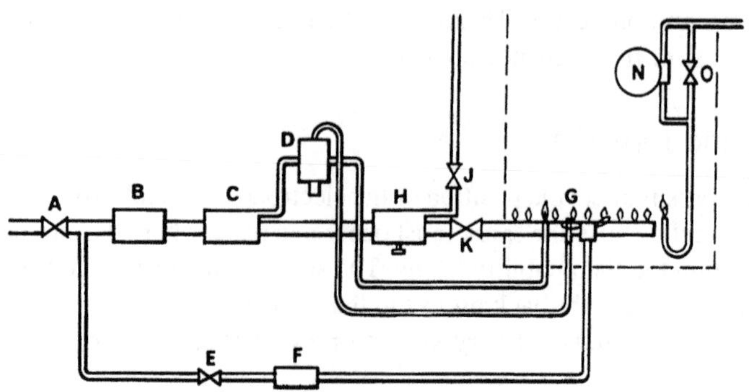

FIGURE 11.59 Pressure operated control system for large boiler.

4. Relay valve D,
5. Burner control cock E.

On the weep line from the relay valve the sequence of the indirect control is:

1. Test cock G,
2. Boiler thermostat H,
3. Air thermostat or room thermostat J,
4. Clock control K,
5. Clock over-ride cock L.

The thermocouple and its pilot burner are shown at M.

A system suitable for a larger appliance is shown in Fig. 11.59. The controls are:

A. Main control cock
B. Main regulator
C. Flame protection relay valve
D. Thermo-electric valve
E. Pilot control cock
F. Pilot regulator
G. Thermocouple and pilot burner
H. Relay valve
J. Test cock
K. Burner control cock
L. Appliance thermostat
M. Weep to and from any remote controls
N. Clock control
O. Clock over-ride cock

A thermo-electric device which could be used on this layout is one with two separate valves linked to the same spindle so that it could control the pilot supply as well as the weep.

On these layouts, the main regulator may be found fitted after the flame protection device and not before it as shown.

ELECTRICALLY OPERATED SYSTEMS

There are many different systems of electric control for gas appliances. They vary from the simple, low voltage types up to the complex, fully sequenced automatic systems which control, ignite and monitor the burner flames. These are described in detail in Volumes 2 and 3.

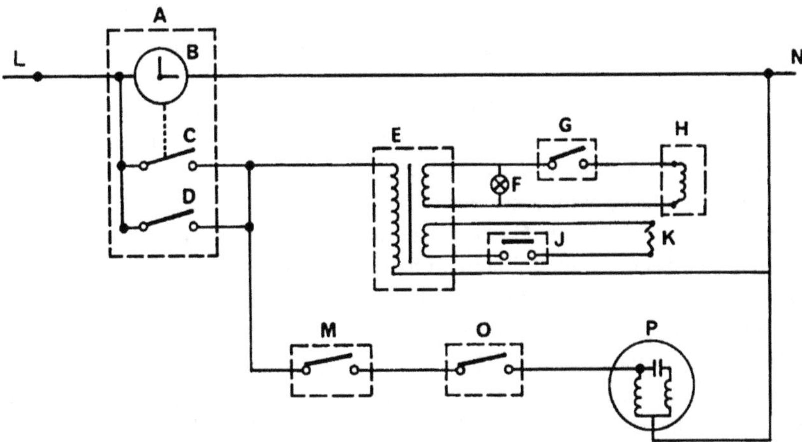

A.	*Clock control*
B.	*Electric clock*
C.	*Clock-operated time switch*
D.	*Clock over-ride switch for continuous operation*
E.	*Transformer (240 V to 24 V for solenoid and 3 V for igniter coil)*
F.	*Indicator light*
G.	*Boiler thermostat*
H.	*Main gas solenoid valve (in multifunctional control)*
J.	*Igniter press-button switch*
K.	*Igniter coil or filament*
L.	*Line connection*
N.	*Neutral connection*
M.	*Programmer 'heating on' switch*
O.	*Room thermostat*
P.	*Circulating pump.*

FIGURE 11.60 Electrically operated control system for central heating boiler.

As an example, Fig. 11.60 shows a schematic wiring diagram of an older type of system (many are still in use), for a domestic central heating boiler. More up-to-date systems are described in Volume 2. The controls are given in Fig. 11.60.

The other control on the boiler would be a multifunctional device which would incorporate:

- flame protection device (this would be thermo-electric and might also operate the ignition switch J);
- constant pressure regulator;
- solenoid valve H.

FAULT DIAGNOSIS

'When all else fails, read the instructions!'

Unfortunately, too many of us adopt this method of working. And it might be a lot better for everybody if we did not. The service engineers who reckon that they can remember it all are probably kidding themselves and trying to impress the customers. Appliances keep getting more and more complex and it would take a genius to remember all the details of the settings and the wiring diagrams. So do not be afraid to use the data sheets and the other material supplied by the manufacturers. It will probably save you quite a lot of time. And the customer is likely to be much more pleased with the result.

If you are faced with a system which is unfamiliar, it often helps to make a line sketch of the layout. Controls can usually be identified by their shape and by what that is connected to them. When you know what they are, make a note of the sequence in which they are fitted.

When you understand the correct sequence of the controls you can begin to look back from the device which is not working to see what can be affecting it.

Check the pressure, gas or electrical, in sequence through the controls. Sometimes it is quicker to check at a halfway point to find out which half contains the fault. And then halfway again.

Many faults can be discovered without the need for elaborate testing procedures or complicated instruments. Start by questioning the customer to get all the information you can. Then use your common sense and your senses.

Look for: signs of obvious damage,
loose wires or connections,
screws missing,
scorch marks or burnt insulation,
displaced electrodes or probes,
badly located pilot burners,
choked weep jets,
soot on burners or thermocouples.

Feel for: overheated solenoids,
transformers or motors loose nuts or connections (when the power's switched off!),
slack belt drives.

Smell for: burning insulation,
overheated coils,
gas leaking from weep connections or vent holes in particular incomplete combustion.

Listen for: chattering or excessive humming,
arcing,
squeaks or grinding noises in motors or motor drives.

Make sure that you have located the fault before you start stripping everything to pieces. The best trouble-shooters are the persons who use their heads before their hands.

One last point. You can sometimes remedy a fault and still not trace the cause of the fault. So it happens all over again.

Always be certain that you have found out not only what happened, but also why it happened. Then you will have ensured that it will not happen again.

Chapter | twelve

Materials and Processes

Chapter 12 is based on an original draft prepared by Mr W. Dale

PROPERTIES OF MATERIALS

In the course of your work you will meet and use many different kinds of materials. They will have been chosen for a particular job because their characteristics make them the most suitable material for that purpose.

If you are to use and choose materials wisely, you will need to be familiar with their characteristics, their advantages and their limitations.

One of the main properties of a material is its strength. That is, its ability to withstand force or resist stress.

Figure 12.1 illustrates the ways in which stress may be applied to materials. These stresses are:

(a) tensile or stretching,
(b) compressive, or squeezing or crushing,
(c) bending, which is both (a) and (b) on either side of a neutral axis,
(d) shear or cutting,
(e) torsion or twisting.

Stress and strain were dealt with in some detail in Chapter 5, and there is no need for them to be studied in any greater depth. But it is necessary to look at the practical applications of this knowledge. For example, you know that:

$$\text{stress} = \frac{\text{load}}{\text{area}}$$

This means that the stress can be increased either by increasing the load or by decreasing the area. So cutting a piece out of a support or beam has the same effect as increasing the load on it. In fact, if the cross-sectional area of a vertical supporting strut was reduced by 1/3, the stress in the strut would increase by 50%.

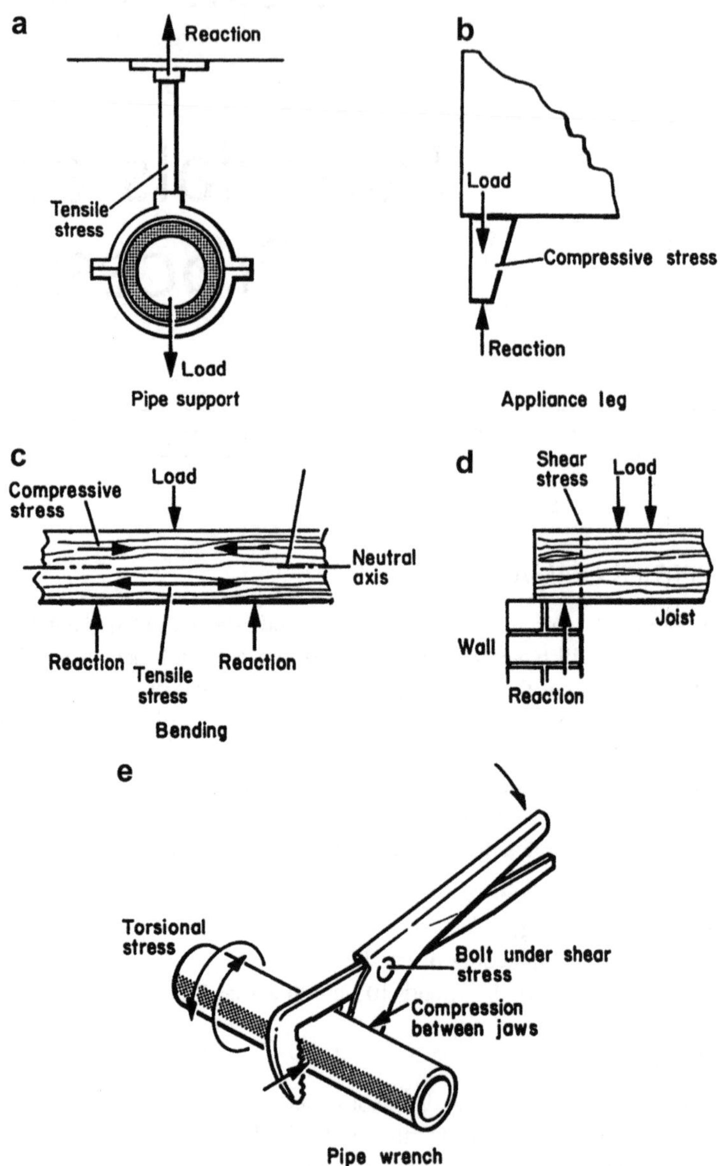

FIGURE 12.1 Types of stress: (a) tensile, (b) compressive, (c) bending, (d) shear, and (e) torsion.

Care must be taken when it is necessary to cut into any beam, joist or structural support. For example, pipes have, on occasion, to be run across the joists in a floor and so the joists must be notched. But notches should not be greater than one sixth of the depth of the joist so as to leave a reasonable factor of safety (Fig. 12.2).

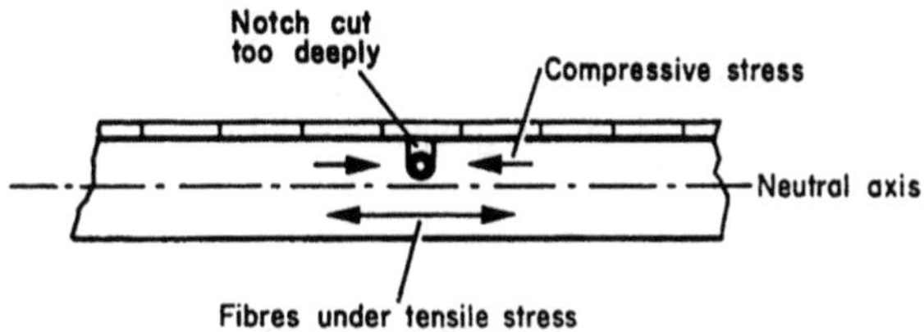

FIGURE 12.2 Notch in joist cut too deeply.

Beams are usually under the greatest bending stress at their centre and under the greatest shear stress at the points where they rest on their supports. So the best place for running a pipe across a floor is just a little way out from the wall.

In addition to strength, materials have a number of other properties which govern their uses. These are:

Malleability: This is the capability of being hammered or beaten into a plate or leaf (Fig. 12.3a).

Ductility: This is the property of many metals of being drawn out into a slender thread or wire without breaking (Fig. 12.3b). Ductile metals may be easily bent.

Elasticity: The ability of a material to return to its original form, or length, after a stress has been removed. The elastic limit is the greatest strain that a material can take without becoming permanently distorted (Fig. 12.3c).

Hardness: The property of a material to resist wear or penetration. This is the requirement of the cutting edge of a tool. The hardest naturally occurring substance known is the diamond.

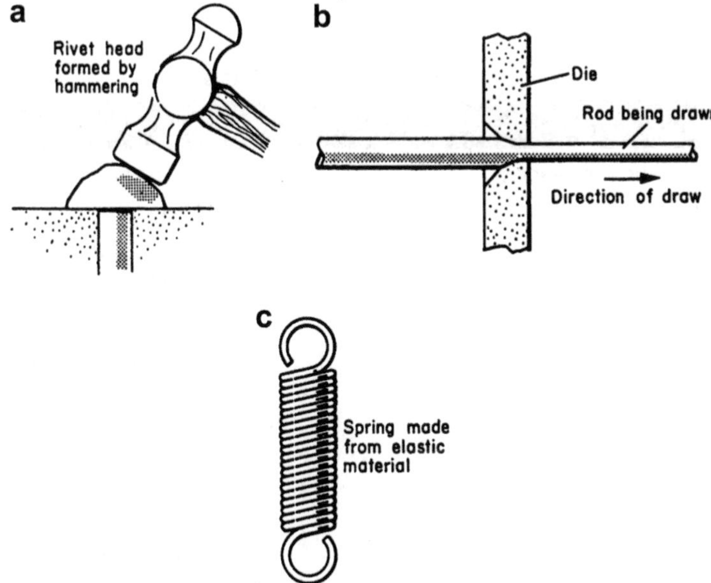

FIGURE 12.3 Properties of materials: (a) malleability, (b) ductility, and; (c) elasticity.

Brittleness: A brittle material is one that is easily broken. Brittleness in metals is usually associated with hardness. So it is often necessary to reduce the hardness of a tool in order to give it greater strength.

METALS

Materials may be divided into different categories in a variety of ways. The simplest division is into:

- metals and
- non-metals.

Metals differ from non-metals in chemical properties as well as in the more obvious physical ones. But there is not a sharp dividing line between the two categories. Some metals, like antimony and selenium, possess characteristics of both classes, and carbon, which is a typical non-metal, behaves like a metal in some respects, including conducting electricity. Table 12.1 gives some of the main differences between the two categories.

Metals themselves can also be subdivided into:

- pure metals and
- alloys.

The pure metals are those which are in their original form and do not have any other metals or materials mixed with them.

Alloys are formed when two or more metals, or a metal and a non-metal are mixed together, usually by melting. An alloy often has useful properties which its

Table 12.1 Characteristics of Metals and Non-Metals

Metals	Non-Metals
Characteristic appearance, 'metallic' sheen or gloss	No particular characteristic appearance
Good conductor of heat	Usually poor conductor of heat, may be insulator
Good conductor of electricity	Usually poor conductor of electricity, may be insulator (except carbon, silicon)
Electrical resistance usually increases as temperature rises	Electrical resistance usually decreases as temperature rises
Density usually high	Density usually low
Melting point and boiling point usually high	Melting point and boiling point usually low

parent metals do not possess. For example, much lower melting point, as in the case of solders. Some of the commoner pure metals and alloys are listed in Table 12.2.

Metals may be further classified as:

- ferrous metals, that is those that consist essentially of iron;
- non-ferrous metals, those that do not contain iron.

The basic differences between the two categories are that, with the exception of the stainless steels, ferrous metals will rust when in contact with water and oxygen. And ferrous metals may be magnetised.

Non-ferrous metals do not rust, although they may corrode under certain circumstances. And they are non-magnetic.

FERROUS METALS

Grey cast iron

This is a mixture of iron with 3.5–4.5% carbon and very small amounts of silicon, manganese, phosphorus and sulphur. The carbon is in the form of 'flake graphite' and it is these flakes which make the material very brittle (Fig. 12.4). The iron fractures easily along the lines of the flakes.

Table 12.2 Commonly Used Metals and Alloys

Pure Metal	Alloy
Aluminium	Brass
Chromium	Bronze
Copper	Chrome-vanadium steel
Gold	Duralumin
Iron	Gun-metal
Lead	Invar
Magnesium	Nickel–silver
Mercury	Pewter
Nickel	Rose's alloy
Platinum	Solder
Silver	
Sodium	Stainless steel
Tin	Type-metal
Tungsten	
Zinc	White-metal (bearing metal)

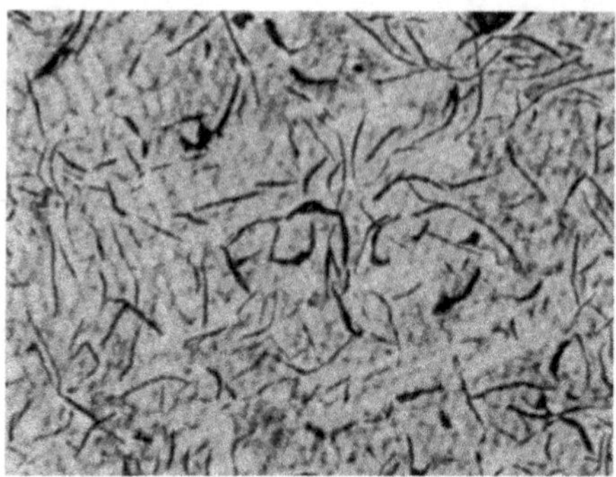

FIGURE 12.4 Micro-section of grey cast iron showing flake graphite.

Cast iron is hard and has a much higher resistance to rusting than steel. It is used for some gas mains, appliance components and tools. It has the advantage of a low melting point (1200 °C), easy casting and machining, and low cost.

Ductile cast iron

This has approximately the same carbon content as grey cast iron. The difference is in the shape, size and distribution of the carbon particles which are changed from flakes into ball-like nodules or 'spheroids' (Fig. 12.5). The change is due to alloying the iron and carbon with magnesium compounds before casting. Subsequent heat treatment at 750 °C changes the brittle iron carbide into softer, ductile ferrite.

Used to make pipes for gas mains, ductile cast iron has the strength and ductility of lower grade steel and the high corrosion resistance of cast iron.

Wrought iron

This is the purest commercial form of iron containing very little carbon. Made from heat-treated cast iron it is rolled or hammered to give it the required grain

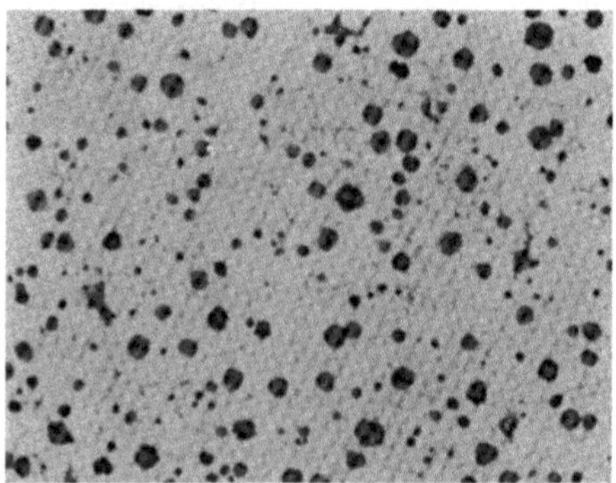

FIGURE 12.5 Micro-section of spheroidal graphite iron showing globular graphite.

structure. It is still used for chains, horseshoes, gates and ornamental ironwork. Because of its high cost it is no longer used for pipes or fittings.

Malleable iron (Malleable cast iron)

Malleable iron is white cast iron in which the carbon has combined with the iron to form iron carbide. This results in a very hard and brittle material which is 'annealed' or softened after being cast into the desired shape.

Most of the ferrous pipe fittings used in conjunction with mild steel pipe in internal installations are malleable iron.

Steel

Steel is another mixture of iron and carbon. Its carbon content lies between that of wrought iron (about 0.05%) and cast iron (about 4%). It may be produced from cast iron by burning out some of the carbon, as in the Bessemer process or in an open-hearth furnace. Or it can be made from wrought iron by adding carbon, as in the crucible process. Low-carbon steel with a carbon content of 0.1–0.25% is called 'mild steel'. It is used for general structural work, pipes, sheets and bars. It can be pressed, drawn, forged and welded easily. But it cannot be hardened.

Steels which contain 0.5–1.5% carbon can be given various degrees of hardness by heat treatment. These are the 'tool steels'.

Table 12.3 gives examples of the percentage carbon content in various steel tools.

Table 12.3 Carbon Steel Tools

Tool	% Carbon
Screwdrivers	0.65
Crowbars	0.75
Cold chisels	0.75
Hammers	0.75–0.95
Pliers	0.75
Vice jaws	0.75
Spanners, wrenches	0.75
Wood saws	0.75–0.85
Wood chisels	0.85–1.00
Screwing dies	1.05
Taps, reamers	1.10
Twist drills	1.15

NON-FERROUS METALS

Aluminium

A soft, light metal, it is used in its pure form for cooking utensils. Rolled into foil it is used for heat insulation, since it reflects radiation.

Aluminium is alloyed with copper, silicon or magnesium to produce tougher metals which can be die-cast to make appliance parts, governor cases and other control devices. It is also used in the motor industry for cylinder heads and pistons.

Antimony

This is a soft, grey, crystalline metal which expands slightly when it solidifies. It is now principally used in alloys of lead or tin, for example type-metal, solders, bearing metals and pewter. Prior to the extensive use of plastics it was used for cheap cast trinkets.

Chromium

A hard, white metal which can take a high polish, it is consequently used to give a plated finish to other metals. In addition it is used in alloys to produce, for example, stainless steel, high-speed tool steels and Nichrome wire for electric elements.

Copper

This is one of the 'coinage' metals (copper, silver, gold), and is still in use as a constituent of British coins. It is a very malleable and ductile metal, a good conductor of heat and electricity and resistant to many forms of corrosion.

Copper is used for pipes and pipe fittings and these may be for gas, water or steam installations. In sheet form it is used for hot-water cylinders, calorifiers and water heater heat exchangers. The sheets are sometimes used for roofing.

It is also used to produce electric wires and in alloys to make nickel–silver, brass and bronze.

Lead

Lead is soft, ductile and malleable, with a low melting point (327 °C) and a high specific gravity (11.34). It is also poisonous and will accumulate in the body, not passing through as most other substances do. The 1986 Model Water Byelaws prohibit the use of lead pipe, fittings etc. for the supply of potable water (Byelaw 9). The use of any pipe made from lead or lead alloy is also prohibited for use in the supply of gas, by the Gas Safety (Installation and Use) Regulations 1998 (Part B Regulation 5(2) (a)).

Lead pipe and fittings etc. were, however, used to supply both water and gas for many years. It will still be encountered in older properties. Methods of connecting acceptable materials to existing lead supplies (gas and water) are dealt with in Volume 2.

Lead is also used in a number of alloys and is an important constituent in many solders; however, only lead-free solders should now be used in fittings on potable water supplies (Byelaw 7).

Nickel

Nickel is a white, malleable and ductile metal which is as hard as iron. It does not tarnish easily and so it is used as a plating metal. Its main use is in alloys with copper and steel, forming 'German silver', invar and armour-plating.

Platinum

This is a rare and expensive metal. It resists heat and corrosion and was used in industry as a catalyst. It is made into gas igniter filaments and used with rhodium as a thermocouple.

Tin

Tin is malleable and ductile and can be rolled to make tin-foil. It is not attacked by organic acids and so it is used to coat steel sheets and form tin-plate. This was used for gas meter cases as well as for food cans.

Tin is used in a number of alloys including solder, pewter and bronze.

Zinc

This is a grey–white metal which resists corrosion in moist air and so is used as a protective coating on iron and steel. This coating is called 'galvanising'. Galvanising is not proof against acids. The other main use of zinc is in alloys with copper to form brass or gun metals.

ALLOYS

Many alloys have already been mentioned. However the principal alloys used in the gas industry are worth studying in a little more detail.

Brass

Brass is an alloy of copper and zinc containing from 10 to 80% zinc and, in some cases, small amounts of tin and lead. The most common brass has about two-thirds copper to one-third zinc. It is a hard-wearing metal which takes a polished finish and can easily be plated, soldered or brazed.

It may be cast, drawn into wires or tubes or stamped. It is used extensively for making cocks, pipe fittings and appliance or control device components.

Solder

Soldering is dealt with in Volume 2, Chapter 1, which covers the methods used and the grades of solder for particular work.

All solders are alloys. You will be mainly concerned with the 'soft solders', which are alloys of tin and lead or tin, lead and antimony.

The melting point of the solder varies with the different percentages of lead and tin, and Fig. 12.6 shows a graph which illustrates this. A solder having about 36% lead and 64% tin has the lowest melting point and changes from solid to liquid between 183 and 185 °C. This is known as the 'eutectic', which is the composition at which the alloy behaves as a pure metal. Adding either more lead or more tin raises the final melting point but the metal still begins to melt at about 185 °C. In its intermediate stage, it becomes 'pasty' or plastic and may be 'wiped' or formed into the required shape and position.

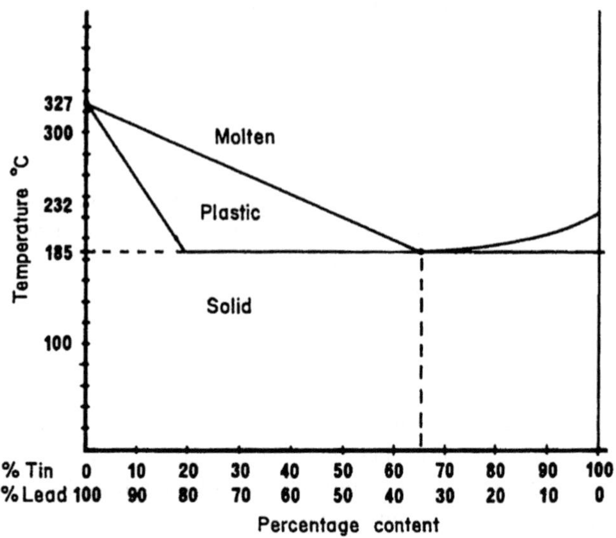

FIGURE 12.6 Tin/lead diagram.

Antimony is added in small amounts to make the solder harder.
Table 12.4 lists the properties and uses of common soft solders.

Lead-free solders

The Water Supply (Water Fittings) Regulations July 1999 recommends the use of nominal lead-free solders in potable water installations. In December 1987 an amendment was made to BS 219: 1977, adding two tin–silver and one tin–copper solders to the standard. Table 12.5 lists the properties and uses of these solders.

Stainless steel

These alloys are now in common use whenever a non-corrosive steel is required. They usually contain between 12 and 30% chromium. Steels containing smaller

Table 12.4 Common Soft Solders

B.S. Grade*	% Composition			Melting Point (°C)		Uses
	Lead	Tin	Antimony	Begins to Melt	Fully Molten	
A	36	64	–	183	185	Electrical connections to copper.
K	40	60	–	183	188	Electronic components and circuits. Capillary joints in copper or stainless steel tubes. Hot-dip coating metals. Soldering brass and zinc.
F	50	50	–	183	212	General engineering work on copper, brass and zinc. Can soldering.
R	55	45	–	183	224	
G	60	40	–	183	234	
H	65	35	–	183	244	Jointing electrical cable sheaths.
J	70	30	–	183	255	
B	47	50	3	185	204	Hot-dip coating ferrous metals.
M	52.3	45	2.7	185	215	Soldering and capillary joints on ferrous metals. Jointing copper conductors.
C	57.6	40	2.4	185	227	General engineering. Heat exchangers. Dip soldering. Jointing copper conductors.
L	66.1	32	1.9	185	243	Plumbing, wiping lead cable sheathing. Dip soldering.
D	68.2	30	1.8	185	248	

*BS 219: 1977.

Table 12.5 Nominal Lead-Free Solders

B.S. Grade*	% Composition			Melting Point (°C)		Uses
	Tin	Silver	Copper	Begins to Melt	Fully Molten	
97S	97	3	–	221	223	For capillary joints in all copper plumbing installations, and particularly in those installations where the lead content of the solder is restricted.
98S	98	2	–	221	230	
99C	99	–	1	227	228	

*BS 219: 1977 (Amendment 23 December 1987).

amounts of chromium but with the addition of vanadium, tungsten or cobalt are used to make very hard-wearing tools, for example drills, dies, gauges, punches and wrenches.

NON-METALS

Plastics

The term 'plastics' covers a wide range of materials which are plastic, or capable of flowing, at some stage of their manufacture but which end up with widely differing characteristics. Plastics may be divided into two groups:

1. thermosetting plastics and
2. thermoplastics.

THERMOSETTING PLASTICS

These undergo an irreversible chemical change when they are heated. They are usually brittle and for some uses may be reinforced with, for example, glass fibre. They include bakelite, casein, melamine and some other formaldehyde resins. Their uses range from electric light fittings, knobs, handles, telephones, clock cases, buttons, toys and radios to plastic foams.

THERMOPLASTICS

These may be repeatedly softened and resoftened by the application of heat and pressure without any chemical change taking place. Examples include most of the natural waxes and resins and substances like cellulose, polystyrene, nylon, polyvinyl chloride (PVC), polyethylene (PE) and their derivatives.

Polyvinyl chloride is a tough, rubber-like material. It is a good electrical insulator and resists attack from many chemicals. It is extruded to form pipes and produced as sheeting in different grades of stiffness. The flexible or 'plasticised' forms are often flammable. Generally only the 'unmodified' (uPVC) is non-flammable. In a stiffer form as 'impact modified' PVC or mPVC, it is used for soil pipes and fittings. These can be joined by the use of solvents. Polyvinyl chloride is commonly used for cable insulation.

Polyethylene covers a range of materials with a waxy appearance, classified according to density:

Low density	(LDPE) 0.91–0.93
Medium density	(MDPE) 0.93–0.94
High density	(HDPE) 0.94–0.97

MDPE and HDPE are used in the manufacture of gas and water distribution pipes and fittings.

Gas distribution pipes are yellow in colour and are used for operating pressures up to 4 bar, depending on the diameter and wall thickness of the pipes and fittings. A specification, on PE pipes, PE joints/fittings and tools/ancillary equipment, were originally produced by British Gas plc: and now are maintained by the "Gas Industry Standards" group.

Gas Industry Standard GIS/PL2-1: 2008 is the specification for polyethylene pipes and fittings for natural gas and suitable manufactured gas. Part 1: General and polyethylene compounds for use in polyethylene pipes and fittings.

Gas Industry Standard GIS/PL2-2: 2008 is the specification for Polyethylene pipes and fittings for natural gas and suitable manufactured gas. Part 2: Pipes for use at pressures up to 5.5 bar.

Gas Industry Standard GIS/PL2-3: 2006 is the specification for Polyethylene pipes and fittings for natural gas and suitable manufactured gas. Part 3: Butt fusion machines and ancillary equipment.

Gas Industry Standard GIS/PL2-4: 2008 is the specification for Polyethylene pipes and fittings for natural gas and suitable manufactured gas. Part 4: Fusion fittings with integral heating element(s).

PE pipes used for the distribution of water are coloured blue. MDPE and HDPE pipes can be joined by melting/fusing them together.

Nylon is used for bearings, hinges, latches and similar components where its hard-wearing, light and non-corrodible properties make it an ideal substitute for metal. It is also used in the form of flexible tubes.

'Neoprene' is the trade name for an early form of synthetic rubber (chloroprene). It resists attack from grease, oil and heat and is used for gaskets, sealing rings, cable insulation and control device diaphragms. There are now a number of similar materials available.

Polystyrene is similar to 'Perspex' and may be either clear or opaque. In unmodified form it tends to be brittle, but high-impact polystyrene is tough.

Polytetrafluoroethylene or PTFE is a waxy material commonly used as a non-stick coating on cooking utensils. In the form of tape, it is also used as a jointing compound on the screw threads of mild steel pipe.

Plastics are composed of 'polymer' molecules. These are large molecules containing carbon and hydrogen atoms with many thousands of the carbon atoms in the form of long chains. Polymer molecules are very much larger than the ordinary molecules of natural materials like ethylene, benzene, chlorine. They are produced by a chemical reaction under very high pressures and high temperatures called 'polymerisation'.

This is an addition process which causes the simple, natural molecules to link together to form long chain groups. For example, ethylene, C_2H_4, becomes chains of CH_2 groups with the carbon forming the links in the chain.

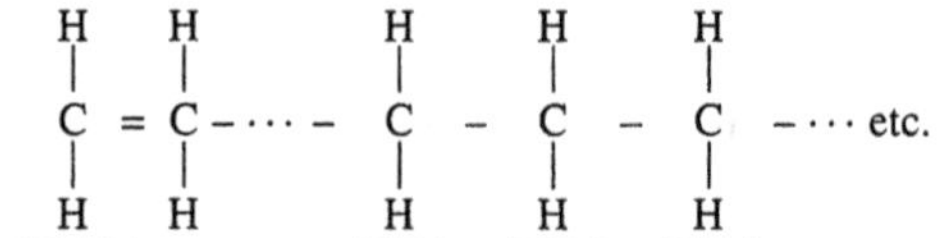

$$\begin{array}{cccccccccc} H & & H & & H & & H & & H & \\ | & & | & & | & & | & & | & \\ C & = & C & -\cdots- & C & - & C & - & C & -\cdots \text{etc.} \\ | & & | & & | & & | & & | & \\ H & & H & & H & & H & & H & \\ (C_2H_4) & & & & (CH_2) & & (CH_2) & & (CH_2) & \\ \text{ethylene} & & & & & & \text{polyethylene} & & & \end{array}$$

Timber

There are many different types of timber in common use and, as the tropical countries continue to develop, new types appear on the market.

Timber may be divided into two categories:

1. softwoods, from conifers and
2. hardwoods, from broad-leaved trees.

Generally the classification is true, but there are exceptions. Some softwoods are harder than some hardwoods, so to a limited extent the names are inaccurate.

Common softwoods are Scots Pine, also known as Red or Yellow Deal, and Douglas Fir, also known as Oregon Pine. They are used for structural work, joists, rafters, floors and also for painted joinery.

Hardwoods include oak, ash, elm, beech, birch, mahogany, teak and the more recently used Iroko, Obeche, Sapele and Afrormosia. These are used for fireplace surrounds, doors and decorative flooring, as well as for furniture.

Timber is likely to shrink or swell with changes in its moisture content, and it is 'seasoned' or dried out after cutting until a specified moisture content is attained. For structural timbers, this is about 20% and for interior joinery 10–15%. In UK, to achieve a moisture content below 20% the timber must be kiln-dried.

Bricks

Bricks are commonly divided into three types:

1. common bricks – for general use,
2. facing bricks – good appearance, suitable for normal external exposure,
3. engineering bricks – dense and strong, made to specifications of absorption and strength.

They may be further classified by:

- quality – internal, ordinary, special;
- finish – sand-faced, rustic, multi-coloured;
- perforation – solid, perforated, hollow, cellular;
- method of manufacture – machine-made (pressed, moulded or wire-cut), hand-made;
- district of manufacture – Staffordshire Blues, Leicester Reds.

An ordinary brick size is 225 mm × 112 mm × 75 mm nominal and 215 mm × 103 mm × 65 mm actual. In addition to bricks, building blocks are used. Blocks are made from a variety of materials. 'Breeze blocks' formed from cement and coke breeze are used for economy of cost in interior walls. Precast concrete blocks are made, Type 'A': density not less than 1500 kg/m^3 and Type 'B': less than 1500 kg/m^3.

Blocks are manufactured from 400 to 600 mm long by 100–225 mm high, their thickness varies from 75 to 225 mm.

Concrete

Concrete is made from cement and an 'aggregate' mixed with water. Aggregates are substances which may be added to cement or plaster to give hardness, resistance to abrasion or heat insulating properties. They are classed as:

- heavy aggregate – sand and gravel, crushed stone;
- lightweight aggregate – clinker, foamed slag, vermiculite.

Heavy aggregates form dense concrete, lightweight aggregates form lightweight concretes. In normal concretes the cement fills the spaces or 'voids' between the pieces of aggregate. No-fines concrete is made from coarse clinker with only enough cement to stick the pieces together. Cellular concretes are made by using materials which foam or give off gas during mixing or by blowing air bubbles into the mix and adding a substance to keep the bubbles fixed there.

With all concrete structures, ducts and sleeves for pipes should be positioned before the concrete is poured. Dense concrete is difficult to cut after it has been formed.

Cement and plaster

Portland cement is available in various forms made to possess particular qualities which include:

- rapid hardening,
- colours,
- water resistant,
- sulphate resistant.

Ordinary Portland cement is made from chalk or limestone with clay and small quantities of gypsum.

For general purposes, including making good holes cut in walls or floors for pipe runs, cement is mixed with sand and water to form a 'mortar'. The proportions are:

- 1 part cement and
- 3 parts sand.

The amount of water used should be kept to a minimum to reduce shrinkage. The sand and cement must be thoroughly mixed in the dry state, before the water is added. The mortar should be used within 2 h of mixing and partially set material should not be remixed.

Plasters are used for the top surfacing or rendering on internal walls and ceilings and for making good any holes cut in them. These are often the calcium sulphate plasters like Plaster of Paris, Keene's or Parian plasters made principally from gypsum. They are corrosive to steel when damp and pipes should be protected from contacting them. Plaster of Paris sets very quickly; other plasters have retarding agents added which give a longer setting period.

Asbestos

Asbestos was extensively used as a building material in the UK from the 1950s through to the mid-1980s. It was used for a variety of purposes and was ideal for fireproofing and insulation. Any building built before 2000 (houses, factories, offices, schools, hospitals etc) can contain asbestos. Asbestos materials in good condition are safe unless asbestos fibres become airborne, which happens when materials are damaged.

The Control of Asbestos Regulations 2006 came into force on 13 November 2006 (Asbestos Regulations – SI 2006/2739). These Regulations bring together the three previous sets of Regulations covering the prohibition of asbestos, the control of asbestos at work and asbestos licensing.

The Regulations prohibit the importation, supply and use of all forms of asbestos. They continue the ban introduced for blue and brown asbestos in 1985 and for white asbestos in 1999. They also continue to ban the second-hand use of asbestos products such as asbestos cement sheets and asbestos boards and tiles; including panels which have been covered with paint or textured plaster containing asbestos. **REMEMBER:** The ban applies to new use of asbestos. If existing asbestos containing materials are in good condition, they may be left in place, their condition monitored and managed to ensure they are not disturbed.

Asbestos is a mineral compound of magnesium, calcium and silica. It is a fibrous mineral found in the following forms:

- crocidolite – blue asbestos,
- chrysotile – white asbestos,
- amosite – brown asbestos.

The material was extensively used in the gas industry for its fire and heat resisting properties. In appliances its uses included gaskets, door seals, insulating etc. On installations, flue pipes, flue jointing compounds, protecting other inflammable materials etc. Since medical research showed the discovery that the tiny fibres from asbestos could cause a form of lung cancer, alternative materials are now used.

ASBESTOS REMOVAL

Most asbestos removal work must be undertaken by a licensed contractor but any decision on whether particular work is licensable is based on the risk. If asbestos is encountered the following points must be borne in mind:

- Avoid making dust, if at all possible.
- Wear the type of dust mask approved for the type of asbestos.
- Wear the recommended protective clothing and either dispose of or have them laundered in the recommended manner.
- Wet the material thoroughly (provided there is no contact with electricity).
- Do not eat, drink or smoke in the working area.
- Collect any dust on dampened newspaper under the work area.
- Do not drill, cut or screw asbestos materials unless there is no alternative. Then use only a hand drill with a sharp bit or a sharp medium-toothed saw.
- Use a dampened cloth to clean up.
- Place debris, dust, contaminated cloths and newspapers, masks etc. in sealed plastic bags and dispose of in the recommended manner.
- Any contaminated clothing (overalls etc.) must be placed in a sealed plastic bag and sent for specialist cleaning.
- Wash hands etc. well, after completion of work.

LIKELY SOURCES OF ASBESTOS

Figures 12.7a–c show examples of where asbestos can be found in gas industry situations.

CORROSION

Corrosion is the eating-away of a metal caused by chemical changes in the composition of its surface. It may be caused by:

- air and water,
- acids,
- electrolysis.

FIGURE 12.7a Asbestos panelling around gas meter. (Picture courtesy of the Health and Safety Executive, HSE)

FIGURE 12.7b Asbestos panelling inside heater cupboard. (Picture courtesy of the Health and Safety Executive, HSE)

FIGURE 12.7c Damaged asbestos lagging on hot water pipes. (Picture courtesy of the Health and Safety Executive, HSE)

Air and water

A combination of air and water may be found in the atmosphere, because the air is usually moist, or in our water supplies, because there is usually some air dissolved in the water.

Apart from stainless steels, any ferrous metals will form a ferric hydroxide, or rust, if in contact with either the atmosphere or water. This particular oxide is porous and crumbly so that it speeds up the rusting by falling away and allowing the oxygen and water to attack new surfaces.

Most metals form a layer of oxide on their surface when exposed to the atmosphere. In the case of copper and lead the layer acts as a protection against any further oxidation or corrosion.

Acids

Some gases in the atmosphere can combine with rainwater to form acids which will corrode any unprotected metal made wet by the rain. For example, carbon dioxide and sulphur dioxide, both of which may be found in flue gases, can form carbonic acid and sulphuric acid, respectively. These gases are found in considerable concentrations in industrial areas.

Other acids may be found in the soil and main and service pipes need to be protected against corrosion from acidic soils like the clays and those containing ashes or clinker.

Electrolysis

This form of corrosion can take place in water systems or in damp soil. Electrolysis or 'galvanic corrosion' requires four things in order for it to take place. These are:

1. an anode – the corroding area,
2. an electrolyte – the means of carrying the electric current (water or soil),
3. a cathode – the protected area,
4. a return path – for the corrosion currents.

An electric current is generated at the anode and flows through the electrolyte to the cathode. The current then flows through the return path back to the anode again (Fig. 12.8).

The principal causes of electrolytic corrosion are as follows:

1. Different metals joined together and both in contact with the electrolyte, for example mild steel radiators and copper pipe in a wet central heating system. Or a steel service pipe connected to a cast iron main. In both

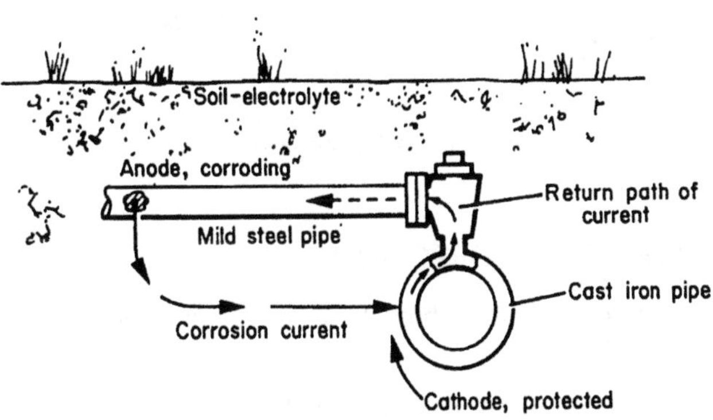

FIGURE 12.8 Electrolytic corrosion, dissimilar metals.

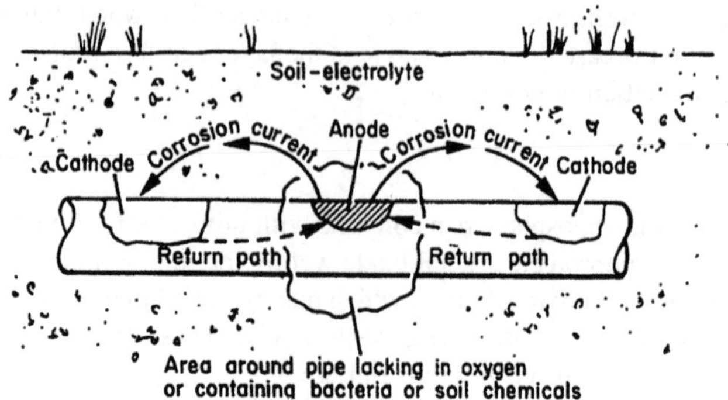

FIGURE 12.9 Electrolytic corrosion, difference in environment.

examples the mild steel is the anode or corroding area. The metals do not have to be completely different to set up electrolysis. It is sufficient to have clean, pure metal at one point and scale, impurities or scarring at another. Corrosion can take place between iron and particles of graphite or carbon in the same metal.

2. Differences in the chemical environment of the metal. For example, in buried pipes, a lack of oxygen or a concentration of soil chemicals or bacteria at the anode point (Fig. 12.9).
3. Stray electric currents. This may occur where bonding is ineffective and a gas service pipe acts as an electrical earth return. It happens on gas mains when in contact with other authorities' plant or electrified railway systems (Fig. 12.10).

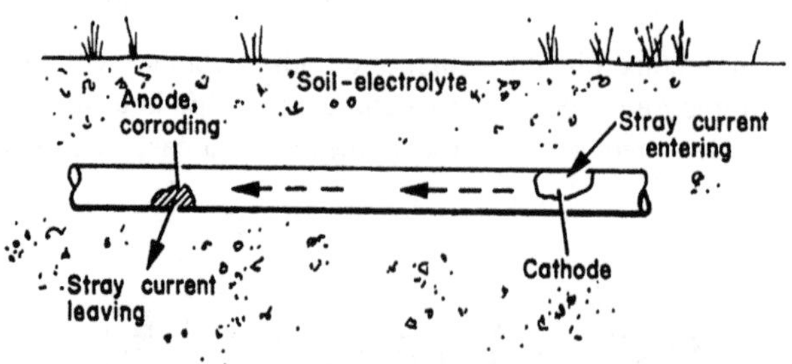

FIGURE 12.10 Electrolytic corrosion, stray currents.

PROTECTION FROM CORROSION

Prevention of corrosion can be brought about in a number of ways. Basically there are two methods:

1. isolate the metal from the corrosive element,
2. provide cathodic protection.

1. *Isolation*

In the case of pipes, this is usually done by providing a coating of:

- coal tar enamel,
- polyethylene sleeving,
- protective tape or bandage,
- a back filling around a buried pipe of alkaline material, for example chalk.

In a correctly designed and installed central heating system, corrosion rarely takes place because there is little or no air dissolved in the water. In areas, however, where the water supply is known to be aggressive, the addition of inhibitors to the circulating water prevents corrosion and the growth of bacteria.

2. *Cathodic Protection*

This ensures that the metal to be protected is always the cathode. So any electric current always flows from the electrolyte to the metal. It can be brought about by:

- sacrificial anodes and
- impressed current systems.

Sacrificial anodes are usually made of an alloy of magnesium because, when joined to mild steel, the magnesium becomes the anode and corrodes away whilst the steel is protected from corrosion.

They are used on buried steel mains and have been fitted to large storage water heaters and water heating systems. Impressed current systems usually employ a DC supply from a transformer/rectifier. The negative connection is made to the pipe and the positive terminal is connected to a bed of special anode material buried some feet from the pipe.

A protected pipeline is isolated by insulating flanges and all joints are electrically bonded. Test points are provided so that the voltage between the pipe and the soil can be checked. The strength of the voltage indicates the degree of protection provided.

PROTECTIVE AND DECORATIVE FINISHES

Paints and lacquers

Stainless steel is one of the few metals which needs no protective coating. When a natural finish is required on copper or brass the metal is cleaned, polished or grained and then given a coat of acetate lacquer. This is frequently 'stoved' or baked on in an oven.

Paints may be applied by dipping, spraying or brushing. An extension of the spraying method is electrostatic attraction in which the paint is attracted to the object by an electric charge. Generally the paints applied by dipping or spraying are subsequently stove-dried and acquire a harder surface than the air-drying paints.

These finishes should generally be cleaned with warm water to which a mild detergent has been added. Do not use domestic scouring powders.

Vitreous enamel

Vitreous enamel is, in effect, a form of glass. It is hard, durable and acid resistant. It must also be capable of withstanding sudden, considerable changes in temperature. It has been used extensively as a finish for cast iron and mild steel appliance panels and components although it has been superseded by synthetic stoved enamels for use where temperatures are not above 95 °C.

Vitreous enamel is made from silica, quartz and felspar, ground and mixed with water to form a 'frit'. The frit is applied to the cleaned metal by dipping or spraying and then dried in an oven. Finally the enamel is fired on at a temperature approaching 800 °C.

Articles are usually given a dark-coloured ground coat followed by one or sometimes two top coats. Metallic oxides are added to the frit to give coloured finishes.

Vitreous enamel should be wiped over whilst still warm to remove dirt or spillage. If this has been burned on, a caustic cleaner may be used occasionally. Fine steel wool impregnated with soap is better than the coarse powder scourers which may damage the enamels if used repeatedly.

Surface conversion finishes

Blueing: There are a number of processes for producing a thin, blue, oxide film on steel. A common method is by heating the metal to about 900 °C and then introducing it to steam. This gives the steel an attractive, corrosion resistant surface and it is used for some tools and appliance components.

It should be cleaned by rubbing over with an oily rag.

Anodising: This is a process of artificially thickening the layer of aluminium oxide on the surface of aluminium or aluminium alloy. It gives extra corrosion resistance and dyes may be introduced to produce attractive colours.

Anodising is commonly applied to aluminium cooking utensils and appliance components. It should be cleaned only with a mild detergent.

Parkerising: This is a method of coating steel with a surface film of phosphate. It is applied by immersing the steel parts in a hot solution of the chemical.

Metallic coatings

Paints: Some paints have a pigment which consists of metallic flakes or powder. Aluminium paint is an example of a paint which provides a form of metallic coating.

Hot dipping: Metals which have a low melting point, like tin or zinc, can be applied to steel by dipping the steel in a bath of the molten metal.

Vapour spraying: This entails melting the metal in an oxy-acetylene flame and spraying it on the surface to be treated in the form of a vapour.

Electro-plating: A film of a metal is deposited on the metal articles to be plated by means of an electric current, the articles are immersed in a suitable solution of electrolyte and connected so that they form the cathode of an electric cell. When current flows metal ions are deposited on the articles.

Plating is commonly used to apply coatings of chromium, nickel and cadmium. It can also be used for tin and zinc coatings.

Galvanising: This is the name given to the coating of iron or steel with zinc. It is usually carried out by hot dipping or electro-plating.

Sherardising: Like galvanising, this is a process for coating steel or iron with zinc. In this case the articles are packed in zinc dust and heated in a furnace to about 400 °C for some hours.

Sherardising is used for small, intricate objects such as screws, nuts and bolts.

HEAT TREATMENT OF METALS

Heat treatment involves a number of processes which change the structure of metals so that they are made harder, softer or tougher. All tools have been treated during manufacture to give them the properties they need for their particular job. If they are heated and cooled again during their use, they may lose their hardness and become useless.

Most metals become harder if they are 'worked', that is hammered or bent. This is particularly true of brass and copper which cannot be hardened by heat treatment. Some aluminium alloys harden up after working by 'age hardening' during the few days immediately afterwards.

The major heat treatment processes are as follows.

Annealing

This is a process of heating and cooling which softens the metal. Although, in a factory, the metal would be heated to an exact temperature suitable to its particular composition, as a rough guide it is sufficient to bring it to a red heat.

If you want to anneal copper pipe, so as to make it softer and easier to bend with a spring, it must be heated to a dull redness. It may then be allowed to cool slowly, but if a supply of water is available, it may be quenched immediately in cold water. Both methods of cooling will have the same effect.

This is not the case with steel. Most steels can be annealed by heating to a cherry red heat and then allowing them to cool slowly in the open air. For some special tool steels, cooling in air is too fast and they would remain hard. They need very slow cooling which may be carried out in the furnace itself.

Normalising

This is similar to annealing and is the treatment given to steel which has been cold worked. The metal is heated to the required temperature, which may be slightly above that used for annealing, and soaked at that temperature for anything from a few hours to several days. It is then cooled slowly.

Hardening

Tool steel can usually be hardened by heating to red heat (about 750 °C) and quenching quickly in water or oil. This makes the metal hard, but brittle and most tools require the hardness to be slightly reduced so that they gain greater strength.

Tempering

Tempering is making steel less hard but tougher. It involves heating the already hardened metal to a particular suitable temperature and then quenching in water or oil. Table 12.6 gives the temperatures for particular tools.

If the metal is cleaned to a bright surface after hardening, it is possible to watch the colour appear and deepen from a pale yellow to a blue as the temperature is increased. The tool is quenched when the appropriate colour is reached.

Case hardening

Mild steel, or steel with a low carbon content, cannot be hardened in the same way as tool steel. To give it a hard surface so that it can resist wear, it is necessary for it to be 'case hardened'. This is done by introducing carbon into the outer layers of the steel to form a 'case' which may be from 0.025 to 0.152 mm thick. The high-carbon case can then be heat treated.

Case hardening or 'carburising' can be carried out in a number of ways, all of which entail heating the steel while surrounding it with a substance from which it can absorb carbon. This is frequently done in a bath of cyanide.

Table 12.6 Approximate Tempering Colours and Temperatures

Colours	Temperature (°C)	Tools
Pale yellow	220	Lathe tools, milling cutters, scribers
Pale straw	230	Taps, dies, reamers
Medium straw	240	Twist drills, centre punches
Dark straw	260	Wood chisels, gouges, planes, hammers, shears
Purple	280	Cold chisels, wood saws
Blue	300	Springs, screwdrivers

THE "CONTROL OF SUBSTANCES HAZARDOUS TO HEALTH" – COSHH REGULATIONS

Why COSHH matters?

Using chemicals or other hazardous substances in the workplace can put peoples' health at risk. So the law requires employers to control exposure to hazardous substances to prevent ill health.

They have to protect both employees and others who may be exposed by complying with the Control of Substances Hazardous to Health Regulations 2002 (COSHH) (as amended). The employer has to put into place the following:

Step 1 Assess the risks.
Step 2 Decide what precautions are needed.
Step 3 Prevent or adequately control exposure.
Step 4 Ensure that control measures are used and maintained.
Step 5 Monitor exposure.
Step 6 Carry out appropriate health surveillance.
Step 7 Prepare plans and procedures to deal with accidents, incidents and emergencies.
Step 8 Ensure that employees are properly informed, trained and supervised.

The Gas industry uses a variety of hazardous materials, a few examples are solvents, blowlamp fuels, descaling solutions and fluxes.

Solvents

Those solvents like acetone, which are similar to, or a form of, alcohol are all highly volatile and flammable. They dissolve organic substances readily and may be used to clean the valves and gas ways of control devices.

Trichloroethylene, which is an excellent grease and dirt remover, is non-flammable. It can, however, dissolve some paints and plastics. Before using any solvent solution make certain of its effect on the material to be cleaned. Solvents should only be used under well-ventilated conditions.

Blowlamp fuels

Propane lamps are in general use and have replaced the earlier and more hazardous petrol and paraffin types – when in use it is essential to ensure that the connection between the nozzle holder and the 'bottle' is completely sound. The lamp must not be placed near any heat source.

When out of use, bottles must be stored in a well-ventilated situation, away from heat and in an upright position.

Methylated spirit lamps or 'mouth lamps' may still be in use in some places. Methylated spirit is another form of crude, wood alcohol and is highly flammable. It burns with a pale blue flame which is very difficult to see in bright sunlight.

Descaling solutions

There are a number of proprietary brands available but, in general, any solution which will dissolve scale in water heaters or systems is acidic. So it must not be allowed to come into contact with galvanised pipes, tanks or cylinders. It might also have an adverse effect on thermostats or automatic valves and it is usual only to pass a descaling solution through the copper heat exchanger or 'heating body' of the water heater. After the scale has dissolved the heater must be thoroughly flushed out.

If it is necessary to dilute an acid solution with water, always add the acid to the water, never the water to the acid. This is because the chemical reaction produces considerable heat. If water is added to concentrated acid, the water may boil and spit out. If acid is added to water, it is immediately diluted and the reaction is much less violent.

The majority of central heating installations currently being installed are replacement boilers that can sometimes lead to the increased problem of 'Dirty' systems. With the ever increasing efficiency of gas appliances, the 'cleanliness' of a system has never been so important. The cleansing of central heating and hot water systems is covered by British Standard BS 7593 '*Code of practice for treatment of water in domestic hot water central heating systems*'. The methods and cleaning agents for flushing are covered in Volume 2, Chapter 10.

Fluxes

Fluxes are used to prevent cleaned metal from oxidising whilst being soldered or brazed. They may be made from borax, tallow, resin, zinc chloride, or mixtures of these or similar substances.

A number of fluxes are corrosive, and joints must be cleaned of any surplus flux when jointing is completed (see Volume 2, Chapter 1).

Chapter | thirteen

Tools

Chapter 13 is based on an original draft prepared by Mr P. Wragg

INTRODUCTION

There are hundreds of different kinds of tools. A service engineer may use up to 140 different items which can be classified as tools. Many tools are the subject of British Standard Specifications.

Some tools can only be used for one specialised job whilst others, like pliers, may be used for a variety of tasks. Some jobs call for a specialised tool, other jobs can be carried out equally well by a number of different tools. An example of the last case is screwing a pipe into a fitting which can be done with a footprint wrench, stillsons or chain tongs (and by a few more less orthodox methods!).

Because of this, the choice of tools for a service engineer's tool kit is largely a matter of opinion, and a subject on which opinions are divided. Each tool has its own particular advantages and disadvantages and everyone has their own preferences and prejudices. There is a limit to a number of tools which a man can reasonably carry, so a choice has to be made.

Engineers are responsible for their tools and in return, are totally dependent on those tools for their livelihood. They should:

- select, from their kit, the appropriate tool for the job,
- ensure that it is in good working order,
- use it correctly,
- transport it safely.

Failure to do this can result in:

- use of more physical effort than is necessary,
- more time spent on the job,
- damage to appliances, customers' property or the engineer.

This chapter serves as an introduction to the tools commonly used for the servicing and installation of gas appliances. Subsequent chapters in Volumes 2 and 3 will deal in more detail with the use of particular tools for cutting, bending and jointing pipes and for servicing appliances.

In many cases, tools are still referred to in imperial rather than metric terms, for example teeth per inch (tpi), ½ lb hammer, 14 in. stillsons etc.

BASIC TOOLS

The tools described in the following sections are those most commonly selected for service work.

Hammers

There are dozens of different kinds of hammers but the ones most frequently used are shown in Figs. 13.1–13.4. They are:

1. cross pane (or 'Warrington'),
2. straight pane,
3. ball pane (or 'engineer's'),
4. club (or 'lump').

The pane is the end opposite to the face of the hammer, and the word was originally spelt 'pein'. Hammer heads are usually made of cast steel and the handles or shafts are ash.

Hammers are named by their weight and their pattern. Those in a service engineer's tool kit are usually cross pane or Warrington hammers in weights of about 700 g (1½ lb) and 250 g (½ lb). Sometimes the ball pane is preferred in place of the larger cross pane. It has the advantage of being suitable for riveting. Straight pane hammers are more frequently used by plumbers than by service engineers. The club hammer is usually of about 1500 g (3 lb).

A 'claw hammer' or carpenter's hammer is sometimes included instead of the large cross pane. The claw is useful for removing nails from floorboards.

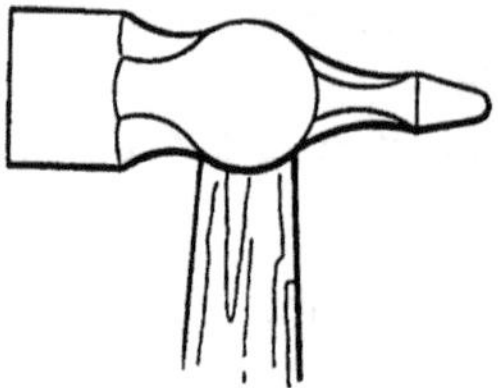

FIGURE 13.1 Cross pane hammer.

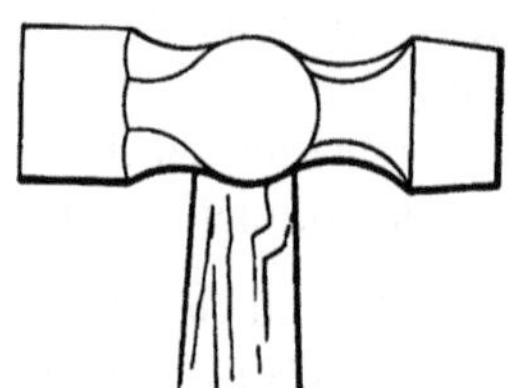

FIGURE 13.2 Straight pane hammer.

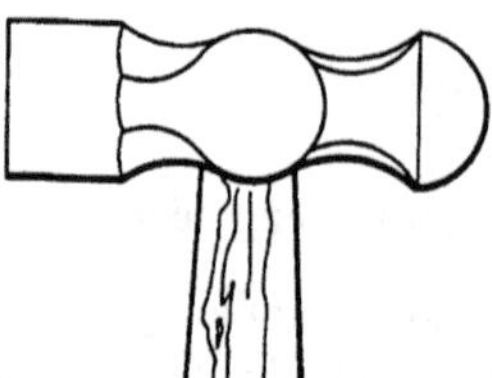

FIGURE 13.3 Ball pane hammer.

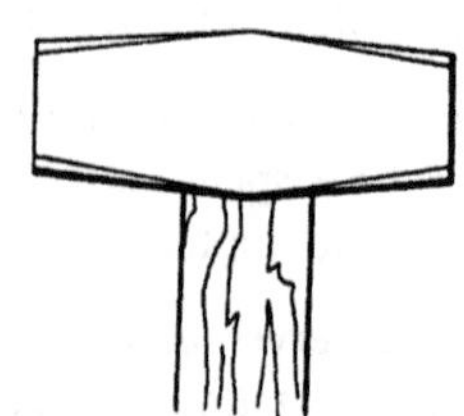

FIGURE 13.4 Club hammer.

Chisels

Cold chisels, so called because they can cut steel when it is cold, are used for cutting metal or, more frequently, for cutting brick or concrete (Fig. 13.5). A Gas Engineers tool kit usually contains one about 230 mm (9 in.) long and 22 mm ($^3/_4$ in.) wide. Sometimes a 300 mm (12 in.) or a 450 mm (18 in.) chisel is required for thick stone walls in old properties.

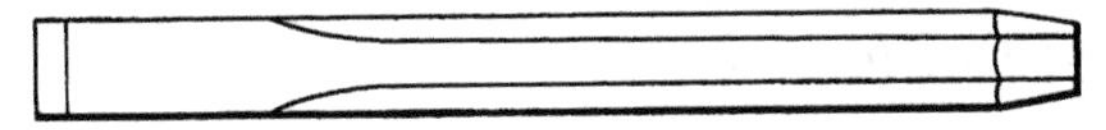

FIGURE 13.5 Flat cold chisel.

Floorboard chisels are similar to cold chisels but have blades usually about 70 mm wide (Fig. 13.6). The blade is parallel so that it can be driven down between floorboards which can then be levered up.

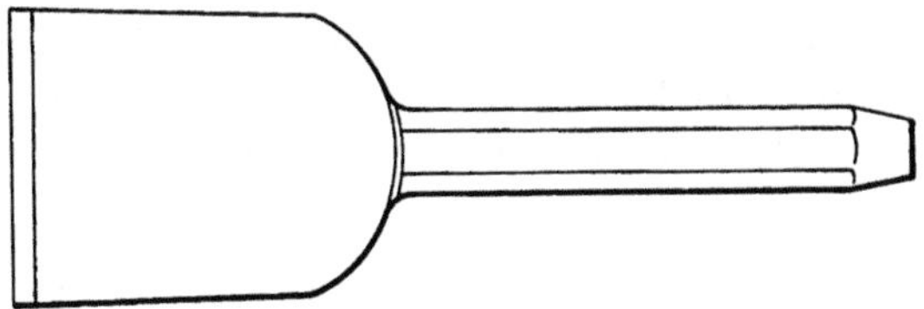

FIGURE 13.6 Floorboard chisel.

Bolsters are very much like floorboard chisels in appearance but the blade is stronger and tapered from 8 mm to about 4 mm before being sharpened off. They are used for cutting bricks or stone slabs and can cut channels in walls. Common size of bolster is 75 mm (3 in.).

Wood chisels are made in a variety of types and sizes. The kit usually has one 250 mm (10 in.) by 12.5 mm ($^1/_2$ in.). This is often all steel, as in Fig. 13.7, because it will be hit with a hammer, and not with a mallet.

FIGURE 13.7 All steel wood chisel.

Figure 13.8 shows a very useful hand-grip protector, being used fitted on a cold chisel. They are available in various sizes to fit all-steel chisels.

Punches

Two punches are usually carried.

The centre punch (Fig. 13.9) is used to mark the centre of holes to be drilled in metal and to provide a means of preventing the drill from slipping. The punch is usually about 100 mm long and the point is tapered at 90.

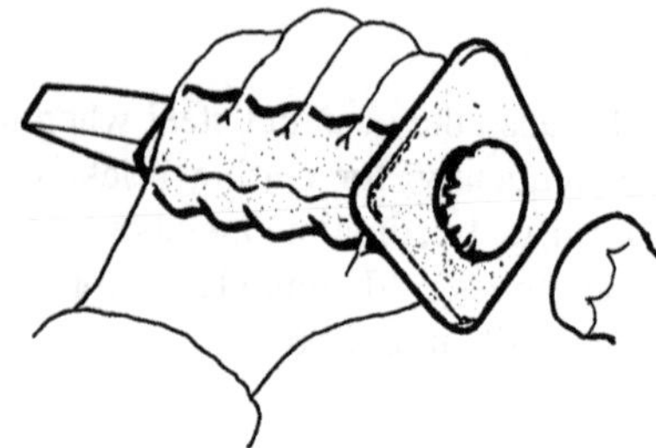

FIGURE 13.8 Hand-grip protector.

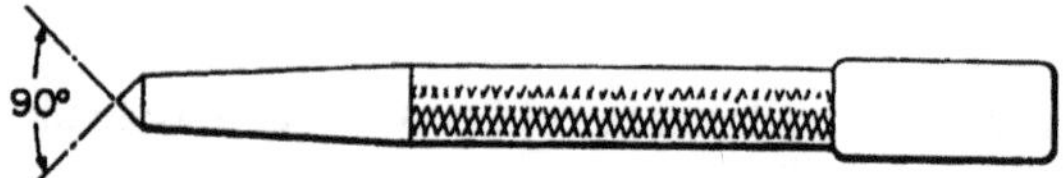

FIGURE 13.9 Centre punch.

The nail punch or brad punch (Fig. 13.10) is for punching nails below the surface of the wood. It can punch nails right through floorboards so that the boards may be lifted easily. It is usually 100 mm long with a 3 mm head which is slightly recessed so that it does not slip off the nail.

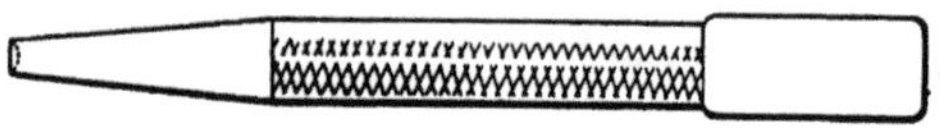

FIGURE 13.10 Nail punch.

Wood saws

Many different kinds of hand saws and floorboard saws have been produced and some are still used by service engineers. A saw in common use is the tenon saw (Fig. 13.11). It may be 250–300 mm long and usually has about 12 tpi (25 mm). It may be used to cut lead pipe as well as wood. All saws have their teeth 'set' that is bent outwards on alternate sides to clear the blade in the cut.

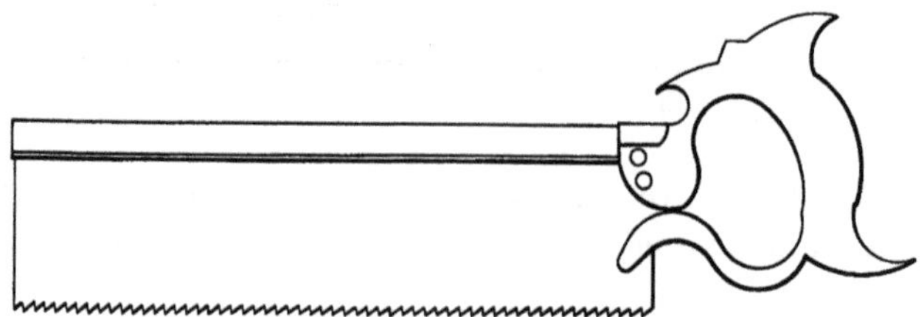

FIGURE 13.11 Tenon saw.

Another very useful saw is the floorboard saw (Fig. 13.12). It is about 280 mm long with 10 tpi (25 mm). It is specially designed to cut along and across the grain of tongue and groove boards. It is equally suitable for use on chipboard

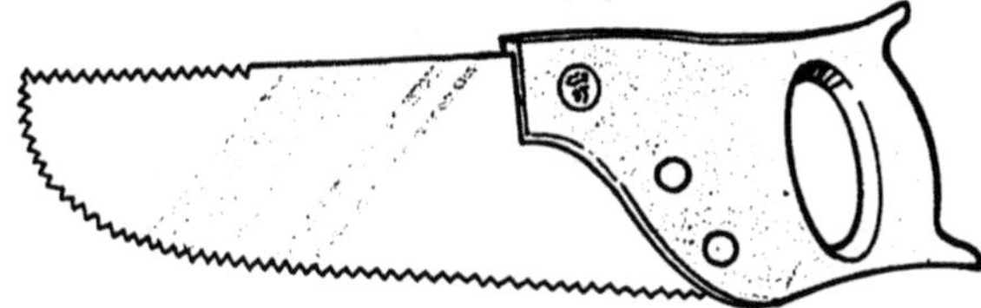

FIGURE 13.12 Floorboard saw.

and as a general purpose saw. The teeth are designed to cut on both the forward and return strokes.

The padsaw (Fig. 13.13) has a metal or plastic handle into which can be fitted either a narrow, tapered blade about 250 mm long and with 10 tpi (25 mm) for cutting wood, or a hacksaw blade for cutting metal. The padsaw can be used in awkward places and the narrow blade can cut out holes in floorboards and skirtings.

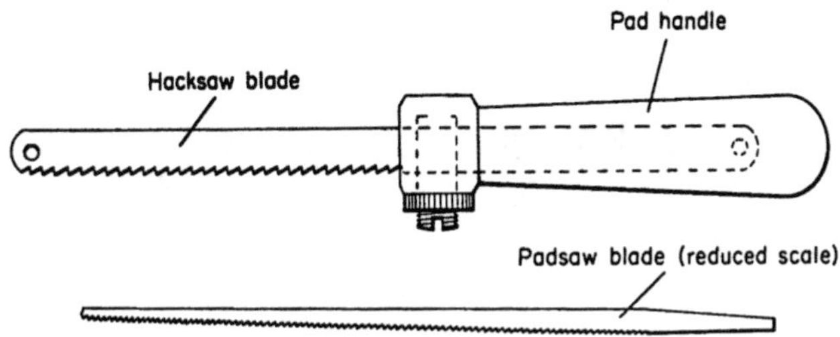

FIGURE 13.13 Padsaw.

Metal saws

The hacksaw (Fig. 13.14) has a pistol grip and the frame is adjustable to take various lengths of blade. Usually 250 or 300 mm (10 or 12 in.) blades are used with 22 tpi (25 mm) for steel pipe and 32 tpi for copper pipe or thin sheet metal.

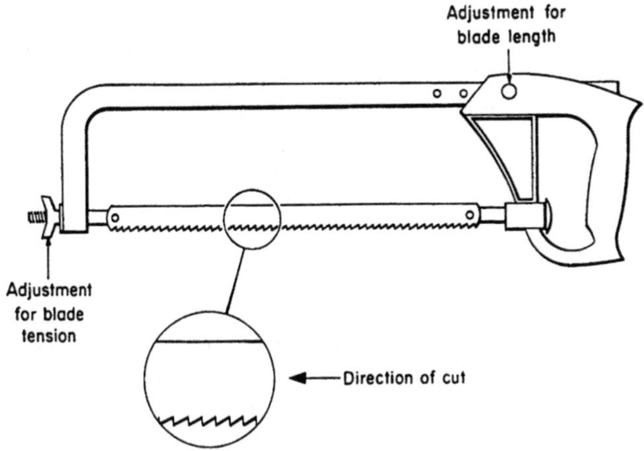

FIGURE 13.14 Hacksaw.

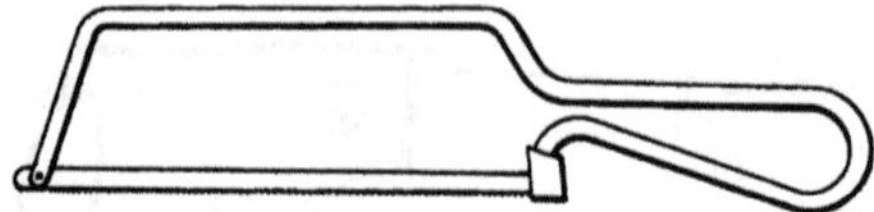

FIGURE 13.15 Junior hacksaw.

It is more usual to have a junior hacksaw (Fig. 13.15) for cutting copper pipe into smaller sizes.

Drilling tools

The carpenter's ratchet brace (Fig. 13.16) is used to hold a variety of drills or 'bits'. It usually has a sweep of 125–200 mm and the ratchet allows it to be used in a corner or close to a wall. The chuck has two jaws and only grips bits with a square tapered shank.

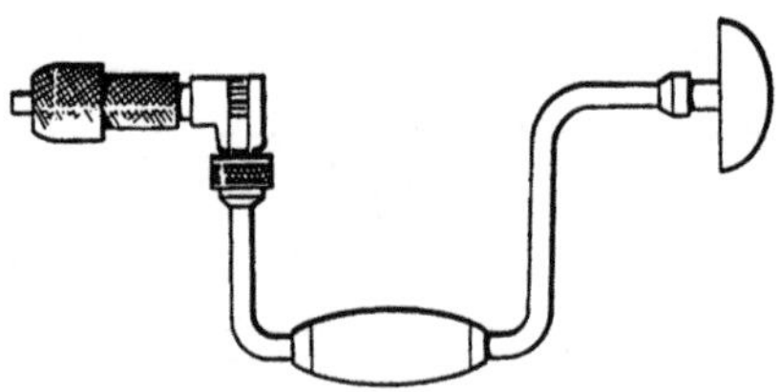

FIGURE 13.16 Carpenter's ratchet brace.

Also sometimes used is a small wheel-brace or engineer's hand brace (Fig. 13.17). This has a three-jaw chuck and holds round-shanked drills. It turns the drills much faster than the carpenter's brace and is used for drilling holes up to about 7 mm in wood, metal or masonry.

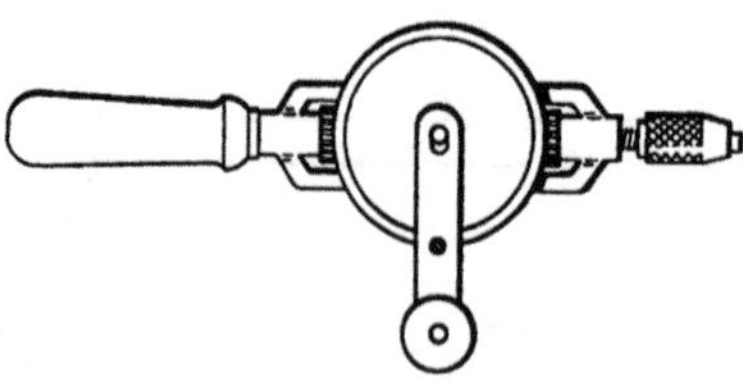

FIGURE 13.17 Wheel brace.

Electric drills are dealt with later in the chapter.

Bits are made in a variety of types and sizes. They can be used for drilling, countersinking, reaming and driving in screws. Most kits now have two wood bits, about 16 and 23 mm. They are usually auger bits (Fig. 13.18) or sometimes the simple flat wood bit or centre bit (Fig. 13.19). Usually included is a rose bit (Fig. 13.20) or the rosehead countersink (Fig. 13.21).

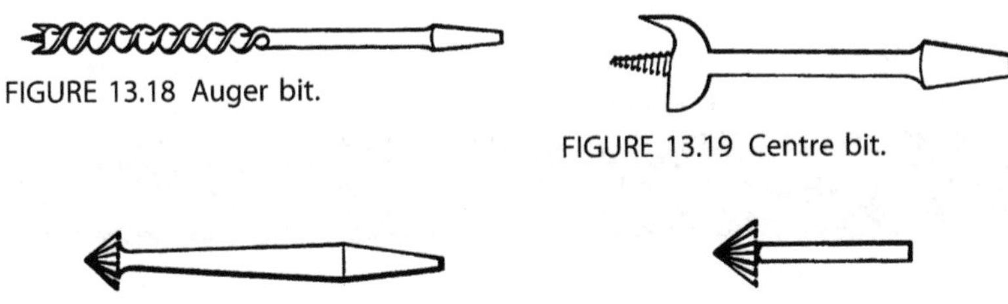

FIGURE 13.18 Auger bit.

FIGURE 13.19 Centre bit.

FIGURE 13.20 Rose bit.

FIGURE 13.21 Rosehead countersink.

Twist drills (Fig. 13.22a), are used in a variety of sizes from 1.5 to 9.5 mm. Often the 9.5-mm size had a square shank (Fig. 13.22b).

Some kits included sets of 'morse' drills which corresponded to the drillings in injectors and burner ports. They were in numbered sizes from No. 1 (5.79 mm) to No. 80 (0.35 mm). These have now been superseded by metric sizes as in Table 13.1.

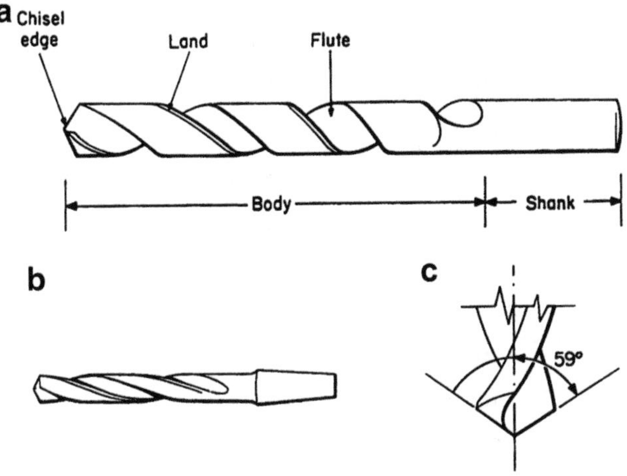

FIGURE 13.22 Twist drills: (a) twist drill, (b) square shank twist drill, and (c) correct point angle for general use.

The angle at the point of a twist drill should be 59 (Fig. 13.22c). Drills may be made of ordinary carbon steel or 'high-speed steel' for faster drilling or harder work.

Tipped drills with a tungsten carbide insert at the point are used for drilling masonry (Fig. 13.23). The usual sizes carried are No. 8, No. 10 and No. 12. These are the sizes of wood screw and wall plug for which the holes are suitable.

Files

Files are made of cast steel with a hard and brittle blade and a softer, tougher tang, onto which fits a handle (Fig. 13.24). Some files are available with a tang formed into a flat handle (Fig. 13.25).

Table 13.1 Metric Equivalents to Morse Drill Sizes (The Sizes Given are for those Drills Recommended for General Use on Gas Equipment)

Morse Drill Number	Millimetre Drill Size	Morse Drill Number	Millimetre Drill Size
18	4.30	47	2.00
19	4.20	48	1.95
20	4.10	49	1.85
21, 22	4.00	50	1.80
23, 24	3.90	51	1.70
25	3.80	52	1.60
26, 27	3.70	53	1.50
28	3.60	54	1.40
29	3.50	55	1.30
30	3.30	56	1.20
31	3.00	57	1.10
40	2.50	58, 59	1.05
41	2.45	60, 61	1.00
42	2.30	62	0.98
43	2.25	63	0.95
44	2.20	64	0.92
45	2.10	65	0.90
46	2.05	78	0.40

FIGURE 13.23 Tipped masonry drill.

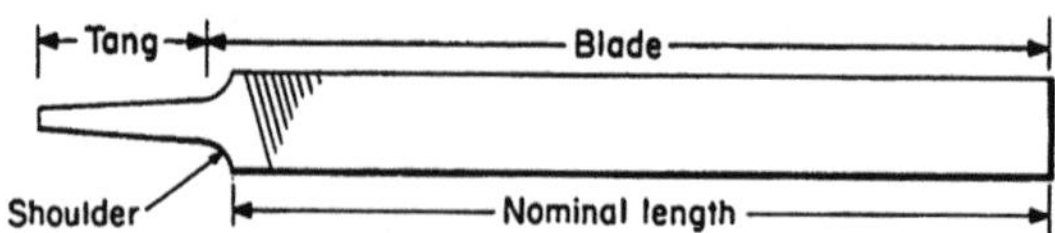

FIGURE 13.24 File.

FIGURE 13.25 Self-handled file.

Strictly, files should be described by their:

- length – 100 mm (4 in.) to 350 mm (14 in.) in 50 mm (2 in.) steps,
- shape and cross-section – flat, hand, half-round, etc. (Fig. 13.26),
- cut – single, double or rasp (Fig. 13.27),
- grade – rough, bastard, second cut, smooth, etc. (Fig. 13.28).

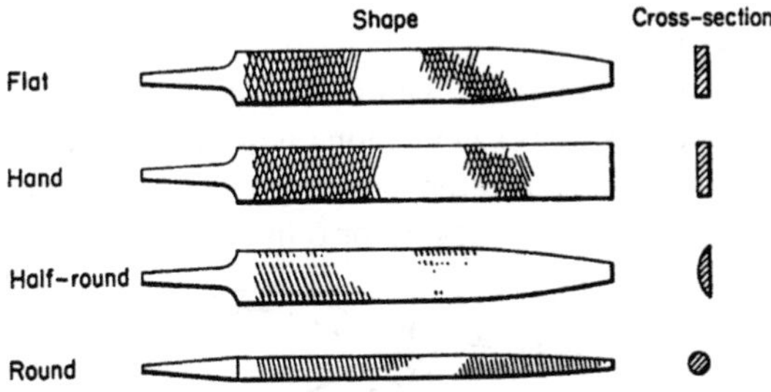

FIGURE 13.26 Shape and cross-section of files.

FIGURE 13.27 Cut of files.

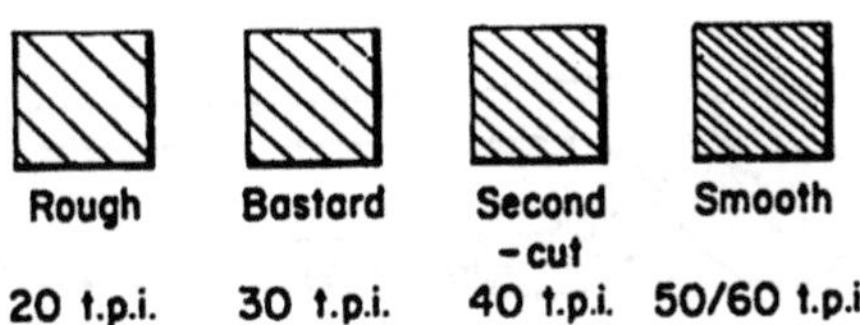

FIGURE 13.28 Grade of files.

In practice it is not generally necessary to state the cut, since this is standard for particular types of files. Flat files are double cut on the face and single cut on the edge. Hand files are similar, but have one edge uncut. They are often termed 'hand safe-edge' files (or HSE for short). Round files are usually single cut and the half-round are double cut on the flat side and single on the curved. A rasp is called a rasp and not a file. The kit usually contains the following files:

- 200 or 250 mm (8 or 10 in.), HSE or flat, bastard or second cut;
- 200 mm (8 in.), half round, bastard or second cut;
- 150 or 200 mm (6 or 8 in.) round, second cut.

Pliers

There are three types of pliers in common use.

1. Combination Pliers
   Pliers of 150 or 180 mm (6 or 7 in.) are used, frequently with insulated grips as shown, when they may be called ‘electrician’s pliers’ (Fig. 13.29).

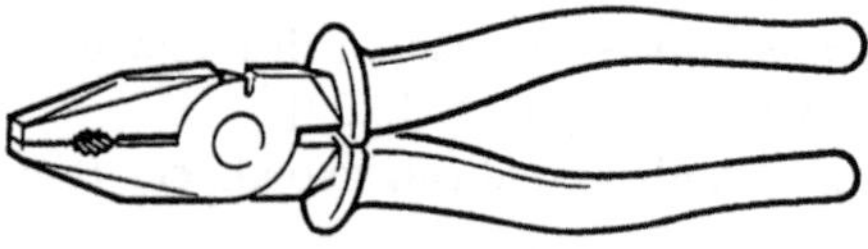

FIGURE 13.29 Combination of pliers, insulated.

2. Snipe Nose Pliers
   These may also be obtained with insulated grips and may be given a variety of names, for example, long nose, taper nose, needle nose, depending on the area and the manufacturer (Fig. 13.30). The common sizes used are 130 or 150 mm (5 or 6 in.).

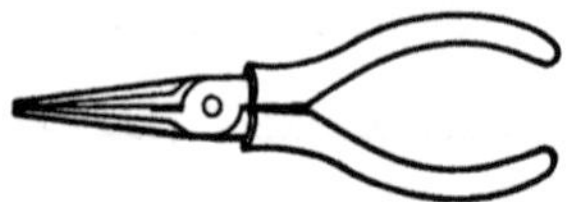

FIGURE 13.30 Snipe nose pliers.

3. Gland Nut Pliers
   These are available in a range of sizes from about 100 (4 in.) to 300 mm (12 in.) and more than one size may be included in the kit (Fig. 13.31). They are extremely useful for many jobs but, like most pliers, they can damage small nuts.

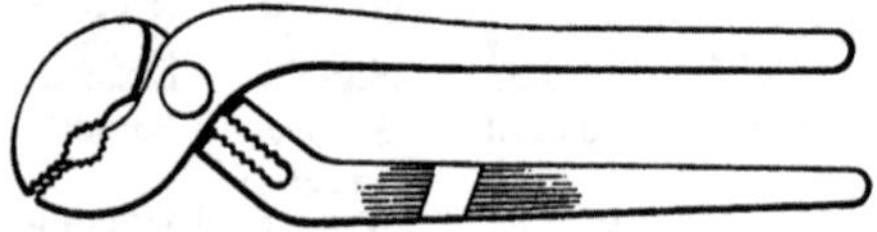

FIGURE 13.31 Gland nut pliers.

Generally the use of special pliers for removing or replacing burner jets is giving way to spanners and keys made specifically for the job.

Wrenches

The following are the types of wrench in common use:

- footprint type (Fig. 13.32),
- stillson (Fig. 13.33),
- chain wrench (Fig. 13.34).

FIGURE 13.32 Footprint pipe wrench.

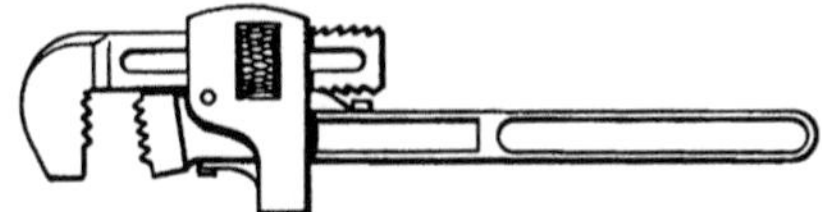

FIGURE 13.33 Stillson pipe wrench.

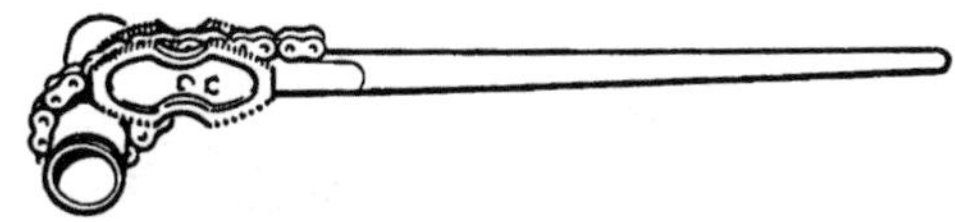

FIGURE 13.34 Chain pipe wrench.

The footprint was, for many years, the type of pipe wrench in general use on domestic installation work. Stillsons or chain wrenches, sometimes called 'chain tongs', were used on pipes over 25 mm (1 in.) diameter and so were reserved for industrial work. They were also used by the gas utility companies for service and main laying., however, most services and mains today are plastic.

The use of the footprint wrench has declined, although the 230 mm (9 in.) size is still used in many areas. Stillson wrenches of 350 mm (14 in.) are included in some kits, and larger sizes are available when required. Chain tongs are still used for large size pipework and often have reversible jaws, as in Fig. 13.34. Lightweight single-jawed grips in 200 and 300 mm sizes (8 and 12 in.) are sometimes used in place of the footprint wrench for ordinary domestic installation work.

Spanners

Spanners may be of the following type:

- adjustable (Fig. 13.35),
- open-ended (Fig. 13.36),
- ring (Fig. 13.37),
- socket (Fig. 13.38),
- box (Fig. 13.39).

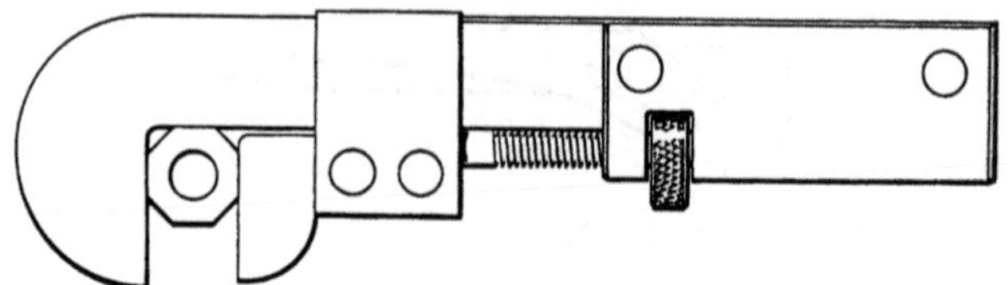

FIGURE 13.35 Adjustable spanner.

FIGURE 13.36 Open-ended spanner.

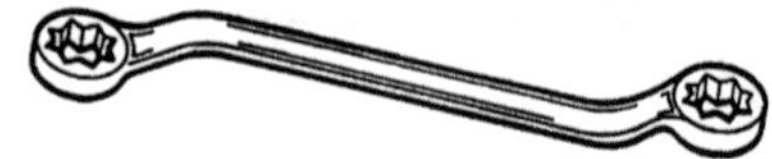

FIGURE 13.37 Ring spanner.

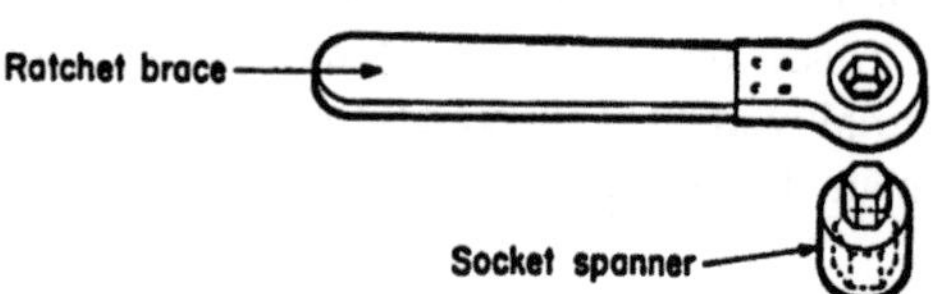

FIGURE 13.38 Socket spanner.

FIGURE 13.39 Double-ended box spanner.

ADJUSTABLE SPANNERS

The 250 mm (10 in.) size is suitable for most domestic meter unions and for assembling pipe fittings and controls, but care must be taken when using this spanner on meter unions as over-tightening can damage the meter washer and cause leakage. There are many different designs of adjustable spanner and the type illustrated is a thin-jawed variety which is frequently chosen.

OPEN-ENDED SPANNERS

These are usually double-ended, with the ends taking consecutively sized nuts. They are described either by the distance across the flats of the nuts (AF sizes) or by the size of screw on which the nut fits. For example:

2 BA (British Association)
5/16 in. BSW (British Standard Whitworth)
8 mm (metric).

so a double-ended spanner could be:

4 × 6 BA
1/4 in. × 5/16 in. BSW
8 mm × 10 mm metric

A number of spanners may be found in the kit, ranging from 3 mm ($^{1}/_{18}$ in.) to 10 mm ($^{3}/_{8}$ in.)

RING SPANNERS

Because they fit completely round the nut, there is less risk of these spanners slipping and they cannot open out with wear as may the open-ended spanners. For this reason, they may be preferred for jobs where nuts must be tightened more securely. Multi-purpose ring spanners will fit nuts from 10 to 22 mm and $^{7}/_{16}$ to $^{13}/_{16}$ in.

SOCKET SPANNERS

Used extensively by motor mechanics, these are a very robust type of spanner. They may be used with a separate lever or brace as illustrated. Alternatively, they are available as a double-ended spanner with two sockets at each end giving a range of sizes, for example 10 and 11 mm × 12 and 14 mm.

BOX SPANNER

In addition to its use as an ordinary spanner, the box spanner can be used on recessed nuts. Small sizes of box spanners are used for fitting burner jets. Box spanners may be double ended and are turned by means of a steel rod called a 'tommy bar'.

Allen keys

The Allen screw (Fig. 13.40a) has a hexagonal recess in the head, into which can be fitted a small hexagonal steel rod, bent near the end to form a key. The Allen keys (Fig. 13.40b) are thus used to turn the screws. A set of seven keys could cover 2–8 mm, or 1/8 in. to 5/16 in. screws. Kits may contain sets of up to ten keys.

Screwdrivers

Like so many other tools, screwdrivers come in many types and sizes. Blades are made of high carbon steel and handles are frequently of plastic in place of

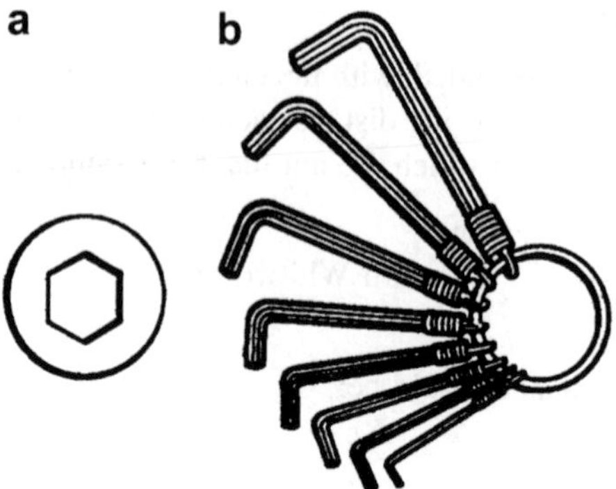

FIGURE 13.40 (a) Allen screw head and (b) Allen keys.

the older, boxwood handles. Some may include a ratchet so that the handle may simply be turned backwards and forwards. Smaller screwdrivers for electrical or servicing work are often in sets with interchangeable blades of various widths and lengths all fitting into a common handle. Screwdrivers are named by their type and the length of the blade. Sometimes the width of the blade is also stated.

CABINET SCREWDRIVER

Used for the bigger screws, this type has a large handle to give more power to turn the screw (Fig. 13.41). It is usually carried in sizes from 150 to 250 mm (6–10 in.).

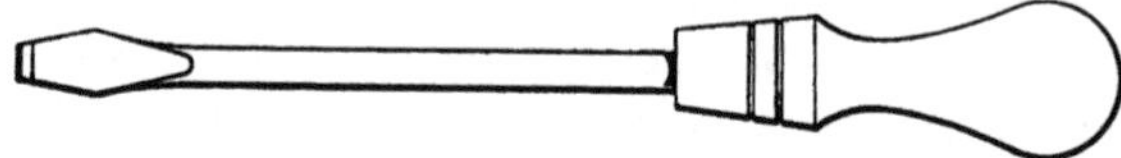

FIGURE 13.41 Cabinet screwdriver.

ELECTRICIAN'S SCREWDRIVER

These are for smaller screws and used in sizes from about 75 mm with a 2-mm wide blade up to 150 mm, with blades about 7 mm wide (Fig. 13.42).

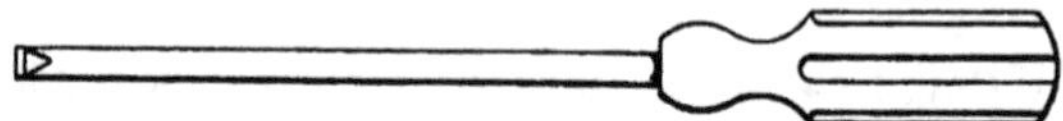

FIGURE 13.42 Electrician's screwdriver.

STUBBY SCREWDRIVER

Also called 'chubby' or 'dumpy', this type is used for larger screws in awkward places (Fig. 13.43). Blades are usually about 25 mm long and in a variety of widths, commonly 5 mm.

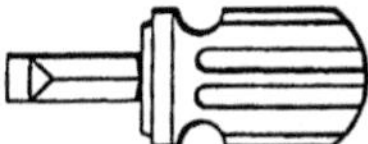

FIGURE 13.43 Stubby screwdriver.

PHILLIPS AND POSIDRIV SCREWDRIVERS

These both have cross-shaped points to fit the cross-slotted head screws (Fig. 13.44b). The Posidriv has superseded the Phillips head, but the Phillips screwdriver has been retained because it will fit both screwheads. The Posidriv is a better fit in its own screw, but it cannot be used on Phillips screws. The end of the Posidriv screwdriver is less sharply pointed than the Phillips, but the essential difference is in the square section between the slots of the cross. This provides a snug fit between the blade and the head. The corners of the square can be clearly seen on the screwhead and are illustrated on the trade-mark. Two sizes of blades are available.

FIGURE 13.44(a) Phillips screwdriver.

FIGURE 13.44(b) Phillips and Posidriv screw heads.

Cross-slotted heads are common on machine screws, but are rarely used on woodscrews, except on mass-produced furniture where power-operated screwdrivers are used.

OFFSET SCREWDRIVERS

Offset or 'cranked' screwdrivers are extremely useful for getting at screws in very awkward places. Two types are shown and the blades may be flat or cross-slotted (Fig. 13.45).

Bradawl and gimlet

Both these tools are used for making holes in wood to start a wood-screw. The bradawl (Fig. 13.46) is usually used on soft wood and the gimlet (Fig. 13.47) on hardwoods.

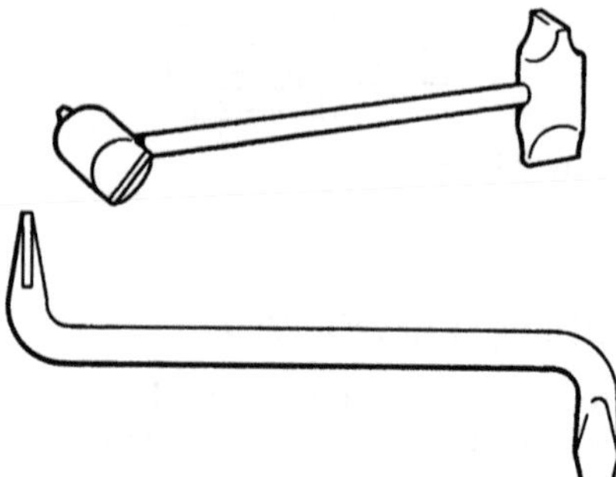

FIGURE 13.45 Offset screwdrivers.

FIGURE 13.46 Bradawl.

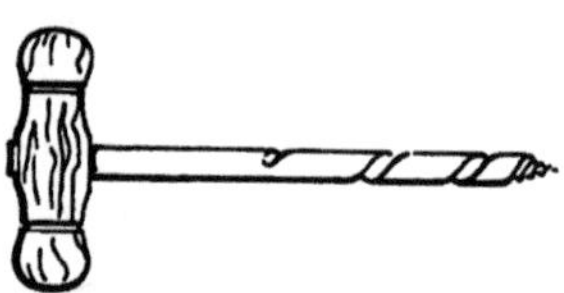

FIGURE 13.47 Gimlet.

The bradawl usually carried has a 38 mm blade which is pinned to the wooden handle. Gimlets may have a metal, wood or plastic handle and are about 100 to 125 mm long.

Taps

Taps are used for cutting internal screw threads, also known as 'female' threads, in metal. There are three taps to a set (Fig. 13.48).

1. Taper tap
   This has no threads near to the end, so that it can enter the hole, and is used to start the thread. If this tap can pass right through the hole, it will cut a complete thread.
2. Second tap
   This has only a few threads ground away at the end and can be used in a 'blind' hole (i.e. one that does not go right through) to cut the thread near to the bottom of the hole, after it has been started by the taper tap.

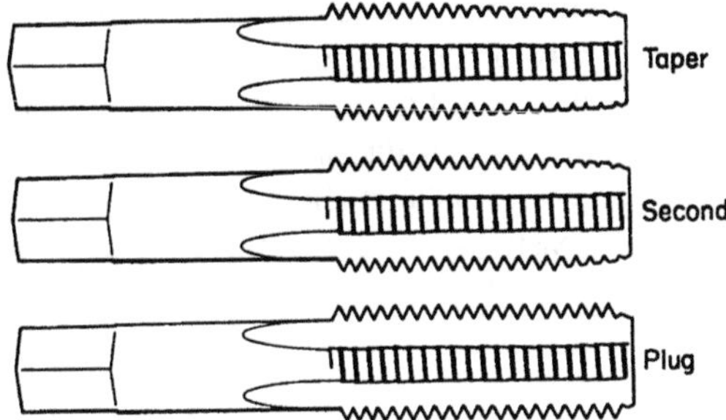

FIGURE 13.48 Taps.

3. Plug tap
   The plug or 'bottoming' tap is used in a blind hole to cut the thread right to the bottom after the other two taps have been used.

Taps are turned by means of a tap wrench, one type of which is shown in Fig. 13.49. Several taps may be included in the kit for cutting or clearing threads for screws, pipes or burner jets.

FIGURE 13.49 Tap wrench.

Dies

Dies are used for cutting external or 'male' threads on metal. They are made in a number of different forms and are held in 'stocks', so that they can be rotated. Figure 13.50 shows circular split dies which are held in the stocks in Fig. 13.51. The split allows some adjustment to be made to the size of thread.

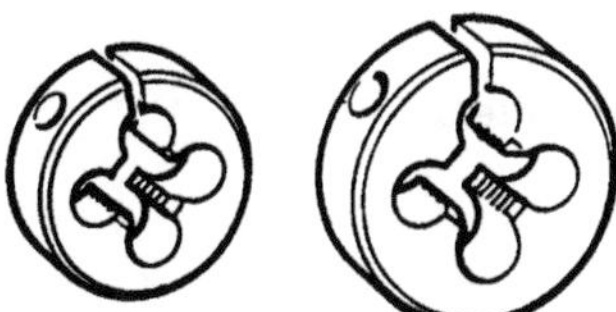

FIGURE 13.50 Circular split dies.

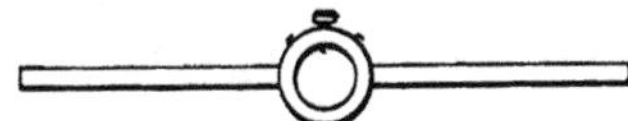

FIGURE 13.51 Circular die stocks.

These dies would be used for cutting threads on metal rods or for clearing threads on screws. 'Die nuts' are also used for thread clearing. They are similar to the circular die but hexagonal, like an ordinary nut and without the split.

Some kits may include a small die or die nut for clearing common sized screw threads. The larger stocks and dies for cutting threads on pipe are generally not part of the basic kit and are dealt with later in the chapter.

Pipe cutters

Pipe cutters operate by the rotation of a hardened steel cutting wheel around the pipe. The wheel is gradually forced through the pipe as the clamp is tightened until the pipe is cut. Although the wheel is made reasonably thin and sharp, the action of cutting produces a burr around the inside of the tube which must be removed by the reamer. Cutters have been used for cutting steel pipe but it is easier and quicker to use a hacksaw which does not leave a burr. Steel pipe cutters are generally no longer used by the service engineer.

The type of cutter which may be included in a Gas Engineers tool kit is the smaller single wheel type for cutting 6–28 mm copper pipe (Fig. 13.52). Using the same type of cutting wheel but the where the clamp is tightened by a spring the modern day equivalent of the 'Pipe Cutter' is the pipe slice. The pipe slice is an ideal tool when removing or breaking into existing pipework in restricted spaces due to its small size. The pipe slice is available in 15 mm, 22 mm and 28 mm sizes.

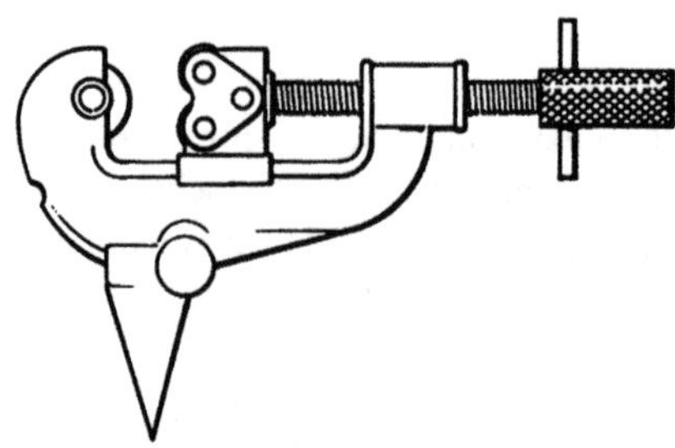

FIGURE 13.52 Copper tube cutters.

Bending springs

Springs are used to support copper pipe whilst it is being bent by hand, to prevent it being flattened or kinked. The springs are made of square-section spring steel (Fig. 13.53). Before use the spring should be lightly lubricated and when bending, the pipe is slightly over-bent and then opened out to release the tension on the spring. The spring is then rotated in the direction required to reduce its diameter by means of a bar through the eye hole. It may then be easily withdrawn.

FIGURE 13.53 Internal bending springs.

External springs are used to support the outer wall of copper pipe and are usually used on small bore, softer copper tube. They are made of round-section spring steel (Fig. 13.54) and are used in a similar manner to internal springs, except that the pipe is inserted into the spring.

FIGURE 13.54 External bending spring.

The sizes of spring usually carried are for 8 and 10 mm (external type) and 12, 15 and 22 mm (internal type). Although still in existence, the use of bending springs is discouraged in favour of mechanical bending machines, whilst small bore springs are acceptable for applications such as hand bending of gas fire tubes, the larger (15 mm and 22 mm) knee bending springs should not be used for health and safety reasons.

Blowlamps and propane torches

Paraffin blowlamps have now been superseded by propane torches and the one in common use is based on the Primus propane bottle No. 2000. To this is fitted a feed tube and valve (Fig. 13.55). More up to date models are fitted with electronic ignition and thermocouples.

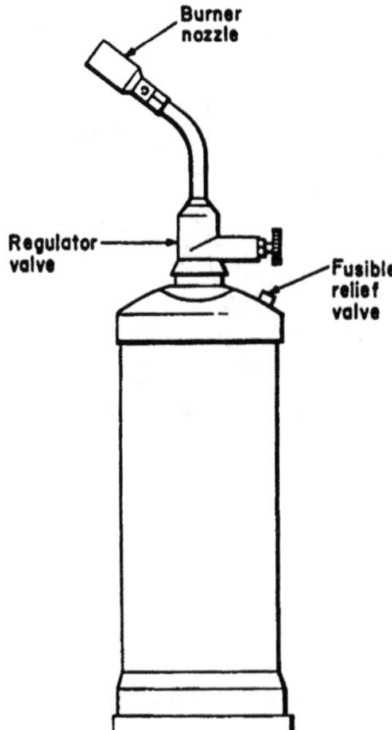

FIGURE 13.55 Propane torch.

Burner nozzles are fitted to the top of the tube and usually two different sized nozzles are carried to deal with lead blown joints and capillary soldered joints (Fig. 13.56).

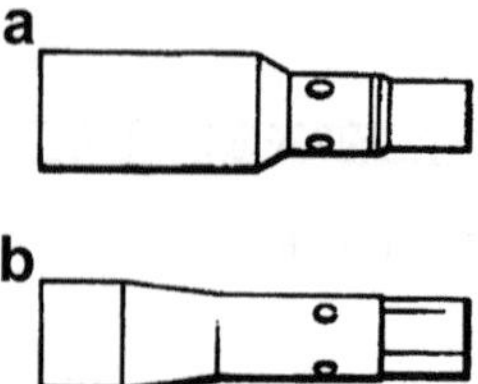

FIGURE 13.56 Burner nozzles: (a) standard flame and (b) needle flame.

Extension hoses are often included to allow the burner to be used in an awkward position where it would not be possible to use the bottle and burner in the normal way.

Using the needle flame burner the bottle will last about 6 h. With the standard burner the bottle will last about 2–3 h.

Fire-resistant mat

Made of several layers of flexible fire proof blanket and often backed with aluminium foil, this 'mat' provides a means of protecting decorations or woodwork when using a blowlamp (Fig. 13.57). It is placed immediately behind the objects being heated, for example, a copper joint being soldered, in situ, up against a wall. Mats are usually about 250 mm × 200 mm.

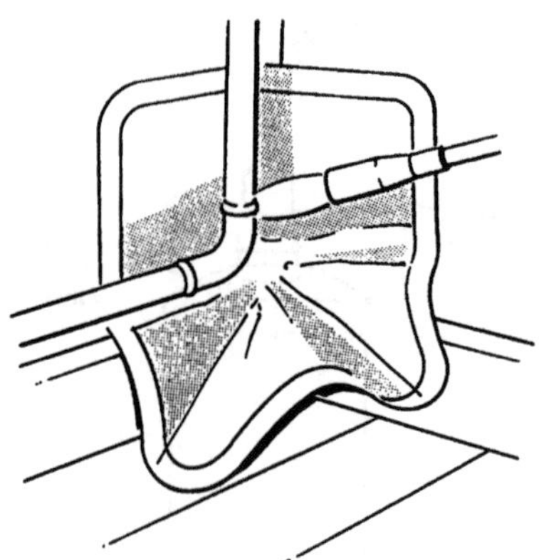

FIGURE 13.57 Fire-resistant mat.

U Gauge

The standard gauge used for general purposes is the 30-mbar gauge (Fig. 13.58). The scale may be in millibars or 'dual' (mbar and in. water gauge). The U tube is housed in a hard, plastic case and the front cover slides off and may be used as

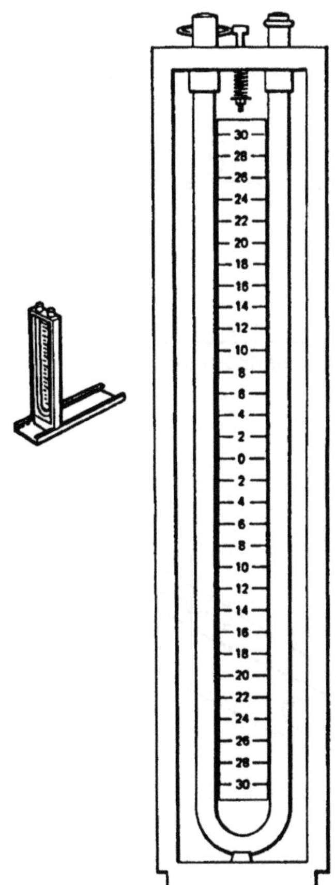

FIGURE 13.58 A 30-mbar 'U' gauge.

a stand. A spring-loaded cap fits over the ends of the tube to prevent spillage of the water when in transit. About 600 mm of rubber tube is supplied to connect the gauge to the point at which pressure is being taken.

Temporary continuity bond

This consists of a 1200-mm long heavy-duty, single-core cable fitted with clamps or insulated crocodile clips at each end (Fig. 13.59). It is used to maintain electrical continuity when disconnecting a meter or a gas supply. It eliminates the risk of fire, explosion or electric shock. The bond is left in position until the pipes or meters have been reconnected.

Spirit level

A 230-mm (9 in.) boat-shaped level, with a die-cast aluminium body and bubbles for checking both horizontal and vertical surfaces, is usually carried (Fig. 13.60).

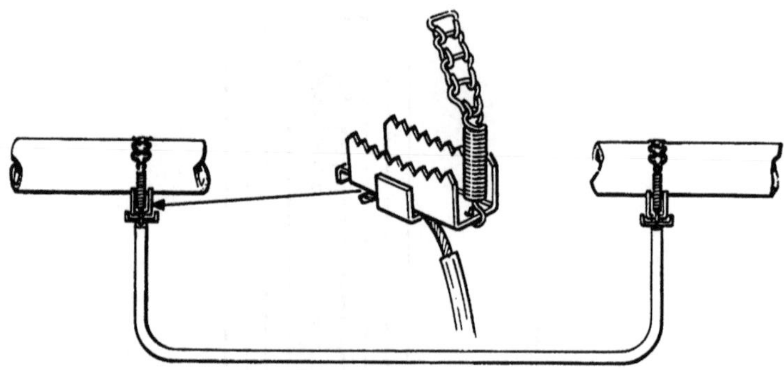

FIGURE 13.59 Temporary continuity bond.

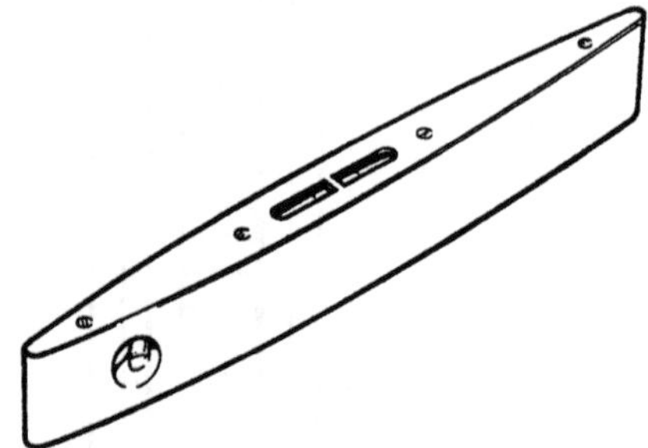

FIGURE 13.60 Boat-shaped spirit level.

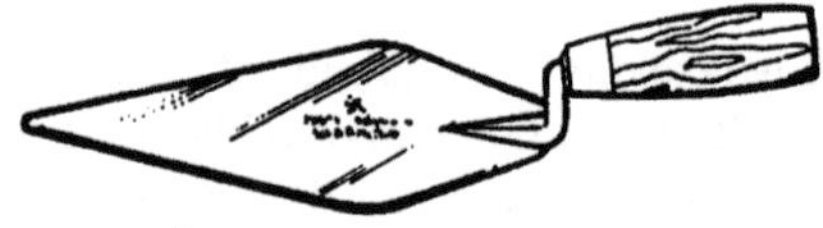

FIGURE 13.61 Pointing trowel.

Trowel

For mixing mortar and making good the holes in walls or concrete floors, a 150-mm (6 in.) pointing trowel is used (Fig. 13.61).

Tinsnips

Tinsnips are like strong, blunt-nosed scissors and are used for cutting thin sheet metal (Fig. 13.62). They are frequently needed, when fitting room heaters, to cut to size the plate which closes off the fireplace opening. The length used is about 280 mm (11 in.).

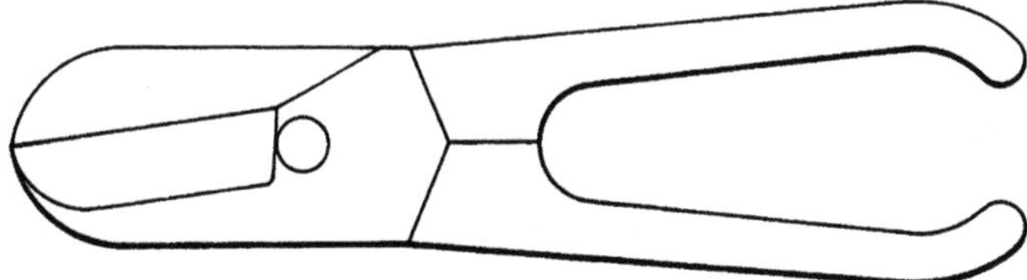

FIGURE 13.62 Tinsnips.

Clasp knife

The kit usually includes a single-bladed knife with a 65–75 mm blade (Fig. 13.63).

FIGURE 13.63 Clasp knife.

Steel rule

The type of rule used varies. It may be the two-fold 600 mm rule (Fig. 13.64a), or the steel tape, usually 3 m (10 ft.) long and often with a dual metric and Imperial scale (Fig. 13.64b).

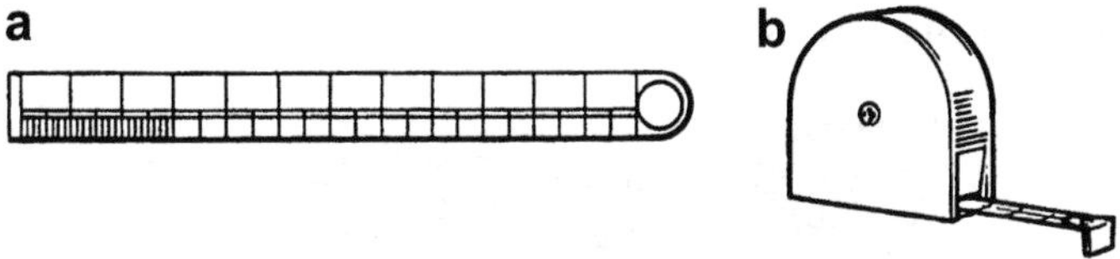

FIGURE 13.64 Rules: (a) folding steel rule and (b) steel tape.

Intrinsic* Safety Torch

This is usually a safe torch, with two 'D' size batteries. The safety torch shown in Fig. 13.65 is safe when used in a gaseous atmosphere and can be opened only with the locking key. (*The term 'Intrinsic' refers to the fact that the electrical switching of the torch is sealed from the outside environment.)

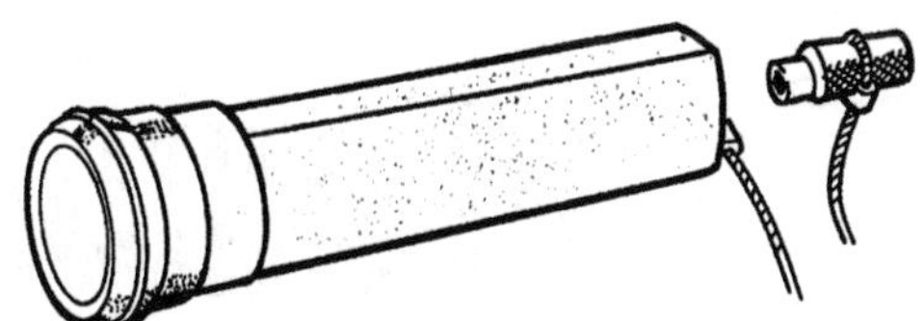

FIGURE 13.65 Safety torch with locking key.

Turnpin

This is a piece of boxwood of about 50 mm diameter and 75 mm long (Fig. 13.66). It is used to open out the end of a piece of lead pipe ready for jointing and as lead piping is no longer used for new gas and water pipework it is unlikely you will come across the need for this tool. The Gas Safety (Installation and Use) Regulations 1998 5(2) states 'Appliances or meters may be connected to existing lead piping using suitable fittings, provided that the piping is in a safe condition, e.g. there is no sign of damage'.

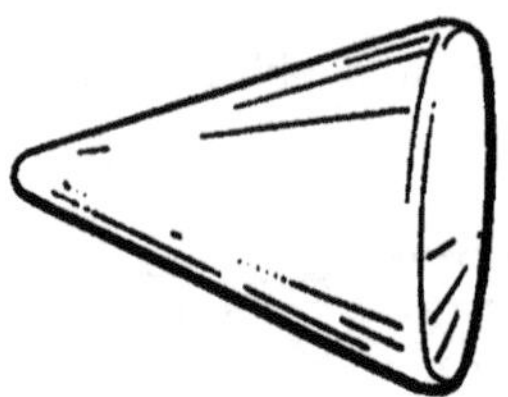

FIGURE 13.66 Turnpin.

Brushes

The kit may contain a number of brushes of different kinds. The following are the most common.

PAINT BRUSHES

The flat paint brush is described by the width of the bristles (Fig. 13.67), 25 mm (1 in.) is the usual size. Circular brushes or 'sash' brushes are in numbered sizes. They are less frequently used.

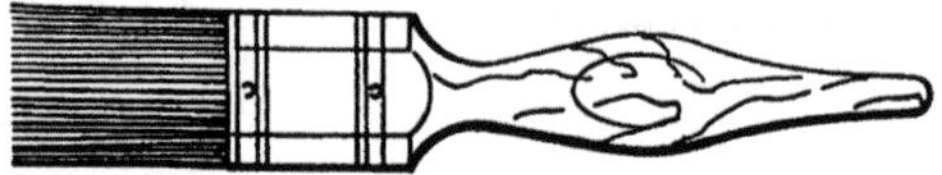

FIGURE 13.67 Flat paint brush.

WIRE BRUSHES

These are included and used for cleaning rust and deposits from burners and appliance components (Fig. 13.68). They are also handy for cleaning files and wrench jaws.

FIGURE 13.68 Wire brush.

DUSTING BRUSHES

A 75-mm (3 in.) dusting brush or 'jamb duster' may be carried for appliance cleaning (Fig. 13.69).

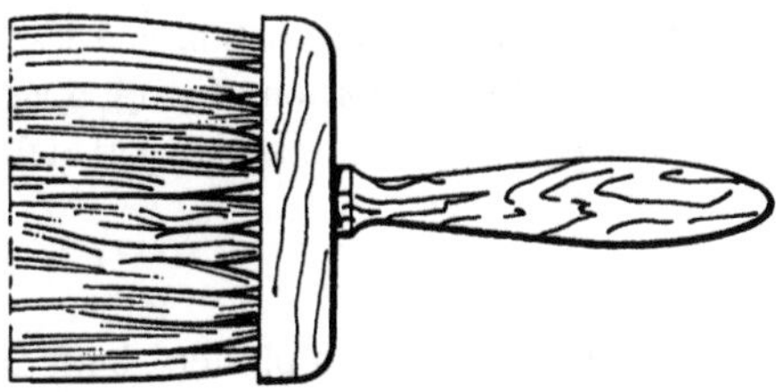

FIGURE 13.69 Dusting brush.

FLUE BRUSHES

These spiral brushes made of wire or bristle may be used for cleaning out the flueways of central heating boilers (Fig. 13.70). Some sets are made with interchangeable heads and with two-piece handles which can be screwed together.

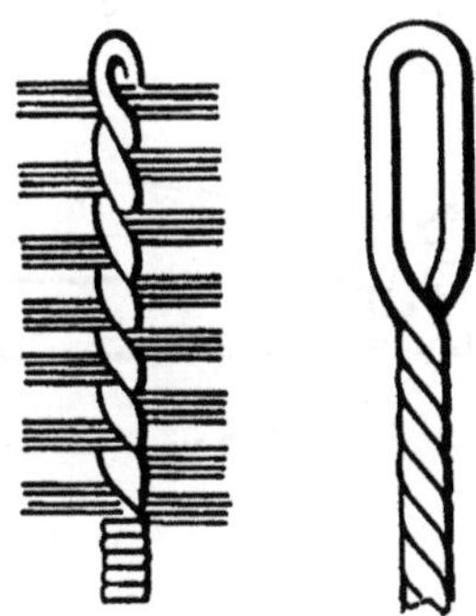

FIGURE 13.70 Flue brush.

Dust sheets

These should be carried to protect customers' property. They could be of plastic or cotton and are usually about 2 m × 1.25 or 2 m.

Vacuum cleaner

Used for clearing combustion chambers and ducting.

Goggles

These should be worn whenever a cold chisel is being used or when using a grinding wheel.

Mirror

Mirrors are usually stainless steel so that they are unbreakable. As part of the gas safety they are used to view the back of a joint which is being made in situ in an awkward position. Mirrors are useful for a number of other purposes when servicing appliances. The Gas Safety (Installation and Use) Regulations 1998 22(1) states 'All joints affected by the portion of work done should be visually inspected to ensure they have been correctly made, as part of the gas tightness test'.

First aid kit

It is useful to carry a small tin of sterilised plasters, burn dressings and one or two small bandages.

Electrical tools

The following tools are often included in the kit for electrical work.

WIRE STRIPPERS

Various types are available, the type shown in Fig. 13.71 will cut and strip insulation from wires, it has an eight-gauge selector and a handle locking device. The handles are insulated.

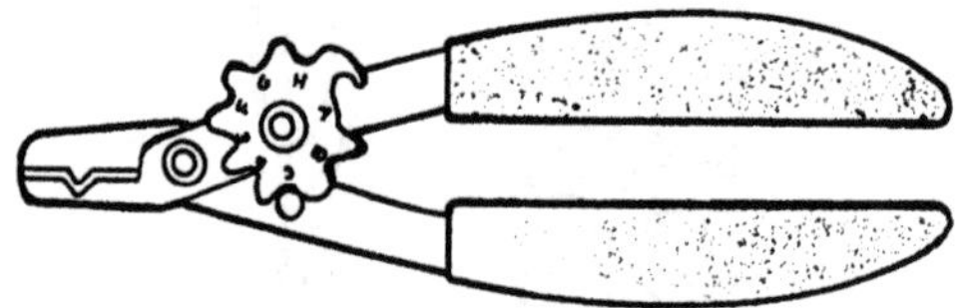

FIGURE 13.71 Wire strippers.

CRIMPING TOOLS

These are designed to fit connectors to the ends of wires. The model shown in Fig. 13.72 also incorporates a bolt cutter and wire stripper.

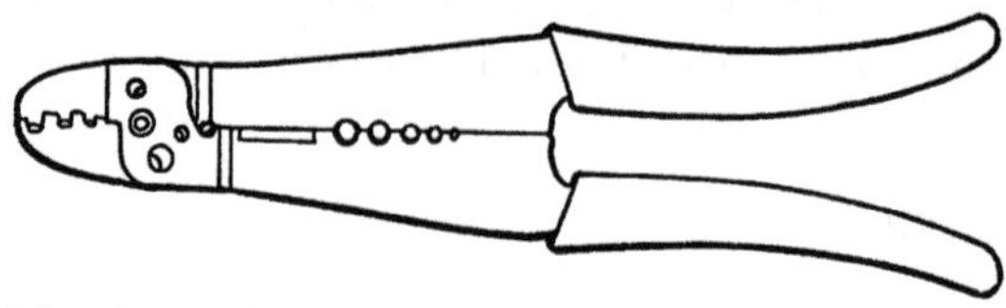

FIGURE 13.72 Crimping tools.

SOLDERING IRONS

Small electric irons are used, mostly for making soldered electric wiring connections. They are from 15 to 40 W and rechargeable types are available.

ELECTRIC CAPILLARY JOINT MAKER

These are used in situations when it is not permissible to use a naked flame.

RING MAIN TESTER

This is used to check a socket before use (Fig. 13.73). It indicates an earth fault, incorrect polarity, a neutral fault or a correct socket.

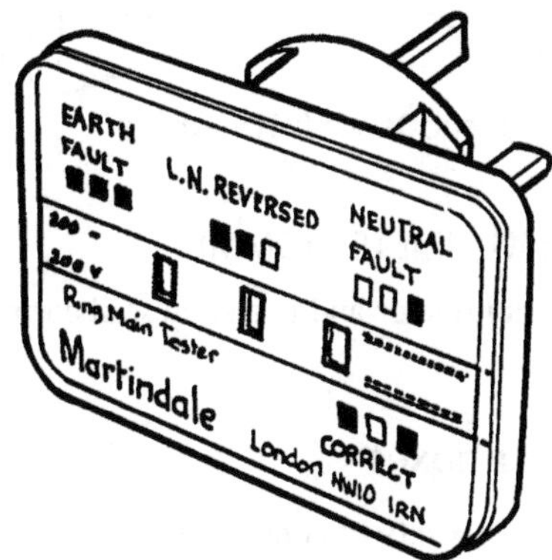

FIGURE 13.73 Ring main tester.

NEON MAIN TESTER

This indicates whether or not a terminal is live (Fig. 13.74).

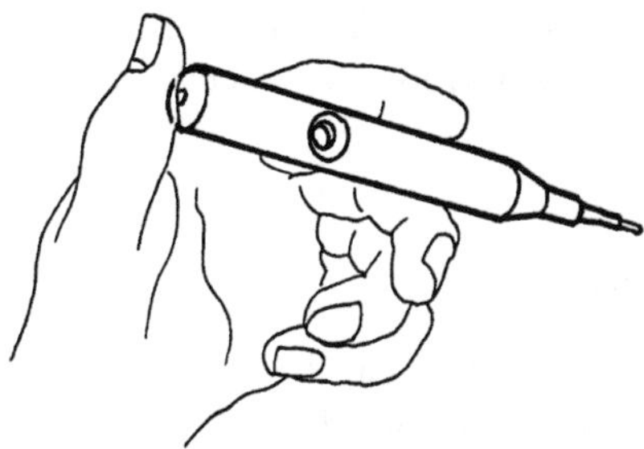

FIGURE 13.74 Neon main tester.

RESIDUAL CURRENT DEVICE (RCD) ADAPTOR

This is an RCCB device used to protect the operative when it is using a portable electric tool (RCCBs are described in Chapter 9) (Fig. 13.75).

MULTIMETER

Designed specifically for use on gas appliances and their control devices. This meter is described in detail in Chapter 9.

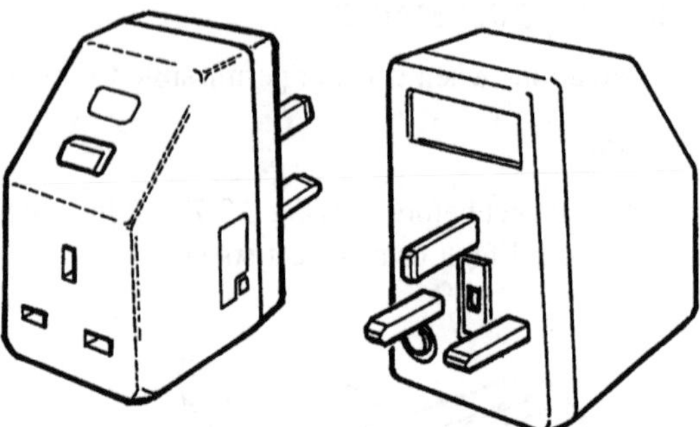

FIGURE 13.75 Residual Current Device (RCD) Adaptor.

FLUKE 75 DIGITAL MULTIMETER (DMM)

This is an alternative multimeter and is also described in detail in Chapter 9.

Tool rolls, bags and boxes

Tools may be carried in a variety of containers. Canvas tool rolls are often used for the smaller tools like drills, taps, punches, small spanners and screwdrivers. Larger tools are generally carried in leather tool bags or sometimes metal boxes.

ADDITIONAL TOOLS

The foregoing list of tools comprises the kit used by the average service engineers in their normal daily work. There are, however, a number of other tools which they might carry or use on particular occasions. These include the following.

Socket forming tool

This is a steel drift which is driven into the end of a piece of copper pipe to open it out so that it forms a socket into which another length of pipe may be soldered (Fig. 13.76). Forming sockets save the cost of pipe fittings. Usually they are confined to sizes 15 and 22 mm.

Tank cutters

This tool is for cutting holes in tanks, cisterns and cylinders so that 'bosses' may be inserted into the holes and pipes connected to them. The type shown in Fig. 13.77 is used in a brace and the cutter is adjusted to the size of hole required. Another type resembles a circular piece of hacksaw blade and is used in an electric drill. These are called 'cup drills'.

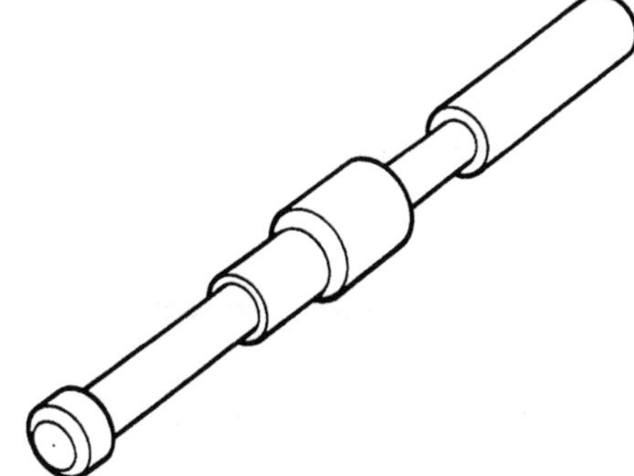

FIGURE 13.76 Socket forming tool.

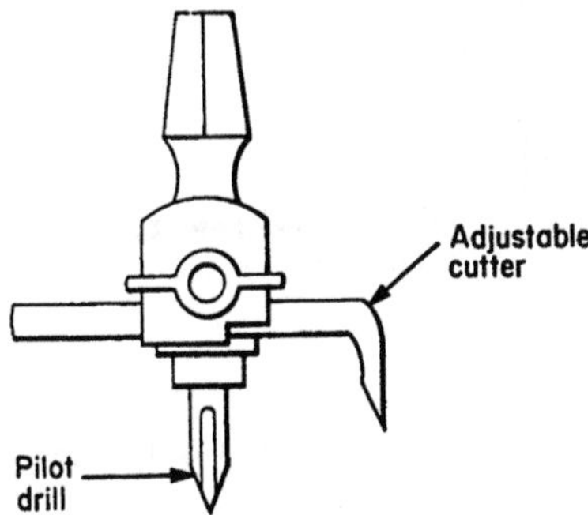

FIGURE 13.77 Tank cutters.

Thermometers

These are used for checking the output temperature of water heaters, cabinet temperatures in refrigerators and centre of oven temperatures in gas cookers (Fig. 13.78). They are mercury-in-glass thermometers in metal or plastic cases and the ranges of temperatures are usually −18 to 115 °C for water and refrigerator temperatures and 100–250 °C for oven temperatures.

More use is being made of the thermistor or electronic thermometer (Chapters 9 and 14) for checking central heating systems. The range is 0–120 °C.

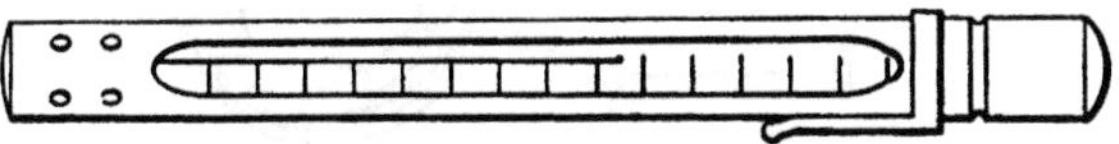

FIGURE 13.78 Water temperature thermometer.

Water flow gauge

Used for checking the rate of flow of water from a water heater, this is also known as a 'weir gauge' (Fig. 13.79). The gauge is held under the spout or tap and the rate of flow is indicated by the height of the water pouring out of the slot in the side of the cup.

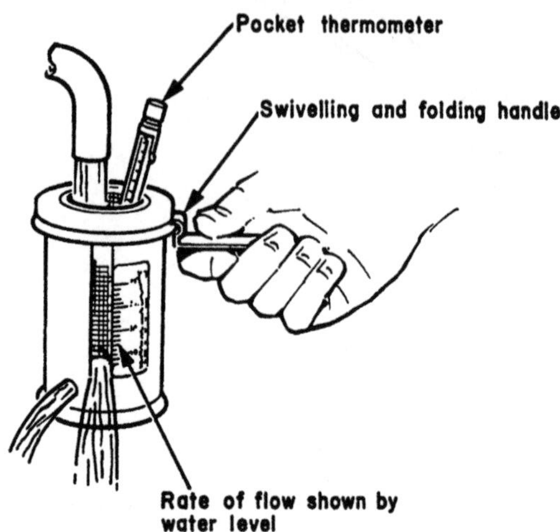

FIGURE 13.79 Water flow gauge.

An alternative method is to use a stop-watch and a container of known quantity, like a bucket or a milk bottle.

Stocks and dies

For cutting threads on steel pipe, sets of stocks and dies are available. Solid dies, as illustrated in Fig. 13.80, are used for pipe sizes DN 8 to DN 25 (3–25 mm). Above this and up to DN 50 (50 mm) the dies may be either solid or adjustable. For sizes larger than DN 50 (50 mm), receder die-stocks or screwing machines are used (see Volume 3, Chapter 1).

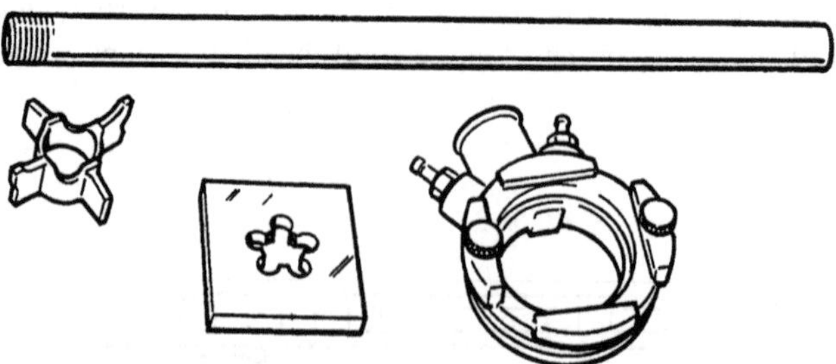

FIGURE 13.80 Stocks and dies.

Pipe vices

Vices are needed to grip the steel pipe which is being cut and threaded (Fig. 13.81). They are usually of the chain type and are attached to a length of wood batten or steel tube. The screw-down type is generally used for larger pipes and may be fitted to portable 'stands' as shown in Fig. 13.82.

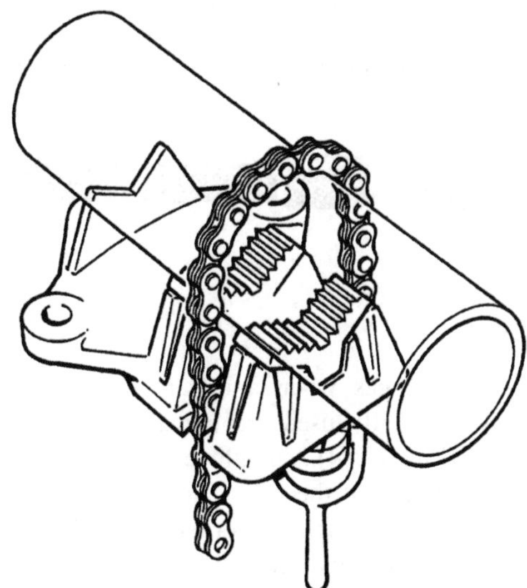

FIGURE 13.81 Chain pipe vice.

FIGURE 13.82 Hinged pipe vice on portable tripod.

Steel pipe in smaller sizes can be threaded by gripping it in chain tongs if no vice is available.

Pin vices

These are made for gripping taper pins so that the pins can be filed to fit in a drilled hole (Fig. 13.83). The pin vice can also be used for holding small drills, taps or reamers and for gripping any other small circular objects which have to be cut or filed. There is considerably less risk of breaking a small morse drill when clearing an injector if the drill is held in a pin vice rather than in a hand drill.

FIGURE 13.83 Pin vice.

Meter control tools

To prevent the escape of gas when a meter control is being greased, or exchanged, the tools generally used are either a 'plastic stopper glove' or a 'Peart Pass stopper' (Fig. 13.84). The glove consists of a transparent plastic bag with a plastic glove set into the side. The bag is put over the control and service pipe and a tourniquet is applied over the end of the bag, sealing it on the service pipe. A hand in the glove can dismantle or remove the control as required. A replacement control or any other materials that may be required are placed in the bag before fitting. The final tightening up or adjustment to the control is carried out after the bag has been removed.

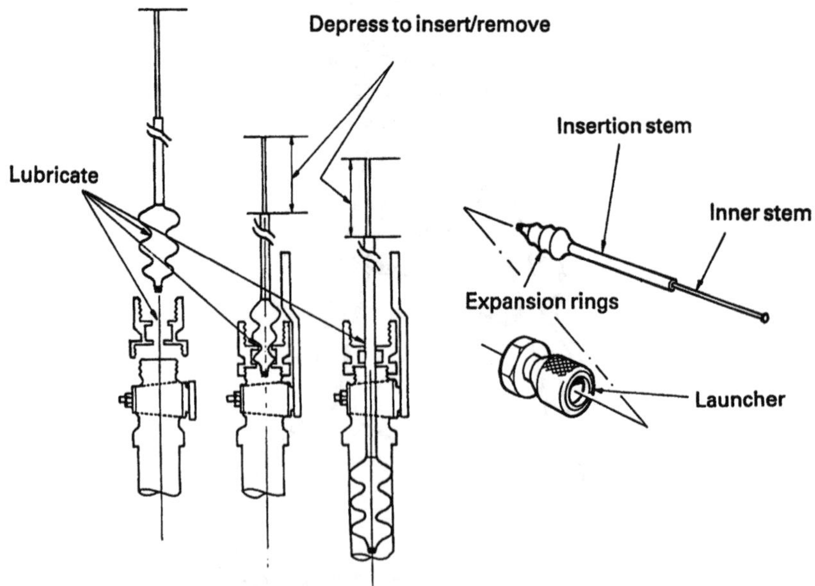

FIGURE 13.84 Peart Pass service pipe stopper.

The Peart Pass stopper (Fig. 13.84) consists of a 'launcher' which is 1×3/4 BS 746, to fit a 3/4 or 1 BS 746 control (the launcher has a rubber membrane through its centre) and a plastic insertion stem fitted with rubber bellows which can be expanded or contracted by the use of an inner stem.

To operate, first lubricate the launcher membrane and the rubber bellows with silicone grease. Depress the inner stem and insert the first pod of the bellows into the launcher. Turn on the meter control. Depress the inner stem and ease the stopper through the barrel until it is located in the service pipe to form an effective seal. Unscrew the launcher and remove. Unscrew the meter control and slide over the stopper stem, taking care not to disturb it.

Reverse the procedure after fitting the new or serviced control. (Note that the greasing or exchanging of meter control taps is the responsibility of the gas transporter and should not be carried out by gas installation or maintenance engineers.)

Testing tee

For reasons of safety no meter shall be installed to an incomplete or unsound installation. The usual way of attaching a U gauge to an installation not connected to a gas supply is by a 'testing tee' and any necessary adaptor. An example of a testing tee and adaptors is shown in Fig. 13.85. It consists of (i) a two-way cock fitted with an R1/4 (BS 21) thread; (ii) a special bush R1/4 by 3/4 BS 746 (meter) thread internally and a 1 BS 746 thread externally and (iii) an R3/4 by 3/4 BS 746 external thread adaptor. With this combination of adaptors, the testing tee can be fitted to a meter union or any carcass point with a BS 21 pipe thread. One connector is for the U gauge tube and the other to blow air into the pipes.

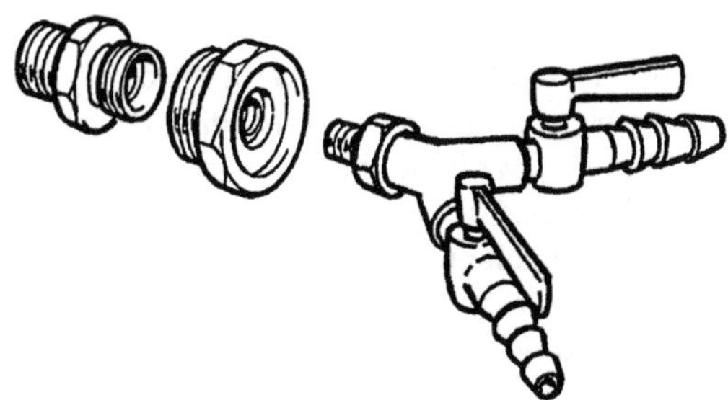

FIGURE 13.85 Testing tee and adaptors.

Other tools

Tools which are made specially for particular appliances will be dealt with in Volume 2 but you may well discover a number of tools which have not been described here. Among these could be jointing tools for lead/gas and water pipes which are no longer in use (see Chapter 12 under 'Lead'). Certain tools that were in use a few years ago have been superseded by new tools and techniques.

Miscellaneous

There is also an assortment of oil cans, grease guns and containers for the sundry lubricants, fluxes and fuels required by the operative.

USE, CARE AND MAINTENANCE OF TOOLS

Although service engineers seldom repair their tools, they do need to keep them generally in good condition and exchange them if they become defective. The following points are worth remembering.

Hammers

To get the full benefit of the leverage, hold it at the bottom of the handle. Make sure the head is accurately and firmly wedged on.

Chisels

All blades need to be kept sharp and, in the case of cold chisels, any 'mushrooms' which form on the heads should also be removed. Grind them off on a wheel or change the chisel. Wood chisels should be protected from damage to the blade when in the tool bag. Keeping them in a tool roll can solve this problem. Always chisel away from the body.

Saws

With hacksaws, use the right size and cut of blade and keep the tension correct. Teeth should point away from the handle. When a blade is blunt or chipped, change it. But preferably not half way through a cut because the new blade may stick and break. Woodsaws can have their teeth protected by a guard. When they become blunt or lose their 'set' change them or have them sharpened and reset.

You are much more likely to cut yourself with a blunt tool than with a sharp one. This is because you have to use a lot more pressure on a blunt tool so it is more likely to jump out of the cut and do some damage. Always put pressure only on the forward stroke when sawing and keep a slow, steady rhythm.

Drills

Small drills may be rotated quickly but large drills and masonry drills should be kept at slower speeds. When drilling a large hole it is easier to drill a small pilot hole first. Do not let drills overheat, they will lose their temper!

Files

Never use a file without a handle, the tangs are often quite sharp. Keep files cleaned out by brushing along the teeth with a file card or wire brush. Any chips of metal which have lodged in the teeth must be removed. This chipping or 'pinning' can be prevented when filing steel by rubbing chalk into the teeth and, with aluminium, by lubricating the file with paraffin.

Wrenches

Keep toothed jaws clean with a wire brush. Do not use toothed wrenches on hexagon heads or nuts when you have a suitable spanner. Do not put a length of pipe on the end to increase the leverage, use a longer or bigger wrench. Keep an eye on the condition of chains, springs and adjusting screws. Do not use splayed or cracked spanners.

Screwdrivers

The blade should be flat and not sharpened too much. It should just fit in the slot of the screw. If the blade becomes worn or chipped it should be reground. Always use the right width of screwdriver. If too big it will damage the surrounding materials, if too small it will tear the screw head. Never use it as a chisel.

Taps and dies

These are both very brittle and the teeth can easily be chipped by rough handling. Small taps must be turned backwards repeatedly to remove the metal cuttings. They are easily broken and must not be forced. Always lubricate the cutting edges of taps and dies using oil or grease for steel and paraffin for aluminium. Brass needs no lubricant. Keep the dies and guides clear of swarf and clean off excess oil. Check that the vice is clean and firmly fixed.

Blowlamps and propane torches

When receiving a new bottle of propane, check:

- a plastic cap has been fitted,
- feed tube connection thread is undamaged and sealing device or 'O' ring is satisfactory,
- needle valve is clean and sound,
- bottle is not corroded or damaged.

Before using, check:

- feed tube connection and needle valve are still undamaged and clean,
- after connecting to the bottle there are no leaks and control valve operates satisfactorily.

In use, ensure:

- torch is not left unattended,
- nothing likely to catch fire is within range of the flame,
- torch is not used in a confined space.

After use remove the feed tube and burner (replace plastic cap to protect the needle valve). If a leak occurs:

- take the torch or cylinder to a safe place in the open air away from sources of ignition and warn anyone in the vicinity,
- a fire extinguisher should be on hand,
- if it ignites on the premises and cannot be carried to safety, call the Fire Brigade and douse it with water to cool the cylinder and minimise damage.

General points to note:

- Propane is heavier than air and will collect in drains, ducts and under floors.
- Bottles should be kept away from sources of heat.
- The relief valve should not be tampered with.
- Feed tubes and burners should only be fitted immediately before the Torch is to be used.
- The feed tube may be removed from the bottle in a confined space only when the valve is properly closed and there is no possible source of ignition.
- When the feed tube is removed the plastic cap must immediately be replaced to protect the needle valve.

General

If your tools get wet, wipe them dry before putting away and periodically wipe them over with an oily rag. This prevents them from rusting and keeps the moving parts lubricated.

And finally, do not forget to keep your first aid tin topped up with plasters.

BENDING MACHINES

Bending machines may be used for copper or steel pipe. They may be necessary when bending pipes which are too large to bend manually or when a lot of bends have to be pre-fabricated, as on a new building estate. There are two main types of bending machines: rotary and ram. *The rotary bender* (Fig. 13.86) holds the pipe at one end by a stop A. The pipe rests in a fixed former B and is bent when the handle D is pulled down, so turning the pipe around the former by means of the guide C. Pipes can be bent through nearly 180° by this method. This type of machine is usually used for copper pipe up to 28 mm and can be used for steel pipe up to size DN 20 diameter. On some of these machines the pressure of the

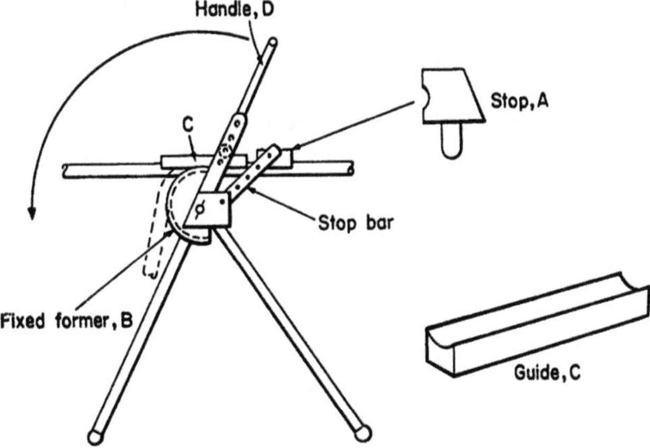

FIGURE 13.86 Rotary pipe bender.

handle roller on the guide is adjustable and this adjustment is critical when bending copper pipe. Excessive pressure will over-compress the pipe and inadequate pressure will produce ripples on the inside of the bend.

The ram bender shown in Fig. 13.87a is operated by a hydraulic pump and is commonly used for steel pipe from size DN 25 up to about DN 50 diameter. Benders for smaller pipes may have a ram operated by a large screw. The pipe is held between two stops A and B, and the former C fits on to the end of the ram D. The pipe is bent when the ram is forced outwards by the hydraulic pressure (Fig. 13.87b). This type of machine produces bends of up to about 90°.

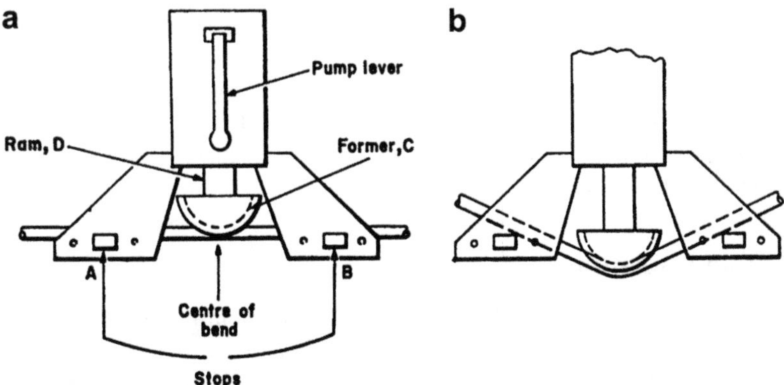

FIGURE 13.87 Hydraulic ram bender: (a) pipe before bending and (b) pipe bent.

POWER TOOLS

Although this title encompasses all kinds of tools driven by all manner of motors, so far as the gas service engineer is concerned it is likely to include only a few tools, all of which are driven by electricity. Pneumatic tools are usually the province of the distribution craftsman. Electrical tools may be of the following:

- mains voltage,
- low voltage, with step-down transformer,
- double insulated,
- all insulated.

Mains voltage

Portable tools, like drills, usually have single-phase universal motors and operate on 240 V. They have a three-core cable with the casing connected to the earth connection. If the earthing is faulty, particularly in wet conditions, the tools can cause a lethal electric shock. Because the earthing in customer premises cannot be guaranteed, it is recommended that double or all insulated portable mains voltage power tools are used. The user should also ensure additional protection by fitting an RCD (Fig. 13.75) in the supply to the tool. Larger workshop tools, like saws, screwing machines and grinders may have small, three-phase motors operating on 415 V.

Low voltage

Transformers are used to step down the mains voltage from 240 to 110 V for tools, or 25 V for hand lamps. Usually both live and neutral connections are fused on the transformer output. Low-voltage tools should be fitted with special plugs so that they cannot be used accidentally on full mains voltage (Fig. 13.88).

Power tools used in commercial and industrial premises should always be of low voltage.

Figure 13.88 is an example of a typical high to low-voltage system. It is important each unit is clearly marked with details of the output voltage and the 'Danger – Electricity' symbol attached, the manufacturer's users instructions should be followed at all times.

Double insulated

These have additional insulation to eliminate risks from faulty earthing. They are tested to 4000 V and may be used on 240 V supplies without an earth lead if they conform to BS 2769 and bear the appropriate symbols (Fig. 13.89).

All insulated

This type is made entirely from shock-proof nylon and does not have a metal case. The chuck spindle is of nylon-glass and electricity cannot be conducted from any part, unless the casing is damaged. It is tested to 4000 V and may be safely used on 240 V without an earth like a double insulated tool.

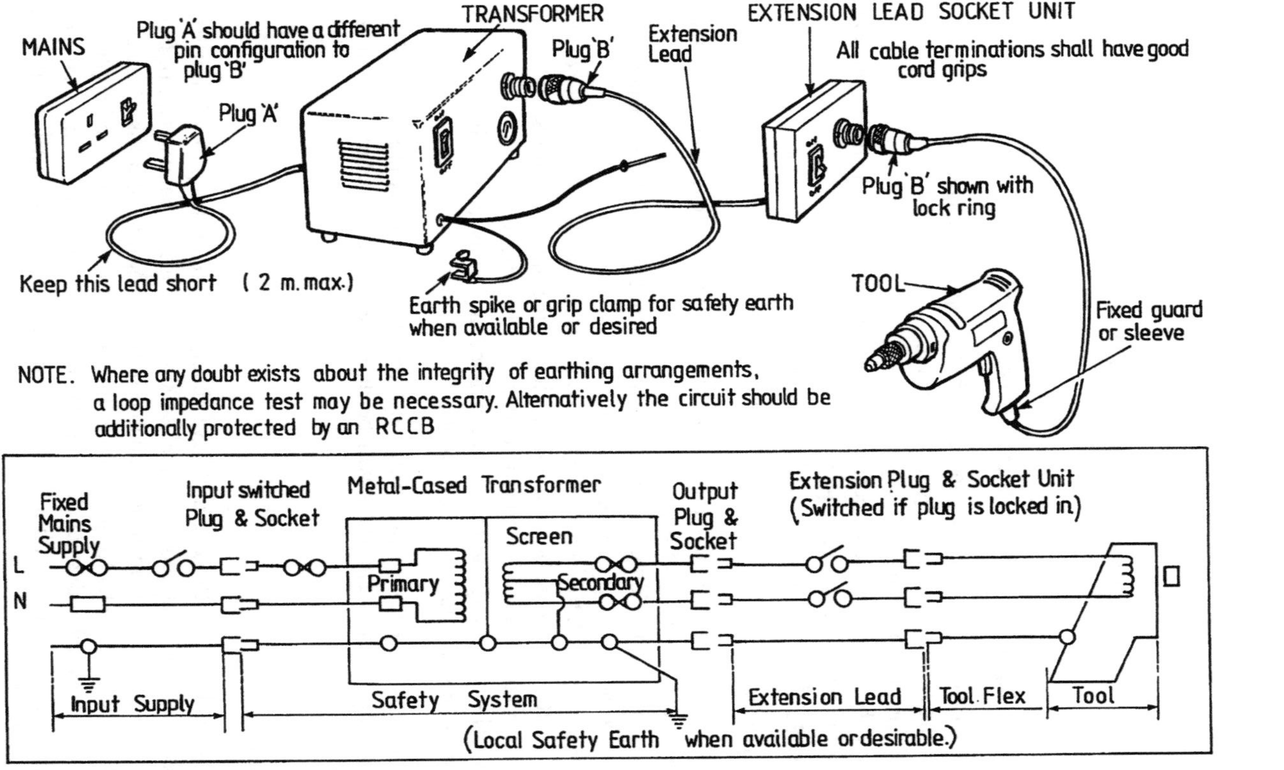

FIGURE 13.88 Installation of a transformer-operated tool.

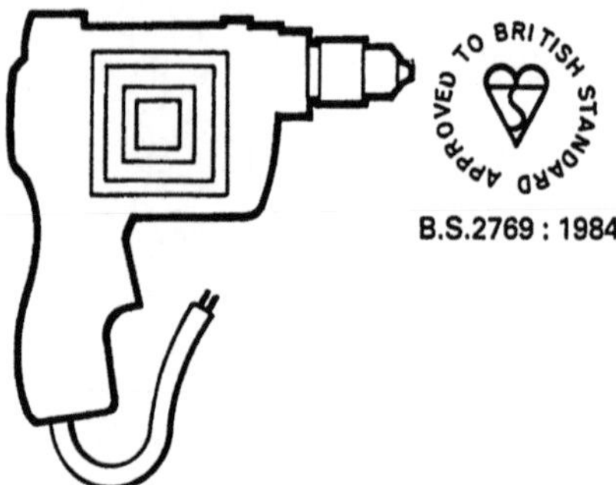

FIGURE 13.89 Symbols for 'all-insulated' and 'double insulated' tools.

Before using any portable power tool check that:

- the voltage on the tool agrees with the supply voltage,
- the casing is not damaged or cracked,
- the cable is sound,
- cable is securely connected to the plug and the tool or the transformer,
- the plugs are not damaged,
- any drills or blades are secured correctly.

When using power tools:

- Wear goggles if there is any danger from flying particles of brick or metal.
- Handle tools carefully.
- Do not lift or drag a tool by its cable, it might loosen the connections.
- Do not apply excessive pressure.
- Do not let the cutter or the tool overheat, it can ruin the temper of the cutter and break down the insulation of the motor windings.
- Do not switch off while the cutter is still in the work.
- Always unplug the tool before making any adjustment to it.
- Do not use the customers' tools.

The following notes on the principal types of power tools are included only as an introduction. The tools will be described in more detail in Volume 3, Chapter 1.

Drills

Portable electric drills generally have chucks to take drills up to 9.5 mm. Two-speed drills rotate at about 900 and 2400 revolutions per minute (rpm). Bench drills for larger holes may be available in workshops. Core drills are used for making holes in walls for the circular flue ducts from room-sealed appliances. They can be used with a standard 850 or 1000 watt, variable speed, electric drill with clutch and fitted with a drive adaptor. A diamond core drill bit is shown in Fig. 13.90.

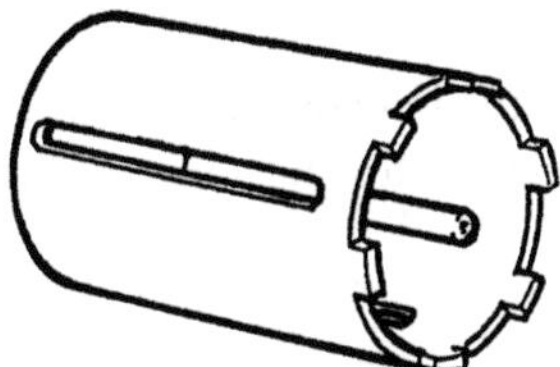

FIGURE 13.90 Diamond core drill bit (Nimbus).

Circular saws

These are now extensively used on timber floors, they are time-saving and can be accurately set to give the correct depth of cut, in order to protect any hidden pipes and cables.

Percussion tools

These tools give the drill fast-hitting blows at the rate of about 50 per second. This action penetrates concrete or similar hard materials which are difficult to cut with the ordinary rotary drill. A special 'impact' type of tungsten carbide-tipped drill is used with percussive tools to produce a smooth, accurate hole.

Grinding machines

These are located in a workshop or depot and may be used for sharpening or reshaping chisels, screwdrivers or drills. Goggles must be worn when grinding. If they become defective, the wheels can burst and fly off in pieces, causing considerable damage. Only specially trained staff should exchange or maintain wheels. If a wheel seems to be faulty switch off, label it 'defective' and report it immediately.

Mechanical saws

The machines used in a workshop for cutting steel bars and mild steel pipe are similar to large mechanical hacksaws. The work should be clamped securely in the machine and excessive pressure must not be used.

Screwing machines

There is a variety of different types of machines for threading large diameter pipe. The cutting heads are often similar to large adjustable dies and have four blades for each size of thread. Some machines incorporate a pipe cutter. The blades must be kept clean and a suitable cutting lubricant should be used to keep them cool and cutting easily. On portable machines, take precautions to protect the cable from damage on site.

VACUUM CLEANERS

To ensure customers property is cared for, a service engineer usually carries a small industrial vacuum cleaner for cleaning combustion chambers and ducting.

CARTRIDGE TOOLS

This tool is like a gun. It shoots a hardened steel fixing stud, like a bullet, into the material to which a fixing is required. Several types of stud are used. Some are similar to nails for fixing wood blocks, others have threaded ends on to which a nut can be screwed. These latter types can support appliances, brackets or pipe clips.

The studs can be fixed into most masonry and also rolled steel joists. They must not be used on very brittle materials like tiles, cast iron or vitreous bricks.

They must also not be used where there is a risk of the stud going right through the structure, as on hollow tiles or no-fines concrete.

The tool is useful when fixings are required for a long run of pipe in a factory or where fixings for appliances are needed in a number of new houses on a building estate.

The following precautions should be taken:

- Ensure material is suitable for studs.
- Use correct strength of cartridge for the material (cartridges are colour-coded to indicate explosive power).
- Do not use near-the-end of a wall, it may break away.
- Do not use existing holes.
- Do not use where there is risk of fire or explosion.
- Only load the tool immediately before use.
- Unload it if it is not required.
- Keep it pointed at the ground, never at anyone.
- Wear goggles and safety helmets and use earplugs in confined spaces.
- Ensure that the tool conforms to BS 4078.

LADDERS

Construction

Ladders are generally made either of wood or of aluminium. They may be single ladders or extension ladders. The traditional wooden, builder's ladder (Fig. 13.91) has rails, or 'stiles', which are half-round and which taper slightly towards the top of the ladder. The flat side of the rails is on the inside of the ladder. The rungs are secured in the rails by tenon joints. The ladder is wider at the bottom and, as the rungs become shorter, it narrows at the top.

Wooden extension ladders are usually of lighter construction and have parallel rails (Fig. 13.92). Construction is similar to the traditional ladder but they are often strengthened by steel tie-bars fitted at about every third or fourth rung and by straining wires inset into the back edges of the rails. Each extension

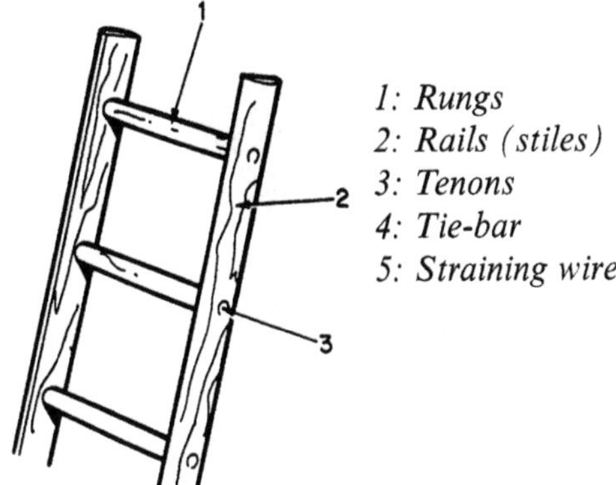

FIGURE 13.91 Builder's ladder.

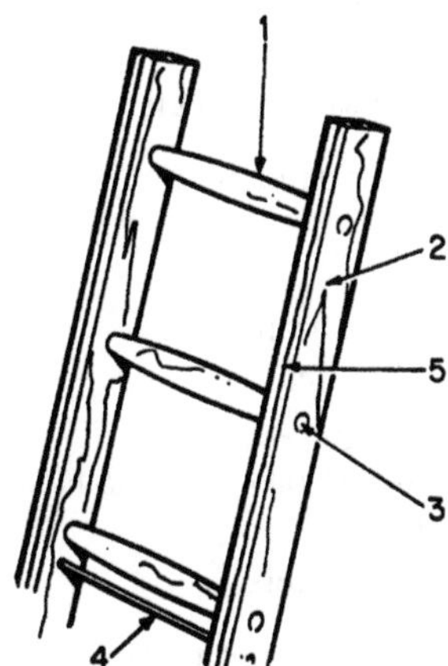

FIGURE 13.92 Extension ladder.

ladder hooks over the rungs of the lower section and they are often extended by ropes and pulleys.

When new, wooden ladders are treated with a clear preservative coating, it prevents the wood from rotting and still allows any cracks or defects to be seen. Ladders should never be painted as this makes it difficult to see any faults which may later occur in the wood.

Aluminium ladders, whatever their type, usually have rectangular, parallel rails and are often fitted with rubber feet to prevent them slipping.

Carrying

When a ladder has to be carried, lift it by the middle and angle it so that the top end is above head height (Fig. 13.93). Take extra care when going around a corner. If it takes two persons, there should be one at each end.

FIGURE 13.93 Carrying a ladder.

Inspecting

All ladders ought to be inspected periodically by an expert, but it is sensible to examine any ladder carefully before you use it.

Check for the following possible defects:

- missing, uneven, loose or defective rungs,
- uneven feet,
- insecure tie-rods,
- split, splintered, or cracked rails,
- temporary repairs or binding,
- excessive twist or warping,
- defective or insecure hooks on extension ladders,
- faulty ropes or pulleys,
- corrosion on aluminium ladders.

Do not use any ladder which is faulty. Put a label on it to warn others and report it immediately.

Selecting

Always choose a ladder which is suitable and safe and particularly one which is the right length for the job. It is an expert's job to lash two ladders together and not something you can learn from experience. If you get it wrong first time, there may not be a second chance. Never try to get greater height by standing a short ladder on a table or box.

Erecting

Make sure that the ground is firm and level and that the top will be resting on something secure. If necessary use a bracing board (Fig. 13.94). With long ladders get some help with the erection. Lay the ladder down, back uppermost, with the foot in the required position. One person then puts a foot on the lowest rung and pulls up on one of the higher rungs. The other person lifts the ladder

FIGURE 13.94 Bracing board across a window.

near the top and walks under it, towards the wall, raising it hand-over-hand on the way. When the ladder reaches a vertical position it becomes the turn of the front person to 'foot' the ladder.

The angle of a ladder is most important. It should be at 75° to the horizontal, that is four lengths upwards for every one length out at the bottom (Fig. 13.95). If the ladder is used to give access to a platform, it should extend at least 3 ft 6 in. (1100 mm) above it. If a ladder is to be used for any length of time it should

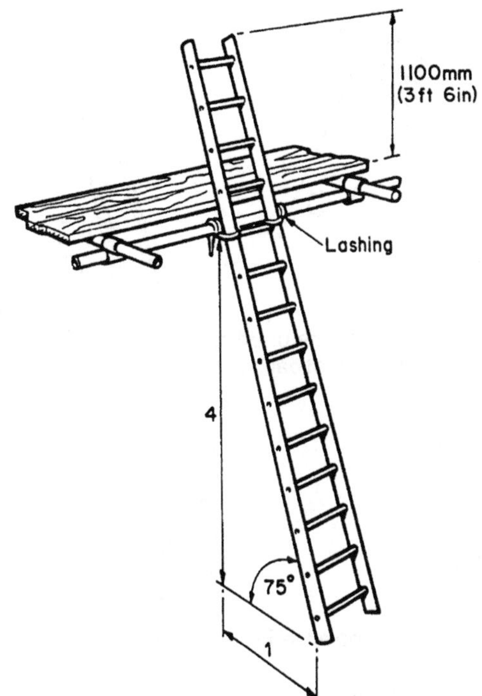

FIGURE 13.95 Ladder correctly positioned and secured.

be securely lashed (Fig. 13.95). Another method of ladder lashing is shown in Fig. 13.96. To carry out the work the following materials are required:

- two 9 m lengths of pre-stretched nylon rope,
- two No. 12 eyelets,
- two No. 12 plastic plugs or
- two 6 mm eyed Rawlbolts.

The method used is as follows: lay the ladder flat on the ground and lash a rope round each stile, in the vicinity of the fourth rung from the top of the ladder (if the rope has an eyelet at one end, it would facilitate easy securing of the rope at the top of the ladder). Erect the ladder in the position required. Fix the eyelets, 1.5 m up from the floor and 1 m out from each stile. Pass the ropes through the eyelets and secure each with a knot. Complete the lashing by securing each rope to the stiles, three rungs from the bottom. Place the ends of the rope in a safe area.

Extension ladders should be extended to the required height after they have been raised. If the ladder has more than two sections, make sure that all sections are extended equally. Check that the hooks are properly engaged before using.

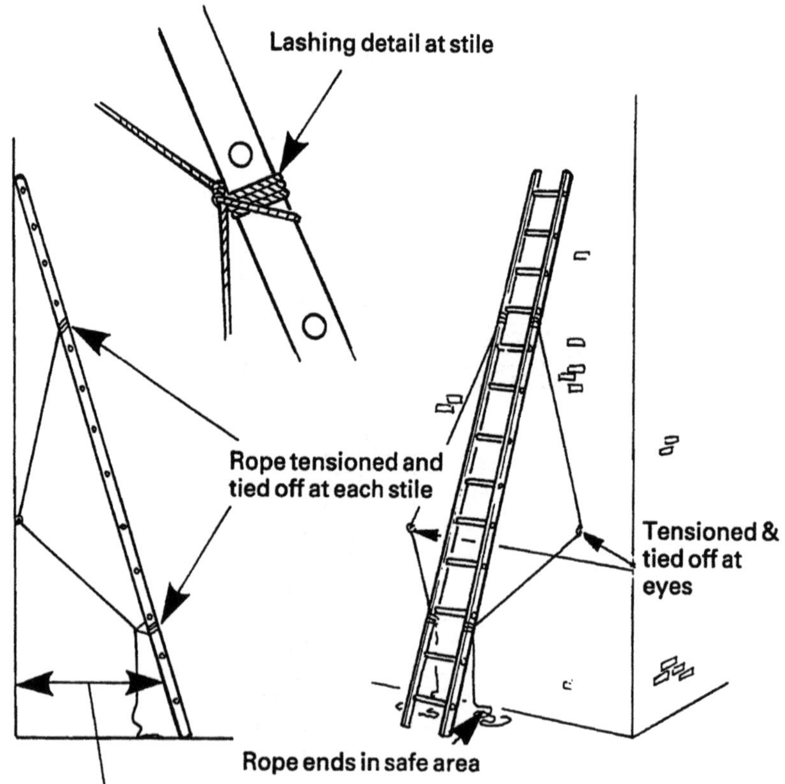

FIGURE 13.96 Ladder lashing.

Working on ladders

The following points should be remembered:

- Climb ladders hand-over-hand so that there is always one hand gripping a rung. Do not slide one hand up the rail. And do the same when you are coming down.
- Use a light rope to get your tools or equipment up or down. Do not try to carry heavy or awkward objects up a ladder and do not put tools loosely in your pocket.
- Do not try to come down a ladder facing outwards.
- Ladders are only made to take one person at a time. Do not work on or climb a ladder at the same time as someone else.
- Make sure that your footwear, preferably with a heel and your clothing is sound. A loose sole or torn trouser leg bottom can cause a fall.
- Do not use metal ladders where there is an electrical hazard.
- Ensure that the ladder will not be hit by passing vehicles.
- If the ladder is in front of a door, lock the door.
- If your feet will be 2 m or more above the ground, lash the top of the ladder, or stake it at the bottom or have someone foot it.
- Do not climb to fewer than three rungs from the top of any ladder.
- In factories take care if working near overhead cables or cranes and do not work over moving machinery.
- Do not lean too far sideways – move the ladder.
- Take extra care in wet or icy conditions.

Working on steps or trestles

Always open them to their fullest extent.

- Where possible, steps should be placed at right angles to the work, that is with either the front or the back of the steps facing the job.
- When using planks avoid excessive spans.
- Do not stand on the top step unless it has been designed as a working platform.
- Do not climb on the rear leg of the steps.

Working on roofs

It may occasionally be necessary to work on a roof or to cross it in order to reach a chimney flue. Flat roofs are usually not a problem, a ladder is often used for access, and tools and materials are brought up by rope. Special equipment is required for hoisting roof-top boilers or air conditioning plant (see Volume 3, Chapters 10 and 11).

Inclined roofs of slates, tiles or shingles call for the use of roof ladders or 'crawling boards' (Fig. 13.97). These lie flat on the roof with the headboard

FIGURE 13.97 Crawling boards on a roof.

bearing on the opposite side of the roof. Gutters and ridge tiles must not be used as supports. The access ladder is usually lashed to the bottom of the roof ladder. Fitting flue liners or a flue and terminal through a sloping roof should not be carried out single-handed.

Asbestos-cement sheet roofs call for special precautions. Although it looks substantial a corrugated, asbestos-cement roof will not support the weight of a man and crawling boards must be laid across the roof so that they are supported by the roof trusses underneath.

When work cannot be carried out safely from roof ladders, a working platform is required. This particularly applies to work on chimneys where a portable scaffold, fixed over the ridge of the roof, can be used.

If there is any risk of slipping, a safety harness should be worn. Precautions must be taken to prevent tools and materials falling from the roof.

SCAFFOLDING

Scaffolding is generally constructed from steel tubes held together by clips or clamps. Its erection is the work of an expert. Small, mobile, tower scaffolds are, however, available which are made up of prefabricated sections which slot one into the top of the other. These can be assembled by anyone with some manual dexterity and are now used by a number of trades who need only a small mobile working platform at a height of about 3–4 m.

Figure 13.98 shows a mobile scaffold which has been constructed from separate steel tubes. The safety rules which are indicated apply equally to any tower scaffold. The following are the rules:

- The height of the working platform should not be more than three times the minimum base width when used outside.
- The height may be increased to three and a half times inside a building.
- The width of the base can be increased by fitting ‘outriggers’.
- The working platform must have handrails and be fitted with toe-boards.

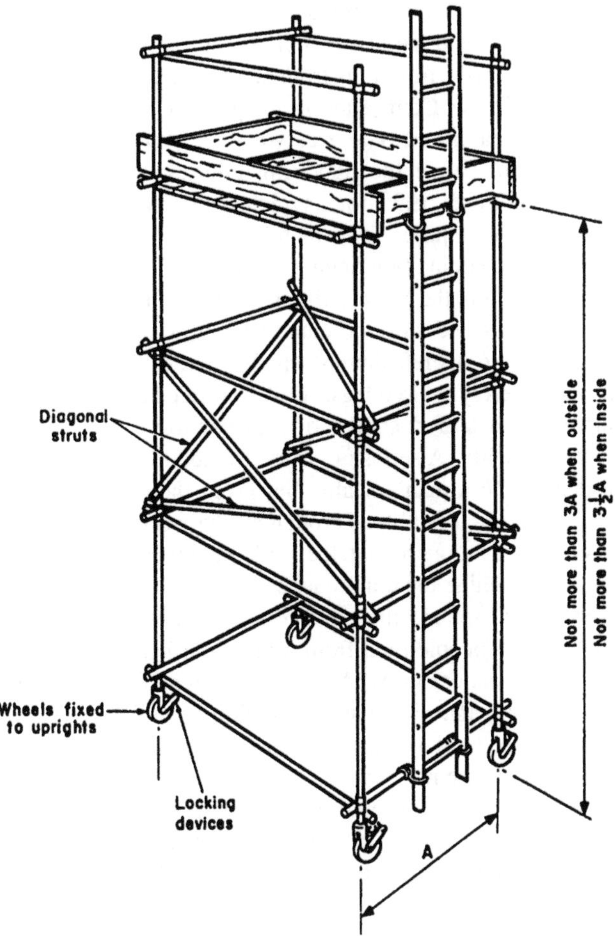

FIGURE 13.98 Mobile scaffold.

- A built-in or securely lashed ladder must be provided for access to the platform.
- The scaffold must be moved by means of the base and not pulled along from the platform. No one should be on the platform while the scaffold is being moved.
- The wheels must be locked when in use to prevent the scaffold moving.
- The scaffold should only be used on firm, level ground.
- To give greater stability, the scaffold could be secured to the building.

WORKING AT HEIGHTS WARNING FROM HSE

The Health and Safety Executive (HSE) has warned employers and senior managers that they must control the risks of working at height in the workplace.

The warning follows the prosecution of the manager of a construction company after an incident led to an employee suffering severe injuries and for failing to control the risks from falls from height; on another site a manager was fined £1500 at the Crown Court after pleading guilty to breaching Section 2 (1) and 3 (1) of the Health and Safety at Work etc Act 1974 and contravening Regulation 3 (1) (b) (ii) of the Reporting Injuries, Diseases and Dangerous Occurrences Regulations 1995 for an incident at the construction site. The builder suffered major injuries when he fell 2 m from an unprotected wall on 6 November 2006 while doing bricklaying work on a housing construction site; he fell onto the floor within the house striking his head on some steelwork and was knocked unconscious and suffered cuts to his head with severe bruising and swelling, a fracture to his left thumb, which has resulted in permanent loss of movement, and severely bruised legs. He also suffered from nerve damage to his right temple and now suffers short-term memory loss.

HSE inspector said: “A series of errors resulted in a tragic incident causing permanent damage to a man’s health, but given the circumstances this could easily have resulted in a fatality. Throughout the work at the construction site there was a complete failure to plan the work, maintain the necessary protection at height or acknowledge the consequence of falls.

Falls from height remain the most common cause of fatal injuries. Latest figures show that 45 people died from a fall from height at work in 2006/07, with 3750 suffering major injury. More than half of all fatalities from falls occur in construction. Companies involved in building, refurbishment or maintenance should ensure that the work is planned properly and sensible measures should be taken so that workers are not exposed to risk.

This illustrates that risks should be properly assessed and the results acted upon to ensure that decisions can be taken on appropriate equipment and working practices to be used so employees are safe.”

PROTECTIVE CLOTHING

The following items may not normally be classified as tools. They are, however, essential for the safety of the service engineers in their work and it seems appropriate to include mention of them at this point.

Safety helmets

These should be worn whenever there is any hazard from overhead or falling objects. They should always be worn on building sites or when carrying out the installation of overhead supplies in factories.

Safety footwear

It is advisable to wear safety footwear at all times while working. Always make sure that your footwear is in good condition, a loose sole or heel can cause a fall.

Eye protection

The eyes should always be protected from hazards as when grinding, chiselling metal or masonry or welding.

Dust masks

An approved dust mask should be worn when the job presents a dust hazard. Special precautions should be observed when dealing with asbestos.

Gloves

Should be worn when handling objects with sharp or abrasive edges or when using acids.

Ear protectors

These may be necessary when using cartridge or percussion tools in confined spaces.

Special protective clothing

Employers are required, by law, to provide special protection when particularly hazardous tasks have to be carried out. Any employee who subsequently ignores his instructions and fails to wear the protection provided risks not only his safety, but also prosecution under the Health and Safety at Work etc. Act 1974. A person causing an accident at work can face a heavy fine or even imprisonment. All building sites and many refurbishment sites now have to enforce the use of protective clothing. Anyone having construction or building work carried out has legal duties under the Construction (Design and Management) Regulations 2007 (CDM 2007), unless they are a domestic client.

A domestic client is someone who lives, or will live, in the premises where the work is carried out. The premises must not relate to any trade, business or other undertaking. Although a domestic client does not have duties under CDM 2007, those who work for them on construction projects will.

WHAT WILL THE REGULATIONS DO?

These Regulations will help you to ensure that your construction project is safe to build, safe to use, safe to maintain and delivers you good value.

Good health and safety planning will also help ensure that your project is well managed and that unexpected costs and problems are minimised.

WHAT DO CLIENTS NEED TO DO?

As a client, you have a big influence over how the work is done. Where potential health and safety risks are low, there is little you are required to do.

FIGURE 13.99 Warning signs. (Pictures courtesy of the Health and Safety Executive, HSE)

Where they are higher, you need to do more. CDM 2007 is not about creating unnecessary and unhelpful processes and paperwork. It is about choosing a competent team and helping them to work safely and efficiently together. Fig. 13.99 shows some of the warning signs a gas engineer may come across when carrying out site work.

Chapter | fourteen

Measuring Devices

Chapter 14 is based on an original draft prepared by Mr B. Hooper

INTRODUCTION

Measuring devices of the types described in this chapter are used for two main reasons:

1. To ensure that appliances are operating at their most efficient rate; to check that the output is of the required standard and quantity and that the fuel is being used as economically as possible, without waste.
2. To make sure that the appliances are installed and operating within the prescribed standards of safety, particularly with respect to operating temperatures, combustion, flueing and risk of explosion.

A number of measuring devices have already been described. Most of them were listed as tools or instruments in Chapter 13 and some have been dealt with in more detail in previous chapters.

They have covered the measurement of:

- length,
- weight or mass,
- volume and rate of flow,
- temperature,
- pressure,
- time,
- electrical potential, current flow and resistance.

There remain four areas of measurement still to be examined:

1. temperature measurement in commercial and industrial applications,
2. air flow measurement,
3. flue gas analysis,
4. leak detection.

The methods used to measure different properties vary considerably. For the simple properties, like length, weight, mass, time, or volume it is usually possible to compare the object to be measured with a known, standard quantity.

The metre rule, the kilogram or pound weight, for example, are simple standard quantities.

For more complex properties, like temperature, or the analysis of gases, the measurement can be achieved by observing their effect on other objects or substances. These effects include the following:

- change of dimensions – expansion or contraction of solids, liquids or gases with temperature;
- change of volume of a gas, as when one of its constituents has been absorbed in a liquid;
- change in the colour or condition of a substance with temperature, presence of a gas or combustion;
- change in the volume of a substance with the absorption of a gas;
- production of an emf by the application of heat;
- conduction of a current by ions in a flame;
- mechanical movements of a bellows, flap or vane with the application of pressure or flow of air or gas.

A number of these effects are used not only to indicate a measured quantity, but also to operate a control device. Some of them were dealt with from this point of view in Chapter 11 and it is now necessary to study their other applications.

TEMPERATURE MEASUREMENT

The various methods of measuring temperature are summarised in Table 14.1.

Liquid in glass thermometers

These simple thermometers (Chapter 5, section on Temperature) are inexpensive but have the disadvantage of being fragile and slow to respond to changes in temperature. The liquids which may be used include the following:

- mercury −35 to +510 °C,
- alcohol −80 to +70 °C,
- toluene −80 to +100 °C,
- pentane −200 to +30 °C.

Liquid-in-steel thermometers

This thermometer consists of three parts (Fig. 14.1):

1. a large bulb containing most of the liquid;
2. a bellows or Bourden tube coupled to a pointer;
3. a capillary tube joining the two other parts.

When the temperature of the bulb is increased, the liquid expands. The increase in volume of the liquid extends the bellows or the Bourden tube and results in a movement on the pointer.

Table 14.1 Methods of Temperature Measurement

Method	Objects Measured	Temperature Range (°C)	Distant Reading	Component Parts
Liquid in glass	Solids, liquids, gases	−200 to +500	Not suitable	Bulb and stem
Liquid in steel	Solids, liquids, gases	0 to +600	Up to 60 m	Bulb, capillary and dial
Gas expansion	Liquids	0 to +550	Up to 60 m	Bulb, capillary and dial
Vapour pressure	Solids, liquids, gases	−20 to +350	Up to 60 m	Bulb, capillary and dial
Metallic expansion	Gases	0 to −400	Not suitable	Bimetal strip or spiral and dial
Electrical resistance wires	Solids, liquids, gases	−240 to +600	Suitable	Element, leads, electrical instrument, source of current
Thermistors	Solids, liquids, gases	0 to +400	Suitable	Element, leads, electrical instrument, source of current
Thermocouples, base metals	Solids, liquids, gases	−200 to +1100	Suitable	Element, leads, electrical instrument
Thermocouples, rare metals	Solids, liquids, gases	0 to 1450	Suitable	Element, leads, electrical instrument
Suction pyrometers	Gases	Depends on thermocouple	Not suitable	Element, leads, suction device, electrical instrument
Total intensity radiation pyrometers	Radiating surfaces	+500 upwards	Suitable	Telescope, leads, electrical instrument
Optical pyrometers	Radiating surfaces	+700 upwards	Not suitable	Telescope, source of current, electrical instrument
Heat sensitive materials, ceramics	Kilns	+600 to +2000	Not suitable	Material made into cones or rings
Heat sensitive materials, crayons	Surfaces	+65 to +670	Not suitable	Crayons or pigments

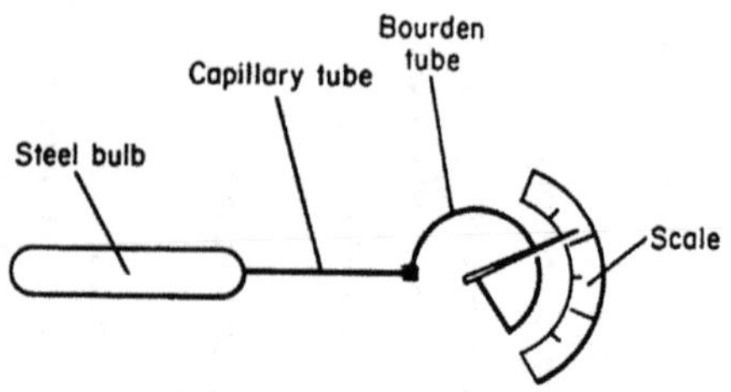

FIGURE 14.1 Liquid-in-steel thermometer.

Mercury was commonly used in these pressure thermometers and the other liquids are now employed for different ranges of temperature.

Gas expansion thermometers

These thermometers consist of the same three components as the liquid expansion types. They are filled with an inert gas, usually nitrogen.

Vapour pressure thermometers

As with the other pressure thermometers, this type has the same components and is operated by pressure from the vapour produced when a liquid is heated. The liquids used include methyl chloride, sulphur dioxide and ether.

Bimetallic thermometers

Similar to the bimetal strips used to operate thermostats (Chapter 11, section on Bimetal Strip Thermostats), these spirals or coils turn a pointer on a dial. They are inexpensive and give a direct reading but they are not very accurate and have a narrow temperature range.

Resistance thermometers

These thermometers operate by the changes in the electrical resistance of a wire brought about by changes in temperature. In order to measure these resistance changes, they use a 'Wheatstone bridge' circuit (Fig. 14.2).

The Wheatstone bridge is a device for determining the ratio of one resistance to another. So it enables a change in the resistance of a wire to be compared with another wire resistance which has not changed. It is named after its developer, Sir Charles Wheatstone, a British electrician and inventor. This bridge circuit may be found in a number of instruments where an accurate measurement of a changing resistance is required.

In Fig. 14.2, P and Q are known resistances, S is an unknown resistance and R is a variable resistance. The resistances are connected up in two loops, abc and adc. A galvanometer is connected across the mid-points of the loops, from b to d, to measure any flow of current.

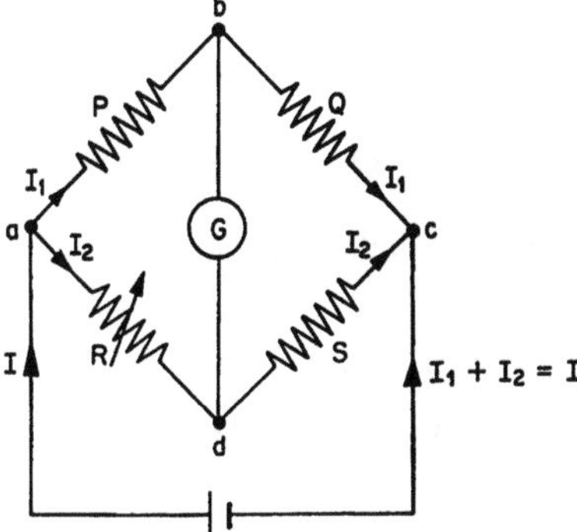

FIGURE 14.2 Wheatstone bridge circuit.

If the resistance R is adjusted so that the galvanometer reads zero, then the voltage at b must be the same as the voltage at d. So the voltage drop through P is the same as the voltage drop through R.

At this point the ratios of the resistances are equal,

$$\frac{P}{Q} = \frac{R}{S}$$

and any alteration of the resistance S will unbalance the circuit so that the galvanometer will again show a current flowing.

In the resistance thermometer (Fig. 14.3) S is the resistance of the wire in the thermometer element, which could be nickel or platinum.
The system is balanced at room temperature and any increases in temperature are shown on the galvanometer, which is calibrated in °C. The theoretical principle of the method of operation is given below, for use if required.

In Fig. 14.2 the current in abc is I_1 and the current in adc is I_2.

When no current is flowing through the galvanometer and b and d are at the same potential then, by Ohm's law *($V = RI$):*

$$\begin{aligned} V_a - V_b &= PI_1, \quad & V_a - V_d &= RI_2 \\ V_b - V_c &= QI_1, \quad & V_d - V_c &= SI_2 \end{aligned}$$

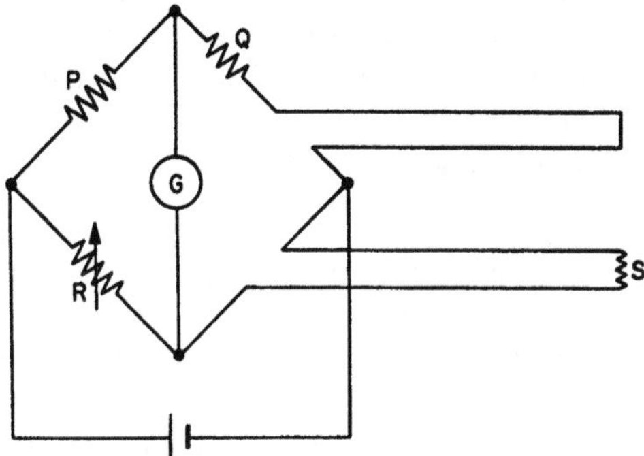

FIGURE 14.3 Resistance thermometer circuit.

Therefore

$$\frac{V_a - V_b}{V_b - V_c} = \frac{P}{Q} \quad \text{and} \quad \frac{V_a - V_d}{V_d - V_c} = \frac{R}{S}$$

But $V_b = V_d$, since no current is flowing. So

$$\frac{V_a - V_b}{V_b - V_c} = \frac{P}{Q} = \frac{R}{S}$$

Therefore

$$\frac{P}{Q} = \frac{R}{S} \quad \text{and} \quad S = \frac{QR}{P}$$

Thermistors

A thermistor (Chapter 9, section on Resistors) is another form of electrical resistance thermometer. The thermistor, which is a non-metallic resistor, shows a decrease in resistance with increasing temperature. It is made in the form of a bead and sealed into a sheath. The leads are then connected to a Wheatstone bridge circuit and the electrical meter is calibrated in °C. A change in temperature alters the resistance of the thermistor and causes a deflection on the meter.

Thermocouples

Thermocouples (Chapter 11, section on Thermo-electric Devices) are probably the most commonly used means of measuring the higher temperatures in commercial and industrial appliances. They are also used in laboratories and by appliance manufacturers for testing domestic appliances. Thermocouples are more accurate than pressure thermometers and respond quickly to changes in temperature. They are used in a number of devices including surface, suction and total radiation pyrometers.

Types of thermocouple in general use are:

1. Base metals

   copper and constantan, maximum temperature 400 °C (Cu and Ni 40% Cu 60%)

   iron and constantan, maximum temperature 850 °C (Fe and Ni 40% Cu 60%)

   chromel and alumel, maximum temperature 1100 °C (Ni 90% Cr 10%, and Ni 94% Al 2%).

2. Rare metals
   Platinum and platinum rhodium, maximum temperature 1400 °C (Pt and Pt 90% Rh 10%).

The maximum temperatures are for continuous use. If a spot reading is made they may be increased by 100–200 °C.

Many modern temperature measuring devices use thermocouples, in varying shapes and sizes, to sense the temperature of a substance. The sensor is then connected, via the electrical leads, to an electronic circuit which in turn produces a digital read-out of the temperature being measured. Such devices are used to measure the temperature of gases/air, liquids, solids, semi-solids and molten metals. Figure 14.4 shows a typical portable electronic thermometer.

In addition to reading temperatures, instruments are available that will also measure humidity (Fig. 14.5).

Suction pyrometers

These pyrometers use a thermocouple or a resistance thermometer as the temperature-sensitive element. The element is situated in a tube through which the gas to be measured is drawn by a suction device.

The main problem when measuring the temperature of gases is that the measuring device is affected by radiation from the walls of the container. The effect is reduced when the velocity of the gas passing the element is increased. Sucking the gas past the element increases its velocity and improves the accuracy of measurement.

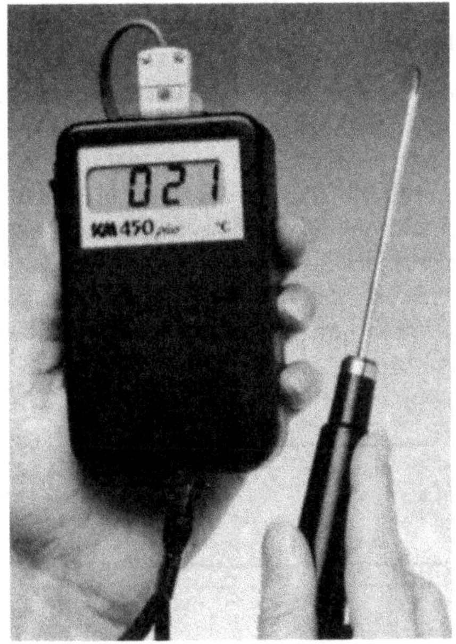

FIGURE 14.4 Portable electronic thermometer (Kane-May Ltd).

FIGURE 14.5 Humidity/temperature meter (Kane-May Ltd).

The suction may be created by an electric fan or by a compressed air or steam ejector. For high temperatures above 1100 °C various types of shield are used and for temperatures up to 1400 °C the pyrometer may be water-cooled.

Radiation pyrometers

Radiation pyrometers use a thermocouple or a 'thermopile', which is a number of small thermocouples mounted together in a circular housing. The radiation from the heated object is focused on to the thermopile by means of a lens (Fig. 14.6) or a mirror (Fig. 14.7). There is no upper limit to the temperature and no physical contact with the heated object. However, the pyrometer must be close enough for direct viewing and the element needs to be protected from corrosive atmospheres, so the choice of locations may be limited. Some pyrometers may incorporate an eyepiece or sighting hole to enable them to be

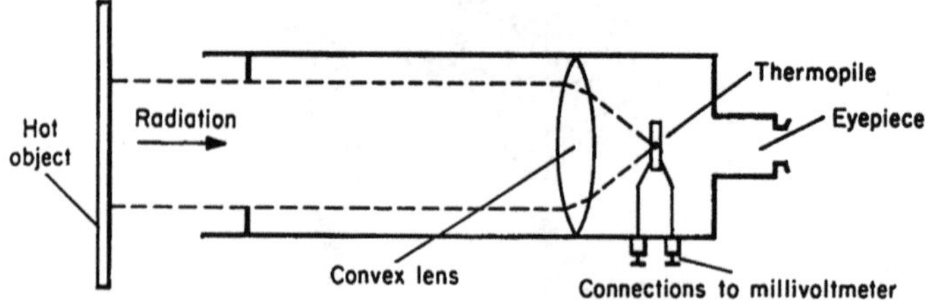

FIGURE 14.6 Radiation pyrometer, lens type.

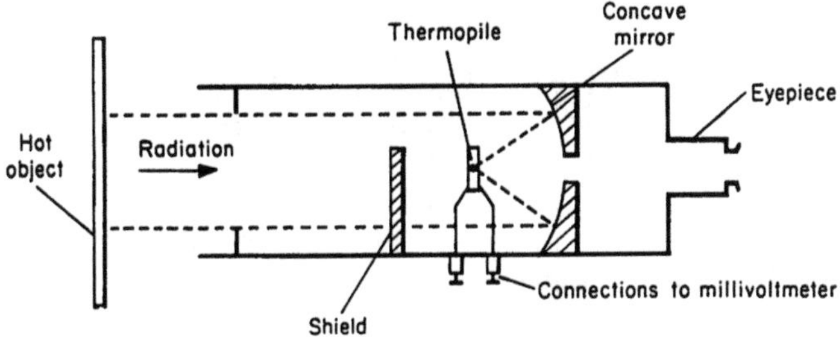

FIGURE 14.7 Radiation pyrometer, mirror type.

directed at the view required. Compensation for ambient temperatures may also be provided.

Optical pyrometers

When measuring the temperature of objects which are incandescent it is possible to compare the light emitted by the object with the light from a controlled light source. This is done in an optical pyrometer (Fig. 14.8). Light from the hot object is brought to a focus in the same plane as a filament A by a lens B. Both the object and the filament are viewed by an eyepiece C.

The filament is heated by a battery D and its temperature and luminosity are controlled by a variable resistance E. When viewed through the eyepiece, the hot object appears as a bright circle against which the filament can be seen (Fig. 14.9). When the resistance is adjusted so that the filament disappears or merges into the background, the current passing through the circuit is proportional to the temperature. The milliammeter measures the flow of current and is calibrated directly in °C. Some devices will now measure the thermal radiation as well as the temperature of a surface. The readings are given on a digital readout and these devices can be connected to a printer recorder giving a continuous hard copy printout for future reference and analysis.

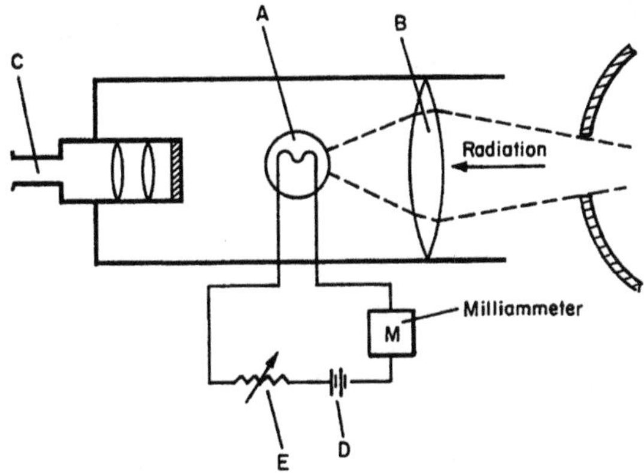

FIGURE 14.8 Optical pyrometer.

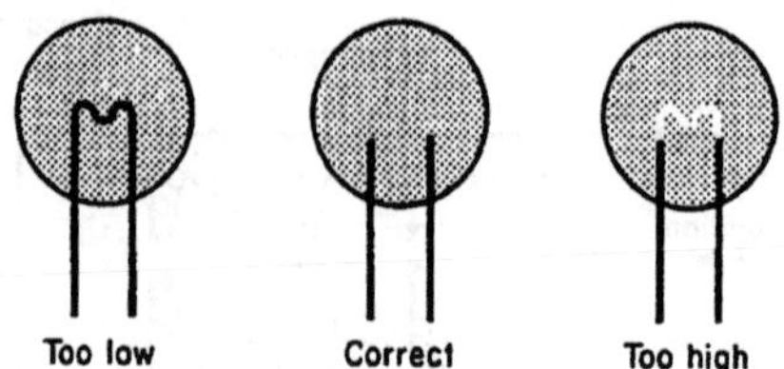

FIGURE 14.9 Disappearing filament.

HEAT SENSITIVE MATERIALS

These may change in condition or colour with increasing temperature. Crayons which change colour in 1 or 2 s are available and cover a temperature range of 65–670 °C. Also used are ceramic materials made in the form of cones or rings which become plastic or melt at specific temperatures. They may be used to indicate temperatures in kilns when firing or glazing pottery.

A range of 'Seger cones' is available, from which three would be chosen (Fig. 14.10). A is for a temperature slightly higher than that required, B is for the required temperature and C is for a temperature just below that required. The three cones are mounted on a tile and placed in the kiln. The operator increases the kiln's temperature until C has melted. At this point he reduces the heat input and continues to raise the temperature slowly until B begins to bend over. If the kiln becomes overheated, this is immediately indicated by A beginning to bend and B beginning to melt.

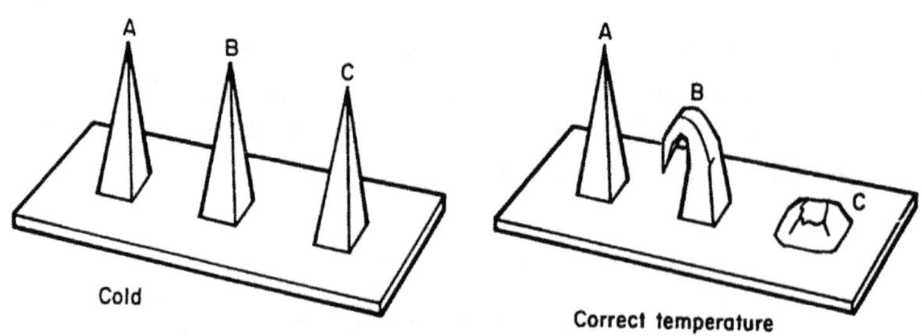

FIGURE 14.10 Seger cones.

AIR FLOW MEASUREMENT

Air velocities may be measured directly, by means of rotating vanes or spring-loaded flaps. These methods are particularly applicable where the air is passing through or emerging from ducts or grilles.

In the case of low air velocities, in rooms or other heated environments, the speed can be determined by the effect of the air movement on the temperature of an object placed in its path.

Kata thermometer

This is an alcohol-in-glass thermometer (Fig. 14.11) graduated from 35 to 37.8 °C. It measures the effective temperature and especially the velocity of air.

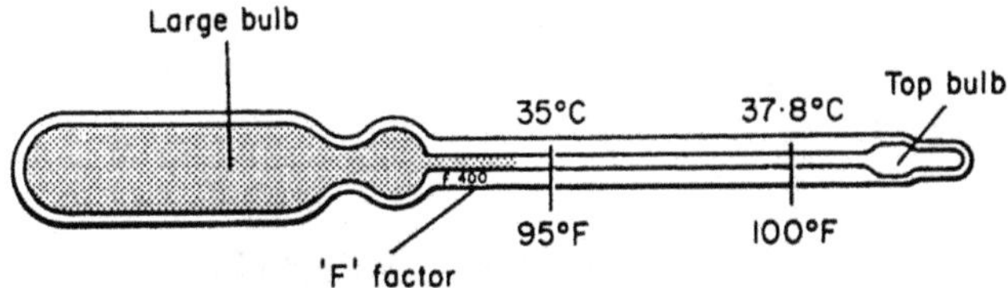

FIGURE 14.11 Kata thermometer.

The large bulb is warmed, in warm water, until the alcohol has expanded up the stem and a little has entered the top bulb. Then the thermometer is dried and placed in the area to be investigated. The time taken for the liquid to cool from 37.8 to 35 °C is noted. This procedure is repeated at least four times. The first reading is ignored and an average taken of the rest.

When the cooling time, in seconds, has been obtained, the cooling power can be calculated from the cooling time and the 'F' factor which is engraved on the stem of the thermometer.

$$\text{cooling power} = \frac{\text{`F' factor}}{\text{cooling time (seconds)}}$$

Using the nomogram in Fig. 14.12 the air velocity can be read off when the cooling power and the air temperature are known.

First, a line from the 'F' factor through the cooling time gives the cooling power. Then a second line from the cooling power through the air temperature gives the air velocity.

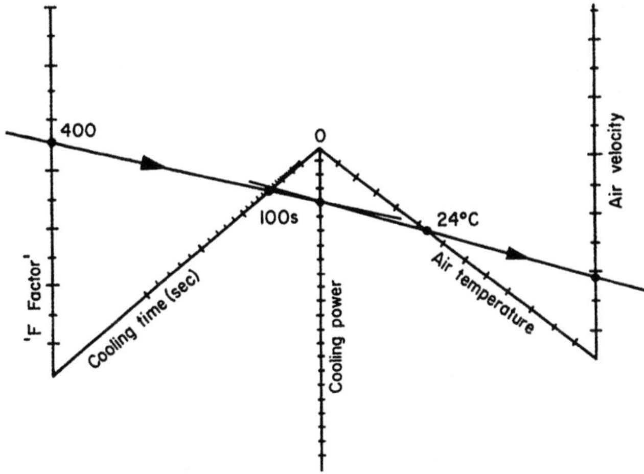

FIGURE 14.12 Nomogram for Kata thermometer.

Vane anemometer

This anemometer (Fig. 14.13) has a set of vanes rotating within an enclosure, similar to the vanes in the rotary gas meter. The vanes are turned by the air stream passing through them and the movement is conveyed by gearing to a dial. A typical instrument would have a range of velocities from 0.5 to 12.5 m/s and could be used to measure the speed of the air emerging from a grille in a warm air heating system.

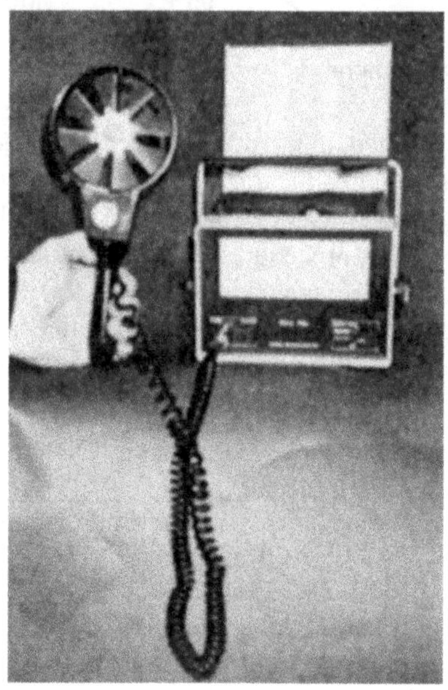

FIGURE 14.13 Vane anemometer.

Velometers

These instruments use a spring-loaded flap which is connected directly to a pointer (Fig. 14.14). The force exerted on the flap is proportional to the rate of air flow and the dial gives a direct reading in velocity. The ranges of measurement could be:

- 0–1.5 m/s,
- 0–15 m/s,
- 5–30 m/s.

Thermo-anemometers

This type of instrument uses temperature measurement to give a precise measurement of airflow velocities. A thermistor is held at a constant temperature and the amount of electrical energy, needed to sustain this temperature, is

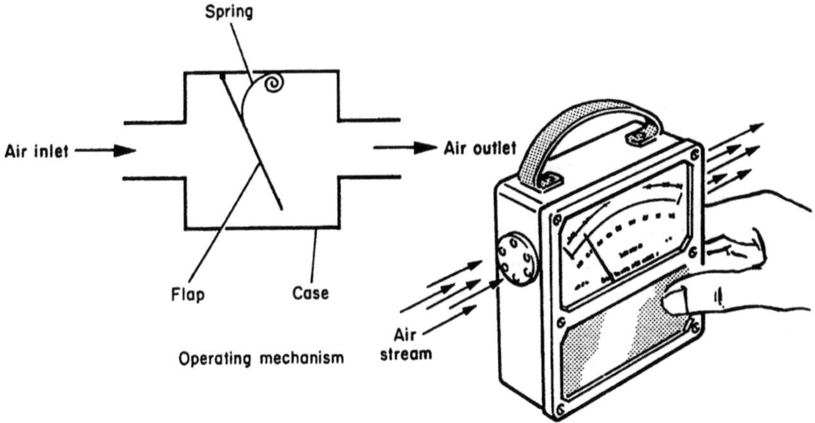

FIGURE 14.14 Velometer.

measured to calculate the air speed passing over the thermistor. Such a device is shown in Fig. 14.15.

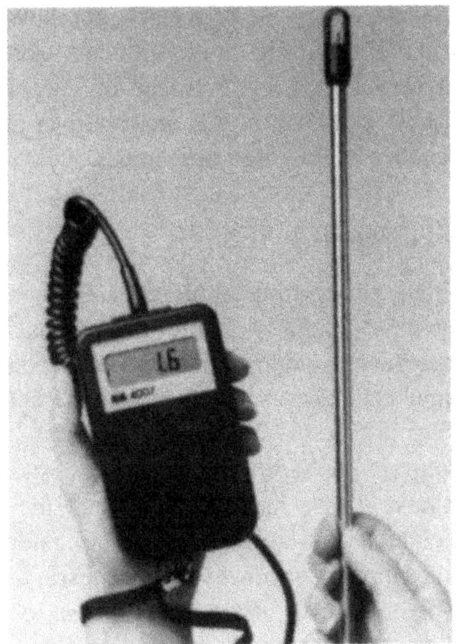

FIGURE 14.15 Airflow meter (Kane-May Ltd).

FLUE GAS ANALYSERS

Flue gas analysis

Flue gas analysis is carried out for a variety of reasons: to assess the combustion of an appliance in regard to its efficiency, its safe operation or when commissioning a new burner or after a service.

Manufacturers use flue gas analysers at the design stage to help to achieve maximum combustion efficiency from their appliances and to ensure safe operation.

The domestic service engineer will use an analyser to check the combustion of an open-flued or room sealed appliance to ascertain if a service of the appliance is required. This type of analyser is known as an appliance performance tester.

The commercial or industrial engineer will use a more sophisticated device to enable him to make adjustments to the air/gas ratio of a burner to obtain maximum efficiency during commissioning or after a service. These devices are known as combustion analysers. On burners with large gas rates, checking the combustion efficiency is carried out on a regular basis with some modern burners having an analyser continually monitoring the flue gases.

Flue gas analyser standards

A suite of new British Standards have been published in order to aid operatives in undertaking portable combustion gas analysis. BS 7967: 2005 Parts 1–3 were released on the 8th December 2005 with Part 1 dealing with reports of smells and fumes, Part 2 guidance on the use of portable combustion gas analysers and Part 3 guide to responding to the measurements obtained from portable combustion gas analysers. Part 4 was released in 2007 and gives a guide for using electronic portable combustion gas analysers as part of the process of servicing and maintenance of gas-fired appliances.

APPLIANCE PERFORMANCE TESTER

In order to ensure that gas-burning appliances are operating safely and at maximum efficiency it is generally recommended that an annual check or service be undertaken. Several organisations operate service contract schemes which include an annual check/service of the appliance and often the associated system.

To comply with current gas safety legislation, any such check/service must include several actions. These include checking the appliance gas rate (working pressure), correct ventilation, operation of safety devices etc. While checking/servicing the appliance, however, unnecessary disturbance or stripping down of equipment, controls, seals etc. can lead to unnecessary expense and inconvenience to the customer and represents inefficiency by the organisation.

In order to help eliminate this unnecessary disturbance/stripping down of an appliance, when checking or servicing central heating appliances, the Performance Tester (Fig. 14.16) has been developed.

The combustion performance of the appliance is tested by analysing the ratio of carbon monoxide (CO) to carbon dioxide (CO_2) from the products of combustion produced by the appliance. This ratio is displayed on a digital readout and if it is within the specified range then the engineer need not strip down

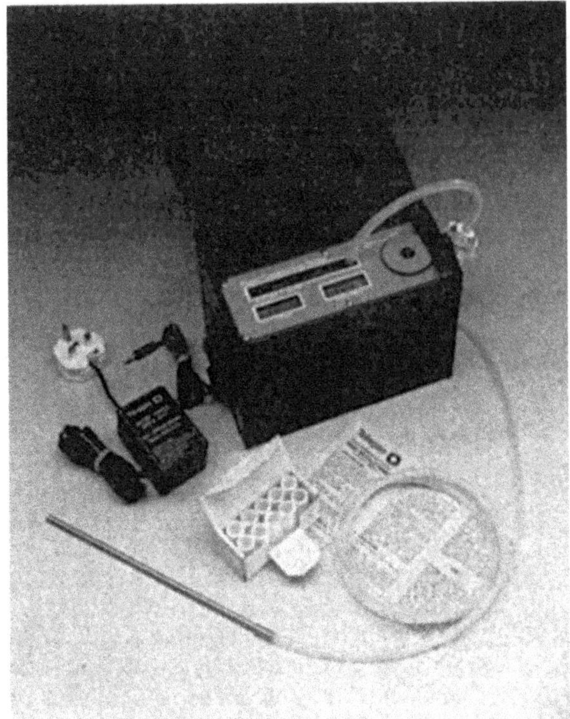

FIGURE 14.16 Telegan/Raco Appliance Performance Tester (Telegan Ltd).

and disturb equipment that is operating safely and efficiently. If, however, the reading is outside the specified range then the engineer will need to carry out certain actions – clean the burner, flueways etc. – to improve the combustion performance of the appliance.

After he has completed this work, the engineer must carry out a second performance test to ensure that the actions taken have brought about the necessary improvements in the appliance performance. Should the results of this second test indicate unsatisfactory performance then further appropriate action will have to be taken.

The tester is supplied and designed to be operated in its carrying case. The case has a rear compartment for storing the filter/water trap, hose probe, filter elements and a battery charger supplied with 240 V AC mains or 12 V DC versions. When conducting combustion tests the filter should be mounted vertically in the retaining strap on the side of the case with arrow pointing upwards to provide proper drainage of the moisture extracted from the incoming gas sample.

Figure 14.17 shows the position of the Tester's controls and displays. The function and operation of these are described in more detail in the User's Manual. Some modern appliance performance testers include a printer to enable the results of tests to be retained.

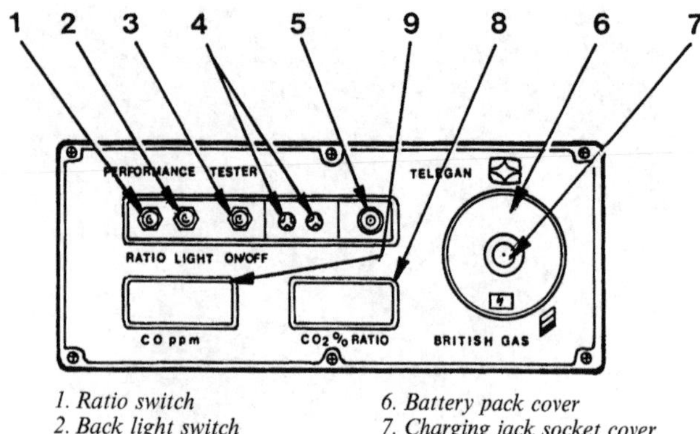

1. Ratio switch
2. Back light switch
3. On-off switcher
4. Exhaust from sensor
5. Recessed spigot for probe
6. Battery pack cover
7. Charging jack socket cover
8. Display 1
9. Display 2

FIGURE 14.17 Telegan Performance Tester.

COMBUSTION ANALYSERS

The equipment illustrated in Fig. 14.18 is based on fuel cell technology and advanced microelectronics. It will measure flue gas temperatures, CO and oxygen (O_2) levels. Carbon dioxide levels are derived from the O_2 reading and all can be displayed at the touch of a button. An electrically operated aspirator draws upon a continuous sample of gases and the probe can traverse the flue and avoid the risks inherent in 'one-shot' sampling.

Modern portable combustion analysers are small enough to fit in the hand (see Fig. 14.18a) and can carry out a wide variety of functions. The device shown is a combination of five instruments:

1. combustion analyser,
2. gas appliance tester,
3. ambient air monitor,
4. pressure meter,
5. temperature thermometer.

The instrument can measure O_2, CO, CO_2 and CO/CO_2 ratio, flue temperatures, surface temperatures or liquid temperatures. It can calculate combustion efficiency for five fuels and differential pressures, all displayed on a digital screen.

The data can be downloaded to a hand-held infrared printer to enable a permanent record to be retained.

LEAK DETECTION

In domestic premises, it is easily possible to prove whether or not a gas leak exists on the internal installation by testing for soundness with a 'U' gauge. But gas could still be leaking into the dwelling from:

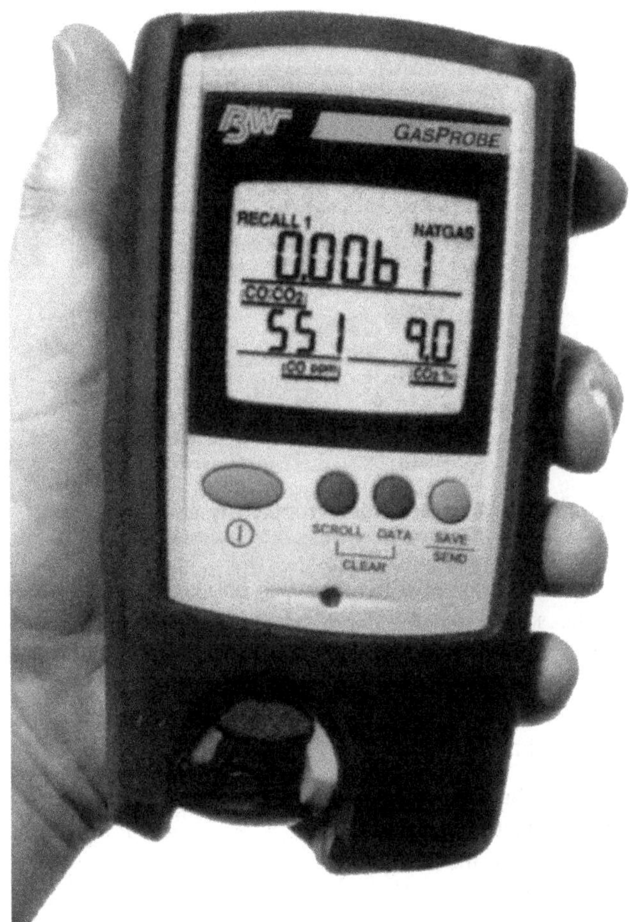

FIGURE 14.18(a) Modern hand-held multi-function combustion analyser (BW Technologies Gas Probe).

- the service pipe within the premises,
- the service pipe or the main outside the premises,
- adjoining properties.

Gas can travel underground for long distances following cracks, ducts and voids until it reaches the open air. It can enter premises by cracks in walls or around the point of entry of the various service pipes or cables. To confirm and locate the source of this leakage a detector is required. The need for confirmation arises from the fact that other substances can smell very much like escaping gas. Apparent gas escapes can be caused by:

- creosote (on newly painted fences),
- sewer gas (from open drains after a long dry spell),
- fuel vapours,
- decaying vegetation.

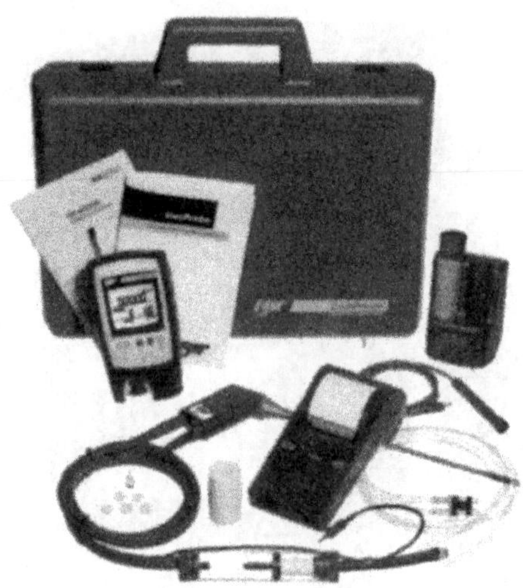

FIGURE 14.18(b) Accessories kit for Gas Probe analyser.

Over the years a number of different types of leakage indicators have been used. Of these, some, like the Short's leakage detector and the palladium chloride indicator, were more effective on town gas containing hydrogen but have become obsolete since the introduction of natural gas.

The types of detectors currently in use are all electrically operated. The methods employed are as follows:

- catalytic combustion,
- thermal conductivity,
- a combination of the two,
- flame ionisation.

Catalytic combustion

This instrument has a catalytic filament connected in one arm of a Wheatstone bridge. The filament is heated by an electric current supplied by dry batteries or re-chargeable cells.

When a flammable gas is introduced into the reaction zone its combustion raises the temperature and so increases the electrical resistance of the filament. This causes an imbalance in the bridge circuit producing a deflection on the meter which indicates the percentage concentration of gas.

Because combustion cannot take place at high gas concentrations, these instruments operate normally in the 0–100% lower explosive limit (LEL) range. Lower explosive limit of 100% corresponds to 5% gas in air.

Thermal conductivity

This instrument also has a filament, heated by current from a dry battery and connected into one arm of a Wheatstone bridge.

In this case, however, the filament is non-catalytic and the gas sample does not burn.

The presence of gas lowers the filament temperature, so unbalancing the bridge circuit. The meter indicates the percentage concentration of gas.

These instruments can detect gas concentrations of up to 100% gas in air.

Flame ionisation

In this type, the gas sample is introduced into a hydrogen flame. The ions resulting from the hydrocarbons in the sample are collected at an electrode. The electric current carried by the flow of ions is approximately proportional to the concentration of gas in the sample and is shown on the meter.

The instrument is extremely sensitive and can be used to detect very small concentrations down to a few parts per million. It requires a cylinder of hydrogen and tends to be heavier and larger than the other devices. The presence of the flame makes it unsuitable for use in enclosed spaces where there would be risk of an explosion.

Because of its characteristics, this type of detector is largely used by the Engineering Department for carrying out district leakage surveys. For this task the instrument is mounted in a patrol vehicle, although portable models are available.

Gascoseeker

This is a portable, multi-purpose gas detector which was developed to replace the diverse range of instruments previously in use. It generally operates on natural gas but specially calibrated instruments can be obtained for use on propane, butane etc. Any instrument used on a gas for which it is not calibrated will give misleading results.

The instrument shown in Figs 14.19 and 14.20 is the Mk 1 Gascoseeker. It has a probe and 1.5 m of flexible hose fitted to the inlet union on the right and an aspirator bulb fitted on the left. Sintered bronze flame arrestors are fitted into the inlet and outlet assemblies.

A water trap with a transparent cover or a hydrophobic filter is provided to prevent the free entry of water. A cotton wad filter, to remove dust etc., is fitted in the inlet of the instrument.

Pellistors are used to sense the gas. They are connected into Wheat-stone bridge circuits which are powered by six 'D size' batteries. Both conductivity and catalytic circuits are included, the circuit in use depends on the range of concentrations being used.

The instrument has three measuring ranges:

1. 0–100% gas in air,
2. 0–100% LEL (or 5% gas in air),
3. 0–10% LEL (or 0.5% gas in air).

The scale required is selected by the rotary switch and is indicated by the appropriate light on the scale plate. The battery condition is also indicated on the lower scale with red and green sectors.

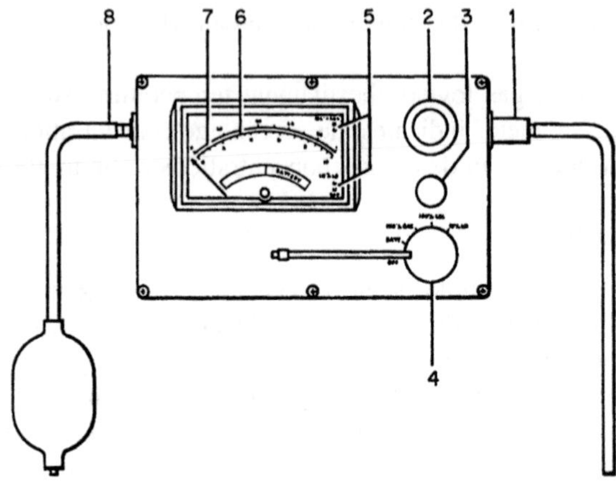

1. Probe connection
2. Water trap
3. Zero adjustment switch (10% L.E.L. range)
4. Range selector switch
5. Range indicator lights and battery test light
6. 10% L.E.L. scale
7. % L.E.L. and % Gas scale
8. Inspirator connection

FIGURE 14.19 Mk 1 Gascoseeker.

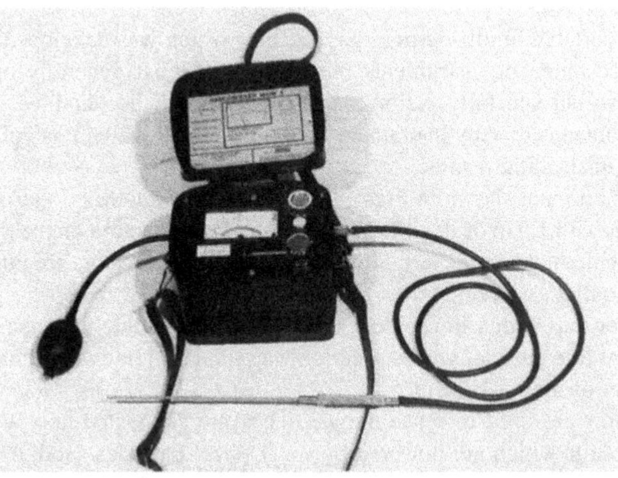

FIGURE 14.20 Mk 1 Gascoseeker with sampling probe and aspirator bulb assembled.

The rotary switch has a locking bar which prevents the cover being replaced while the instrument is switched on. A zero adjustment for the 0–10% LEL range completes the controls.

The Gascoseeker is approved by the British Approval Service for Electrical Equipment in Flammable Atmospheres (BASEEFA) as a safe instrument for use in enclosed spaces, providing it is kept in the PVC carrying case while in use. It is, however, not safe to be used in the presence of acetylene or excessive oxygen, that is near containers of compressed or liquid gases.

Instrument checks

Like all instruments, the Gascoseeker should be checked before use each day to ensure that it will operate satisfactorily and give accurate readings. The checks to be made are as follows.

1. *Leakage.* Check the aspirator bulb, tube and sampling probe for leaks. Ensure that the non-return valve in the bulb is working satisfactorily.
2. *Batteries.* Raise the bar on the range switch and turn the switch clockwise until the light marked BAT comes on. The meter needle should enter the green sector on the scale. If it stays in the red sector, change the batteries.
3. *Zero.* Set the range switch to 100% GAS and aspirate the bulb four or five times with the probe in fresh air. The pointer should be within a small division of the zero.

Turn the switch to the 100% LEL range and check the zero by again aspirating the bulb four to five times.

Finally turn to the 10% LEL range, aspirate and adjust the zero by the adjusting knob.

Sampling

Set the instrument to the 100% Gas range.

Place the end of the probe in the atmosphere to be tested, aspirate the bulb four or five times and read the upper scale as 100% gas in air while the bulb is still expanding.

If the reading is 5% gas or below, select the 100% LEL range and repeat the procedure, reading the upper scale as 100% LEL.

If the reading is then 10% LEL or below, select the 10% LEL scale and repeat the measurement, reading the lower scale.

After taking readings, purge the instrument with fresh air and switch off.

When sampling, care must be taken to prevent water entering the instrument. Water is likely to be found at the bottom of a bar-hole made in the road or pavement. If the bulb takes longer than usual to expand, it may be because water is being drawn up the probe. If water is seen to enter the trap, it must immediately be removed, the trap dried out and a new filter inserted.

The Mk 2 Gascoseeker is illustrated in Figs 14.21 and 14.22. The instrument is a development of the Mk 1 and has two measuring ranges:

1. 0–100% gas in air,
2. 0–100% LEL (or 5% gas in air).

On the gas in air range the operating principle of the detection system is thermal conductivity. The % LEL uses the operating principle of catalyst reaction. The instrument has a built-in pump which is brought into operation when the instrument is switched on. Readings are given on an LCD display providing both analogue and digital representation of the gas being measured. This display

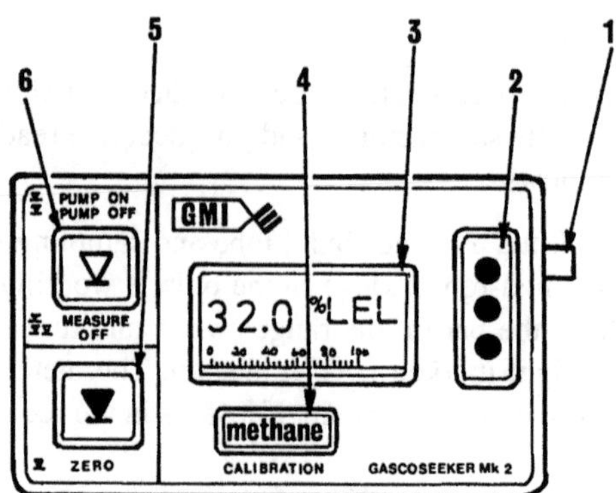

1. Sample line connector
2. Socket connection for 'microcal' (calibration check)
3. Analogue and digital display
4. Calibration display (type of gas instrument is calibrated to measure)
5. Zero button
6. Measure (on/off) button

FIGURE 14.21 Mk 2 Gascoseeker.

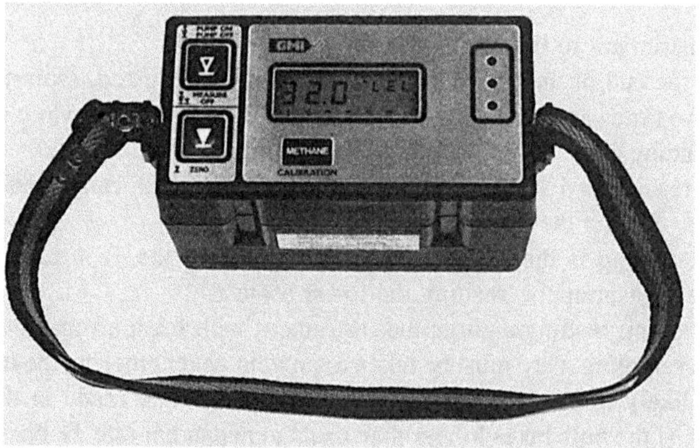

FIGURE 14.22 Mk 2 Gascoseeker with digital and analogue display showing a reading of 32% LEL.

is illuminated under poor lighting conditions. The instrument performs a self check and energises all the segments in the display window in a few seconds, prior to being ready for use.

The sample tube is connected to the right-hand inlet nozzle and the sample passes through a charcoal filter and a hydrophobic filter to prevent the ingress of foreign bodies and water.

The precautions which must be taken when dealing with gas escapes are to be found in Volume 2, Chapter 3.

Appendix | one

SI Units

The standards describing the essential features of the International System of Units (SI) are BS 5555: 1981 and ISO 1000: 1981. Both standards are entitled: SI units and recommendations for the use of their multiples and of certain other units.

The International System of Units is founded on the following seven base units:

Quantity	Name of base SI unit	Symbol
length	metre	m
mass	kilogram	kg
time	second	s
electric current	ampere	A
thermodynamic temperature	kelvin	K
amount of substance	mole	mol
luminous intensity	candela	cd

In addition to the base units, there are two units which may be regarded as base or derived units, they are called 'supplementary units' and are shown below:

Quantity	Name of supplementary SI unit	Symbol
plane angle	radian	rad
solid angle	steradian	sr

Other units are derived from the base and supplementary units, they are termed 'derived units' and are shown below:

Quantity	Special name of derived SI units	Symbol	Expressed in terms of base or supplementary SI units or in terms of other derived SI units
frequency	hertz	Hz	1 Hz = 1 s^{-1}
force	newton	N	1 N = 1 kgm/s^2
pressure, stress	pascal	Pa	L Pa = 1 N/m^2
energy, work, quantity of heat	joule	J	1 J = 1 Nm
power	watt	W	1 W = 1 J/s
electric charge, quantity of electricity	coulomb	C	1 C = lA.s
electric potential, potential difference, tension, electromotive force	volt	V	1 V = 1 J/C
electric capacitance	farad	F	1 F = 1 C/V
electric resistance	ohm	Ω	1 Ω = 1 V/A
electric conductance	siemens	s	1 S = 1 Ω^{-1}
flux of magnetic induction, magnetic flux	weber	Wb	1 Wb = l V.s
magnetic flux density, magnetic induction	tesla	T	1 T = 1 Wb/m^2
inductance	henry	H	1 H = 1 Wb/A
Celcius temperature	degree Celcius	°	1 °C = 1K
luminous flux	lumen	lm	1 lm = 1 cd·sr
illuminance	lux	lx	1 lx = 1 lm/m^2

Appendix | two

Conversion Factors

The standard which deals with interconversion from one unit of measurement to another for quantities which are in general use in the gas industry is BS 350 Conversion factors and tables.

Part 1: 1974 Basis of tables. Conversion factors incorporate a 1983 amendment and Part 2: 1962 Detailed conversion tables had a supplement added in 1967 giving additional tables for SI conversions. Part 2 was withdrawn in 1981 but the supplement is still valid and has been retained.

CONVERSIONS IMPERIAL TO METRIC UNITS

1 inch	= 25.4 mm	= 2.54 cm
1 foot	= 304.8 mm	= 0.3048 m
1 yard	= 914.4 mm	= 0.9144 m
1 mile	= 1609 m	= 1.609 km
1 square	= 645.2 mm^2	= 6.452 cm^2
1 square foot	= 929.0 cm^2	
1 square	= 0.8361 m^2	
1 acre	= 4047 m^2	= 0.4047 ha
1 cubic inch	= 16390 mm^3	= 0.01639 dm^3
1 cubic foot	= 0.02832 m^3	= 28.3 dm^3
1 cubic yard	= 0.7647 m^3	
1 pint	= 0.5683 l	= 568.3 ml
1 quart	= 1.137 l	
1 gallon	= 4.546 l	= 4.546 kg of water
1000 gallons	= 4.546 m^3	
1 grain	= 0.06480 g	
1 ounce	= 28.35 g	
1 pound	= 453.6 g	= 0.4536 kg
1 ton	= 1016 kg	= 1.016 tonne
0.1 inches water gauge	= 0.25 mbar	
1.0 inch water gauge	= 2.5 mbar	
1 atmosphere	= 1.01325 bar	
British thermal unit (Btu)	= 1055 J	
1000 Btu	= 1.055 MJ	

1 therm	= 105.5 MJ	
1000Btu/h	= 0.2931 kW	= 1.055 MJ/h
1 horsepower	= 745.7 W	= 2.685 MJ/h
1 Btu/s ft^3	= 0.037 96	
1 Btu/ft^3(dry)	= 0.037 23	
1 Btu/lb	= 2326 J/kg	
1lbf	= 4.448 N	

METRIC TO IMPERIAL UNITS

1 millimetre	= 0.03937 in	
1 metre	= 39.37 in	= 3.281 ft = 1.094 yd
1 kilometre	= 0.624 mile	= 1094 yd
1 square millimetre	= 0.001550 in^2	
1 square metre	= 1.196 yd^2	
1 acre	= 119.6 yd^2	= 0.0247 acre
1 hectare	= 11960 yd^2	= 2.471 acre
1 cubic millimetre	= 0.000061 in^3	
1 cubic decimetre	= 61.02 in^3	= 0.035 31 ft^3
1 cubic metre	= 35.31 ft^3	= 1.308 yd^3 = 220 gal
1 millimetre	= 0.001760 pint	
1 litre	= 1.760 pint	= 0.2200 gal
		= 0.035 ft^3
1 gram	= 0.035 27 oz	= 15.43 grains
1 kilogram	= 2.205 1b	
1 tonne	= 2205 lb	= 0.9842 ton
1 millibar	= 0.4 in wg	
1 bar	= 14.50 lbf/in^2	
1 joule	= 0.000 947 8 Btu	
1 megajoule	= 947.8 Btu	
100 megajoule	= 0.9478 therm	
1 kilowatt	= 3.6 MJ/h	= 3412 Btu/h
		= 1.341 horsepower
1 kilowatt hour	= 1 kWh	= 3.6 MJ
1 megajoule per hour	= 0.2778 kW	= 947.8 Btu
1 MJ/m^3	= 26.34 Btu/ft^3	
1 kgf	= 2.205 lbf	

PREFIXES

The following prefixes, with significance, name and symbol as shown below, are used to denote decimal multiples or submultiples of (metric) units. These prefixes developed in conjunction with the metric system are now authorised as 'SI prefixes'.

To indicate multiples

$\times 10^{18}$	exa	E	=	1000	000	000	000	000	000
$\times 10^{15}$	peta	P	=	1000	000	000	000	000	
$\times 10^{12}$	tera	T	=	1000	000	000	000		
$\times 10^{9}$	giga	G	=	1000	000	000			
$\times 10^{6}$	mega	M	=	1000	000				
$\times 10^{3}$	kilo	k	=	1000					
$\times 10^{2}$	hecto	h	=	100					
$\times 10$	deca	da	=	10					

Examples

gigajoule	(GJ)
megawatt	(MW)
kilopascal	(kPa)
hectonewton	(hN)
decavolt	(daV)
decigram	(dg)
centiampere	(cA)
millisecond	(ms)
micrometre	(am)

To indicate submultiples

$\times 10^{-1}$	deci	d	=	0.1					
$\times 10^{-2}$	centi	c	=	0.01					
$\times 10^{-3}$	milli	m	=	0.001					
$\times 10^{-6}$	micro	μ	=	0.000	001				
$\times 10^{-9}$	nano	n	=	0.000	000	001			
$\times 10^{-12}$	pico	p	=	0.000	000	000	001		
$\times 10^{-15}$	femto	f	=	0.000	000	000	000	001	
$\times 10^{-18}$	atto	a	=	0.000	000	000	000	000	001

Those multiples in most common use are mega, kilo, milli and micro. Instead of writing lots of 0s or noughts, it is easy to use the powers of ten given in the first column.

For example, $1000 = 10 \times 10 \times 10$ or 10^3. The superscript 3 is the same as the number of noughts in 1000. Similarly,

$$0.001 = \frac{1}{10 \times 10 \times 10} \text{ or } \frac{1}{1000}, \text{ that is } 10^{-3}$$

The superscript -3 is one more than the number of noughts after the decimal point 0.001.

Index

Absolute zero, 3, 120
Acceleration, 115
Adhesion, 156, 157
Aerated flame, 26
Aeration, 9
  air shutter, 87, 88
Air blast burners, 83
Air flow measurement, 447, 448
Air movement, 126
Air required for combustion, 9, 14, 19, 44
Air temperature, 126
Aldehydes, 24, 25, 33
Allen keys, 399
Alloys, 369
Alternating current, 255
  frequency, 256
  power factor, 287
  rectification, 273
  three-phase, 259
  voltage, 257
Altitude, effect on pressure, 167
Aluminium, 45, 365, 368
Ambient temperature, 296
Ammeter, 275
Anemometer, vane, 450
  thermo, 450
Aneroid barometer, 154
Aneroid gauge, 161
Angular velocity, 114
Annealing, 383
Anode, 243
Anodising, 382
Antimony, 368
Appliance performance testers, 451, 452
Approved Document L1, 308
Abestos, 376
  likely sources of, 377
  regulations, 376
  removal, 377
  working precautions, 376
Atmospheric burners, 99, 100
Atmospheric pressure, 151
  effect of altitude, 167
  measurement of, 156
  standard conditions, 152
Atmospheric sensing devices *see* Oxygen depletion systems
Atom, components of, 14, 15, 16
Atomic structure, 15, 16, 240
Automated reading system, 222
Automatic changeover (LPG cylinders), 67, 68, 75, 203
Automatic touch hold, 278
Axial flow turbine meter, 214

Baffles, 91
Bar (pressure unit), 151
Barometer, 153
  aneroid, 156
Battery, 242
Beams, stresses in, 135
Bending machines, pipe, 422
Bending moments in joists and beams, 135
Bending spring, 404
Bimetal strip thermostat, 336
Bimetallic devices, 321
Blowlamp, 451, 469
  fuels, 231
Blow-off, 31
Blueing, 382
Boilers, 302, 308
Bourden tube gauge, 161
Boyle's Law, 207
Bradawl, 401
Brass, 369
British Gas multimeter, 276
British thermal unit, 111
Bricks, 375
Brittleness, 374
Brushes, 410
Building blocks, 375
Burner
  body tapered, 84, 90, 92, 108
  components, 83
  examples of, 106
  head, 90
  location, 102
  ports, 84, 90, 91, 93
  pre-mix, 98
Butane, 10, 42–51, 63, 64, 77, 78
Butane/air mixture, 10
  combustion equations, 18, 19, 20
  constituents, 20, 21, 23
  oxygen requirement, 20
  products of combustion, 22, 23
  properties of, 12
  Butane cabinet heater, 201
  Butylene, 42, 43
Byelaws for water pipework, 368

Calorific value, 6, 12
Calorifiers, 304, 305, 368
Capacitive reactance, 582, 583
Capacitor, 263
Capillary attraction (Capillarity), 157
Carbon deposits, 33
Carbon dioxide, 10
  flame/atmospheric effect, 45
Carbon monoxide, 6
  flame/atmospheric effect, 45
  poisoning, 34
Carbon monoxide: carbon dioxide
  ratio, 35
Carbon steel tools, 393, 399
Carburising, 39
Cartridge tools, 428
Case hardening, 384
Cast iron, 365
Catalytic combustion burner, 100
Catalytic rich gas, 11, 12
Cathode, 243

Cathodic corrosion protection, 381
Causes of incomplete combustion, 40
Celsius scale, 118–120
Cement, 376
Centigrade scale, 118, 119
Central heating unit burner, 109
Centre of gravity, 143
Charger unit, 233, 234
Charles' Law, 209
Chemical symbols, 4
Chisels, 389, 420
Choke (electrical), 286
Chromium, 368
Circuit breakers, 271
  miniature, 271, 273
Circular saws, 427
Clasp knife, 409
Clothing, protective, 436
Cock, 311
  faults, 313
Coefficient of cubical expansion, 49
Cohesion, 156
Combustion, 14
  analyzers, 451
  equation for, 17
  flameless, 100
  incomplete, 33
  LPG, 45
  products of, 21, 22, 23, 105
  standards, 35
Comfort conditions, thermal, 125, 310
Commercial butane, 50, 51, 71
Commercial propane, 50, 51, 71
  combustion equations 20, 21
  constituents, 10
  oxygen requirements, 20
  products of combustion 22, 23
  properties, 12
Commutator, 258
Compounds, 16
Concrete, 375
Condensation, 125
Conduction heat loss, 292
Conduction of heat, 288
Conductors, electrical, 240
Conservation of energy, 112
Constant pressure regulators, 184
  compensated, 188
  need for compensation, 187
Constituents of gases, 10
Containers (LPG), 31, 53, 81
Contents gauge, 72
Continuity bond, temporary, 285, 453
Control of pressure, 181
  constant pressure regulators, 184
  district variations, 181
  high-to-low pressure regulators, 192
  installing regulators, 194
  LPG regulators, 194
  methods, 128
Control systems, 357
  electrically operated, 358
  pressure operated, 357
Controlled atmospheres, 38
Controls
  in domestic appliances, 311
  multifunctional, 322
Control of Asbestos Regulations 2006, 376
Convected heat flow, 296
Convection, 288, 293
  currents in room, 294
  forced, 293, 301
  in liquid, 293
Cooker
  hotplate burner, 106, 107
  oven thermostat, 334
  temperatures, 121
Copper, 368
Corrosion, 378
  acid, 379
  atmospheric, 378
  galvanic, 379
  protection from, 380
COSHH Regulations, 429
Coulomb, 243
Crimping tools, 412
Critical pressure, 46, 48
Critical temperature, 46, 48
Cross-bonding to earth, 285
Cubic feet, 5, 6
Current, electric, 241
  induced magnetically, 254
  magnetic effect, 252
Cylinder, hot water storage, 295, 304
Cylinders, LPG, 45, 47, 53, 55, 64
  automatic changeover, 67, 68, 75
  design and construction, 55
  regulators, 46, 54, 67, 74, 75, 78
  valves, 57
  vapour offtake, 58, 61, 65

Deflectors, 84, 103
Density, 3, 6, 7
Department of Trade and Industry, 6
Descaling solutions, 386
Deuterium, atomic structure, 15
Dew point, 50, 125
  butane/air, 51
Dielectric, 263
Dies, 416, 421
Diffusion, 2, 3, 4
Digital multi-meter, 414
Direct current, 243
Discharge rates (pipe), 178
Distribution of electricity, 259
Domestic internal installation, electricity, 260
Drilling tools, 392, 420
Drills and bits, 420, 426
Dry meter, 216
Dry pressure gauges, 161
Ductility, 363
Dust, explosive, 25
Dust sheets, 411
Dusting brushes, 411
Dynamics, 130
Dynamo, 258

E6 meter, 228
Earthing, 280
Elastic limit, 135
Elasticity, 363
Electric capillary joint maker, 413
Electric drill, 426
Electric motors, 254, 264

Electric shock, 279
Electrical circuits
resistances in series, 247
resistances in parallel, 248
Electrical energy, 111, 240–247
Electrical test meters, 277
Electrical power, 244
Electrical supply coding, 259
Electrical tools, 412
Electricity, 240
distribution, 259
domestic internal installation, 260
generation, 258
Electrolyte, 243
Electrolytic corrosion, 379
Electromotive force, 243, 247
Electrons, 14, 15
flow of, 242
Electronic pulse/coil ignition, 355
Electro-plating, 383
Elements, 16
Emergency/meter control, 312, 313
Emissivity, 299
Energy, 111
conservation of, 112
electrical, 240
forms of, 111
Ethane, 5, 10
Evaporation, 125
Excess air, 24
Explosion, 25

Fahrenheit scale, 118, 119
Families of gases, 8
comparison of properties, 12
Family 1 gas, 8
Family 2 gas, 8
Family 3 gas, 8
Farad, 264
Fault diagnosis, 101, 264, 359
Fault voltage-operated circuit breaker, 282
Ferrous metals, 365
Filament ignition/igniter, 37, 350, 352
Files, 393, 395, 421
Filling connection (LPG tank), 71
Fire-resistant mat, 406
First aid kit, 412
Fixed aeration, 87
Flame front, 28
Flame kernel, 38
Flame lift, 30, 31, 32
Flame protection devices, 320
flame rectification, 332
oxygen depletion systems, 316
photo-electric, 323
thermal backflow prevention device (TTB), 319
thermo-electric, 319
vapour pressure, 312
Flame speed, 10
and aeration, 24, 30
Flames, 26, 29–38
fault diagnosis by, 101
oxidising effect, 38, 39
post-aerated, 26
pre-aerated, 26
reducing effect, 38
Flammability, 24
limits, 10, 44, 49, 50
Flash tube, 351
Fleming's right-hand rule, 254
Flow measurement, 448
Flue brushes, 411
Flue gas analysis, 451
Flue Gas Analyser Standards, 452
digital multi-meter, 277
series 2 model, 278
Fluxes (jointing), 385
Force, 116
Forced-draught burners, 83, 99
Frequency, 256
Fuse, 231
Fusion of ice, latent heat of, 124

Galvanising, 369, 383
Gas bills, 238
Gas fire burner, 90, 101, 102, 108
Gas laws, 207
Gas measurement *see* Measurement of gas
Gas modulus, 9
Gas pressure, 169
Gas quality regulations, 181
Gas rate (appliance), 101
measurement of, 175
Gas Safety (Installation and Use) Regulations 1998, 102, 284, 285, 312, 315, 327
Gas Transmission System, National, 169
GASCARD, 233, 234
Gascoseeker, 506
Gases, 1
comparison of properties, 12
constituents of, 10
families of, 9
molecules in, 2
toxic, 5, 6, 13
Gimlet, 401
Goggles, 413
Graham's law of diffusion, 3
Grinding machine, 427

Hammers, 388, 420
Handling, 141
Hardening metals, 384
Hardness, 363
Harmonized cable core colors, 259
Health and Safety Executive, 82
Heat, 120
Heat energy, 111
rates, 127
Heat exchangers, 201–304
Heat loss from buildings/walls, 306
Heat sensitive materials, 448
Heat transfer *see* Transfer of heat
Heat treatment of metals, 383
Heating systems
circulating pressure, 165
hydrostatic pressure, 163
water pressure, 165
Henry, 286
Hertz, 256
High explosive limit, 10
Home energy rating, 309
Home Information Pack, 308
Hooke's Law, 135
Hot dipping, 383

Hot wire filaments, 37
How is Gas Delivered, 169
Humidity, 127
Hydrogen
atomic structure, 14, 15
flame/atmospheric effect, 38
Hydrostatic pressure, 163

Ice, latent heat of, 124
Ignition, 36
Ignition devices, 350
Ignition system, direct spark, 347
Ignition temperatures, 12
Impedance, 281, 286, 287
Inclined gauge, 159
Incomplete combustion 30, 40
chilling the flame, 33, 35
oxygen starvation, 35
products of, 33
Indexes, meter, 218
cubic feet, 224
cubic metres, 224
dial-type, 225
liquid crystal display (LCD), 225
older types, 220, 225
reading, 218
R5, 226
Induction, magnetic, 254
Inductive reactance, 286
Inductor, 286
Inferential meters, 212
Injector, 27, 85, 86
comparison of characteristics, 86
eccentric, 85
orifice size vital, 86
Institution of Gas Engineers, meter specification, 217
Gas Industry Standard for Polyethylene pipes and fittings, 373
Insulation (electrical), 241
Intermittent pilot systems, 347
Internal gauze, 84, 91, 104
Inverse square law, 300
Ions, 242
Iron, types of, 365
Itron unit, 227

Jets, pre-aerated, 96
Joint maker, electric capillary, 413
Joule, 112, 117

Kata thermometer, 449
Kelvin, 120
Kilogram, 113
Kilowatt, 117
Kilowatt-hour, 6, 112, 117, 238, 273
Kinetic energy, 111
Kinetic Theory, 2
Kirchhoff's Law, 251

Lacquering, 381
Ladders, 428
Latent heat, 123, 124
Lead, 368
Leak detection, 455
catalytic combustion, 456
flame ionisation, 456
thermal conductivity, 456
Levers, 132
Lifting, 141
kinetic, 142, 143
Light emitting diode (LED), 227
Lighting-back, 29, 30
Limit of proportionality, 135
Likely sources of asbestos, 377
Lint, composition of, 103
Linting, 103
Liquefied Petroleum Gas (LPG), 41
air systems, 51
boiling points, 46, 48
calorific value, 50
chemical properties, 44
coefficient of cubical expansion of liquid, 49
combustion, 45
containers, 31, 53, 81
dew point, 50
distribution, 51, 53, 54
flammability, limits of, 49
latent heat, 48
odorisation of, 49
physical properties, 46
pressure standard, minimum, 181
regulators, 194
relative density, 48
sources, 41
special grades, 44
specific volume, 48
typical properties, 44, 45
vapour pressure, 46
what is? 41
Liquefied Natural Gas (LNG), 8
Liquid crystal display (LCD), 225
Liquids, 2
Low pressure cut-offs, 316
Lower explosive limit, 10
LPG Association, 79
Luminous flame, 26

Magnetic flux, 252
Magnetism, 251
Main cock, 312
Mains ignition transformer, 355
Malleability, 363
Malleable iron, 367
Manufactured gases, 8
properties of, 12
Masonry drill, 326, 394
Mass, 113
Materials, properties of, 361
Matrix burner, 97
Measurement of appliance gas rates, 175
Measurement of gas, 207
Measuring devices, 439
Mechanical advantage, 138–140
Mechanical energy, 112
Mechanics, 129
Melting points of metals, 121
Meniscus, 156
Metals, 364
Meter control cock, 312
tools, 387
Meter (electrical), 274
Meter (gas), 212
faults, 237
Index (cubic feet), 175, 177
Index (cubic metres), 175, 177
reading indexes, 218
regulations for, 235
test conditions, 236
test dial, 193, 194

Meter (gas) (*continued*)
test drum, 193, 194
see also names of gas meters
Metering orifices, 95
Methane, 10, 13
molecular structure, 15
Micronic unit, 227
Millibar, 151
Mirror, 412
Mixing tube, 27, 84–109
Mixtures, 16
Modifing appliances, 104
Modulation of gas supply, electronic, 349
Modulus of elasticity, 135
Molecules, 1, 2, 3, 4, 15
Moment of a force, 130
Multifunctional controls, 347–350
Multimeter, 275, 413

Naphtha, 8, 12
Natural draught burners, 83
Natural gas, 10, 11, 12
combustion equations, 16, 23
constituents, 10, 42
oxygen requirements, 19
pressure standard, minimum, 181
products of combustion, 20–22
properties, 12
Natural rubber, 45, 78
Neat flame, 26
Neon main tester, 413
'Neoprene', 374
Neutrons, 14, 15
Newton, 117
Nickel, 369
Niting, 312
Nitrogen, 5, 9, 10
Non-aerated flame, 26
Non-ferrous metals, 365
Non-metals, 372
Normalising metals, 384
Nucleus, atomic, 14
Nylon, 373

Odour, 5, 49
Ohm, 245
Ohmmeter, 275
Ohm's law, 245
Optical pyrometer, 447
Orifice, flow of gas through, 174
Origins of Gas Pressure, 169
Oxidation, 38
Oxygen, 4, 5, 6, 9, 10
flame/atmospheric effect, 39
requirement of gases, 19, 20
Oxygen depletion systems, 326
operation, 327
servicing, 328

Paint brushes, 410
Paints and lacquers, 381
metallic, 382
Parkerising, 382
Pascal (unit), 174
Pascal's apparatus, 151
Peart pass stopper, 418, 419
Percussion tools, 427
Performance testers for appliances, 451
Phase relationships, current and voltage, 285, 286
Piezo-electric effect, 240
Piezo-electric igniter, 396, 397
Pilot flames 36, 390
pre-aerated, 96
Pin vices, 418
Pipe cutters, 404
Pipes, gas flow in, 171
diameters, 173
estimating sizes, 177
factors in pressure loss, 172
Pipe slice, 404
Pipe vices, 417
Pipework, internal installation, 177
Pitot tube, 212
Plaster, 376
Plastics, 373
Platinum, 369
Platinum filament, 37
Pliers, 396
Poles formula, 174
Polyethylene (PE), 373, 381
Polymer molecules, 374
Polystyrene, 374
Polytetrafluoroethylene (PTFE), 374
tape, 46, 374
Polyvinyl chloride (PVC), 374
Ports, 27
Positive displacement meters, 214, 215
Post-aerated burner, 96, 97
with retention flame, 31
Post-aerated flames, 26
Potential difference, 244–246, 271
Potential energy, 111
Power, 117
Power factor (alternating current), 225–257
Power tools, 432
insulated, 424
Pre-aerated burner, 83, 101, 103
bar burner, 84
ring burner components, 83
with retention flame, 33
Pre-aerated flames, 26
flame front, 28, 29
inner cone, 27
outer mantle, 28, 29
reaction zone, 28, 29
temperatures of, 28, 29
zones of, 28
Pre-mix burner, 90, 98, 99, 107, 108
Prepayment meters, 230
quantum, 231–232
Pressure absolute, 154
atmospheric, 151
control *see* Control of pressure
effect of altitude, 167
in fluids, 148
loss in pipes, 179
measurement of, 177
recorders, 162
units, comparative, 151
Pressure regulator, 56, 75, 78
Pressures, lift and reversion, 32
Primary air, 27–28, 30, 40
Primary air port, 83, 87, 89, 103
Propane, 5, 8, 9, 10, 12, 42–82
see also Commercial propane
Propane torch, 405, 421

Propylene, 5, 10, 42, 43
Protective clothing, 436
Protective multiple earthing, 280
Protons, 14, 15
Pulley systems, 139
Punches, 489
Pyrometers, 118, 446, 447

Quantum meter, 231
Quantum system, 232
 charger unit, 234
GASCARD 233, 234
SERVICE CARD, 234

R5 Index, 221
 anti-tamper device, 234
 automated reading system, 222
 fischer socket, 222
 reed switch, 222
 security label, 217
 telecom socket, 222
Radian, 114
Radiant burners, 99
Radiant heat, 127
 flow, 299
Radiant heating appliances, 297, 298
Radiation, 288, 297
 environmental, 126
Radiation pyrometer, 446
Radiator, hot water, 305
Rat-tail flames, 97
Reaction zones 28, 29, 31
Rectifiers, 273
Reduction, chemical, 38
Refrigerator
 temperatures, 121
 thermostat, 332
Regulators
 auxillary diaphragm, 183
 'chattering', 206
 commissioning, 204
 compensation requirement, 183
 components, 185
 construction, 190, 191
 faults, 205
 high-to-low pressure regulators, 192
 'hunting', 206
 impulse pipe, 189
 installation, 194
 loading, 190
 LPG, 46, 54, 56, 67, 74, 78, 221
 meter, 194, 199
 safety, 76
 service, 184
 servicing, 205
 valve, 197
Relative density, 6, 48
Relay, electrical, 270
Relay valves, 335
 fail safe, 319
Residual current circuit breaker, 316
Residual current device adaptor, 413
Resistance, 245
Resistor, 265
 colour coding, 256
 letter/number coding, 268
Retention flames, 26, 31, 32
Retention ports, 84, 94, 95, 110
Reversion pressure, 23
Rheostat, 265
Right-hand rule (magnetic flux), 252–254
Ring main circuit, 260-1
Ring main tester, 413
Roofs, working on, 433
Room-sealed appliances, 104
Room temperatures, 121, 126
Root mean square voltage and current, 257
Roots type meter, 212, 213
Rotary displacement meter, 212, 213

Safety consideration and emergency action, 77–81
Safety relief valve, 47, 61, 70
Safety torch, 409
Saturated air, 125
Saws, mechanical, 427
Saws, metal, 391
Saws, wood, 390
Scaffolding, 343
Screwdrivers, 399, 421
Screwing machines, 427
Secondary air, 27
Seasonal Efficiencies of Domestic Boilers in the UK (SEDBUK), 308
Seger cones, 448
Sensible heat, 123, 124
Service regulator, 196
Sherardising, 383
Sine wave, 256
Single column pressure gauge, 159
Slings, 141
Smartcard, 231–233
Socket forming tool, 414
Solar power, 113
Solder, 370
 lead-free, 370, 372
Soldering irons, 412
Solenoid, 253, 263, 282
Solenoid valves, 270
Solids, 1
Solvents, 385
Soot, 33
Spanners, 397
Spark ignition, 37, 38, 347
Specific gravity, 6
Specific heat, 122
Specific heat capacity, 122
Specific volume, 49
Speed, 114
Spirit level, 407
Stainless steel, 370
Standing pressure, 172
Statics, 130
Steel, 367
Steel rules, 409
Stefan-Boltzmann Law, 300
Stefan's constant, 300
Stocks and dies, 463
Stopper glove, plastic, 418
Strain, 133
Stratification, heat, 295
Stress, 133, 315, 354
 in beams, 136
Substitute natural gas (SNG), 11
 characteristics, 13
 pressure standard, minimum, 86
Substitute natural gas plant, 11
Suction pyrometers, 445
Symbols, chemical, 5

Tank cutters, 414, 415
Tanks (LPG), 48, 67, 69, 72
 contents gauge, 72
 filling connection, 71
 outlet (service) connection, 71

Tanks (LPG) (*continued*)
  pressure gauge, 71
  pressure relief valve, 64, 71
  temperature gauge, 71
Tap, gas, 311
  faults, 313
Taps, thread cutting, 402
Target burner, 97
Temperature,
  ambient, 296
  measurement of, 118, 119
scales of, 119–120
Tempering, 384
  colours and temperatures, 384
Temporary continuity bond, 285
Testing tee, 419
Therm, 6, 111, 112
Thermal backflow prevention device (TTB), 329
  operation, 329
  servicing, 330, 331
Thermal conductance, 291
Thermal conductivity, 289, 291
Thermal cut-offs, 315
Thermal efficiency, 128
Thermal transmittance
  coefficient, 306
Thermistor, 266, 444
Thermocouple, principle of, 323
Thermometer, 118, 120
  bimetallic, 442
  gas expansion, 442
  kata, 449
  liquid in glass, 440
  liquid in steel, 440
  pyrometers, 445
  resistance, 442
  vapour pressure, 442
Thermoplastics, 373
Thermosetting plastics, 372
Thermostat, 332
  bimetal rod-type, 333
  bimetal strip type, 336
  cooker oven, 335
  liquid expansion, 337
  refrigerator, 338
  vapour pressure, 339
Thorpe rotary meter, 214
Throat restrictor, 84, 89
Timber, 374
Tin, 369
Tinsnips, 408
Tools, use, care and maintenance, 420
Tool rolls and bags, 414
Torch, 409
Toxic gases, 5, 13
Trade and Industry, Department of, 6
Transfer of heat, 288
Transducer, 244, 245
Transformer, 260, 263, 280
Trowel, 408
Turnpin, 410

'U' gauge, 158, 159, 406, 407
U6 meter, 217–231
  prepayment, 230
  with £1 coinplate, 231
  prepayment token meter, 232
Ultrasonic UE6 meter, 221
  advantages of, 227
  construction of, 221
  index, 229
  liquid crystal display (LCD), 225
  measurement, principle of, 225
U6-E6 comparison, 228
  optical communication port, 227
'U' values for building components, 306, 307
Unit construction meter, 216–221

Vacuum cleaner, 411, 428
Value added tax, 239
Valve, gas, 332
Valves
  ball, 313
  butterfly, 313, 315
  electro-hydraulic, 341
  gate, 314
  motorised, 340
  power operated, 339, 349
  relay, 317
  solenoid, 340
  zoning, 341
Vane anemometer, 450
Vapour, explosive, 455
Vapour offtake, 65, 66
Vapourisation, latent heat of, 124
Vapouriser, 73
Velocity, 113–115
Velocity ratio, 139
Velometer, 450
Ventilation heat loss, 306
Venturi, 84, 85, 87, 89, 90
  energy changes within, 90
  flow of gas through, 175
Viscosity of LPG, 53
Vices
  pin, 418
  pipe, 417
Vitreous enamel, 382
Volt, 243
Voltage alternating current, 264
  out-of-phase, 319–21
Voltmeter, 275
Volume, variation with pressure, 208

Water flow gauge, 416
Water heater burner, multipoint, 109
Water pipework byelaws, 368
Water pressure
  circulating, 165, 166
  heads, 165
  hydrostatics, 163
Water temperatures, 121
Water vapour, 7, 125, 127
  flame/atmospheric effect, 39
Watt, 117, 244
Wavelength of radiant heat, 227
Weight, 113, 114
Wet meter, 215, 230
Wheatstone bridge, 443
Wire brushes, 410
Wire strippers, 412
Wiring regulations, 280, 284
Wobbe number (index), 7
Work, 113
Working at Heights, 435
Working pressure, 172
Warning signs, 438
Wrenches, 396, 421
Wrought iron, 363

Young's modulus, 135

Zinc, 56, 242, 369

Milton Keynes UK
Ingram Content Group UK Ltd.
UKHW020009071024
449327UK00031B/2709

9 780367 659349